Film Synthesis and Growth Using Energetic Beams

MATERIALS RESEARCH SOCIETY
SYMPOSIUM PROCEEDINGS VOLUME 388

Film Synthesis and Growth Using Energetic Beams

Symposium held April 17-20, 1995, San Francisco, California, U.S.A.

EDITORS:

H.A. Atwater
California Institute of Technology
Pasadena, California, U.S.A.

J.T. Dickinson
Washington State University
Pullman, Washington, U.S.A.

D.H. Lowndes
Oak Ridge National Laboratory
Oak Ridge, Tennessee, U.S.A.

A. Polman
FOM-Institute for Atomic and Molecular Physics
Amsterdam, Netherlands

MATERIALS
RESEARCH
SOCIETY

PITTSBURGH, PENNSYLVANIA

Single article reprints from this publication are available through
University Microfilms Inc., 300 North Zeeb Road, Ann Arbor, Michigan 48106

CODEN: MRSPDH

Published by:

Materials Research Society
9800 McKnight Road
Pittsburgh, Pennsylvania   15237
Telephone (412) 367-3003
Fax (412) 367-4373
Homepage http://www.mrs.org/

Library of Congress Cataloging in Publication Data

Film synthesis and growth using energetic beams : symposium held April 17-20,
    1995, San Francisco, California, U.S.A. / editors, H.A. Atwater, J.T. Dickinson,
    D.H. Lowndes, A. Polman
        p. cm.-(Materials Research Society symposium proceedings ; v. 388)
        Includes bibliographical references and index.
        ISBN 1-55899-291-X (alk. paper)
        1. Thin films-Congresses. 2. Ion implantation-Congresses.   3. Laser
    beams-Industrial applications-Congresses. I. Atwater, H.A. II. Dickinson, J.T.
    III. Lowndes, D.H. IV. Polman, A. V. Series: Materials Research Society
    symposium proceedings ; v. 388.
TA418.9.T45F54   1995   95-31836
621.3815'2-dc20   CIP

Manufactured in the United States of America

CONTENTS

*Invited Paper

*Invited Paper

*Invited Paper

*Invited Paper

*Invited Paper

PART IV: BEAM-INDUCED DEFECTS, CHEMICAL EFFECTS, AND CHARACTERIZATION

*Invited Paper

PREFACE

The symposium "Film Synthesis and Growth Using Energetic Beams" was held on April 17-20, during the 1995 MRS Spring Meeting in San Francisco, California. It covered a variety of topics, all related to the synthesis and growth of thin films using laser deposition and ion-beam techniques. A common theme in all presentations was the interaction of energetic beams (with energies from eV to MeV) with surfaces, and how energetics and kinetics affect film synthesis and growth.

The symposium consisted of three and one-half days of oral sessions and one evening poster session. Fourteen speakers were invited to review the status of the field and report about recent discoveries. In addition, 55 contributed oral presentations and 24 poster presentations were given. Both experimental and theoretical approaches were discussed. A panel discussion addressing "Issues and Challenges for Manufacturing Using Energetic Beams" surveyed four beam synthesis technologies, including pulsed-laser deposition, highly ionized sputter deposition for ULSI interconnects, separate isolation by implantation of oxygen (SIMOX), and reactive atom source growth of III-V and II-VI semiconductor visible diode materials. The symposium also was enlivened by two heavily attended sessions on nitrides growth. Of the presented papers, 65 have been accepted for publication in this proceedings. The speakers and attendees represented a broad range of academic institutions, national laboratories and industrial organizations. With contributions from 16 different countries, the symposium was quite international in nature as well.

The first section of this proceedings focuses on pulsed-laser deposition. Fundamental issues pertaining to the generation of laser ablation plumes, temperature distributions and collisional effects are described. A significant part of this section is devoted to the growth of superconducting and ferro-electric oxides, semiconductors and other materials by laser ablation. In Section II, ion-assisted pulsed-laser deposition, pulsed-ion deposition, applications of hyperthermal beams, and aspects of surface dynamics are described. The inclusion of an ion beam with the ablation process leads to some unique modifications in the thin film growth mechanisms, and hence, film properties. Likewise, the collision of high-mass metal cluster ions with substrates shows promise for growth of novel structures.

Section III is devoted to exciting new developments in the growth of optoelectronic materials, nitrides, and carbon films using a variety of techniques. These papers focus on a number of applications of considerable current interest. Section IV discusses the effects of beam-induced defects on growth and surface morphology, chemical effects occurring during growth, and characterization of film growth and film properties.

Overall, this proceedings volume represents the most recent developments in the field of film synthesis and growth using energetic beams. It shows that this is a lively field in which much remains to be learned and subsequently brought into practice. It is the organizers' hope that this symposium has helped to stimulate this development.

H.A. Atwater
J.T. Dickinson
D.H. Lowndes
A. Polman

July 1995

ACKNOWLEDGMENTS

We wish to thank all of the speakers, authors, and referees for their contributions to the success of the symposium and these proceedings.

It is our pleasure to acknowledge with gratitude the financial support provided for this symposium by: Hauzer Techno Coating Europe B.V., High Voltage Engineering Europa B.V., Lambda Physik, Lumonics, Inc., Martin Marietta Energy Systems, and Spectra-Physics.

We also wish to thank Mary Dawes, Department of Physics, Washington State University, for her able assistance in putting together these proceedings.

MATERIALS RESEARCH SOCIETY SYMPOSIUM PROCEEDINGS

MATERIALS RESEARCH SOCIETY SYMPOSIUM PROCEEDINGS

Volume 374—Materials for Optical Limiting, R. Crane, K. Lewis, E.V. Stryland, M. Khoshnevisan, 1995, ISBN: 1-55899-276-6

Volume 375—Applications of Synchrotron Radiation Techniques to Materials Science II, L.J. Terminello, N.D. Shinn, G.E. Ice, K.L. D'Amico, D.L. Perry, 1995, ISBN: 1-55899-277-4

Volume 376—Neutron Scattering in Materials Science II, D.A. Neumann, T.P. Russell, B.J. Wuensch, 1995, ISBN: 1-55899-278-2

Volume 377—Amorphous Silicon Technology—1995, M. Hack, E.A. Schiff, M. Powell, A. Matsuda, A. Madan, 1995, ISBN: 1-55899-280-4

Volume 378—Defect- and Impurity-Engineered Semiconductors and Devices, S. Ashok, J. Chevallier, I. Akasaki, N.M. Johnson, B.L. Sopori, 1995, ISBN: 1-55899-281-2

Volume 379—Strained Layer Epitaxy—Materials, Processing, and Device Applications, J. Bean, E. Fitzgerald, J. Hoyt, K-Y. Cheng, 1995, ISBN: 1-55899-282-0

Volume 380—Materials—Fabrication and Patterning at the Nanoscale, C.R.K. Marrian, K. Kash, F. Cerrina, M. Lagally, 1995, ISBN: 1-55899-283-9

Volume 381—Low-Dielectric Constant Materials—Synthesis and Applications in Microelectronics, T-M. Lu, S.P. Murarka, T.S. Kuan, C.H. Ting, 1995, ISBN: 1-55899-284-7

Volume 382—Structure and Properties of Multilayered Thin Films, T.D. Nguyen, B.M. Lairson, B.M. Clemens, K. Sato, S-C. Shin, 1995, ISBN: 1-55899-285-5

Volume 383—Mechanical Behavior of Diamond and Other Forms of Carbon, M.D. Drory, M.S. Donley, D. Bogy, J.E. Field, 1995, ISBN: 1-55899-286-3

Volume 384—Magnetic Ultrathin Films, Multilayers and Surfaces, A. Fert, H. Fujimori, G. Guntherodt, B. Heinrich, W.F. Egelhoff, Jr., E.E. Marinero, R.L. White, 1995, ISBN: 1-55899-287-1

Volume 385—Polymer/Inorganic Interfaces II, L. Drzal, N.A. Peppas, R.L. Opila, C. Schutte, 1995, ISBN: 1-55899-288-X

Volume 386—Ultraclean Semiconductor Processing Technology and Surface Chemical Cleaning and Passivation, M. Liehr, M. Hirose, M. Heyns, H. Parks, 1995, ISBN: 1-55899-289-8

Volume 387—Rapid Thermal and Integrated Processing IV, J.C. Sturm, J.C. Gelpey, S.R.J. Brueck, A. Kermani, J.L. Regolini, 1995, ISBN: 1-55899-290-1

Volume 388—Film Synthesis and Growth Using Energetic Beams, H.A. Atwater, J.T. Dickinson, D.H. Lowndes, A. Polman, 1995, ISBN: 1-55899-291-X

Volume 389—Modeling and Simulation of Thin-Film Processing, C.A. Volkert, R.J. Kee, D.J. Srolovitz, M.J. Fluss, 1995, ISBN: 1-55899-292-8

Volume 390—Electronic Packaging Materials Science VIII, R.C. Sundahl, K.A. Jackson, K-N. Tu, P. Børgesen, 1995, ISBN: 1-55899-293-6

Volume 391—Materials Reliability in Microelectronics V, A.S. Oates, K. Gadepally, R. Rosenberg, W.F. Filter, L. Greer, 1995, ISBN: 1-55899-294-4

Volume 392—Thin Films for Integrated Optics Applications, B.W. Wessels, D.M. Walba, 1995, ISBN: 1-55899-295-2

Volume 393—Materials for Electrochemical Energy Storage and Conversion—Batteries, Capacitors and Fuel Cells, D.H. Doughty, B. Vyas, J.R. Huff, T. Takamura, 1995, ISBN: 1-55899-296-0

Volume 394—Polymers in Medicine and Pharmacy, A.G. Mikos, K.W. Leong, M.L. Radomsky, J.A. Tamada, M.J. Yaszemski, 1995, ISBN: 1-55899-297-9

Prior Materials Research Society Symposium Proceedings available by contacting Materials Research Society

Part I

Pulsed-Laser Deposition

DYNAMICAL MODELING OF LASER ABLATION PROCESSES

J. N. Leboeuf*, K. R. Chen*, J. M. Donato†, D. B. Geohegan‡, C. L. Liu‡,
A. A. Puretzky‡, and R. F. Wood‡
*Fusion Energy Division
†Computer Science and Mathematics Division
‡Solid State Division
Oak Ridge National Laboratory, Oak Ridge, TN 37831-8071

ABSTRACT

Several physics and computational approaches have been developed to globally characterize phenomena important for film growth by pulsed laser deposition of materials. These include thermal models of laser-solid target interactions that initiate the vapor plume; plume ionization and heating through laser absorption beyond local thermodynamic equilibrium mechanisms; gas dynamic, hydrodynamic, and collisional descriptions of plume transport; and molecular dynamics models of the interaction of plume particles with the deposition substrate. The complexity of the phenomena involved in the laser ablation process is matched by the diversity of the modeling task, which combines materials science, atomic physics, and plasma physics.

I. INTRODUCTION

The laser ablation technique for pulsed laser deposition of thin films is reported to be conceptually and experimentally simple [1] and yet has proven extremely successful at growing high-quality films of very complex materials, such as high-temperature superconducting compounds. The physics ingredients that come into play are also quite complicated given that they involve laser–solid interactions at the target, plasma formation off the target, vapor/plasma plume transport toward the deposition substrate with its associated hydrodynamics and atomic physics, as well as plume–solid interactions at the deposition substrate.

We have taken a global physics and computational approach to the laser ablation process that relies on thermal models to describe laser–solid interactions; on kinetic models of plasma formation in the plume; on an assorted variety of hydrodynamic, gas dynamic, and collisional, models of plume transport; as well as on molecular dynamics methods to treat plume–substrate interactions. We have chosen to concentrate mostly on silicon to validate our models. The application of our physics results does however go beyond silicon, given the universality of many experimental observations, such as plume splitting for instance [2], for a wide variety of laser-ablated materials, be it carbon, copper, yttrium or YBCO. For laser–target interactions, the issues that we have addressed with our thermal models cover the vaporization rate as a function of laser fluence and the effect of the pressure at the liquid surface on the vaporization threshold. For plasma formation, the likelihood of vapor breakdown during the laser pulse has been tackled with nonequilibrium, kinetic rate equations models. As far as plume transport is concerned, the differing character of plume dynamics in near-vacuum and in the presence of a higher pressure background gas [3] has been treated with our gas dynamic and collisional models. In the area of plume–substrate interactions, the possibility of film damage by highly energetic plume particles has been investigated with the embedded atom method of molecular dynamics.

II. LASER–TARGET INTERACTIONS

Thermal models of laser–solid interactions have been successfully applied to laser annealing of semiconductors [4]. As implemented in the one- and two-dimensional (1- and 2-D) Laser8 computer programs [5], these models solve the enthalpy ($\rho h = \rho e + P$, with ρ the density, h the enthalpy, e the internal energy, and P the pressure) diffusion equation using finite differences

$$\frac{\partial}{\partial t}(\rho h) = \frac{\partial}{\partial x}\left(\kappa \frac{\partial}{\partial x} T\right) + S \ .$$

3

The laser energy input is specified at each instant of time and each point in space through the source term S related to the intensity of the laser pulse Φ:

$$S(x,t) = [1 - R(x,t)]\Phi(t)\,\alpha\,e^{-\alpha x} .$$

For silicon, the absorption coefficient α is taken to be $\alpha = 1 \times 10^6$ cm^{-1} in both the solid and liquid phases; the reflectivity R is set at 0.58 for the solid phase and 0.69 for the liquid phase; and the profile of the thermal conductivity κ is as in Ref. 5. Phase transitions, which may be time-delayed, are handled through the state array concept that determines the state of each cell in space and time according to the appropriate (enthalpy h, temperature T) state diagram. In addition to the solid-liquid transition appropriate for laser annealing, we have extended the process to include the liquid-vapor transition necessary to model the initial stages of laser ablation. The extended thermal model also includes the effect of the pressure at the liquid surface P_s on the vaporization temperature through the Clausius-Clapeyron equation:

$$T_v = \left[\frac{1}{T_0} - \frac{\ln(P_s/P_0)}{\Delta H} \right]^{-1} ,$$

where T_0 is the vaporization temperature at atmospheric pressure P_0 and ΔH is the latent heat.

Results from calculations with the 1-D version of the Laser8 computer program are displayed in Fig. 1. The depth of vaporization and maximum recession speed at atmospheric pressure as a function of laser energy density are displayed on the right. Both are linear with energy density and indicate that the vaporization threshold is around 4 J/cm^2. The vaporization temperature and threshold laser energy density for onset of vaporization as a function of pressure at the liquid surface P_s are displayed on the left. Both go like log(P_s) and indicate that a lower vaporization temperature and a lower threshold laser energy density are obtained in near-vacuum than at atmospheric pressure in accordance with the Clausius-Clapeyron equation. In particular, the predicted threshold laser energy density decreases from ~4 J/cm^2 at atmospheric pressure to ~1.5 J/cm^2 at a pressure of 1 mTorr.

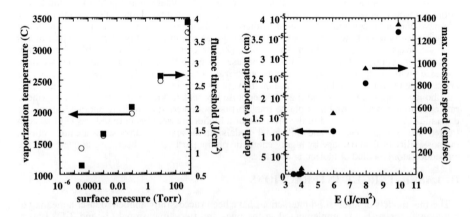

Fig. 1 Results from 1-D Laser8 for silicon: vaporization temperature and laser energy density threshold for vaporization as a function of surface pressure (left) and vaporization depth and maximum recession speed as a function of laser fluence at a liquid surface pressure of 1 atm (right).

The thermal model allows us to specify the initial conditions for the plume transport models to be described next.

III. PLUME FORMATION AND TRANSPORT

Plasma formation

Before considering gas dynamic and collisional models of plume transport from the target to the deposition substrate, let us briefly discuss the issue of plasma formation. From local thermodynamic equilibrium considerations, it has proven difficult to ionize the silicon vapor plume at the typical vaporization temperature of 0.3 eV. Even for a neutral density as low as 10^{15} cm^{-3}, an ionization fraction of only 0.001 is obtained according to the Saha equation. The lack of electrons makes absorption of laser light by the vapor unlikely. According to the best formulas available for electron-neutral inverse Bremstrahlung [6], electron-ion inverse Bremstrahlung [6], and plasma resonance absorption [7], the first is the largest of the three but at least 9 orders of magnitude smaller than the absorption coefficient in the solid and liquid phase for the most optimistic ionization fraction at the nominal vaporization temperature of 0.3 eV. The last is not important for KrF lasers since the critical plasma density required is close to solid density and therefore much greater than the expected plume densities at the end of the laser pulse.

These considerations have prompted us to investigate nonequilibrium processes for plume ionization and absorption of laser light by the plume. Our breakdown model for 248-nm KrF laser light interacting with silicon within a pulse length of ~40 ns is described in detail in a companion paper [8] and will only be briefly reviewed here. It is based on a kinetic model of breakdown first applied by Rosen et al. [9] to coupling of 1-μs pulses of 0.35-μm XeF laser radiation to aluminum alloys. Rate equations for densities and temperatures of electrons, excited state neutrals, ground state neutrals and singly charged silicon ions are evolved in time. These include dilution upon vapor expansion and resupply of particles from vaporization according to the Clausius-Clapeyron equation at the normal evaporation temperature and various density and temperature acceleration and deceleration mechanisms whose rates are taken from Zel'dovich and Raizer [10].

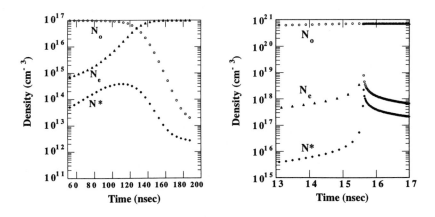

Fig. 2 Results from kinetic model of breakdown for silicon: ground state neutral (N_0), excited state neutral (N^*), and electron (N_e) densities in the plume as a function of time for laser energy densities of 5 J/cm^2 (left) and 30 J/cm^2 (right).

The key ingredient of the model is the two-stage process of electron impact excitation from the ground state to the excited state for neutrals at an energy of 5 eV and the subsequent electron impact ionization and photoionization of the excited state neutral to a singly charged silicon ion, which requires an additional 3.1 eV. This is to be contrasted with the jump from ground state neutral to singly charged ion at an energy of 8.1 eV, which is therefore more difficult to achieve with energies of 5 eV per photon in the KrF laser pulse. Mechanisms for absorption of laser light now include photoionization of the excited state [10], in addition to electron-neutral and electron-ion inverse Bremstrahlung [6]. Results from the solution of these zero-dimensional rate equations are shown in Fig. 2 where the evolution in time of the densities of electrons, ground state neutrals, and excited state neutrals is displayed for a laser energy density of 5 J/cm2 on the left and 30 J/cm2 on the right. Acceleration of the electron density is clear in both cases from an initial state given by the Saha equation for the priming electrons and a Boltzmann distribution for the excited state silicon neutrals. Production of a significant number of electrons and therefore plume breakdown do occur within the laser pulse at the higher energy density of 30 J/cm2. However, breakdown does not occur within the laser pulse for the lower energy density of 5 J/cm2. We will compare our transport modeling results with experimental observations at low fluences for which particulates or clusters are absent. Because ionization is difficult to achieve from a theoretical point of view at these low fluences, we have concentrated on neutral fluids for our models of plume transport.

Plume transport

Experimental observations have shown marked differences between plume expansion in vacuum and in the presence of a higher pressure background gas. These observations are common to a wide range of ablated materials including silicon, carbon, yttrium, and high-temperature superconducting compounds such as YBCO. Ablation in high-pressure ambient gases results in shock waves and expansion fronts propagating through the background gases. Time-of-flight measurements also show two components in the ion probe signals, an energetic component that propagates at vacuum speed and another that is more or less significantly slowed down depending on the pressure of the background gas [2]. We have applied several hydrodynamic, gas dynamic, and collisional models to study plume expansion in vacuum and in a higher pressure background gas. Results from some of these models will be described starting from more qualitative but higher dimensional ones and proceeding to the more detailed and flexible 1-D ones.

To investigate the gross hydrodynamic features of plume transport in vacuum and in background gas, we have used a 2-D gas dynamic model that solves conservation equations for mass density (ρ), momentum ($\rho \vec{v}$), and energy ($\rho e + 1/2\rho v^2$):

$$\frac{\partial}{\partial t}\rho = -\vec{\nabla} \cdot (\rho \vec{v}) \ ,$$

$$\frac{\partial}{\partial t}(\rho \vec{v}) = -\vec{\nabla} \cdot \left(P\vec{I} + \rho \vec{v}\vec{v}\right) \ ,$$

$$\frac{\partial}{\partial t}\left(\rho e + \frac{1}{2}\rho v^2\right) = -\vec{\nabla} \cdot \left\{\vec{v}\left[\left(\rho e + \frac{1}{2}\rho v^2\right) + P\right]\right\} \ .$$

They are augmented by the following equation of state for internal energy e and pressure P:

$$P = (\gamma - 1)\rho e = \frac{\rho}{M}k_B T \text{ with } \gamma = 5/3.$$

This model is numerically implemented using finite differences in space and the Rusanov scheme [11] in time. Results from the 2-D model are displayed in Fig. 3 for plume expansion in vacuum (left) and in background gas (right). Contours of total density (plume plus background) are shown as time progresses from top to bottom. It is evident that strong shocks are generated as the plume expands in presence of the background gas. The plume does in fact snowplow the background gas, giving rise to the crescent feature at the leading edge, which is clearly seen in Fig. 3.

6

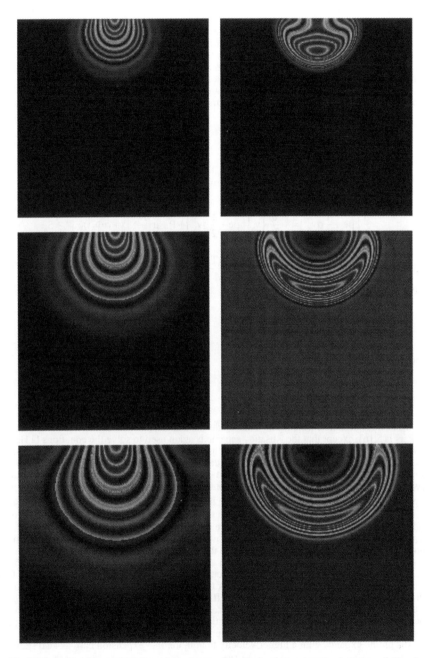

Fig. 3 Results from 2D gas dynamic model: contours of density for plume expansion in vacuum (left) and in background gas (right) as a function of time (top to bottom).

We note that the gross characteristics of the expansion in vacuum and in background gas are remarkably similar to the light emission patterns experimentally detected using a gated CCD camera [3]. We have also produced similar features [12] using a novel 2-D particle-in-cell hydrodynamic model [13,14] of plume transport in which particles representing elements of the fluid are followed in space and time in the pressure gradient field.

Characteristics of plume expansion in vacuum and in background gas have been examined in more detail using a 1-D gas dynamic model that solves the same conservation equations as the 2-D model, but includes ionization through the Saha equation, and energy input through laser light absorption. The set of equations used in the 1-D model are as follows:

$$\frac{\partial}{\partial t}\rho = -\frac{\partial}{\partial x}(\rho v_x) + S_\rho \ ,$$

$$\frac{\partial}{\partial t}(\rho v_x) = -\frac{\partial}{\partial x}\left(\rho v_x^2 + P - \mu\frac{\partial v_x}{\partial x}\right) \ ,$$

$$\frac{\partial}{\partial t}\left(\rho e + \frac{1}{2}\rho v_x^2\right) = -\frac{\partial}{\partial x}\left\{v_x\left[\left(\rho e + \frac{1}{2}\rho v_x^2\right) + P - \mu\frac{\partial v_x}{\partial x}\right]\right\} + \alpha\Phi(t)e^{-\alpha x} + S_e \ ,$$

with μ denoting the viscosity and $P = (1+\eta)\rho\dfrac{k_B T}{M}$. The ionization fraction η is determined by a simultaneous solution of the Saha equation:

$$\frac{\eta^2}{1-\eta} = \frac{M}{\rho}\left(\frac{2\pi m k_B T}{h^2}\right)^{\frac{3}{2}}e^{-\frac{I_p}{k_B T}} \ ,$$

and the equation of state:

$$\rho e = \frac{\rho}{M}\left[\frac{3}{2}(1+\eta)k_B T + \eta I_p\right]$$

where I_p is the ionization potential. This model is similar to the 1-D models of Vertes and coworkers [15,16], except that the Rusanov scheme is again used to solve the equations [11]. Our model also contains source terms for mass density and energy input denoted by S_ρ and S_e, respectively, in the mass density and energy equation. These source terms allow us to start the calculations with a clean slate and input mass and energy into the transport model according to the results from calculations of laser–target interactions using the Laser8 computer program. The mass density source is then given as $S_\rho = n_{liq}Mv_{rs}$ and the energy one as $S_e = n_{liq}Mk_B T v/(\gamma-1)$, with v_{rs} the recession speed and n_{liq} the liquid density. For the duration of the laser pulse, these terms provide a dynamic source of mass and energy into the system.

Results from 1-D gas dynamic calculations of silicon plume transport in vacuum with the dynamic source effect and of calculations where the density and temperature profiles at the end of the laser pulse are taken as initial conditions and allowed to freely expand are displayed in Fig. 4. These calculations were performed without ionization of the vapor and without absorption of laser light by the vapor plume. The time evolution of the pressure at the solid surface with the source effect and for free expansion are shown on the left, while the plume front position as a function of time in both of these cases is shown on the right. It is clear from Fig. 4 that high pressure at the surface is maintained for a longer time due to continuous ablation for the duration of the laser pulse compared to the free expansion case where the pressure at the surface rapidly drops as 1/t. As a result, the plume expands with a higher velocity than in the free expansion case as also shown in Fig. 4 where the plume front speed is $\sim 1 \times 10^6$ cm/sec with the dynamic source effect compared to ~ 5 × 10^5 cm/s in the case of free expansion. More details on the dynamic source effect for plume transport in vacuum, including analytical expressions for the steady-state density profile and maximum front velocity, can be found in a companion paper [17].

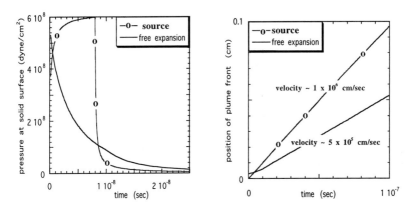

Fig. 4 Results from 1-D gas dynamic model with dynamic source effect from continuous ablation of material within the laser pulse: pressure at the solid surface with dynamic source effect and for free expansion (left) and position of plume front in these same two cases (right) as a function of time.

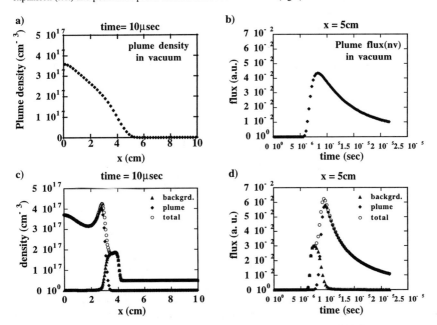

Fig. 5 Results from 1-D gas dynamic calculations of plume free expansion in vacuum and in background gas: a) density profile of plume in vacuum at 10 μs; b) plume flux in vacuum as a function of time at 5 cm from the target; c) total, plume, and background density profiles at 10 μs; d) total, plume, and background fluxes as a function of time at 5 cm from the target.

Plume free expansion in vacuum and in a higher pressure background gas has also been studied with our 1-D gas dynamic model, albeit for silicon expanding into a silicon background. For the calculations with background gas presented in Fig. 5, the background density n_b was set at $n_b/n_p = 5 \times 10^{-3}$ compared to the plume density n_p. Comparison of the density profiles in vacuum and in background gas displayed in Fig. 5 shows that the plume snowplows the background gas, which is pushed ahead of the plume, while the expansion of the plume is slowed down in presence of the background gas. The particle fluxes displayed in Fig. 5(d) show that the snowplowed background gas reaches the probe at 5 cm from the target at plume vacuum speed, with the plume arriving later at the detector because of slowing down from interaction with the background gas.

The results of pursuing the time evolution further are shown in Fig. 6 for a calculation with dynamic source effect and a silicon background gas pressure of 200 mTorr. The density profile is displayed as a function of distance from the target at four different times in the calculations up to 500 μs. Snowplowing of the background gas at the leading edge (a); rarefaction of the plume (b); slowdown and turnaround of the plume peak, the peak between target and front, by the snowplowed and piled-up background gas at the leading edge (c); and the subsequent reflection of the plume peak from the target (d) lead to multiple shocks between target and front, as observed in experiments performed in high background gas pressure [2].

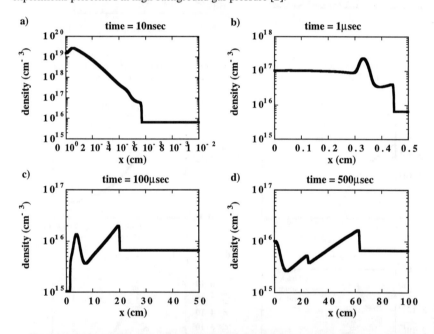

Fig. 6 Results from 1-D gas dynamic calculations of plume expansion in background gas at a pressure of 200 mTorr: density profile as a function of distance from the target at various times during the evolution.

Other models of plume–background interactions are also being considered in which the background and the plume are treated as two distinct species, but which include collisional interactions between them. One such model [18] treats each species using a set of fluid equations for density and momentum:

$$\frac{\partial}{\partial t} n_s + \frac{\partial}{\partial x}(n_s v_s) = 0 \quad ,$$

$$\frac{\partial}{\partial t}(n_s v_s) + \frac{\partial}{\partial x}\left[\frac{P_s}{m_s} + v_s(n_s v_s)\right] = -\frac{n_s}{m_s}\sum_{s'}\frac{m_s m_{s'}}{m_s + m_{s'}}C_{ss'}n_{s'}(v_s - v_{s'}) \ ,$$

with a collisional drag term between species s (plume) and species s' (background) included on the right-hand side of the momentum evolution equations. The strength of the interaction is denoted by $C_{ss'}$ which is taken to be constant for the calculations presented here. Note that the drag term is momentum conserving when summed over the two species. The pressure is taken to be $P_s = n_s k_B T_s$. Results from a calculation with this two-species (silicon in silicon) model are displayed in Fig. 7 where total, plume and background density profiles at a particular time and fluxes at 5 cm from the target as a function of time are shown. It is clear from the density profiles that the plume has penetrated through the background gas and that the background gas is being dragged along by the plume. The fluxes show that the plume component gets to the probe first at essentially vacuum speed, with the dragged background gas reaching the detector some time later.

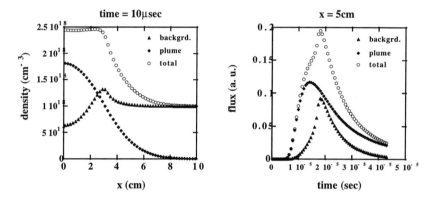

Fig. 7 Results from hydrodynamic model with two interpenetrating species: background, plume, and total densities (left) and fluxes (right).

Other approaches that show promise in elucidating the phenomena that take place when the plume interacts with a background gas include a model first proposed by Koopman and Goforth [19] and a scattering model we have recently developed. The model of Koopman and Goforth relies on a fully ionized layer of background gas to scatter the plume ions through ion-ion collisions. It is only by including this interaction that two velocity components are observed in the plume fluxes detected 5 cm from the target in the numerical calculations. Our scattering model includes plume-plume, plume-background, and background-background collisional interactions with cross sections that depend on the difference of the velocities to various inverse powers. The calculated fluxes at a certain distance from the target indicate that background or plume get to the probe first depending on whether the background gas is lighter, such as helium, or heavier, such as argon, than the silicon plume, with velocities exceeding plume vacuum speed in the helium case under conditions appropriate for elastic collisions.

IV. PLUME–SUBSTRATE INTERACTIONS

Time-of-flight measurements [2], as well as the model calculations presented here, clearly show that material reaches the deposition substrate at speeds equal to and even exceeding plume speed in vacuum even in the presence of a background gas, followed by a slower moving component whose velocity depends on background pressure. It is expected that the fast-traveling component may cause film damage due to the high kinetic energy of these particles. In the case of deposition of copper films, time-of-flight measurements indicate that the fast-moving particles can have kinetic energies in excess of 200 eV, while the energies of the slow-moving particles are < 100 eV under normal ablation conditions.

We have performed atomistic calculations using the embedded atom method of molecular dynamics to assess the effect of particle kinetic energies on the film quality by pulsed laser deposition from a copper target. A copper particle above the Cu(100) surface is allowed to hit the surface at velocities corresponding to kinetic energies ranging from 100 eV to 200 eV. The most probable point on the surface for the incoming particle to hit was found to be the open crystal channel along the <100> direction on Cu(100). Trajectories of the particles and substrate atoms are recorded as a function of time. As seen in Fig. 8, the critical energy for penetration of a copper particle into the Cu(100) surface is found to be ~150 eV. At this energy, the particle almost touches the surface atoms but does not penetrate. Figure 8 also shows that an energetic copper particle with a kinetic energy of 200 eV can however penetrate the first layer of the Cu(100) surface. The hopping distances for this 200 eV particle are large initially and gradually decrease as the particle loses its kinetic energy. It finally comes to rest after hopping about 5 A along the <110> surface channel. It is therefore possible for 200 eV particles to cause permanent mechanical damage according to these molecular dynamics calculations.

Critical Kinetic Energy for Penetration: 150 eV for Neutrals Bombardment of Cu(100)   **200 eV Neutrals Bombardment of Cu(100) Trajectory of the Energetic Particle**

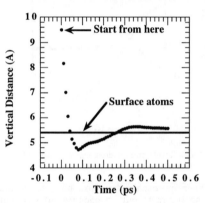

Fig. 8 Results from molecular dynamics calculations using the embedded atom method of film damage by energetic particles: vertical trajectories of a copper plume particle with kinetic energies of 150 eV (left) and 200 eV (right).

V. SUMMARY

We have described a global approach to modeling of the laser ablation process with several physics and computational models applied to laser–target interactions, plume transport in vacuum and in background gas, and plume–substrate interactions. The remaining challenges include extending models so far applied to silicon to more complex materials, that is, progress from elemental materials to compounds. They also entail continued quantitative comparisons with experiments, as well as model integration.

ACKNOWLEDGMENTS

This research is sponsored at the Oak Ridge National Laboratory (ORNL) by Laboratory Directed Research and Development (LDRD) Funds and by the Division of Materials Sciences, U. S. Department of Energy, under contract DE-AC05-84OR21400 with Martin Marietta Energy Systems, Inc. K. R. Chen and Chun-Li Liu are supported by an appointment to the ORNL Research Associate Program administered jointly by the Oak Ridge Institute for Science and Education and ORNL. Alex Puretzky is on assignment to the ORNL from the Institute of Spectroscopy, Russian Academy of Sciences, Troitsk, Russia.

REFERENCES

1. J. T. Cheung, Pulser Laser Deposition of Thin Films, ed. D. B. Chrisey and G. K. Hubler (Wiley, New York ,1994) pp. 1-22.
2. D. B. Geohegan, A. A. Puretzky, J. N. Leboeuf, K. R. Chen, C. L. Liu, R. F. Wood, J. M. Donato, and D. Lowndes, Paper in these proceedings.
3. D. B. Geohegan, Appl. Phys. Lett. 60, 2732 (1992).
4. R. F. Wood and G. A. Geist, Phys. Rev. Lett. 57, 873 (1986).
5. R. F. Wood and G. A. Geist, Phys. Rev. B34, 2606 (1986).
6. A. Anders, A Formulary for Plasma Physics (Akademie-Verlag, Berlin, 1990), pp. 159-187.
7. T. W. Johnston, and J. M. Dawson, Phys. Fluids 16, 722 (1973).
8. C. L. Liu, J. N. Leboeuf, R. F. Wood, D. B. Geohegan, J. M. Donato, K. R. Chen, and A. A. Puretzky, Paper in these proceedings.
9. D. I. Rosen, J. Mitteldorf, G. Kothandaraman, A. N. Pirri, and E. R. Pugh, J. Appl. Phys. 53, 3190 (1982).
10. Ya. B. Zel'dovich, and Yu. P. Raizer, Physics of Shock Waves and High-Temperature Hydrodynamic Phenomena (Academic Press, New York and London, 1966).
11. G. A. Sod, Journ. Comput. Phys 27, 1 (1978).
12. C. L. Liu, J. N. Leboeuf, R. F. Wood, D. B. Geohegan, J. M. Donato, K. R. Chen, and A. A. Puretzky, Beam-Solid Interactions for Materials Synthesis and Characterization, ed. D. E. Luzzi, T. F. Heinz, M. Iwaki, D. C. Jacobson (Mater. Res. Soc. Proc. 351, Boston, Mass., 1994).
13. J. N. Leboeuf, T. Tajima, and J. M. Dawson, Journ. Comput. Phys 31, 379 (1979).
14. F. Brunel, J. N. Leboeuf, T. Tajima, J. M. Dawson, M. Makino, and T. Kamimura, Journ. Comput. Phys. 43, 268 (1981).
15. A. Vertes, P. Juhasz, M. D. Wolf, and R. Gijbels, Scanning Microscopy 2, 1853 (1988).
16. L. Balazs, R. Gijbels, and A. Vertes, Analytical Chemistry 63, 314 (1991).
17. K. R. Chen, J. N. Leboeuf, D. B. Geohegan, J. M. Donato, R. F. Wood, and C. L. Liu, Paper in these proceedings.
18. P. W. Rambo, and J. Denavit, Journ. Comput. Phys. 98, 317 (1992).
19. D. W. Koopman, and R. R. Goforth, Phys. Fluids 17, 1560 (1974).

MECHANISMS OF EXCIMER LASER INDUCED POSITIVE ION EMISSION FROM IONIC CRYSTALS

J. T. DICKINSON, J-J. SHIN, and S. C. LANGFORD
Department of Physics, Washington State University, Pullman, WA 99164-2814

ABSTRACT

The energy distributions of positive ions produced by exposing single-crystal MgO to pulsed 248 nm excimer laser light at fluences of 200-1200 mJ/cm^2 were determined by combined quadrupole mass spectrometry and time-of-flight techniques. The dominant ionic species is Mg^+, although small amounts of Mg^{2+}, MgO^+, and Mg_2O^+ are also observed. In particular, the Mg^+ and Mg^{2+} energy distributions each show two broad peaks, with the energies of the Mg^{2+} peaks at significantly higher energies. Ion trajectory simulations (accounting for Coulomb forces only and assuming no surface relaxation) suggest that Mg^{2+} adsorbed at sites directly atop surface F-centers (oxygen vacancies with two trapped electrons) would be ejected upon photo-ionization of the F-center. The experimentally observed Mg^{2+} kinetic energies agree well with the energies predicted by the simulation.

INTRODUCTION

Laser desorption of positive ions have been observed from many wide band gap materials, but in most cases the precise ion desorption mechanism is not clear.[1-3] Recent studies of single-crystal MgO (E_g = 7.9 eV) irradiated with 248 nm (5 eV) excimer laser (KrF) radiation have shown that lattice defects play an important role in desorption and ablation processes.[4-7] In this work, quadrupole mass spectrometry is used to determine the energy distributions of photodesorbed Mg^+ and Mg^{2+} from MgO. Comparing these results with ion trajectory simulations, we show that these energy distributions are consistent with the emission of cations via "Coulomb explosions" from sites atop surface F-centers following photo-ionization of the F-center by the laser. This mechanism is similar to the Knotek-Feibelman mechanism for photostimulated desorption,[8] except in our case emission requires lattice defects prior to exposure. This allows for desorption at much lower photon energies (here 5 eV as opposed to the 525 eV required for core-hole production and emission of O^+ from MgO surfaces[9]), and it results in desorption of cations (as opposed to anions only in x-ray stimulated desorption).

EXPERIMENT

Arc-fused, single crystal MgO, 99.99% pure with respect to metallic impurities, was obtained from Tateho Chemical Industries, cleaved into 0.1 × 1 × 2 cm^3 samples, and polished to a 1-μm surface finish with diamond paste. Experiments were conducted in a diffusion pumped, liquid nitrogen trapped vacuum chamber at a pressure of about 10^{-4} Pa. Sample irradiation was performed with a Lambda Physik Lextra 200 excimer laser with a pulse width of 30 ns at 248 nm (KrF). The beam energy was measured with a Gentec ED 500 joule meter. Ion TOF measurements and mass spectra were obtained with a UTI 100C quadrupole mass spectrometer (QMS) with the ionizer removed. The region between the sample and entrance to the QMS was field free. Thus the only perturbation to the direct flight of ions to the QMS detector (a CEM mounted at the exit aperture of the mass filter) was a small radial oscillation in the quadrupole mass filter; for the ion energies observed here, this oscillation had a negligible effect on the measured TOF. Time-resolved signals at individual masses were acquired to determine TOF distributions.

15

The ion energy distribution was estimated from TOF signals accompanying the first few laser pulses on previously unexposed surfaces. The energy distribution may be obtained by transforming the TOF distribution

$$\frac{dN}{dE} \quad = \quad -\frac{dN}{dt}\left(\frac{dE}{dt}\right)^{-1} \quad = \quad I[E(t)]\,\frac{t^3}{md^2} \quad , \tag{1}$$

where E is the kinetic energy of an ion detected at time t after the laser pulse ($E = mv^2/2 = md^2/2t^2$), $I(t)$ is the ion intensity as a function of time, m is the ion mass, and d is the distance between the sample and the detector. To facilitate numerical analysis, the TOF signal was smoothed by performing a least-squares fit of the TOF signal, where energy distribution E(t) was empirically expressed as a sum of Gaussians.

To predict the ion energies associated with emission from various defect configurations, ion trajectory simulations were performed by summing the electrostatic forces due to the individual ions in a simulated MgO cluster and integrating the resulting equation of motion using a second order Taylor approximation.[10] Computations with clusters of different sizes showed little improvement in the computed trajectory when the cluster size was increased beyond a $19 \times 19 \times 10$ ion array. Trajectories computed without image charge corrections typically yielded ion energies within 2% of the (repulsive) Madelung binding energy of the initial defect configuration; this difference is a reasonable estimate of errors due to the finite cluster size and computational inaccuracies. The ion energy simulations reported below incorporated an image charge term to account for the dielectric response of the bulk. Since the image charge reduces the accelerations along the trajectory, the addition of the image charge term should not degrade the numerical accuracy of the trajectory simulation.

RESULTS

A typical set of TOF signals at a mass to charge ratio of 12 amu/e (Mg^{2+}) accompanying the first five laser pulses (1.2 J/cm^2 per pulse) on previously unexposed, polished MgO appear in Fig. 1. The results of fitting Eq. 1 to the experimental data are also shown. The TOF signal accompanying the first pulse is slightly distorted relative to the signals accompanying the remaining pulses. In particular, the first pulse produces a number of high energy ions. At this fluence, the second and successive pulses on polished surfaces consistently produce two broad peaks which are well described by the model Gaussian energy distributions. The position and breadth of these peaks are remarkably consistent from pulse to pulse, until the decreasing signal intensities no longer provide adequate statistics.

Typical TOF distributions at a mass to charge ratio of 24 amu/e (Mg^+) accompanying the first five pulses incident on a previously unexposed, polished surface at a fluence of 1.1 J/cm^2 appear in Fig. 2. Again, two peaks were observed. Figure 2 also shows the results of a least squares fit of Eq. 2 to the corresponding experimental TOF distributions.

As in the case of the Mg^{2+} emissions, the Mg^+ emission accompanying the first pulse is more intense than the emission accompanying the following pulses. The first pulse also produces a larger number of faster (higher energy) and slower (lower energy) ions. Previous experience suggests that the low energy ions produced by the first laser pulse are accompanied by a "cloud" of electrons (at least until they enter the mass filter), and thus correspond to a weakly bound plasma. The TOF behavior of emissions accompanying the first pulse on "fresh" portions of the surface is highly variable, reflecting the operation of strongly nonlinear ion emission and acceleration mechanisms at fluences well below threshold for breakdown. At higher fluences and particle densities, this plasma becomes much more strongly bound and is remarkably stable against perturbation by applied electric fields; the plasma also displays the bright green luminescence characteristic of excited neutral and ionic Mg, due to collisions with energetic electrons. Although the emissions accompanying the first pulse (and at higher fluences) are of considerable importance, in this work we focus on emission during subsequent pulses, which is unaffected by laser-particle interactions after emission. Again, weak ion

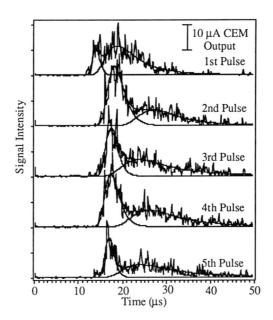

FIG. 1.   Mg^{2+} TOF distributions from polished MgO at a fluence of 1.2 J/cm^2 per pulse. The least squares fits to the TOF data used in the energy distribution calculation are shown by the fine lines, while the data is shown by the dark lines.

emissions are readily observed after exposing the surface to many tens of laser pulses at this fluence.

Despite considerable pulse-to-pulse variation in the Mg$^+$ and Mg^{2+} intensities, both emissions gradually become weaker over the first several pulses. This is consistent with the gradual removal or annealing of defects involved in the emission process (a "clean up" effect). Using a fast electrometer to obtain higher sensitivities, measurable ion emission is observed for at least several tens of pulses. Ultimately, sustained emission requires that any surface defects consumed by emission processes be replenished, either during laser exposure or in the interval between successive pulses. Unfortunately, the time response of typical electrometers is inadequate for TOF measurements on these time scales.

DISCUSSION

The low photon energies (5 eV) and fluences ($\lesssim$ 1 J/cm^2) employed are insufficient to produce excitons across the band gap of MgO (7.9 eV). Therefore the photon absorption responsible for electron and ion emission must be associated with lattice defects. In MgO, electrons trapped at oxygen vacancy defects absorb strongly at 5 eV and are therefore candidates for absorption centers. These vacancies readily accommodate either one electron (F$^+$-center) or two electrons (F-center). Both the F- and F$^+$-centers absorb strongly at 5 eV at surface and bulk sites. Recent calculations suggest that the ground state of these centers lie about 1.5 eV below the bottom of the conduction band,[11] so that the absorption of a 5 eV photon at surface defects would be sufficient to photo-ionize the center. This is consistent with the observation of electron emission due to single-photon absorption at the fluences employed in this work.[12]

The loss of charge by F-centers in the near surface region would reduce the binding energy of nearby Mg$^+$ ions, leading us to propose that the principle ion emission mechanism is

17

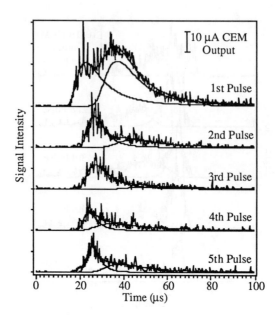

FIG. 2.   Mass selected Mg$^+$ TOF signal from polished MgO at a fluence of 1.1 J/cm^2 per pulse. The least squares fits to the TOF data used in the energy distribution calculation are shown by the fine lines, while the data is shown by the dark lines.

the Coulomb repulsion of weakly bound Mg$^+$ ions due to the photo-ionization of nearby F-centers or comparable defects. The relevant photo-ionization events include

$$F_s^0 + h\nu \rightarrow F^+ + e^- \qquad (2a)$$
$$F_s^+ + h\nu \rightarrow F^{2+} + e^- \quad , \qquad (2b)$$

where the subscript s indicates a surface defect.

The binding energies of Mg$^+$ and Mg^{2+} at surface and surface defect sites were estimated using Madelung energy arguments.[13] For this purpose, F-centers were treated as point charges at the defect lattice site, and lattice relaxation effects were neglected. In this approximation, F-center photo-ionization reduces the binding energy of nearby Mg ions by $q_{ion}\Delta q_F/r$, where q_{ion} is the charge of the Mg ion, Δq_F is the change in the charge of the F-center, and r is the distance between the F-center and the ion. Defect configurations where the electrostatic binding energy of the cation was less than the energy available from a photo-ionization event were taken as candidates for further analysis.

These arguments suggest that emission is possible for a large number of defect configurations. However, ion trajectory simulations rule out most of these configurations as potential emission centers.[14]   Magnesium ions with a single, fully charged (-2e), nearest neighbor anion were not desorbed in any tested defect configuration involving a single F-center (either singly or doubly ionized). The electrostatic potential of a nearest neighbor anion produces a local potential minimum (about 1 eV deep) from which the adjacent cation cannot escape, despite its negative (global) binding energy. The only simple defect configuration without a nearest neighbor anion is an Mg ion adsorbed on a terrace site directly atop an F-center, as indicated schematically in Fig. 3. This site symmetric about the surface normal, and would produce highly directed emissions normal to the surface. Recent angular distribution measurements indicate that ions emitted at fluences of 1-2 J/cm^2 from cleaved MgO, polished MgO, and thin film polycrystalline MgO, are in fact strongly directed along the surface normal.[14] The estimated ion kinetic energies associated with these defect configurations appear in Table II, along with the average observed kinetic energies determined from the TOF data.

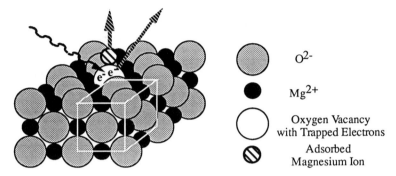

Fig. 3.   A schematic diagram of the proposed defect configuration responsible for low-fluence ion emission under 248 nm laser irradiation. An Mg$^+$ or Mg^{2+} ion is initially adsorbed atop a surface F-center. Subsequent photo-ionization of this F-center produces a strong repulsive electrostatic force on the cation, resulting in emission.

Table II.   Comparison of the average peak energies obtained from experimental TOF signals from polished material with ion energies predicted by idealized trajectory simulations.

| | Mg$^+$ | | Mg^{2+} | |
	Low Energy	High Energy	Low Energy	High Energy
Experiment	2.7 eV	7.7 eV	6.7 eV	19.2 eV
Simulations	4.4 eV	11.2 eV	6.0 eV	19.6 eV

The agreement between the measured and predicted Mg^{2+} energies is excellent, particularly considering the approximations in the ion trajectory simulations and in the modeling of the ion TOF signals. We attribute the poor agreement in the case of Mg$^+$ to lattice relaxation effects (where an adsorbed Mg$^+$ ion relaxes away from the surface); lattice relaxation would be stronger for Mg$^+$ than Mg^{2+}, since Mg$^+$ lacks the full charge of the lattice. Nevertheless, the ratio of the high and low peak energies for Mg$^+$ determined from TOF data is similar to the ratio of the predicted energies from the trajectory simulations.

CONCLUSIONS

Irradiating previously unexposed MgO surfaces with pulsed, 248-nm laser radiation at fluences below 1 J/cm^2 produces significant positive ion emissions. TOF measurements of Mg^{2+} emissions show two distinct components which we attribute to Coulomb ejection from two distinct electrostatic environments; the measured ion energies are those expected when Mg^{2+} is emitted from a site atop a surface F-center which is photo-ionized by the laser. The corresponding Mg$^+$ energies are somewhat lower than expected on the basis of ion trajectory models for a singly charged ion emitted from the high symmetry site over the vacancy defect.

Previous observations have shown that the ion emissions under these conditions are highly directed along the surface normal,[14] consistent with electrostatic ejection from this defect configuration. Both surface F-centers and adsorbed ions appear to be produced by deformation processes accompanying cleavage, polishing, and abrasion.

We emphasize that deformation-induced defects can play a dominant role in laser-material interactions. Treatments like mechanical polishing, which enhance the optical quality of a surface may, in fact, strongly increase interactions in the UV portion of the spectrum. Localized deformation on cleaved surfaces can also produce measurable interactions. Although uv irradiation might be used to improve the resistance of many surfaces to laser interactions via "clean up" effects, it appears that this approach would be ultimately limited by the thermal transport of defects from the bulk. Chemical etching or thermal annealing treatments may be required to further improve the resistance of MgO to laser interactions. Ion emission intensities can be a sensitive probe of surface defects which could find use in surface characterization and preparation.

ACKNOWLEDGMENTS

This work was supported by the Department of Energy under Contract DE-FG06-92ER14252 and the National Science Foundation Surface Engineering and Tribology Program Grant No. CMS-9414405.

REFERENCES

1. E. Matthias, H. B. Nielsen, J. Reif, A. Rosen, and E. Westin, J. Vac. Sci. Technol. B 5, 1415 (1987).
2. E. Matthias and T. A. Green, in *Desorption Induced by Electronic Transitions, DIET IV*, edited by G. Betz and P. Varga (Springer, Berlin, 1990), pp. 112-127.
3. L. L. Chase, A. V. Hamza, and H. W. H. Lee, in *Laser Ablation: Mechanisms and Applications*, edited by J. C. Miller and R. F. Haglund, Jr. (Springer, Berlin, 1991), pp. 193-202.
4. R. L. Webb, S. C. Langford, L. C. Jensen, and J. T. Dickinson, Mat. Res. Soc. Symp. Proc. 236, 21 (1992).
5. R. L. Webb, L. C. Jensen, S. C. Langford, and J. T. Dickinson, J. Appl. Phys. 74, 2323 (1993).
6. J. T. Dickinson, L. C. Jensen, R. L. Webb, M. L. Dawes, and S. C. Langford, Proc. Mater. Res. Soc. 285, 131 (1993).
7. J. T. Dickinson, L. C. Jensen, R. L. Webb, M. L. Dawes, and S. C. Langford, J. Appl. Phys. 74, 3758 (1993).
8. M. L. Knotek and Peter J. Feibelman, Phys. Rev. Lett. 40, 964 (1978).
9. Richard L. Kurtz, Roger Stockbauer, Ralf Nyholm, S. Anders Flodström, and Friedmar Senf, Phys. Rev. B 35, 7794 (1987).
10. R. W. Stanley, Am. J. Phys. 52, 499 (1984).
11. Andrew Gibson, Roger Haydock, and J. P. LaFemina, Appl. Surf. Sci. 72, 285 (1993).
12. J. T. Dickinson, L. C. Jensen, R. L. Webb and S. C. Langford, in *Laser Ablation: Mechanism and Applications-II*, AIP Conf. Proc. 288, 13-25 (1993).
13. J. Magill, J. Bloem, and R. W. Ohse, J. Chem. Phys. 76, 6227 (1982).
14. J. T. Dickinson, S. C. Langford, J. J. Shin, and D. L. Doering, Phys. Rev. Lett. 73, 2630 (1994).

COLLISIONAL EFFECTS OF BACKGROUND GASES ON PULSED LASER DEPOSITION PLASMA BEAMS

DAVID B. GEOHEGAN* AND ALEX A. PURETZKY**
*Oak Ridge National Laboratory, P.O. Box 2008, Oak Ridge, TN 37831-6056.
**Institute of Spectroscopy, Russian Academy of Sciences, Troitsk, Russia.

ABSTRACT

The penetration of energetic pulsed ablation plumes through ambient gases is experimentally characterized to investigate a general phenomenon believed to be important to film growth by pulsed laser deposition (PLD). Under typical PLD conditions involving background gases, the ion flux in the ablation plume is observed to split into distinct fast and slow components over a limited range of distances.[1,2] The fast component is transmitted with near-initial velocities and high kinetic energies, potentially damaging to growing films at these distances. Formation of the second, significantly-slowed component correlates with the bright contact front[3] formation observed[1,4] in fast ICCD imaging studies. This general effect is explored in detail for the case of yttrium ablation into argon, a single-element target into an inert gas.[5] Time-resolved optical absorption spectroscopy and optical emission spectroscopy are employed to simultaneously view the populations of both excited and ground states of Y and Y^+ for comparison with quantitative intensified-CCD photography of the visible plume luminescence and ion flux measurements made with fast ion probes during this phenomenon. These measurements confirm that, in addition to the bright significantly-slowed front which has been described by shock or drag propagation models[1], a fast-component of target material is transmitted to extended distances for some ambient pressures with near-initial velocities.

INTRODUCTION

The ability of energetic ablation plasma plumes or 'beams' to penetrate low-pressure background gases during PLD is a key processing advantage. However, the plume/background-gas interaction is a mixture of gas-dynamic and thermal effects which result in nontrivial variations of key processing parameters with distance, namely: kinetic energy, density, and temperature. Reported here are combined diagnostic measurements of silicon and yttrium ablation plume penetration through inert background gases during a key transitional regime in which the ion flux is split into distinct fast and slowed components. This apparently general phenomenon occurs over a limited range of distances at ambient pressures, including those sometimes used for PLD (as reported for YBCO ablation into O_2).[1,2,6,7] This "plume-splitting" is significant because a 'fast' component of ions can arrive at the ion probe (or substrate) with little or no delay compared to propagation in vacuum (e.g., Y^+ kinetic energies in this study were up to 250 eV, and Si^+ up to 215 eV). Kools recently employed a Monte Carlo simulation involving elastic collisions to prove this possibility.[8] However, at longer distances this 'fast' component is completely attenuated, and a single, slowed distribution of ions is observed. This 'fast' component is easily overlooked in imaging studies because enhanced plume luminescence occurs in the slowed distribution.

However, the transmitted flux is readily observed with fast ion probes which provide convenient measurements of the magnitude and time-of-flight of the ion current (flux) in plasma plumes propagating through background gases.[1,2,5-7,9] Figure 1 serves to review the ion probe measurements and illustrate the splitting effect.

Figure 1(a) shows the ion flux arriving at d = 5 cm along the normal to a silicon target following 2.8 J/cm^2 KrF-laser irradiation in vacuum and various helium ambient pressures. As in the case of YBCO penetrating through O_2,[1,2,6,7] the ion flux in these Si/He and Si/Ar studies was attenuated exponentially with background pressure and/or distance in general agreement with the simple scattering model (effective cross sections in both cases $\sim 1 \times 10^{-16}$ cm^2) utilized for YBCO.[1,2,6,7] At long distances and long time delays, ion probe and fast imaging measurements

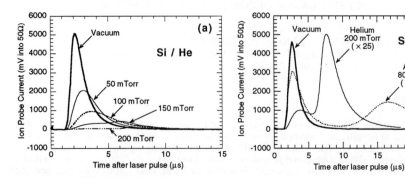

Figure 1: Ion probe current waveforms measured d = 5 cm along the normal from a silicon target following 2.8 J/cm^2 KrF-laser irradiation in vacuum and background helium and argon: (a) Slowing and attenuation vs. helium ambient pressure (unscaled), (b) Comparison of fast and slow ion flux arrival at d = 5 cm through helium (200 mTorr) and argon (80 mTorr) with that in vacuum. In addition to the slowed material, a fast component of ions arrives in both cases with little or no delay compared to the propagation in vacuum.

respectively record a decelerating, stable shock-structure which propagates in accordance with drag or shock models.[6,7,9] However, for some pressure/distance combinations, *two* distinct ion flux distributions are resolvable as shown in Fig. 1(b).

Figure 1(b) shows scaled ion fluxes measured at d = 5 cm following laser ablation of silicon in helium (200 mTorr) and argon (80 mTorr, dashed curve). Despite the high background pressures, a 'fast' component of ion flux arrives at the detector position at d = 5 cm with little or no apparent slowing. A 'slow' component of material arrives much later, in coincidence with the arrival of the bright shock structure (as will be shown for yttrium into argon, below).

Variations of the shape of the 'fast' component with background gas, distance, and pressure hold clues to the fundamentals of the scattering processes at play. These will be presented elsewhere, as will the various computer models and simulations which have been undertaken. In this paper, it is necessary to prove that the 'fast' component contains fast plume ions (and not, for example, exclusively background gas ions which have been swept-up, or 'snowplowed' in front of the target material). In addition, *quantitative* imaging will be introduced in support of this hypothesis.

EXPERIMENTAL RESULTS

In this paper, time- and spatially-resolved optical absorption and emission spectroscopy are applied to determine the composition of the 'fast' and 'slow' propagating plume components for single-component target ablation (yttrium) into an inert gas (argon). These measurements are correlated with quantitative imaging and ion probe measurements. Although the splitting effect appears quite general, the yttrium/argon system was chosen because optical absorption spectroscopy of both Y and Y$^+$ was simultaneously possible[10] and because argon is chemically inert.

The experimental apparatus has been described previously.[4–7,10] A cylindrical lens (f_L = 500 mm) focused the apertured 248-nm beam from a Questek 2960 KrF-excimer laser (25 ns FWHM pulse width) to a horizontal line (2.0 cm × 0.075 cm) on the face of a sanded yttrium (99.99%) pellet at an incidence angle of 30°. For optical absorption spectroscopy, a pulsed (1.5 μs FWHM) Xe lamp beam (width 0.5 mm) was passed parallel to the target surface and just above a wire ion probe tip (floating bias -70 V with respect to its shield). The expansion of the ablated yttrium formed a vertical fan which expanded little along the horizontal lamp beam axis (toward the camera

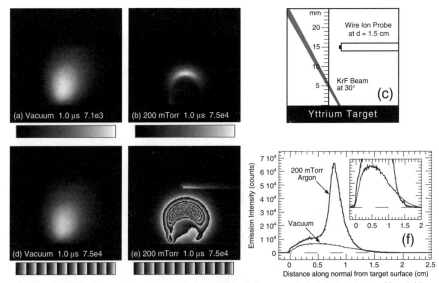

Figure 2: Gated ICCD photographs of the total visible optical plasma emission during $\Delta t = 1.00 - 1.05$ μs following 0.8 J/cm^2 KrF-laser irradiation of yttrium metal into (a) 1×10^{-6} Torr and (b) 200 mTorr argon. The peak of the grayscale palette is normalized to 7100 and 75000 counts, respectively. (c) Schematic indicating the side-on view of the yttrium target, mm-scale, and wire-ion probe position at $d = 1.5$ cm. (d),(e) The same photographs are replotted with a 10-grayscale palette and equal normalization to 75000 counts for direct comparison. A line-profile of the emission intensity from the irradiated spot along the target normal (f) shows the correct relative scaling between the vacuum and background images. The inset in (f) more clearly shows the same data replotted on a 0-8000 count scale to illustrate the small, but resolvable emission in 200 mTorr argon out to and beyond 1.5 cm.

or spectrometer), so the absorption path length remained at approximately 2 cm for the two distances investigated here ($d = 1.5, 2.5$ cm). Optical detection utilized a 1.33-meter spectrometer (McPherson 209, 1800 g/mm holographic grating) outfitted with an intensified, gated diode array (Princeton Instruments IRY-700RB, 5-ns resolution) and photomultiplier tube (Hamamatsu R955). Fast unfiltered imaging was performed with a gated, (ICCD), lens-coupled camera system (Princeton Instruments) with variable gain, 200–820 nm response, and variable gating above 5-ns.

Typical gated images of the visible yttrium-plume luminescence at $\Delta t = 1.0$ μs in vacuum and 200 mTorr argon are shown in Figs. 2(a) and 2(b). The free expansion in vacuum [Fig. 2(a)] shows luminescence extending to 2.0 cm, while the luminescence in argon [Fig. 2(b)] appears not to reach the ion probe at $d = 1.5$ cm. However, this is a result of the scaling of the intensity data to accommodate the bright shock structure [over 10 times brighter than the peak in Fig. 2(a)]. When normalized to the same peak intensity, and replotted in Figs. 2(d) and 2(e) with a 10-grayscale palette, the same two images reveal luminescence extending well past the ion probe to nearly the same distances as in vacuum. A line profile along the normal to the irradiated spot [Fig. 2(f)] correctly portrays the relative intensities.

The ion probe currents collected in vacuum and 200 mTorr argon at $d = 1.5$ cm are shown in Fig. 3(a). At this distance and pressure, the diminishing 'fast' component of the ion flux is still an appreciable fraction of the total signal, and is temporally well-resolved from the 'slow' component. In order to investigate the composition of the plume during the splitting of the ion flux, optical absorption spectra were obtained at $d = 1.5$ cm at three times (1.3 μs, 2.1 μs, and 2.9 μs) corresponding to times during the first peak, shoulder, and second peak of the 200 mTorr ion flux in Fig. 3(a) [see diagram, Fig. 3(b)]. This spectral region for absorption was chosen because both

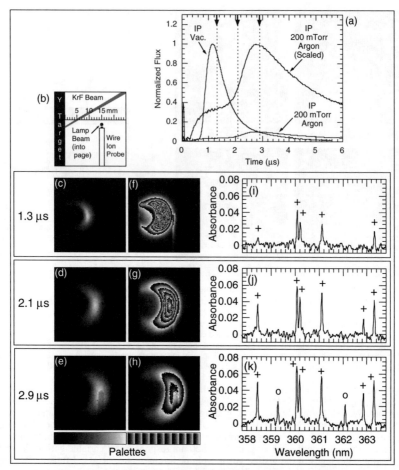

Figure 3: (a) Ion probe current pulses measured d = 1.5 cm along the normal to an yttrium target following KrF-laser irradiation (0.8 J/cm^2) in vacuum (1 × 10^{-6} Torr) and 200 mTorr argon. The bimodal 200 mTorr waveform is plotted in correct magnitude relative to the vacuum pulse and replotted scaled by a factor of 10. The peak of the vacuum pulse corresponds to a current flux of 25 A/cm^2. Imaging and optical absorption spectroscopy were performed at the three times indicated in 200 mTorr argon. (b) Schematic of the side-on view of the Y-target and focusing sheet of KrF-irradiation, wire ion probe at d = 1.5 cm, and position of Xe-lamp beam (propagating into the page) used for optical absorption spectroscopy just above the wire probe tip. (c) Gated ICCD photographs of the total visible optical plasma emission during 50 ns gates at the indicated time delays following KrF-laser irradiation of yttrium metal into 200 mTorr argon. The times correspond to the first peak, shoulder, and second peak of the ion probe waveform in (a). The images represent equal exposures: (c)–(e) utilize the grayscale palette and are independently normalized to 10,000, 20,000, and 38,000 counts, respectively, while (f) - (h) utilize the 10-grayscale palette and are all normalized to 38,000 counts. (i)–(k) Time-resolved optical absorption spectra (~ 2 cm optical path length, d = 1.5 cm, 100-spectra averaged with 200 ns gates centered at the imaged times) in 200 mTorr argon show that ground-state Y$^+$ ion lines (indicated by +) dominate absorption at (c) 1.3 μs and (b) 2.1 μs (first peak, shoulder of ion probe signal): 358.45, 360.07, 360.19, 361.10, 362.87, 363.31 nm. Y neutral absorption becomes (indic. by o) noticeable in the second peak of the ion probe signal as indicated in (k) 2.9 μs, at 359.29, 362.09 nm.

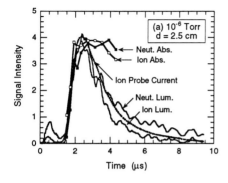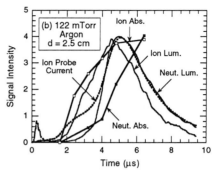

Figure 4: Normalized ion probe current (IP), luminescence (lum.) intensities of Y*(410.2 nm) and Y^{+*}(363.3 nm), and optical absorbencies (abs.) of Y(362.1 nm) and Y^+(363.3 nm) at d = 2.5 cm following 0.8 J/cm^2 KrF-laser irradiation of yttrium in (a) vacuum (1×10^{-6} Torr) and (b) 122 mTorr argon. Scaling factors: (a) lum: Y* (/3), Y^{+*}($\times$ 1), abs: Y ($\times$ 70), Y^+($\times$ 94), IP (mV/81), (b) lum: Y* (/23), Y^{+*}(/12.5), abs: Y ($\times$ 100), Y^+($\times$ 70), IP (mV/81).

ground-state neutral (Y) and ground-state ionic (Y^+) populations could simultaneously be investigated (ionic lines indicated by + and neutral lines indicated by o). In addition, quantitative imaging was performed at the same three times and these images are again plotted in two representations (see figure caption) in Fig. 3(c)–3(e) and Fig. 3(f)–3(h).

At 1.3 μs [Fig. 3(c) and 3(f)], very weak optical emission is present at the probe position during the first peak of the ion flux signal. However, the optical absorption spectrum [Fig. 3(i)] reveals strong Y^+ absorption [lines indicated by + in Figs. 3(a)–3(c)] at this time, confirming the penetration of target ions through the background gas at 'vacuum' velocities. At 2.1 μs [Fig. 3(d) and 3(g)] the steep gradient of plume luminescence has reached the probe at 1.5 cm (shoulder of the ion probe current, Fig. 3(a)). The peak luminescence is 1.9 times smaller than at 1.3 μs, and the optical absorption spectrum is still dominated by Y^+ absorption. During the second peak of the ion probe current at 2.9 μs, the probe is immersed in the brightest region of luminescence [Fig. 3(e) and 3(h)] which is half the peak intensity of that at 2.1 μs. In addition to the ion lines, ground-state neutral yttrium becomes noticeable [lines indicated by o in Fig. 3(k)].

At longer distances, a similar situation exists at lower pressures. Figure 4 shows a comparison of ion probe current, optical emission intensity (of Y* and Y^{+*}) , and optical absorbance (of Y and Y^+) time dependencies at d = 2.5 cm following KrF-laser irradiation of yttrium in (a) vacuum (1×10^{-6} Torr) and (b) 122 mTorr argon. In vacuum [Fig. 4(a)], ground-state neutrals and ions are found (through optical absorption) in coincidence with excited ions and neutrals (from optical emission) on the leading edge of the ion probe current. In background argon [Fig. 4(b)], excited and ground-state Y^+ ions dominate the first peak of the ion probe waveform, while excited and ground-state neutrals appear in the delayed, second distribution.

DISCUSSION AND CONCLUSIONS

In summary, simultaneous application of the four diagnostic techniques confirms that a component of target material is transmitted through low-pressure background gases with little or no delay compared to vacuum. Weak optical emission from this 'fast' component originates principally from yttrium ions. Ground state and excited neutrals, which are present during the 'fast' vacuum distribution, appear delayed (or depopulated) at similar times in background gases and are correlated principally with the bright luminescence at the slowed, second peak of the ion probe signal.

The data are consistent with scattering over extended distances in which ions and atoms in the ablation plume undergo collisions with mean-free-paths ~1 cm at 200 mTorr. Plume material is retarded and lost from the 'fast' vacuum velocity distribution due to collisional momentum transfer with the background gas (as well as other retarded plume material). This slowed material becomes observable as a distinct, second distribution only for a limited range of distances and pressures. Alternate explanations have been proposed, including ion-ion interactions with pre-ionized background gas atoms near the target surface,[11-13] however the imaging data implies interactions continuing at long distances from the target surface as evidenced by the plume luminescence.

The luminescence represents populations of short-lived excited states (radiative lifetimes ~10 ns which can be populated by a variety of collisional processes, including three-body recombination of ions with electrons, electron-impact excitation of ground and low-lying states of neutrals and ions, and collisional or radiative deactivation of Rydberg states. The luminescence intensity does not necessarily correspond to the overall plume density, but is determined by velocity- or temperature-dependent rate constants and their effects on the population kinetics of luminescent levels. The enhanced luminescence in the slowed 'shock region' is explained generally by favorable population kinetics there. The depletion of ground- and excited-state neutrals observed in the 'fast' distribution is more likely the result of conversion to neutral Rydberg states or ions than a preferential slowing of neutrals. A detailed experimental description of the plume luminescence and implications for plume temperatures and density-redistributions, as well as computer simulation of the above effects, will be presented elsewhere.

ACKNOWLEDGMENTS

The authors gratefully acknowledge many helpful discussions with C.-L. Liu, J. N. Leboeuf, K.-R. Chen, J. Donato, R. F. Wood, D. P. Norton, and D. H. Lowndes. This work was supported by the Division of Materials Sciences, U.S. Department of Energy under contract DE-AC05-84OR21400 with Martin Marietta Energy Systems, Inc.

REFERENCES

1. D. B. Geohegan, p. 28 in *Laser Ablation: Mechanisms and Applications*, ed. by J. C. Miller and R. F. Haglund, Springer-Verlag, Heidelberg, (1991).
2. D. B. Geohegan, Ch. 5. in *Pulsed Laser Deposition of Thin Films*, D. B. Chrisey and G. K. Hubler (eds.), Wiley, New York (1994) and references cited therein.
3. Roger Kelly and Antonio Miotello, (Ibid).Ch. 3.
4. D. B. Geohegan, *Appl. Phys. Lett.* **60**, 2732 (1992).
5. D.B. Geohegan and A.A. Puretzky, submitted to *Appl. Phys. Lett.* . (1995).
6. D. B. Geohegan,*Thin Solid Films* . **220**, 138 (1992).
7. D. B. Geohegan, p. 73 in *Laser Ablation of Electronic Materials: Basic Mechanisms and Applications*, ed. by E. Fogarassy and S. Lazare, North Holland (1992).
8. J. C. S. Kools, *J. Appl. Phys.* **74**, 6401 (1993).
9. D. B. Geohegan, pp. 165–185 in *Excimer Lasers*, NATO ASI Series E: Applied Sciences Vol. 265, L. D. Laude (ed.). Kluwer, Netherlands (1994).
10. D. B. Geohegan and D. N. Mashburn, *Appl. Phys. Lett.* **55**, 2345 (1989).
11. R. R. Goforth and David W. Koopman, *Phys. Fluids* **17**, 698 (1974).
12. David W. Koopman and R. R. Goforth, *Phys. Fluids* **17**, 1560 (1974).
13. David W. Koopman, *Phys. Fluids.* **15**, 1959 (1972).

LASER-SOLID INTERACTION AND DYNAMICS OF THE LASER-ABLATED MATERIALS

K. R. CHEN, J. N. LEBOEUF, D. B. GEOHEGAN, R. F. WOOD, J. M. DONATO, C. L. LIU, and A. A. PURETZKY
Oak Ridge National Laboratory, P. O. Box 2009, Oak Ridge, TN 37831-8071, USA

ABSTRACT

Rapid transformations through the liquid and vapor phases induced by laser-solid interactions are described by our thermal model with the Clausius-Clapeyron equation to determine the vaporization temperature under different surface pressure condition. Hydrodynamic behavior of the vapor during and after ablation is described by gas dynamic equations. These two models are coupled. Modeling results show that lower background pressure results lower laser energy density threshold for vaporization. The ablation rate and the amount of materials removed are proportional to the laser energy density above its threshold. We also demonstrate a dynamic source effect that accelerates the unsteady expansion of laser-ablated material in the direction perpendicular to the solid. A dynamic partial ionization effect is studied as well. A self-similar theory shows that the maximum expansion velocity is proportional to c_s/α, where $1 - \alpha$ is the slope of the velocity profile. Numerical hydrodynamic modeling is in good agreement with the theory. With these effects, α is reduced. Therefore, the expansion front velocity is significantly higher than that from conventional models. The results are consistent with experiments. We further study the plume propagates in high background gas condition. Under appropriate conditions, the plume is slowed down, separates with the background, is backward moving, and hits the solid surface. Then, it splits to be two parts when it rebounds from the surface. The results from the modeling will be compared with experimental observations where possible.

I. INTRODUCTION

Materials processing with lasers is considered to be a technique which will have great impact on materials science and engineering in the 90's and beyond. One of the main techniques for laser materials processing is Pulsed Laser Deposition (PLD) for thin film growth. While experimentalists search for optimal approaches for thin film growth, a systematic effort in theory and modeling of various processes during PLD is needed. In this paper, we will study three physics issues important to PLD: laser-solid interaction, dynamic source effect for an accelerated expansion, and background gas effect on the dynamics of ablated materials.

II. LASER-SOLID INTERACTION

We have extanded a laser-annealing model, "Laser8" [1], to include the vapor production stage. The model employs finite-difference method to solve the heat diffusion equation with enthalpy, temperature, and a state diagram. It can handle phase transitions, which may or may not be in equilibrium, through the state array according to the state diagram. For the phase transition from liquid to vapor, the vaporization temperature is determined by the pressure at the liquid surface, P_s, according to the Clausius-Clapeyron equation:

$$T_v = [\frac{1}{T_o} - \frac{ln(P_s/P_o)}{\Delta H}]^{-1} , \qquad (1)$$

where T_o is a known reference vaporization temperature at a reference pressure, (P_o), and ΔH is the latent heat for vaporization. Here, we have assumed that the materials emitted from the surface before vaporiation due to other mechanisms such as ionic emission maintain an equilibrium pressure with the background. In this model, the surface pressure is given as a constant for boiling cases or by the surface pressure from the gas dynamic model.

When the surface pressure is at one atmosphere, the corresponding boiling temperature is 3267 C for silicon. We calculate the interaction of the silicon solid with a 40 ns (FWHM)

KrF (248 nm wavelength) laser. The modeling results give the time history of the surface temperature for the cases of the laser energy density, E, being 3.8 and 3.9 J/cm², respectively. With 3.8 J/cm², the silicon surface never reaches the boiling temperature and no material is vaporized while it reaches the boiling temperature and about 4.4×10^{-8}cm thin layer of silicon becomes vapor with 3.9 J/cm². So, the laser energy density threshold for boiling of silicon in air with KrF laser is about 3.9J/cm². This is consistent with experimental measurements, which show that the silicon surface remains liquid with 1.9 J/cm²[2] and that visible surface damage by vaporization (frozen-in ripples and craters) was found not to occur until $E \sim 4.5$J/cm² of XeCl (308 nm wavelength) laser[3].

For the surface pressure being at 1 mTorr, the corresponding boiling temperature is 1615 C. From the modeling, we know that the laser energy density threshold is about 1.4J/cm². The experimental data of D. B. Geohegan shows that his ion probe receives signal at 1.5J/cm² or higher and no signal is detected at lower laser energy density. Again, the modeling result agrees with experimental measurement.

Fig. 1 shows the laser energy density threshold and the boiling temperature versus the surface pressure for silicon. For the surface pressure below 3.1×10^{-5}Torr, the boiling temperature is equal to the melting temperature; that is, there is no liquid phase to exist and the solid can become vapor directly. The laser energy density required to reach this temperature is about 0.7J/cm². For the surface pressure higher than that, the higher is the pressure, the higher is vaporization temperature. The modeling results show that the laser energy density threshold is linearly proportional to $\log P_s$. We also know the amount of material removed under different laser energy density. Fig. 2 show that both the maximum speed of surface recession due to vaporization and the depth of vaporization are linearly proportional to the laser energy density above its threshold at different background pressures.

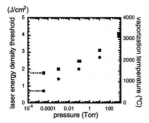

Fig. 1. laser energy density threshold(•) and vaporization temperature(square)vs. pressure

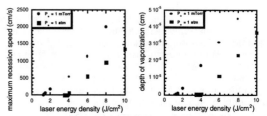

Fig. 2. maximum recession speed (left) and depth of vaporization (right) vs. laser energy density.

III. DYNAMIC SOURCE EFFECT FOR ACCELERATED EXPANSION

For laser ablation in materials research, the quality of film deposited critically depends on the range and profile of the kinetic energy and density of the ablated plume[6, 7, 8]. When it is the advantage of pulsed laser deposition to have high kinetic energy, plume being too energetic have been observed to cause film damage. It has long been an important conclusion[4]-[5],[11, 12] that the maximum escape velocity of an original stationary gas has a limit, which for an ideal gas is $c_s\sqrt{2/(\gamma - 1)}$ for a steady expansion and $2c_s/(\gamma - 1)$ for an unsteady expansion, where c_s is the initial sound speed and γ is the ratio of specific heats. However, experimental measurements always show that, at low laser fluence in which the laser energy absorbed by the plume is negligible, the expansion front is a factor of 2-3 faster than predicted from unsteady adiabatic expansion with typical vaporization temperature[6, 9, 10, 11]. The effect of a Knudsen layer[14] was studied in order to explain the higher expansion velocity. It gives a velocity of $4u_k$[11, 12], where $u_k < c_s$ is the Knudsen layer velocity. This front expansion velocity is still too low. The inability to explain the experimental observation through gas dynamics has prompted a suggestion[13] of increased vapor temperature due to violent interactions inside the target such as phase explosion[16, 17].

We will demonstrate a dynamic source effect that accelerates the unsteady expansion significantly faster than predicted from conventional models in the direction perpendicular to the target surface. An effect of dynamic partial ionization that increases the expansion in all directions is also studied. This may explain the historical puzzle observed in laser ablation experiments. As in previous work[9]-[13], we are interested here in a laser fluence range high enough for hydrodynamic theory to be applicable but low enough for the absorption of the laser energy by the plume to be weak so that we can compare with free expansion models that do not include absorption.

In free expansion model the plume is given in a reservoir at $t = 0$. When the gate of the reservoir is removed, the gas adiabatically expands forward and a rarefection wave moves with the sound speed from the gate to the back wall in a period of time (t_r) during which the wall pressure remains constant. Then the wall pressure drops quickly. The average velocity gained per particle after the wall pressure drop is $c_s/\sqrt{\gamma}$. The maximum expansion velocity is $2c_s/(\gamma - 1)$ [4]-[5],[11, 12].

In our model the same material is treated as a source dynamically into the system after $t = 0$. For the plume pressure, P, below its thermodynamic critical pressure and with low plume viscosity, we may assume that the plume behaves as an ideal gas such that $P = n(1 + \eta)k_B T$, where n (T) is the density (temperature) of the plume, η is the ionization fraction, and k_B is the Boltzmann constant. We use Euler's equations to model the plume dynamics and the Saha equation to determine the ionization fraction[15]:

$$\frac{\partial}{\partial t}(n) = -\frac{\partial}{\partial x}(nv) + S_n\delta(x - x_s) , \qquad (2)$$

$$m\frac{\partial}{\partial t}(nv) = -\frac{\partial}{\partial x}(P + mnv^2) , \qquad (3)$$

$$\frac{\partial}{\partial t}(E) = -\frac{\partial}{\partial x}[v(E + P)] + S_E\delta(x - x_s) , \qquad (4)$$

$$\frac{\eta^2}{1 - \eta} = \frac{2}{n}\frac{u_+}{u_o}\left(\frac{2\pi m_e k_B T}{h^2}\right)^{3/2}e^{-\frac{U_i}{k_B T}} , \qquad (5)$$

where v is the plume velocity, $E = mne + mnv^2/2$ is the plume energy density, $e = (1+\eta)(k_B T/m)/(\gamma - 1) + \eta U_i$ is the plume enthalpy, U_i is the ionization potential, u_+ and u_o are the electronic partition functions, m is the mass of the plume atom, m_e is the electron mass, h is Plank's constant, $S_n = n_{liq}v_{rs}$ is the density source, $S_E = n_{liq}v_{rs}k_B T_v/(\gamma - 1)$ is the energy source, n_{liq} is the liquid density, v_{rs} is the recession speed of the target surface due to ablation, and T_v is the vapor temperature. Here we take the small Knudsen layer limit, use $v = 0$ at the surface, and let S_n and S_E be constant. Because $c_s >> v_{rs}$, the surface recession on the plume expansion can be neglected[18]; i.e., $x_s = 0$.

A self-similar theory for the dynamic source effect. For simplicity and comparision with the free expansion results, our analysis considers the gas to be neutral, which is a good approximation for $T_v \ll U_i$. With an energy source, the system is not adiabatic near the surface. We expect self-similar expansion, except for early times and a transition region near the surface (δx). The self-similar variable is $\xi \equiv x/v_m t$, where v_m is the maximum expansion velocity. The source boundary conditions at $\xi = \delta \equiv \delta x/v_m t \ll 1$ are given by the constants $n = n_\delta$, $T = T_\delta$, and $v = v_\delta$. The downstream boundary conditions are $n = 0$, $v = v_m$, and $T = 0$ at $\xi = 1$ for expansion in vacuum. We assume a velocity profile of $v = v_m[\alpha + (1 - \alpha)\xi]$, where α is determined by the flow properties $(1 \geq \alpha \geq 0)$. We transform the independent variables from (x, t) to ξ. From Eqs.(2) and (3) we obtain the density profile $n = n_\delta(1 - \xi)^{(1-\alpha)/\alpha}$ and the pressure profile $P = n_\delta v_m^2 m \alpha^2(1 - \alpha)/(1 + \alpha)(1 - \xi)^{(1+1/\alpha)}$. So the temperature profile is $k_B T/m = v_m^2\alpha^2(1-\alpha)/(1+\alpha)(1-\xi)^2$. From mass, momentum, and energy conservations, we know the relations of v_m, α, n_δ, and T_δ. We note that if $\alpha = (\gamma - 1)/(\gamma + 1)$ is used, as for adiabatic cases, the analytical theory can recover the previous results[11, 12] of free expansion with a Knudsen layer. Figure 3 shows the normalized maximum expansion velocity as a function of α for a monatomic gas, $\gamma = 5/3$. The rapid rise of the maximum expansion velocity at $\alpha \leq 0.1$ is due to the $1/\alpha$ dependence. The value $v_m/c_\delta = 4$ for $\alpha = 1/4$ corresponds to the case of adiabatic expansion with a Knudsen layer[11, 12]. The flow at δ is sonic for the case of adiabatic expansion with the Knudsen layer. Figure 3 also shows that the flow at $\xi = \delta$ is subsonic (supersonic) for $\alpha < 1/4$ ($\alpha > 1/4$). The local temperature in terms of T_v is not sensitive to α.

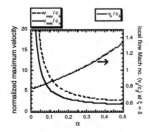

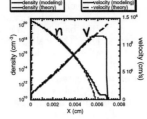

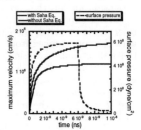

Fig.3. The normalized maximum expansion velocity and the local flow Mach number as a function of α for $\gamma=5/3$ from the self-similar theory. In the free expansion models, $\alpha=0.25$.

Fig.4. The plume density and velocity profiles at $t = 5$ ns from the self-similar theory and numerical hydrodynamic modeling. Here, we have $\alpha=1/14$. The corresponding results from the free expansion models are $\alpha=1/4$, $v_m=3$ c_s.

Fig. 5. The histories of the surface pressure, and the maximum velocity from the velocity profile with and without the Saha equation for ionization. The maximum expansion velocity from the free expansion model is 5.5×10^5 cm/s.

Numerical hydrodynamic modeling. The Rusanov scheme[19] is used to solve Euler's equations, Eqs.(2)-(4); the nonlinear calculation of T and η is done with the Newton-Raphson method[20]. We use the logarithm of Eq.(5) for numerical stability. The system size is 1,000 spatial grids, Δx. The initial adaptive size is 10^{-5}cm, which is required for numerical convergence. New vapor is added into the first cell near the surface perturbatively. This limits the size of time step according to $n_{liq} v_{rs} T_v \Delta t \ll n_1 T_1 \Delta x$, where the subscript 1 represents the first cell.

The typical physical parameters are as follows. The system is initiated with a uniform background gas with its density $n_{bg} = 1 \times 10^{10}$ cm^{-3} and its temperature $T_{bg} = 293$ °K, which give a pressure $P_{bg} \sim 0.3\mu$Torr. A constant source of vapor is specified for 6 ns with a temperature $T_v = 7000$ °K, given by the Clausius-Clapeyron equation, and the target recession speed is $v_{rs} = 1 \times 10^3$ cm/s, which are typical for the ablation of silicon with laser fluence of a few J/cm^2 [1]. We choose the mass of both source and background gas to be 28 a.m.u., a solid density of 5.01×10^{22} cm^{-3}, an ionization potential of 1.3×10^{11} erg (8.1 eV), $u_+ = 6$, and $u_0 = 15$. These parameters correspond to silicon. The normalized results should also be applicable to different materials. We use $\gamma = 5/3$ appropriate for monatomic gas. Thus, $c_s = 1.85 \times 10^5$ cm/s. We note that the conventional free expansion model for no background gas (vacuum) gives a maximum expansion velocity of 5.55×10^5 cm/s.

We first study the case without the Saha equation (no ionization; i.e., $\eta = 0$). Figure 4 shows the profiles of density and velocity at $t = 5$ ns, at which time the expansion is almost steady state. From the simulations, we observe that the expansion develops self-similarly after 0.1 ns. The front position is at $x = 0.0069$ cm at $t = 5$ ns. The maximum expansion velocity at this time is defined to be the ratio of the front position and the time; i.e., $v_m = 1.38 \times 10^6$ cm/s or $7.46 c_s$. From the slope of the velocity profile, we estimate $\alpha = 1/14 = 0.07143$, which gives $v_\delta = 9.85 \times 10^4$ cm/s. Thus, $\delta x = 6.4 \times 10^{-5}$ cm. The simulation also shows that $n_\delta = 4.7 \times 10^{20}$ cm^{-3} and $T_\delta = 3693$ °K. The analytical maximum expansion velocity is $7.42 c_s$. Also, $n_\delta = 5.07 \times 10^{20}$ cm^{-3} and $T_\delta = 2836$ °K. The analytical profiles in the figure are given by $n = n_\delta(1-x/0.0069\,\mathrm{cm})^{13}$ and $v = v_m/14+(13/14)(x/5\,\mathrm{ns})$ from the self-similar theory. Although the profiles at the shock front are flattened due to the finite background pressure (not included in this analytical theory), the overall profiles and scalings are in good agreement with the analytical theory. Figure 5 shows how the dynamic source causes the surface pressure to rise quickly and approach a saturation level of about $6 \times 10^8\,\mathrm{dyne/cm^2}$, or 600 atmospheres. Then the surface pressure exponentially drops after the source is terminated at $t = 6\,\mathrm{ns}$. Meanwhile, the maximum velocity from the velocity profile rises and saturates due to the surface pressure and the nonadiabatic unsteady expansion. The maximum velocity at $t = 10\,\mathrm{ns}$ is about $1.22 \times 10^6\,\mathrm{cm/s}$. When we use the Saha equation (the more physical case), we find that the surface pressure remains unchanged and the maximum velocity is about 40% higher. It reaches $1.70 \times 10^6\,\mathrm{cm/s}$ or $9.2\,c_s$ at $t = 10\,\mathrm{ns}$. As discussed earlier, the higher maximum velocity is an effect due to dynamic partial ionization as a result of increased energy channeled into directed motion. This effect is reduced when the vapor temperature is lower; it gives only about a 6% increase when $T_v = 3500\,°\mathrm{K}$, for example.

IV. PLUME DYNAMICS IN BACKGROUND GAS

Laser ablation experiments have shown that the plume propagation in background gas can lead to the stopping of the ablated materials. In some cases, the materials can even go backward and several reflected shocks within the plume are evidenced.

With the hydrodynamic modeling, we have simulate the plume dynamics in following parameters: the recession speed of solid silicon surface is 100 cm/s lasting for 6 ns with a vapor temperature 7000 K and the background gas density is $6.6 \times 10^{15}\mathrm{cm^{-3}}$ with room temperature (i.e., the background pressure is 200 mTorr). Fig. 6 shows the plume dynamics at different times. At t=10ns as shown in Fig. 6(a), the background gas has been snowplowed. Also, the temperature and ionization fraction rise at the shock front. Fig. 6(b) show that the relative higher pressure at shock front has split the plume and background. This couples with the rarefaction of the plume to begin pushing the main body of the plume (2nd peak) backward and, thus, to slow it down. As a result, the velocity of the second peak is decreased toward zero. By t=100 μs, the velocity has become negative; that is, the second peak moves backward as inidcated in Fig. 6(c). The backward moving plume eventually hits the target surface, rebounds, and moves forward again. The resultant plume splits as shown on Fig. 6(d).

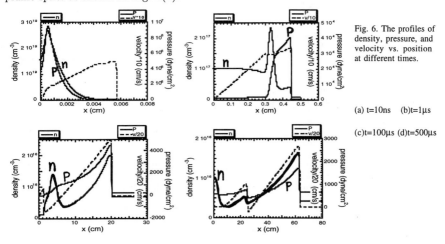

Fig. 6. The profiles of density, pressure, and velocity vs. position at different times.

(a) t=10ns   (b)t=1µs

(c)t=100µs (d)t=500µs

We have also checked the scaling law of the turnover position of ablated plume. The numerical modeling results show that the turnover position of ablated plume is inversely proportional to the gas pressure and is proportional to the amount of ablated materials.

V. SUMMARY

Both a thermal model for studying laser-solid interaction and a hydrodynamic model for the dynamics of laser-ablated materials have been developed. It is shown that lower background pressure results lower laser energy density threshold for boiling, which is consistent with experimental measurements. Both the recession speed of the surface due to vaporization and the vaporization depth are proportional to the laser energy density above its threshold. We have treated the laser-ablated materials as a dynamic source, which is closer to experimental condition, instead of an initial constant source as in free expansion models. It is demonstrated that the dynamic source and partial ionization effects can dramatically increase the front expansion velocity, which becomes significantly higher than predicted from conventional free expansion models, while the average momentum in the direction perpendicular to the solid surface is moderately increased. Since the expansion is accelerated mainly in the perpendicular direction, it should become more nonsymmetric and forward-peaked. Two dimensional model would be required to study the resultant plume profile and dynamics away from the target surface. The profiles and scalings from numerical hydrodynamic modeling are in good agreement with our self-similar theory. The results are consistent with experimental observations. For plume propagation in a background gas, our results show that the background gas acts on the main body of the rarefying plume, tends to slow it down, and in some cases even results in backward going materials.

ACKNOWLEDGE

This work is supported by the Oak Ridge National Laboratory (ORNL) Director's Research and Development Funds, and the Division of Materials Sciences, U. S. Department of Energy, under contract DE-AC05-84OR21400 with Martin Marietta Energy Systems, Inc. Chen and Liu were supported in part by appointments to the ORNL Research Associate Program administered jointly by the Oak Ridge Institute for Science and Education and ORNL.

References

[1] R. F. Wood and G. A. Geist, *Phys. Rev. Lett.* 57, 873 (1986).
[2] G. E. Jellison, Jr., D. H. Lowndes, D. N. Mashburn, and R. F. Wood, *Phys. Rev. B* 34, 2407 (1986).
[3] D. H. Lowndes, *et. al.*, *Appl. Phys. Lett.* 41, 938 (1982).
[4] L. D. Landau and E. M. Lifshitz, *Fluid Mechanics*, Addison-Wesley:New York, 357 (1959).
[5] Ya. B. Zel'dovich and Yu. P. Raizer, *Physics of Shock Waves and High-Temperature Hydrodynamic Phenomena*, Academic Press:New York (1966).
[6] J. F. Ready, *Effects of High-Power Laser Radiation*, Academic Press:Orlando Florida, (1971).
[7] D. B. Chrisey and G. K. Hubler, *Pulsed Laser Deposition of Thin Films*, John Wiley & Sons:New York, (1994).
[8] R. K. Singh, O. W. Holland, and J. Narayan, *J. Appl. Phys.* 68, 233 (1990a).
[9] R. Kelly and R. W. Dreyfus, *Surf. Sci.* 198, 263 (1988).
[10] R. Kelly and R. W. Dreyfus, *Nucl. Instrum. Methods in Phys. Res. B* 32, 341 (1988).
[11] R. Kelly, *J. of Chem. Phys.* 92, 5047 (1990) and references therein.
[12] R. Kelly, *Phys. Rev. A.* 46, 860 (1992).
[13] R. Kelly, A. Miotello, A. Mele, A. Giardini, J. W. Hastie, P. K. Schenck, and H. Okabe, submitted to *Surf. Sci.* Also, private communication with R. Kelly.
[14] C. J. Knight, *AIAA J.* 17, 519 (1979).
[15] See,for example, A. Vertes, in *2nd Int'l Conf. on Laser Ablation: Mechanisms and Application-II*, edited by J. C. Miller and D. B. Geohegan, AIP Conf. Proc. No. 288, 275 (1994).
[16] S. Otsubo, T. Minamikawa, Y. Yonezawa, A. Morimoto, and T. Shimizu, *Japanese J. of Appl. Phys.* 29, L 73 (1990).
[17] R. K. Singh, D. Bhattacharya, and J. Narayan, *Appl. Phys. Lett.* 57, 2022 (1990).
[18] G. Weyl, A. Pirri, and R. Root, *AIAA J.* 19, 460 (1981).
[19] G. A. Sod, *J. of Comput. Phys.* 27, 1-31 (1978).
[20] W. H. Press, B. P. Flannery, S. A. Teukolsky, and W. T. Vetterling, *Numerical Recipes (FORTRAN Version)*, 267 (1989).

PLANAR LASER-INDUCED FLUORESCENCE DIAGNOSTICS OF PULSED LASER ABLATION OF SILICON

D. G. GOODWIN,* D. L. CAPEWELL,* AND P. H. PAUL**
*Division of Engineering and Applied Science, California Institute of Technology, Pasadena, CA 91125
**Combustion Research Facility, Sandia National Laboratories, Livermore, CA

ABSTRACT

Planar laser-induced fluorescence has been used to acquire time sequence images of ground-state, neutral Si and SiO during laser ablation of an Si target in vacuum and in the presence of a background gas at a fluence of 3–4 J/cm^2. The SiO images, taken in air, strongly suggest that the observed SiO is created through reaction of silicon with oxygen at the contact front as the plume expands.

INTRODUCTION

Pulsed laser deposition (PLD) is an attractive method to deposit a wide range of thin films [1]. The physics leading to plume formation is complex, resulting in a rich variety of plume phenomena, depending on target material, fluence, and the presence or absence of a background gas. Time-resolved imaging can shed light on some of the complex processes in the plume. Several different methods have been employed to image PLD plumes, including emission, dye laser absorption, shadow photography, and planar laser-induced fluorescence (PLIF). (See Geohegan [2] for a review of these studies.)

Planar laser-induced fluorescence is particularly useful, since it involves no line-of-sight averaging, is species-specific (like absorption) and can detect non-luminous ground state species. A number of groups have used single-point or one-dimensional laser-induced fluorescence for plume studies [3, 4, 5]. However, there are only two reports of two-dimensional planar LIF imaging of a laser-ablation plume [6, 7].

In the present paper, we present some results of a PLIF study of the ablation of an Si target into vacuum and air. We have acquired time sequence images of ground-state Si and, in the experiments with air, ground-state SiO. The images of SiO are, to our knowledge, the first reported of a reactive intermediate species generated during pulsed laser deposition.

EXPERIMENTAL

The experiments were carried out in a stainless steel vacuum cube with quartz windows. The experimental setup is shown schematically in Figure 1. The target was a silicon wafer and was rotated continuously during the experiments. The ablation laser was an injection-seeded 10 Hz excimer laser running on KrF with a nominal energy per pulse of 175 mJ. The excimer laser beam was focused onto the target at normal incidence using a single 500 mm focal length aberration-corrected doublet lens. The spot size on the target was approximately 2.5 mm diameter, resulting in a fluence of 3–4 J/cm^2.

An excimer-pumped dye laser was used for the fluorescence excitation. Coumarin 480 dye was used for both the Si and SiO measurements, with a nominal energy per pulse of 1 mJ and a bandwidth of 0.15 cm^{-1}.

The dye laser beam was transformed into a collimated sheet nominally 1 cm wide using a telescope made from a −64 mm cylindrical lens and a +500 mm spherical lens. The laser sheet was brought to a soft waist of thickness 0.02 cm.

Mat. Res. Soc. Symp. Proc. Vol. 388 © 1995 Materials Research Society

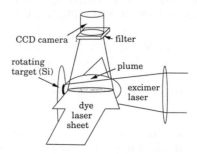

CCD camera — filter

rotating
target (Si) — plume

excimer
laser

dye
laser
sheet

Figure 1: Schematic of experimental setup used for PLIF measurements.

A gated single-microchannel-plate intensified CCD camera oriented perpendicular to the dye laser sheet and to the target was used to collect the fluorescence. A 105 mm f/4.5 UV-Nikor lens was used to image the fluorescence onto the image intensifier. The gate width was set at 50 ns. The camera was operated in an RS-170 video format and the image recorded using a frame grabber in a 486 computer.

After the experiment, each image was corrected for the dye laser sheet intensity variation ($< 40\%$). Typically either 20 or 40 images were acquired for each experimental condition and averaged to reduce statistical noise.

For the Si fluorescence experiments, the dye laser output was doubled in BBO to excite the Si $3p^2\ ^3P \rightarrow 4s\ ^1P^o$ transition near 243.88 nm. A 290 ± 5 nm bandpass filter was used to collect fluorescence from the $3p^2\ ^1D \leftarrow 4s\ ^1P^o$ transition near 288.1 nm. For the SiO fluorescence experiments, an overlap of the $Q1(35)$ and $R1(42)$ transitions in the SiO $A^1\Pi \leftarrow X^1\Sigma^+(0,0)$ system near 235.25 nm ($f_{00} = 0.13$) was excited. A 260 ± 10 nm filter was used to detect fluorescence from the (0,4) band as well as portions of emission from the (0,3) and (0,5) bands. A 1 mm thick piece of Schott UG1 filter glass was also used to increase the rejection of scattered KrF laser light. For both Si and SiO detection, the short upper-state lifetimes make collisional quenching negligible; consequently, the LIF signal is proportional to the number density in the lower level.

For the SiO measurements, significant luminous emission was detected in the 260 ± 10 nm filter bandpass which interfered with the LIF signal. To overcome this, images were acquired with the laser on and with the laser off and subtracted to remove the emission interference. No interfering emission was observed with the 290 ± 5 nm filter used for the Si experiments.

RESULTS

Si Fluorescence Measurements

Ablation into vacuum

Shown in Figure 2 are normalized Si atom spatial distributions resulting from ablation in vacuum for delay times between the ablation laser and dye laser of up to 4 μs. In these plots, the target is at the left, and the horizontal axis dimensions are in cm. The intensity scale is normalized to the maximum value in each image.

As is usually observed for ablation into vacuum, the Si plume forms a narrow jet perpendicular to the target. In the images at 0.2 and 0.6 μs, the plume appears to consist of two

34

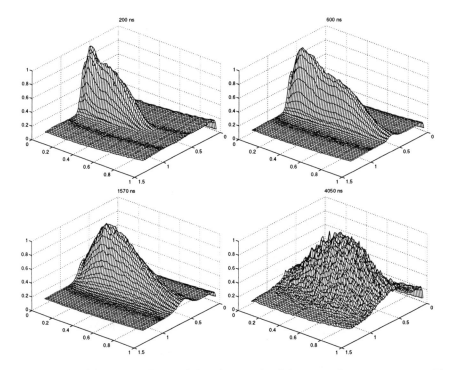

Figure 2: Neutral Si atom LIF spatial distributions for delay times from 0.2 to 4.0 μs. The target is on the left.

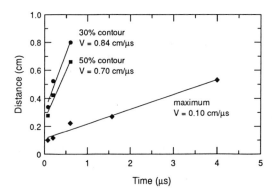

Figure 3: Axial position vs. time for Si LIF intensity maximum and for contours of 30% and 50% of maximum intensity.

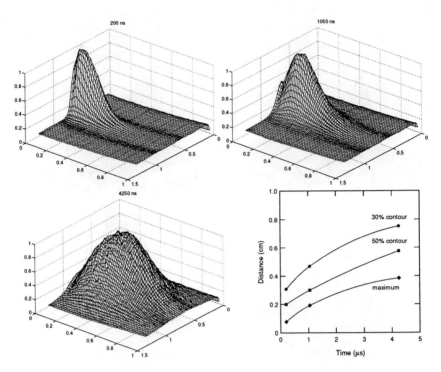

Figure 4: Neutral Si atom LIF spatial distribution for expansion into 100 mTorr of air. Lower right: position vs. time of maximum and 30% and 50% intensity contours.

components. A fast component is seen at the forward edge of the plume, which is no longer apparent by 1.57 μs, having left the image field. Similar two-component plumes have been observed by others [2]. While the mechanism for producing a two-component plume is still unclear, the fast Si atoms may result from fast ions which are neutralized by recombination or charge exchange.

As shown in Figure 3, the region of maximum Si LIF intensity moves away from the target at close to 10^5 cm/s. However, the fast component, as measured by the points on the forward edge where the intensity has dropped to 50% or 30% of the maximum, is traveling 7–8 times as fast.

Ablation into air

Shown in Figure 4 are Si atom LIF distributions resulting from ablation into 100 mTorr of air. As expected, the plume is broadened compared to the results in vacuum. No fast component is observed, and the region of maximum LIF intensity moves at 10^5 cm/s. It is interesting to note that this velocity is very similar to the slow, neutral component seen in vacuum.

Emission images of PLD plumes at the same pressure and similar fluence [2] usually show strong forward peaking in the intensity. In contrast, the present LIF images do not show

 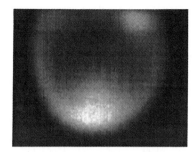

Figure 5: Detection in the 250–270 nm spectral range at 2.4 μs in 1 Torr of air. Left: emission only; Right: emission + SiO LIF. The target is at the top of the image, the image height is 0.96 cm, and the width is 1.27 cm.

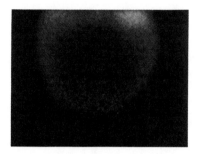 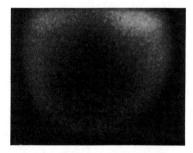

Figure 6: Emission-subtracted SiO LIF image at 2.0 μs (left) and 3.75 μs (right). Background gas is 1 Torr of air.

peak signal at the forward edge. These results show that the high emission intensity at the leading edge is not associated with high neutral density. It is likely that the forward peaking in emission is due to a high temperature or electron density at the leading edge, resulting in collisional population of excited states.

SiO Fluorescence Measurements

Images of ground-state SiO were acquired at a pressure of 1 Torr in air. The interference from emission in this spectral band (250 – 270 nm) is seen in Figure 5, which shows the measured signal at 2.4 μs with the dye laser off (left) and with the dye laser on (right). The emission-only image is similar to emission images acquired during YBCO ablation [2].

The emission-subtracted SiO PLIF images obtained for ablation into 1 Torr of air are shown in Figure 6 for delay times of 2 and 3.75 μs. SiO fluorescence is seen around the entire perimeter of the plume at the contact front between the expanding Si plume and the air. These images strongly suggest that the SiO we observe is created by gas-phase chemistry at the contact front between the Si and air.

The SiO density is observed to be greatest in the wings of the plume near the target. There are several possible explanations for this. It may be that the temperature near the leading edge of the plume is too high for SiO formation. (If the gas were in chemical equilibrium, at 1 Torr little SiO could form for temperatures much above 4000 K.) Alternatively, the observed distribution may simply be the result of the stretching of the interface region as the plume expands. Since the stretching will be greatest on-axis and least near the target, this could account for the observed distribution.

SUMMARY

Planar laser-induced fluorescence images have been acquired of ground-state Si and ground-state SiO during laser ablation of Si into vacuum and air at a fluence of 3–4 J/cm^2. In vacuum, the plume is ejected in a narrow jet, with a fast component with velocity approaching 10^6 cm/s and a larger slow component traveling at approximately 10^5 cm/s. In the presence of 100 mTorr of air, the plume is broader, and the entire plume travels with velocity of order 10^5 cm/s. The SiO images show that SiO is localized at the contact front between the ablated Si and the background gas, suggesting that the observed SiO is formed due to gas-phase chemistry.

ACKNOWLEDGMENTS

PHP acknowleges support from the United States Department of Energy, Office of Basic Energy Sciences, Chemical Sciences Division. DLC and DGG acknowledge partial support from a National Science Foundation Presidential Young Investigator grant.

References

[1] *Pulsed Laser Deposition of Thin Films*, edited by D. B. Chrisey and G. K. Hubler (John Wiley and Sons, Inc., New York, 1994).

[2] D. B. Geohegan, *ibid.* pp. 115–165.

[3] D. Fried, S. Jodeh, and G. P. Reck, J. Appl. Phys. **75**, 522 (1994).

[4] D. L. Pappas, K. L. Saenger, J. J. Cuomo, and R. W. Dreyfus, J. Appl. Phys. **72**, 3966 (1992); R. W. Dreyfus, J. Appl. Phys. **69**, 1721 (1991); C. E. Otis and R. W. Dreyfus, Phys. Rev. Lett. **67**, 2102 (1991). R. W. Dreyfus, Appl. Phys. A **55**, 335 (1992).

[5] T. Okada, N. Shibamaru, Y. Nakayama, and M. Maeda, Appl. Phys. Lett. **60**, 941 (1992); T. Okada, N. Shibamaru, Y. Nakayama, and M. Maeda, Jpn. J. Appl. Phys. **31**, L367 (1992); T. Okada, Y. Nakata, W. Kumuduni, and M. Maeda, Appl. Surf. Sci. **79/80**, 136 (1994); Y. Nakata, K. A. Kumuduni, T. Okada, and M. Maeda, Appl. Phys. Lett. **64**, 2599 (1994); W. Kumuduni *et al.*, J. Appl. Phys. **74**, 7510 (1993).

[6] M. A. Cappelli, P. H. Paul, and R. K. Hanson, Appl. Phys. Lett. **56**, 1715 (1990).

[7] D. P. Butt, P. J. Wantuch, and A. D. Sappey, J. Amer. Ceram. Soc. **77**, 1411 (1994).

VELOCITY SELECTION OF LASER ABLATED METAL ATOMS
BY A NOVEL NON-MECHANICAL TECHNIQUE

MARIO E. FAJARDO AND MICHEL MACLER*
Emerging Technologies Branch, Propulsion Directorate, Phillips Laboratory,
OLAC PL/RKFE, 10 E. Saturn Blvd., Edwards Air Force Base, CA 93524-7680.
*AFMC PL/NRC Post-doctoral Research Associate.

ABSTRACT

We present the results of experiments on velocity selection of fast laser ablated Al, Ga, and In atoms by a novel, non-mechanical, technique. Pulses of atoms with broad velocity distributions are produced by laser ablation of a single component pure metal target in vacuum. After a delay of ~ 1 µs, there exists a strong one-to-one correlation between atomic velocity and distance traveled from the ablated surface. Thus, a second pulsed laser, delayed by ~ 1 µs and crossed at a right angle to the atomic beam, can be used to photoionize only those atoms with unwanted velocities, *i.e.*: atoms moving too fast or too slow to be hidden behind an opaque mask placed ~ 1 cm from the ablated surface. The photoions, and any ions surviving from the ablation event, are subsequently deflected from the beam by a static magnetic field. By a fortunate coincidence, Al, Ga, and In atoms all have very large single photon photoionization cross sections at 193 nm, the output wavelength of the ArF excimer laser; thus, well over 95% of the unwanted atoms can be easily photoionized and rejected. We have demonstrated velocity selected Al, Ga, and In atom fluxes equivalent to $\Phi \sim 10^{11}$ atoms/(cm^2-eV-pulse) at a working distance of 10 cm.

INTRODUCTION

Pulsed laser ablation of solid targets is an increasingly popular technique for the deposition of a wide variety of thin film materials [1]. There are many fundamental and applied studies underway of *e.g.*: laser/surface interactions, plume plasma hydrodynamics, plume composition *vis-à-vis* ions/neutrals/clusters/particulates, internal and kinetic energy content of ablated species, *etc.*, and of how all these factors ultimately affect the processes of thin film deposition and growth. Unfortunately, progress towards this ultimate goal is hampered by the bewildering complexity of the phenomena involved, a predicament that has been likened to studying "a tornado in a garbage can." We believe that any modification of "standard" laser ablation techniques which result in a simplification of this situation, and/or in improved parametric control over deposition conditions, will prove to be highly valuable to the thin film deposition community.

As a case in point: we have been employing pulsed laser ablation of metal targets as a source of metal atoms for matrix isolation spectroscopy (MIS) studies for several years now [2-6]. We prepare our MIS samples by codepositing the products of a laser ablated plume along with a large excess of an inert matrix host gas onto a cryogenically cooled substrate. These experiments have lead us to hypothesize that the incident kinetic energy (KE) of the ablated metal atoms plays a key role in determining the atomic isolation efficiency of the matrix deposition process, and in the formation of novel metal atom trapping site structures.

However, this hypothesis is largely speculation, as we have not measured *in situ* (in our MIS apparatus) the actual composition of these laser ablated plumes, or the kinetic energy distributions (KEDs) of the ablated atoms. Furthermore, the laser ablation process can produce a mixture of metal atoms, clusters, and ions with rather broad KEDs, in some cases with KEs tailing out to several tens of electron volts. Thus, we undertook an effort to better characterize the chemical identity and KEDs of the laser ablated species produced under our experimental conditions, and to find out if we could gain improved control over these properties [7,8]. This effort culminated last year in our demonstration of velocity selection of fast laser ablated Al atoms by Temporally And Spatially Specific PhotoIonization (TASSPI) [9]; a novel, non-mechanical technique which provides a clean, stable, compact, intense, and tunable source of fast monoenergetic metal atoms for MIS experiments in particular, and thin film depositions in general.

In this manuscript, we briefly describe the TASSPI technique, and present the highlights of our recent results on velocity selection of Ga and In atoms; the interested reader is directed to ref. 9 for details of our Al atom work. A more detailed presentation of all our TASSPI related work is currently in preparation [10].

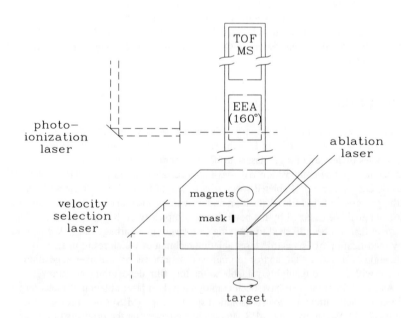

Fig. 1: *TASSPI Experimental Diagram.*

EXPERIMENTAL

Fig. 1 shows a schematic of our experimental setup, as configured for velocity selection by TASSPI. The metal ablation targets are mounted on a rotatable rod within a vacuum chamber pumped directly by a small turbomolecular pump to ~ 10^{-6} Torr. The ablated plumes are generated by an excimer laser beam (XeCl, $\lambda = 308$ nm) incident at an angle of 45° from the surface normal, and focused to a ≈ 0.003 cm² spot so as to cut a circular track on the rotating target. The laser ablated plume products pass through a magnetic field region ($|\mathbf{B}|$ = 2.8 kG), located 2.5 cm from the ablation target surface, which deflects ionic species (e.g.: Ga^+ and In^+ ions with KE ≤ 15 eV) from the atomic beam axis. The beam axis is defined by the ablation spot on the target and a 0.2 cm diameter isolation aperture located ≈ 6 cm from the target surface, from which we estimate the final beam divergence as ≈ 40 mrad.

We characterize the chemical composition and KEDs of the species emerging through the isolation aperture with an electrostatic energy analyzer/time-of-flight mass spectrometer (EEA/TOFMS) system, using pulsed photoionization (ArF, $\lambda = 193$ nm) of neutrals [8,9]. As we have discussed before [8,9], the apparent redundancy of this detection scheme (which requires that the time of flight of neutrals from the ablation target surface to the photoionization region be appropriate to the photoion pass energy selected by the EEA) allows us to detect the presence of non-atomic M^+ ion precursors in the atomic beam. The photoions which transit the EEA are accelerated to 2.0 keV, separated by mass in a 1 m TOF tube, and detected by a low gain microchannel plate detector.

Velocity selection by TASSPI entails photoionizing those metal atoms with unwanted velocities using another ArF excimer laser ("velocity selection laser" in fig. 1) and rejection of the resulting photoions by the magnetic deflection field. We have previously reported measurements of Al atom KEDs made close to and far from the ablated target surface which demonstrate that, under our mild ablation conditions, the Al atom KEDs are established on a sub 100 ns timescale [7,8]. Thus, after a delay of ~ 1 μs, there exists a strong one-to-one correlation between atomic velocity and distance traveled from the ablated surface (i.e. the atoms have sorted themselves out with the faster atoms traveling farther from the surface). A simple opaque mask of width Δx, placed at a distance x from the target surface, serves to protect those metal atoms hidden behind it at the arrival time of the velocity selection laser (t_{VS}). These metal atoms, with mean velocity v = x/t_{VS}, will not be photoionized, and hence will not be deflected from the atomic beam by the magnetic rejection field. The mean velocity can thus be set by adjusting x, and/or by adjusting t_{VS}, within limits set by the requirement that the fastest metal atoms to be rejected still be in the velocity selection region at t_{VS}. The surviving metal atoms have a spread of velocities, Δv (full width at half maximum, FWHM), and a corresponding spread of kinetic energies, ΔKE, all related by:

$$\frac{\Delta KE}{KE} \approx 2\frac{\Delta v}{v} \approx 2\left[\left(\frac{\Delta x}{x}\right)^2 + \left(\frac{\Delta t_{vs}}{t_{vs}}\right)^2\right]^{\frac{1}{2}}$$

in which Δt_{VS} is taken as the sum of the duration of the velocity selection laser pulse and the jitter between the ablation and velocity selection pulses (Δt_{VS} ≈ 50 to 100 ns).

RESULTS AND DISCUSSION

We have previously shown in our Al experiments that maintaining the ablation laser intensity (I_{abl}) below $\approx 1 \times 10^8$ W/cm² greatly reduces the number of metal clusters and particles formed in the ablation process [8,9]. For Ga and In targets, we achieved the same results by keeping I_{abl} below $\approx 4 \times 10^7$ W/cm².

By varying the fluence of the ArF photoionization laser up to about 500 mJ/cm² we were able to saturate the intensities of both Ga⁺ and In⁺ signals. By fitting these saturation curves we can estimate the atomic photoionization cross sections at 193 nm as $\sigma_{PI}(Ga) = 2.7(\pm 0.6) \times 10^{-17}$ cm² and $\sigma_{PI}(In) = 1.7(\pm 0.6) \times 10^{-17}$ cm², in good agreement with previously published values [11]. These measurements confirm the velocity selection laser fluences, F_{VS}, required for efficient implementation of the TASSPI process.

Fig. 2 shows the results of a velocity selection experiment on laser ablated Ga

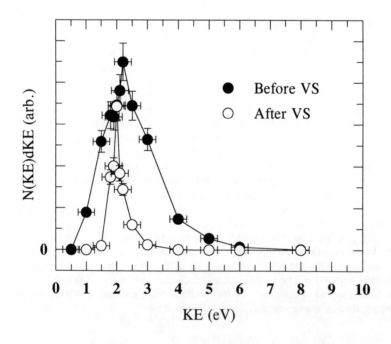

Fig. 2: *Velocity Selection of Ga Atoms by TASSPI. The closed circles show the nascent Ga atom KED, the open circles show the TASSPI effect. The error bars represent the estimated $\pm 1\ \sigma$ limits. The experimental conditions were: $I_{abl} = 3.4 \times 10^7$ W/cm², $F_{VS} = 500$ mJ/cm², $x = 0.90$ cm, $\Delta x = 0.16$ cm, $t_{VS} = 3.8$ µs.*

atoms. The nascent Ga atom KED peaks near 2.2 eV and shows a FWHM of ≈ 1.8 eV. The velocity selected Ga atom KED peaks at 2.0 eV with a FWHM of 0.5 eV. The transmission of the velocity selector at the 2 eV peak is very nearly 100%, and the integrated transmission is ≈ 20 % of the original laser ablated flux. Quartz crystal microbalance (QCM) measurements indicate a velocity selected Ga atom flux of Φ_{Ga} ≈ 8x10^{10} atoms/(cm^2 eV pulse) at a distance of 10 cm from the ablation target surface.

Fig. 3 shows the results of a velocity selection experiment on laser ablated In atoms. The nascent KED peaks near 2.8 eV and shows a FWHM of ≈ 2.4 eV. The velocity selected KED peaks near 3 eV with a FWHM of 0.9 eV. The transmission at the 3 eV peak is very nearly 100%, and the integrated transmission is ≈ 20 % of the original atomic flux. QCM measurements yield a velocity selected In atom flux of Φ_{In} ≈ 6x10^{10} atoms/(cm^2 eV pulse) at a working distance of 10 cm.

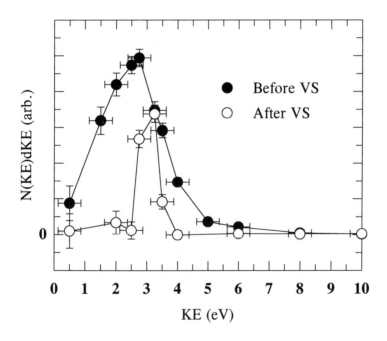

Fig. 3: *Velocity Selection of In Atoms by TASSPI. The closed circles show the nascent In atom KED, the open circles show the TASSPI effect. The error bars represent the estimated ± 1 σ limits. The experimental conditions were: I_{abl} = 3.4x10^7 W/cm^2, F_{VS} = 530 mJ/cm^2, x = 0.90 cm, Δx = 0.16 cm, t_{VS} = 4.0 μs.*

CONCLUSIONS AND FUTURE DIRECTIONS

We have demonstrated a new, non-mechanical technique for performing velocity selection of fast laser ablated Al, Ga, and In atoms. Actually, the relatively low velocities of the heavy Ga and In atoms allow the use of conventional velocity selection schemes, but we found switching between group IIIB metals very convenient with the TASSPI method.

We are currently pursuing velocity selection of lighter species such as fast Li and B atoms. Unfortunately, the single photon photoionization cross section for Li is smaller than that for Al by over two-orders of magnitude [11], and the ionization potential for B atoms is almost 6.7 eV [12], corresponding to a presently inconvenient photon wavelength of 185 nm. It may be that the implementation of TASSPI in these systems will require the use of multi-photon ionization techniques, perhaps involving tunable pulsed lasers able to effect resonance enhancement via intermediate states [12].

We hope to ultimately apply a TASSPI source to help answer some of the questions arising from our MIS work mentioned above. We will begin with co-depositions of fast Al atoms and molecular hydrogen at 2 K, examining the KE dependence of the atomic isolation efficiency and the formation of Al_xH_y reaction products [6].

We also hope that the TASSPI technique will find its way into other thin film deposition experiments, at least as a research tool. Three simultaneous modifications: increasing the ablation laser spot size while maintaining I_{abl} constant, increasing the diameter of the beam defining aperture, and increasing the magnetic rejection field strength, should allow for pure atomic fluxes of over 10^{12} atoms/(cm^2 eV pulse) at a 10 cm working distance. Excimer lasers operating at ~ 1 KHz repetition rates would then provide ~ 10^{15} atoms/(cm^2 eV s). Ultimately, TASSPI may help elucidate the roles of incident atomic KE, beam composition, *etc.* in the formation of mixed semiconductors, nonlinear optical materials, and other thin films produced by laser ablation techniques.

REFERENCES

1. Laser Ablation in Materials Processing: Fundamentals and Applications, B. Braren, J.J. Dubowski, and D.P. Norton, eds. (Mater. Res. Soc. Proc. **285**, Boston, MA, 1993).
2. M.E. Fajardo, P.G. Carrick, J.W. Kenney III, J. Chem. Phys. **94**, 5812 (1991).
3. M.E. Fajardo, J. Chem. Phys. **98**, 110 (1993).
4. S. Tam and M.E. Fajardo, J. Chem. Phys. **99**, 854 (1993).
5. R.A. Corbin and M.E. Fajardo, J. Chem. Phys. **101**, 2678 (1994).
6. M.E. Fajardo, S. Tam, T.L. Thompson, and M.E. Cordonnier, Chem. Phys. **189**, 351 (1994).
7. M. Macler and M.E. Fajardo, pp. 105-110 in ref. 1.
8. M. Macler and M.E. Fajardo, Appl. Phys. Lett. **65**, 159 (1994).
9. M. Macler and M.E. Fajardo, Appl. Phys. Lett. **65**, 2275 (1994).
10. M. Macler and M.E. Fajardo, manuscript in preparation.
11. R.D. Hudson and L.J. Kieffer, Atomic Data **2**, 205 (1971).
12. V.S. Letokhov, Laser Photoionization Spectroscopy, (Academic, Orlando, FL 1987).

In Situ Monitoring and Model Simulation
of BaTiO$_3$ Pulsed Laser Thin Film Deposition

ALBERT J. PAUL, PETER K. SCHENCK, DAVID W. BONNELL, JOHN W. HASTIE

National Institute of Standards and Technology, Gaithersburg, Md 20899

ABSTRACT

We have applied Intensified Charge Coupled Detector (ICCD) imaging to real-time, *in situ* monitoring of excited vapor species emission during the pulsed excimer (248 nm wavelength) laser deposition (PLD) of BaTiO$_3$ ferroelectric thin films. Molecular beam mass spectrometry (MS) was used for species identification and velocity distribution analysis. Gasdynamic model simulations of the plume formation and transport process are tested by comparison with the ICCD and MS results. Particular attention is given to the time scale during and immediately following the laser pulse (0 – 460 ns), and also to the effects of added O$_2$. Atomic oxygen was found to be the predominant species present at the leading edge of the vapor plume. Other, slower species found included neutral Ti, TiO, Ba, and BaO.

INTRODUCTION

The PLD technique is emerging as an alternative approach to more conventional methods for the production of stoichiometrically complex thin films such as BaTiO$_3$[1]. High energy pulsed laser light is focused onto a target, generating a vapor plume that is transported gasdynamically and is condensed onto a substrate. The vapor pressures developed at the target surface typically lie in the range, 10^5 - 10^7 Pa (~ 1 - 100 bar). Current understanding of the detailed dynamics of the evaporation/ablation process is limited, and new or improved diagnostics and models are needed to clarify the gas-collision and laser-gas interaction events occurring in the near-target region of the expanding plume[1,2]. Experimental tests of most reported gasdynamic models under pre- or interim-expansion (near-field) conditions have been limited by the lack of direct observations within the short time frame of the early expansion (0 - 100 ns)[1]. In this work we applied ICCD imaging to the temporal and spatial analysis of the early stages of the laser plume expansion in the near-field. The results are used to test a three-dimensional gasdynamic simulation of plume expansion. Post-expansion (far-field) MS–determined velocity distributions and temperatures were used to assign gasdynamic model parameters, from which the near-field nature of the plume was simulated.

MODEL SIMULATION

Our model simulation approach has been discussed earlier[1,2] and is based primarily on the work of Dawson, *et al*[3], Singh and Narayan[4], and Predtechensky and Mayorov[5]. In the present three-dimensional simulations, we continue to utilize the isothermal expansion approximation to model the plume region during the laser pulse[4]. Following the ending time ($t = \tau$) of the laser pulse, the plume is considered to expand adiabatically (*see Eq. 1*). To model this expansion process with a background gas present, we adopted Predtechensky's approach[5], converting his radial formulation to Cartesian coordinates assuming equivalent equations of motion in each direction (*see Eq. 2*). The adapted model correctly reduces to the Dawson model[3] when the background gas density $\rho_0 \approx 0$. Each individual species was considered independently of other species of differing mass. Thus, an <u>effective</u> combined pressure at the species-specific plume "front", $P_i(t)$ for a species i at any time $t > \tau$, was treated using:

45

$$P_i(t) \approx P_o + P_{s,i} \left(\frac{X_o Y_o Z_o}{X_i(t) Y_i(t) Z_i(t)} \right)^{\gamma} \qquad ; \qquad (1)$$

$P_{s,i}$ is the maximum partial pressure of species i at the target surface, P_o is the static background pressure; the coordinates $X_i(t)$, $Y_i(t)$, and $Z_i(t)$ are the Cartesian extents of the plume "front," an ellipsoidal "boundary" where the species vapor density has fallen to 60.65 percent of its maximum at t; X_o, Y_o, and Z_o are the plume extent coordinates at the end of the isothermal expansion stage[4]; γ is the gasdynamic factor (the effective plume heat capacity ratio). The equation of motion for the plume front location of species i in the z-direction (perpendicular to the target surface) is:

$$\frac{d^2 Z_i}{dt^2} = \frac{X_i(t) Y_i(t) P_i(t) - X_i(t) Y_i(t) P_o \left(\frac{dZ_i}{dt} \right)^2}{(M_i + [X_i(t) Y_i(t) Z_i(t)] P_o)} \qquad ; \qquad (2)$$

M_i is the total vaporized mass (in gm) of species i; ρ_o is the mass density of the background gas, determined from P_o and the calculated volume at the end of the laser pulse. The time evolution of the plume extent was simulated using a step-wise Runge-Kutta method to obtain time-dependent solutions of the model equations of motion. The model for the centerline (z-axis) number density (n) of the plume during adiabatic expansion ($t > \tau$) at distance z is given by[4]:

$$n(z,t) = \frac{N_T}{2^{0.5} \pi^{1.5} X(t) Y(t) Z(t)} \times e^{\left(\frac{-z^2}{2 Z(t)^2} \right)} , \qquad (3)$$

where N_T is the total population of all species produced during the isothermal stage.

The required model input parameters included: the initial plume coordinates that we obtained from direct measurements of the laser footprint at the target; the pressure and temperature of the vapor at the target, taken from estimates based on MS data and deposition rates[6]; the measured pressure of any added background gas (usually, $P_o \ll P_s$); the pre-expansion plasma temperature (treated as an adjustable parameter); γ taken as 1.3, based on MS observations of the expanded plume[6]; and the time interval for the isothermal expansion phase of the simulation (assumed to be the duration of the laser pulse, $\tau \sim 30$ ns, in this study).

EXPERIMENTAL

Apparatus for *In Situ* Monitoring

The apparatus details have been given elsewhere (ICCD[1,2]; MS[1,7]). The essential features are as follows. The ICCD camera provides time-resolved (to 5 ns, both in frame duration {gate width}, and time of frame relative to the beginning of the laser pulse) two-dimensional images. These images may be linked together to form sequential "movie"-like representations of the PLD plume emission history. The MS has a direct path to the target through differentially pumped apertures. This geometry allows one to define a molecular beam with a composition that is representative of the expanded plume. The total path length from the target to the MS ion source is $\ell = 47.4 \pm 0.1$ cm. Time-resolved molecular beam analysis of mass-selected species provides post-expansion intensity I ($\propto$ beam number density) vs. time-of-arrival (TOA) profiles of all major vapor species. These profiles can also be transformed into flux distributions ($I' = I \times \ell/t$) for velocity analysis, as shown in Figs. 1 and 2.

RESULTS AND DISCUSSION

TOA profiles and ICCD images were obtained for $BaTiO_3$ using excimer laser irradiation at 248 nm wavelength and ~ 13 J cm^{-2} fluence. The MS intensity (I) data were converted to flux (I') by scaling each intensity point by the factor ℓ/t to correct for the v discrimination of the number-density-sensitive MS ion source.

Velocity and Temperature Analysis

We have determined[1] that the MS-derived velocity (v, measured here in the Z-direction) distributions for individual species are best represented as a full-range Maxwellian velocity distribution (beam *flux*) that includes a net (center-of-mass) flow velocity, u:

$$f(v) \quad \propto \quad v^3 \exp\left[\frac{-m}{2kT_b}(v-u)^2\right] \quad , \quad (4)$$

where $-\infty < v < \infty$, T_b is the beam temperature, k ($= 1.38 \times 10^{-16}$ erg K^{-1}) Boltzmann's constant, and m, the species atomic/molecular mass. The values of T_b and u were obtained from non-linear fits of this expression to our experimentally-derived TOA distributions, as shown in Fig. 2. Likewise, the mass dependence of t_{max}, the time of peak *flux* in Eq. 4, can be used to obtain T_b, as well as to provide a stringent test for thermal equilibration among the different molecular weight (M) species present, as shown in Fig. 3. We derive the following relationship between T_b and the *slope* of the fitted line (in $\mu s/M^{1/2}$, e.g., as in Fig. 3):

$$T_b \quad = \quad 4.01 \times 10^3 \left\{\frac{mol \cdot K}{erg}\right\} \frac{\ell^2}{(slope)^2} \left[\frac{t_{max}}{t_{mp}}\right]^2 \left(1 - \frac{u t_{max}}{\ell \times 10^6}\left[\frac{t_{mp}}{t_{max}}\right]\right) \quad , \quad (5)$$

where T_b is in K, ℓ is in cm, u is in cm s^{-1}, and t_{max} is in μs. The transformation term, t_{mp}/t_{max} (i.e., the ratio of the time corresponding to the most probable velocity to the TOA of peak flux), is a function of the flow velocity and temperature:

$$\frac{t_{mp}}{t_{max}} \quad = \quad \left(1 \pm \left[1 + \left(\frac{20kT_b}{mu^2}\right)\right]^{1/2}\right) \Bigg/ \left(1 \pm \left[1 + \left(\frac{12kT_b}{mu^2}\right)\right]^{1/2}\right) \quad . \quad (6)$$

T_b is obtained by iteration of Eqs. 5 and 6.

Model Simulations

A representative model simulation fit to the O-atom data of Fig. 1 is indicated in Fig. 4. Here, the model pre-expansion plume temperature (T_p) was adjusted until the model post-expansion temperature (T_b) agreed with the MS-determined value and the resultant distribution was then calculated. The experimental t_{max} values for all the neutral species could be satisfactorily simulated by the model, e.g., as shown for O in Fig. 4. However, no reasonable set of input parameters allowed the model to simulate the observed *shape* of the distribution. We believe that one source of this discrepancy is due to not including the flow velocity term (u) in the simulation, with the consequence that the model assumes the vapor density maximum to

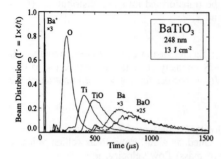

Figure 1. TOA species flux (distribution) curves (arb. units) for MS-sampled PLD beam. $P_o \lesssim 10^{-5}$ Pa (chamber background). Measurement uncertainties are: $I' \pm 0.01$; $t \pm 5$ μs.

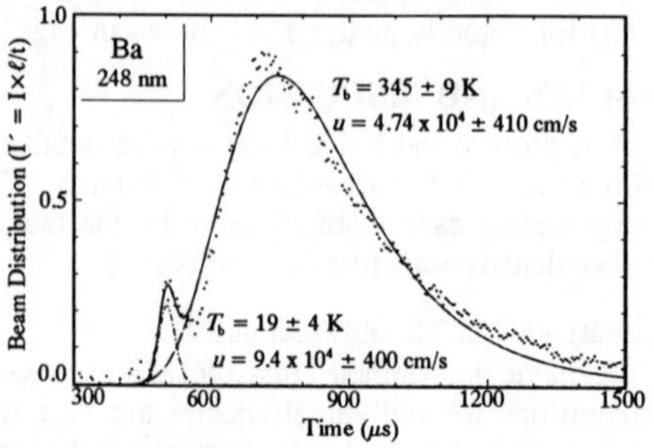

Figure 2. Comparison of TOA distribution (points) for Ba (from Fig. 1) with fitted (sum of two; includes u term — *see text*) full-range Maxwell distributions.

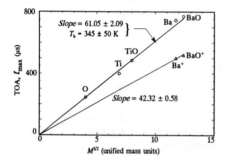

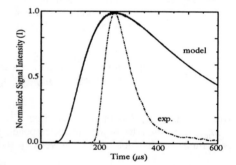

Figure 3. Plot of t_{max} *vs* $M^{1/2}$ for species sampled from BaTiO$_3$ plumes (from Fig. 1). The symbols denote fast (triangle) and slow (circle) peaks in the TOA curves of Fig. 1. Uncertainties: $t_{max} \lesssim \pm 20$ μs; MS-observed $M \pm 0.1$ umu.

Figure 4. Comparison of far-field model simulation of O-atom density *vs.* time with MS data (I) for Fig. 1 conditions. Model density derived using Gaussian density expression (*see* Eq. 3), with $z = \ell = 47.4$ cm. Model parameters: laser footprint $0.1 \times 0.063 \pm 0.005$ cm; $T_s = T_p = 6700$ K, $P_s = 5$ MPa (50 bar), $P_o = 5$ Pa (5×10^{-5} bar), $\gamma = 1.3$.

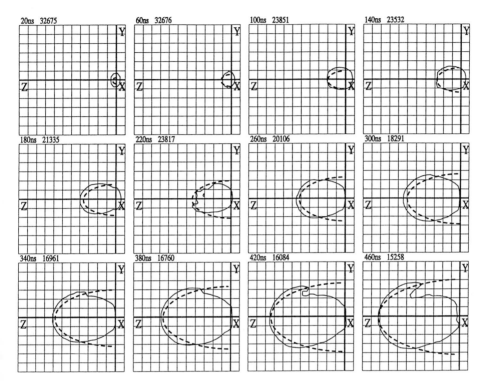

Figure 5. Contour plots (solid curves) from sequential ICCD images of visible emission (500–800 nm) from laser-irradiated BaTiO$_3$ at 248 nm, ~ 13 J·cm^{-2} fluence, with P_o ~ 13 Pa (O$_2$). Grid scale is 1 mm. Individual image intensities have been scaled to the frame maximum (*see text*); relative intensity maxima (in digitizer counts) for each frame are indicated beside the frame time shown at the upper left of each frame. Frame times are ± 1 ns and are referenced to the beginning of the laser pulse, whose duration was nominally 30 ns. Contours outline 60.65 percent of emission intensity maximum in each frame. The camera "shutter" (frame duration) gate width was nominally 10 ns; images were collected at 20 ns (nominal) intervals as the sum of eight generated plumes, but only the 40 ns interval frames are shown here. Dashed curves are two-dimensional projections in the ICCD camera plane of model simulations of the ICCD image contours. As the long dimension of the laser footprint was oriented at 45° to the image plane, a linear transformation of the model $X(t)$, $Y(t)$ coordinates was made to coincide with the experimental X, Y observation coordinates noted on each frame. Uncertainties in experimental contour location estimated as ± 0.2 mm in observed X and Y.

Model conditions used were: laser footprint 0.16×0.04 cm; plasma temperature $T_p = 75$ eV (~750,000 K); $T_s = 6700$ K; $P_s = 5 \times 10^6$ Pa (50 bar); $P_o = 13$ Pa; $\gamma = 1.3$.

remain at the origin (laser impact point) whereas both the MS data and the ICCD images indicate motion of the maximum. Explicit incorporation of the flow term into the model is under investigation. Also, the model assumes continuum flow at all distances, while the MS is located considerably downstream of the transition to molecular flow — explicit inclusion of the transition to molecular flow is also currently being considered.

Mass and optical emission spectral observations have indicated the outer plume extent to be predominantly composed of atomic oxygen. Model simulation of the 60.65 percent atomic oxygen density contours in the plume yielded an expansion time/distance history similar to the image data when high model plasma temperatures and target vapor densities were assumed, as indicated in Fig. 5. The presence of strong continuum plasma emission close to the surface, which persisted through the 180 ns frame and exceeded the collisional emission maximum, distorted the "normal" plume contours at 100–180 ns. As the collisional emission is more representative of the species transport the model describes, the frames at 100, 140, and 180 ns have been corrected to eliminate the effect of the continuum emission maximum. Other effects may also be present, but are not considered in present expansion models. These effects include ion-electron recombination and surface charging[8], and the phase explosion mechanism discussed elsewhere for AlN PLD[9].

In summary, these results clearly indicate the distinctly differing energetics of those components of the plume responsible for observed emission (where $T_p \sim$ 750,000 K, as in Fig. 5), in contrast to the lower energetics of the primarily non-emitting ground state species components (where $T_p \sim$ 6700 K, as in Fig. 4) observed *via* MS in the far-field molecular beam. This aspect, of various plume components with widely differing energetics, will be discussed in future work. Also needing further consideration is the question of the most appropriate transition period, τ, between the isothermal and supersonic adiabatic expansion regions of the expansion model. The current assumption, that the laser pulse termination time determines τ, is a rather arbitrary convenience, as plume densities and velocities generally appear to attain a supersonic expansion state well before the end of the laser pulse under typical film deposition conditions.

REFERENCES

1. J.W. Hastie, D.W. Bonnell, A.J. Paul, & P.K. Schenck, "Gasdynamics and Chemistry in the Pulsed Laser Deposition of Oxide Dielectric Thin Films," in *Gas-Phase and Surface Chemistry in Electronic Materials Processing*, T. Mountziaris, G. Paz-Pujalt, F. T.J. Smith & P. Westmorland (Eds), MRS Proceedings **334** (MRS Pittsburgh, PA 1994), p. 305-316.
2. P.K. Schenck, J.W. Hastie, A.J. Paul, & D.W. Bonnell, "Imaging and Modeling of Pulsed Laser Thin Film Deposition Plumes," in *Laser-Induced Thin Film Processing*, J.J. Dubowski (Ed), SPIE Proceedings **2403**, 26-38 (1995).
3. J. Dawson, P. Kaw, & B. Green, *Phys. Fluids*, **12(4)**, 875-882 (1969).
4. R.K. Singh & J. Narayan, *Phys. Rev. B*, **41(13)**, 8843-8859 (1990).
5. M.R. Predtechensky & A.P. Mayorov, *Appl. Supercond.*, **1(10-12)**, 2011-2017 (1993).
6. J.W. Hastie, D.W. Bonnell, A.J. Paul, J. Yeheskel, & P.K. Schenck, *High Temp & Matr. Sci.* (formerly *High Temp. Sci.*), **33**, 135-169 (1995).
7. J.W. Hastie, A.J. Paul, and D.W. Bonnell, "Free-Jet Mass Spectrometry of Laser Ablation Plumes in Thin Film Deposition," in *Specialists Workshop on Applications of Free-Jet Molecular Beam Sampling*, *NREL Report CP433-7748*, T. Milne (ed), (NTIS, Washington, DC, 1995).
8. R.M. Gilgenbach, C.H Ching, J.S. Lash, & R.A. Lindley, *Phys. Plasmas*, **1**, 1619-1625 (1994).
9. R. Kelly, A. Mele, A. Miotello, A. Giardini, J.W. Hastie, P.K. Schenck, & H. Okabe, *Nucl. Instr. Meth.*, in press (1995).

PULSED LASER DEPOSITION OF CADMIUM STANNATE, A SPINEL-TYPE TRANSPARENT CONDUCTING OXIDE

JEANNE M. McGRAW*, PHILIP A. PARILLA**, DOUGLAS L. SCHULZ**, JEFFREY ALLEMAN**, XUANZHI WU**, WILLIAM P. MULLIGAN*, DAVID S. GINLEY**, TIMOTHY J. COUTTS**
*Colorado School of Mines, Golden, CO 80401
**National Renewable Energy Laboratory, 1617 Cole Boulevard, Golden, CO 80401

ABSTRACT

We present the first report of the synthesis of Cd_2SnO_4 films by pulsed laser deposition. Controlling the substrate temperature and the ambient atmosphere allowed for the synthesis of films ranging from amorphous to crystalline with some crystalline films exhibiting strong texture. Highly transparent films with large mobilities were obtained for both the amorphous and crystalline films. Sheet resistances of 15.5 Ω/square and mobilities of up to 44.7 $cm^2/V.s$ were observed. Typical carrier concentrations showed the crystalline films to be degenerate with carrier concentrations of 5×10^{20} cm^{-3} while amorphous films had carrier concentrations lower by about half. Band gaps for the films ranged between 3.1-3.8 eV. These films are attractive candidates for TCO applications in thin film photovoltaic devices, flat panel displays, electrochromic windows, and as plasma filters for thermophotovoltaic devices.

INTRODUCTION

Transparent conducting oxides (TCO) are vital to a number of key new technologies. In photovoltaic (PV) cells and flat panel displays, TCOs act as a transparent electrical contact. Building glazing applications such as electrochromic windows and heat mirrors use TCOs as critical elements. Thermophotovoltaic (TPV) applications can utilize TCOs as an infrared (plasma) filter which allows usable radiation to pass through to the cell and reflects the unusable radiation back to the source thereby increasing energy efficiency.

The dominant design criteria for most TCO applications are high transparency in the visible range, high conductivity, high mobility, and a plasma edge near 3 μm. A material with a band gap of 3.0 -3.5 eV will be transparent to light from 350-400 nm to its plasma edge which is suitable for most solar cell and flat panel applications. In thermophotovoltaic (TPV) cells, the position of the plasma edge is crucial for the proper filtering of the heat radiation.

Cubic spinel Cd_2SnO_4 is a TCO which has a small effective mass and a typical range of carrier concentrations from 10^{18} to 10^{20} cm^{-3} resulting in a plasma edge suitable for TPV applications. Cadmium stannate has been studied in the past using numerous deposition techniques.[1,2] To the best of our knowledge, this paper reports the first Cd_2SnO_4 thin films grown by pulsed laser deposition (PLD). Our films exhibit outstanding transparency, mobility, and resistivity making Cd_2SnO_4 films and derivatives thereof suitable toward a wide range of applications.

EXPERIMENTAL

Equipment

The PLD system used for the deposition of Cd_2SnO_4 films consists of a 5-cross 6" diameter sphere. A turbo pump backed by a mechanical pump evacuates the system to a base pressure $<3 \times 10^{-6}$ Torr. Ambient gas flow is controlled with a Flo-Box by Sierra Instruments, Inc., with MKS Baratron feedback. Vitrosil optical grade quartz substrates (ESCO) are mounted onto a 2" diameter resistance heater (U.S. Gun) with silver paint (SPI Supplies). A thermocouple imbedded in the heater monitors substrate temperature. The 1" diameter x 1/4" hot-pressed Cd_2SnO_4 target (Cerac, Inc., 99.5% pure) is mounted on a 1" diameter stainless-steel platen with epoxy (Ultra seal). The excimer laser (Lambda Physik EMG50) is charged with KrF (248 nm). Spot size of the beam is approximately 1 x 2 mm with ~35 mJ/pulse. The incoming beam hits the target at 45° incidence and the distance between the target and the substrate is fixed at ~7 cm.

Growth Conditions

For all depositions, laser energy density was 1-2 J/cm^2, repetition rate was 5 Hz, chamber pressure was 0.4 Torr of ambient gas flowing at 50 sccm, and deposition time was 15 minutes. Base pressure was established before the substrate heater was turned on. High-purity O_2 and N_2 were employed as background gases. When temperature reached ~50°C below the intended temperature, the ambient gas flow was initiated. After the temperature stabilized, the deposition commenced. The target rotated during the run and a new target surface was used for each run. After 15 minutes, the laser was turned off, the gas flow was stopped, the chamber was evacuated, and finally the heater was turned off. The substrate cooled to room temperature before being removed from the chamber. The substrate temperatures and ambient gas for the various samples are listed in Table I.

Table I: Deposition Conditions. For all samples, repetition rate was 5 Hz, chamber pressure was 0.4 Torr at a flow rate of 50 sccm Ablation time was 15 minutes.

Sample	Ox25	Ox400	Ox500a	Ox500b	Ox550	N500
Substrate Temp, °C	25	400	500	500	550	500
Ambient Gas	O_2	O_2	O_2	O_2	O_2	N_2

Film Characterization

Film thickness was determined using a Dektak^3 profilometer. The scanning electron microscope (SEM) micrographs were taken on a Perkin-Elmer PHI 670 Auger Nanoprobe. Crystalline phase development was determined by x-ray diffraction (XRD) θ/2θ using a Rigaku DMAX-A instrument with a rotating Cu Kα anode. Sheet resistance measurements were performed using an Alessi Industries 4 point probe tip together with a Keithley Instruments voltmeter and current source. Gold wire and indium solder were employed as contacts for the Hall measurements taken on a Physical Electronics Industries, Inc., Hall meter with an electromagnet operating at 6 kGauss. Optical properties in the ultraviolet-visible range were determined using a Varian Cary 2300 UV-Vis spectrophotometer with air as background for normal transmissivity and an integrating sphere for reflectance measurements. In the infrared (IR) region, a Nicolet 550 medium resolution Fourier Transform Infrared (FTIR) spectrophotometer using air as the background was used for normal transmittance measurements and a Nicolet 800 high resolution FTIR spectrophotometer using gold on silicon as the reference was used for total reflectance measurements.

RESULTS AND DISCUSSION

The films were specularly smooth as observed by SEM. The amorphous films were absolutely featureless at 15,000x, with the exception of a few pinholes. At 10,000x, the Ox500a crystalline film shows a slight surface roughness, also with an occasional pinhole. This surface roughness is consistent with XRD grain size calculations discussed below.

X-ray diffraction data of the as-deposited films are shown in Figure 1. Films Ox25 and N500 were both amorphous The emergence of two very small peaks corresponding to the (311) and (222) peaks of cubic spinel Cd_2SnO_4 shows partial crystallization of the Ox400 sample.[4] All films grown in oxygen at ≥500°C were crystalline with the cubic spinel structure.

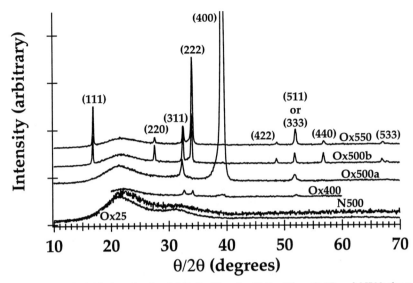

Figure 1. XRD θ/2θ data for the Cd₂SnO₄ films by PLD. Films Ox25 and N500 show amorphous patterns, Ox400 shows little crystallinity, and films Ox500a, Ox500b, and Ox550 give cubic spinel patterns. Note that film Ox500a is highly (400) textured.

The Ox500b and Ox550 samples have no obvious texture, whereas the Ox500a film exhibits strong (400) texturing. In the textured film, there is a slight asymmetry to the (400) peak which may be due to the (111) CdO peak (Intensity = 100, JCPDS PDF# 5-640). From the (400) peak of this sample, the grain size is calculated to be ~875 Å by the Scherrer method.[5] X-ray diffraction characterization of the target showed it to have the orthorhombic Cd₂SnO₄ structure (JCPDS PDF# 31-242).

Table II summarizes the crystalline phase development as well as film thickness, sheet resistance (R_s), carrier concentration, mobility, and resistivity data. The resistivity of the Ox25 film was too high for our 4 point probe and Hall measurement apparatuses. The best R_s value observed is 15.5 Ω/square for the crystalline Ox500a film. For comparison, sheet resistance of ITO is typically 2-5 Ω/square.[2] Carrier concentrations are approximately $2\text{-}5\times10^{20}$ cm⁻³, which confirms that these are degenerate films. Interestingly, the crystalline films, including the (400) textured film all have similar carrier concentrations, mobilities, and resistivities. Mobilities range from about 30 to 45 cm²/V·s. Note that the amorphous N500 film has the highest mobility. Generally, crystalline materials have mobilities 1-2 orders of magnitude higher than amorphous materials due to the long range order in the crystal lattice.[3] Both oxygen vacancies and cadmium interstitials have been identified as possible doping mechanisms in Cd₂SnO₄.[6] Additional work is needed to explore this phenomenon.

The Ox500a and Ox500b films were grown under nominally identical conditions. Two films grown under these conditions exhibited the (400) texturing whereas three other films exhibited an XRD pattern similar to the Ox500b film. The reason for the difference in the crystal phase development is being investigated. We suspect that we were operating near a "texturing threshold" where small variations in the deposition parameters were sufficient to cause the texturing to be present or absent. Prime suspects are variations in the laser fluence and aging effects in the targets. In future experiments, these and other parameters will be more tightly controlled in a new PLD system. This new system consists of a new deposition chamber, a more powerful laser (440 mJ/pulse), and provides better monitoring of the in situ

Table II: Film Properties. Thickness and R_s measurements were taken at both ends of the film, and averaged. Hall measurements were performed at room temperature.

Sample	a=amorph or c=crystal	Avg. Thickness Å	R sheet R_s Ω/Square	Carrier Conc. n x 10^{20} cm^{-3}	Mobility μ $cm^2/V \cdot s$	Resistivity ρ x 10^{-4} Ω-cm
Ox25	a	3800	--	--	--	--
Ox400	a	1500	55.5	2.5	29.7	8.5
Ox500a	c (400)	2750	15.5	4.2	30.9	4.9
Ox500b	c	2000	30.5	5.5	28.7	3.9
Ox550	c	1500	34.0	5.6	34.9	3.2
N500	a	1410	41.2	2.9	44.7	4.8

deposition conditions. It will provide improved film thickness uniformity and allow the effects of laser fluence to be investigated.

Both the carrier concentration and the mobility strongly influence the onset of reflectance at the plasma frequency. The FTIR total reflectance spectra, Figure 2, show that this onset is much sharper for the crystalline films than for the amorphous films. Inspection of the carrier concentrations and mobilities in Table II suggest that the carrier concentration plays a stronger role in the onset of IR reflectance. As expected, the plasma edges for the crystalline, higher carrier concentration films occur at smaller wavelengths than the amorphous, lower carrier concentration films. The Ox25 sample did not show a sharp onset of reflection, nor was it very reflective. This is consistent with the film being highly resistive and supports the 4 point probe and Hall measurements.

The UV-Vis normal transmissivity vs. wavelength spectra are shown in Figure 3. For clarity, only three films are plotted. The crystalline Ox500b and Ox550 films follow the (400) textured Ox500a film closely. Substrate absorption is responsible for the low transmissivity at 2700 nm. The lack of free carriers in the amorphous, highly resistive Ox25 film results in a high transmissivity throughout the visible wavelength range as expected.[6] The amorphous Ox400 film with a carrier concentration of 2.5 x 10^{20} cm^{-3} lies between the Ox25 and crystalline

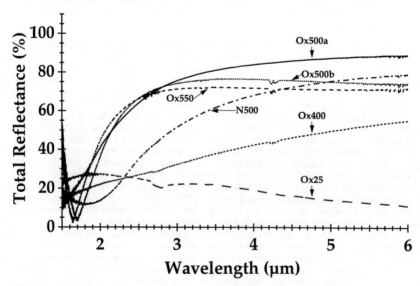

Figure 2. FTIR total reflectance data versus wavelength for the Cd_2SnO_4 films.

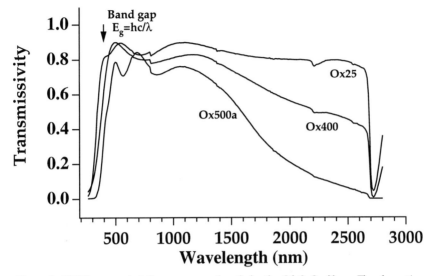

Figure 3. UV-Vis transmissivity versus wavelength for the Cd_2SnO_4 films. The absorption edge of the band gap at shorter wavelengths is clearly visible for all the films. Note that the free carrier absorption is absent in the highly resistive film Ox25 at longer wavelengths.

Ox500a films. The Ox500a has the highest carrier concentration of 4.2 x 10^{20} cm^{-3} and transmissivity falls off sooner than the films with lower carrier concentrations due to free carrier absorption.[7]

UV-Vis transmissivity measurements showed the band gap to be 3.1 eV for the crystalline Ox500a film and 3.8 eV for the amorphous Ox25 film.. All the films exhibited a fairly sharp band edge. Interestingly, the highly resistive Ox25 film is at the highest energy, the crystalline films are grouped together at the lowest energy, and the two amorphous films Ox400 and N500 (not shown in the figure) fall in the middle. This is in opposition to what is expected for the Moss-Burstein effect[8,9] since the amorphous materials have a lower carrier concentration. Nozik[3] reported that the Moss-Burstein effect occurred in both crystalline and amorphous Cd_2SnO_4; however, further studies are necessary to determine if this shift is indeed the Moss-Burstein effect.

Listed in Table III are electrical data from the literature for Cd_2SnO_4 films grown by different methods. Film Ox550, grown in this study, is the PLD film used for comparison in the table. Resistivities, with the exception of the Ox400 film, are lower for the PLD films and all mobilities of films grown in this study are higher. Clearly, the Cd_2SnO_4 films grown by PLD are superior to films grown by the other methods.

Table III. Comparison of film Ox550 grown in O_2 at 550°C by PLD to Cd_2SnO_4 films grown by other deposition methods.

Phase	Resistivity $\Omega\cdot$cm	Carrier Conc. cm^{-3}	Mobility cm^2/V·s	Deposition Method	Reference
Cd_2SnO_4	4×10^{-3}	1.0×10^{20}	14.4	electroless deposition	[10]
Cd_2SnO_4	6.2×10^{-4}	5×10^{20}	22	sputtering	[11]
Cd_2SnO_4	2.6×10^{-3}	1.2×10^{20}	20	sputtering	[3]
Cd_2SnO_4	3.2×10^{-4}	5.6×10^{20}	35	PLD	This paper

SUMMARY

We have demonstrated pulsed laser deposition of spinel-type Cd_2SnO_4. In these initial experiments, the as-deposited films exhibited excellent optical and electrical properties. Both crystalline and amorphous films are highly conductive. A sharp onset of reflectance makes cubic spinel cadmium stannate a good candidate for further study in thermophotovoltaic plasma filter applications. The high transmissivity in the visible range and the high mobility and carrier concentration make it potentially suitable as a transparent contact layer in flat panel display and photovoltaic applications. Studies will continue on Cd_2SnO_4 as well as other TCO materials grown in the aforementioned new PLD system which we expect to result in optically and electrically better films.

ACKNOWLEDGMENTS

The authors gratefully acknowledge Amy B. Swartzlander for SEM micrographs and John Webb for assistance with FTIR measurements. This research was funded by the Department of Energy (Contract No. DOE/CH10093-264).

REFERENCES

1. J. L. Vossen In *Physics of Thin Films* 1977; Vol. 9; pp. 1.
2. K. L. Chopra, S. Major and D. K. Pandya, Thin Solid Films **102**, 1 (1983).
3. A. J. Nozik, Phys. Rev. B **6**, 453 (1972).
4. L. A. Siegel, J. Appl. Cryst. **11**, 284 (1978).
5. B. D. Cullity, Elements of X-ray Diffraction (Addison-Wesley, Reading, Mass., 1978).
6. G. Haacke, W. E. Mealmaker and L. A. Siegel, Thin Solid Films **55**, 67 (1978).
7. I. Hamberg and C. G. Granqvist, J. Appl. Phys. **60**, R123 (1986).
8. E. Burstein, Phys. Rev. **93**, 632 (1954).
9. T. S. Moss, Proc. Soc. London **B76**, 775 (1954).
10. Raviendra, D. and J.K. Sharma, J. Appl. Phys.,**58**(2), 838 (1985).
11. Miyata, N., et al., J. Electrochem. Soc.,**127,** 918 (1980).

FORMATION OF ARTIFICIALLY-LAYERED THIN-FILM COMPOUNDS USING PULSED-LASER DEPOSITION

DAVID P. NORTON, B. C. CHAKOUMAKOS, D. H. LOWNDES, AND J. D. BUDAI
Oak Ridge National Laboratory, P. O. Box 2008, Oak Ridge, TN 37831-6056

ABSTRACT

Superlattice structures, consisting of $SrCuO_2$, $(Sr,Ca)CuO_2$, and $BaCuO_2$ layers in the tetragonal, "infinite layer" crystal structure, have been grown by pulsed-laser deposition (PLD). Superlattice chemical modulation is observed for structures with component layers as thin as a single unit cell (~3.4 Å), indicating that unit-cell control of $(Sr,Ca)CuO_2$ growth is possible using conventional pulsed-laser deposition over a wide oxygen pressure regime. X-ray diffraction intensity oscillations, due to the finite thickness of the film, indicate that these films are extremely flat with a thickness variation of only ~20 Å over a length scale of several thousand angstroms. Using the constraint of epitaxy to grow metastable cuprates in the infinite layer structure, novel high-temperature superconducting structural families have been formed. In particular, epitaxially-stabilized $SrCuO_2/BaCuO_2$ superlattices, grown by sequentially depositing on lattice-matched (100) $SrTiO_3$ from $BaCuO_2$ and $SrCuO_2$ ablation targets in a PLD system, show metallic conductivity and superconductivity at T_c(onset) ~70 K. These results show that pulsed-laser deposition and epitaxial stabilization have been used to effectively "engineer" artificially-layered thin-film materials.

INTRODUCTION

Since the discovery of the high-temperature superconductivity (HTSc),[1] intensive efforts have revealed numerous families of layered crystal structures containing copper oxide layers,[2] with superconducting transition temperatures as high as 135 K for the Hg-containing cuprates.[3] Typically, bulk synthesis techniques have been the primary tool in the search for new HTSc materials. Recently, high-pressure synthesis methods have played a prominent role in investigating metastable cuprate phases.[4-7] However, thin-film growth methods offer unique advantages for the atomic engineering of new HTSc materials through the ability to form artificially layered crystal structures. Moreover, the surfaces of single-crystal substrates provide an "atomic template" that can be used to stabilize epitaxial films in metastable crystal structures. Advances in the understanding of epitaxial thin-film growth of the cuprates have heightened interest in the possibility of creating artificially layered materials.[8-10] The cuprate superconductors are ideal candidates to work on, because their crystal structures are composed of chemically distinct layer modules, most important of which is the square CuO_2 layer.

A prelude to the formation of artificially-layered cuprates has been the epitaxial stabilization of infinite layer $(Ca,Sr)CuO_2$ single-crystal thin films.[11-14] The infinite layer $(Ca,Sr)CuO_2$ structure type, consisting of square CuO_2 layers alternately stacked with square layers of alkaline earth atoms, is the simplest cuprate with the essential structural features for HTSc.[15] Indeed, the infinite-layer structure can be viewed as a fundamental building unit of all of the HTSc cuprates. Bulk synthesis of $(Ca,Sr)CuO_2$ with the tetragonal, "infinite layer" structure generally requires the use of high pressure and high temperature bulk processing techniques.[16-21] However, recent experiments show that this metastable compound can be epitaxially stabilized at less than atmospheric pressure, resulting in the growth of tetragonal $(Ca,Sr)CuO_2$ single crystal thin films of the infinite layer defect perovskite structure by pulsed-laser deposition (PLD) over a wide range of growth conditions.[11-14, 22-24] However, in order to synthesize artificially-layered HTSc structures, the growth of $(Ca,Sr)CuO_2$ must be controlled at the unit-cell level (~3.4 Å).

In this paper, we show that unit-cell control of $(Ca,Sr)CuO_2$ film growth is possible using conventional PLD. We have successfully grown $SrCuO_2/(Ca,Sr)CuO_2$ superlattice structures utilizing a PLD multi-target system operating at an oxygen pressure of 200 mTorr.[25] X-ray

Mat. Res. Soc. Symp. Proc. Vol. 388 © 1995 Materials Research Society

diffraction peaks attributed to the superlattice structures are observed, even for $SrCuO_2/(Ca,Sr)CuO_2$ superlattice structures with $SrCuO_2$ and $(Ca,Sr)CuO_2$ layer thicknesses of a single unit cell (~3.4 Å). The x-ray diffraction data also reveal finite-thickness oscillations in the x-ray intensity, which is indicative of films with extremely flat surfaces. The growth of superlattice structures by PLD is made possible, in large part, by this surface flatness. In addition, we also report the synthesis of novel artificially-layered HTSc compounds, grown as $SrCuO_2/BaCuO_2$ crystalline superlattice structures.[26] In showing that the growth of infinite layer materials in superlattice structures can be controlled on the unit cell scale using conventional PLD with no in situ surface monitoring, this work offers the exciting possibility of greatly broadening the conditions under which new artificially-layered HTSc phases can be formed.

EXPERIMENTAL CONDITIONS

The films were grown on (100) $SrTiO_3$ substrates utilizing conventional multi-target PLD. Polycrystalline, orthorhombic $(Ca,Sr)CuO_2$ and cubic $BaCuO_2$ ablation targets were mounted in a multi-target carousel. The $(Ca,Sr)CuO_2$ targets were made by solid state reaction of high-purity $SrCO_3$, $CaCO_3$, and CuO which were pressed and fired at 1025°C. Powder x-ray diffraction confirmed complete decomposition of the carbonates. The $BaCuO_2$ target was prepared using high-purity $BaCuO_2$ powder. (100) $SrTiO_3$ substrates were cleaned with solvents prior to being mounted with silver paint on the substrate heater. The KrF excimer laser ablation beam was focused to a 1 cm horizontal line and vertically scanned over the rotating targets to improve film thickness uniformity. The focused laser energy density was approximately 2 J/cm^2, and the substrates, heated to 500–625°C, were placed 10 cm from the ablation targets. Film growth was carried out in 200 mTorr of oxygen. Before growing the $SrCuO_2/BaCuO_2$ layered structures, a 9.0 nm thick $SrCuO_2$ buffer layer was grown to initiate epitaxial growth of the infinite layer structure on the $SrTiO_3$ surface. Total film thickness varied from 90 to 120 nm, corresponding to 60 or more superlattice periods. After deposition, the films were cooled at ~80°C/min in 200 mTorr of oxygen, with the pressure increased to 760 Torr at 375°C for the $SrCuO_2/BaCuO_2$ superlattices. The structure and epitaxy of the films were investigated by x-ray scattering measurements obtained using both two-circle and four-circle diffractometers with monochromated $CuK\alpha$ x-ray sources. In addition, more sensitive synchrotron x-ray measurements were obtained for one film using beamline X14 at the National Synchrotron Light Source (NSLS) operating with a Si(111) monochromator and a Ge(111) analyzer set near the $CuK\alpha$ wavelength.

$SrCuO_2/(Ca,Sr)CuO_2$ SUPERLATTICES

The growth of superlattice structures by PLD requires a well-calibrated and controlled growth rate. Initial estimates of the film thickness per laser shot were obtained by growing single-component $(Ca,Sr)CuO_2$ films and measuring their thickness. However, for superlattice growth, the growth rate and/or sticking coefficient of each component layer can deviate from the values obtained from single-component film growth. For instance, growth of a single unit cell of $SrCuO_2$ on a $Sr_{0.2}Ca_{0.8}CuO_2$ surface involves slightly different surface chemistry than growth on $SrCuO_2$, and can lead to a different growth rate. Thus, the final growth rate calibration was obtained from the superlattice satellite peak locations in the x-ray diffraction data.

For $SrCuO_2/(Ca,Sr)CuO_2$ superlattices with the infinite layer crystal structure, the x-ray diffraction data should show the infinite layer (00ℓ) peaks, along with superlattice satellite peaks due to the additional chemical modulation from the superlattice structure. The nomenclature used to describe the nominal superlattice structure is M × N, where M is the number of unit cells of the first component, and N is the number of unit cells of the second component. Figure 1 shows the x-ray diffraction pattern, obtained using a rotating anode x-ray source, for a 2 × 1 $SrCuO_2/Sr_{0.2}Ca_{0.8}CuO_2$ superlattice structure with 80 periods grown at 600°C. The pattern shows peaks due to both the infinite layer crystal structure as well as to the superlattice Sr/Ca chemical modulation. Based on the location of the superlattice peaks, the superlattice chemical periodicity is 9.28 Å. The fundamental peaks yield an average lattice parameter of c ~3.38 Å. A rocking curve through the infinite layer (002) reflection for this film yields a mosaic FWHM of ~0.11°, an indication of the high degree of alignment.

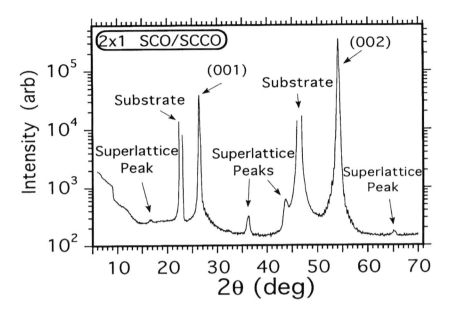

Fig. 1 X-ray diffraction (Cu Kα radiation) data for a 2×1 SrCuO$_2$/Sr$_{0.2}$Ca$_{0.8}$CuO$_2$ (SCO/SCCO) superlattice showing diffraction peaks due to the Sr/Ca chemical modulation of the superlattice thin-film structure. From the superlattice peak locations, the superlattice period is 9.28.

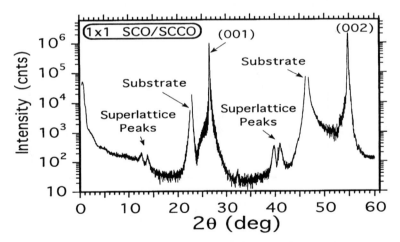

Fig. 2 X-ray diffraction data for a 1×1 SrCuO$_2$/Sr$_{0.2}$Ca$_{0.8}$CuO$_2$ (SCO/SCCO) superlattice showing diffraction peaks due to the Sr/Ca chemical modulation of the superlattice thin-film structure. The superlattice peaks are split due to the fact that the chemical periodicity is slightly incommensurate with the "infinite layer" structural periodicity. From the superlattice peak locations, the superlattice period is 7.0 Å.

Figure 2 shows the x-ray diffraction pattern for a 1×1 $SrCuO_2/Sr_{0.2}Ca_{0.8}CuO_2$ superlattice. Even for this structure, with $SrCuO_2$ and $Sr_{0.2}Ca_{0.8}CuO_2$ layers only 1 unit cell thick, x-ray diffraction peaks attributed to the superlattice are observed indicating control of the film growth at the infinite layer unit cell level of ~3.4 Å. The superlattice modulation periodicity for the 1×1 structure is 7.0 Å and the average lattice parameter is c~3.35Å. For $SrCuO_2$ and $Sr_{0.2}Ca_{0.8}CuO_2$, the nominal ideal lattice constants are 3.45 and 3.25 Å, respectively. Thus, for both the 1×1 and 2×1 $SrCuO_2/Ca_{0.8}Sr_{0.2}CuO_2$ structures, the chemical modulation periodicities are within 8% of that for the nominal structure. In addition, the average c-axis lattice parameters are in excellent agreement with the values[11] for the infinite layer films with the same average stoichiometries.

The quality of the superlattice structures is sensitive to the growth temperature as shown by x-ray diffraction. While 4×4 $SrCuO_2/Sr_{0.2}Ca_{0.8}CuO_2$ superlattices grown at 500 and 600°C had superlattice peaks that are similar in both intensity and width, the 2×2, 2×1, and 1×1 $SrCuO_2/Sr_{0.2}Ca_{0.8}CuO_2$ superlattice structures grown at 500°C had superlattice peaks significantly lower in intensity and broader than for similar structures grown at 600°C. This suggests that the growing surface is significantly smoother at 600°C than at 500°C due to increased surface diffusion. In addition, this result indicates that bulk interdiffusion is not a significant factor at temperatures up to 600°C, as interdiffusion of the layers would tend to make the diffraction peaks broader and lower in intensity at the higher temperature.

It is interesting that superlattice structures with component layer thicknesses as small as 3.3 Å can be obtained with a relatively simple thin-film growth system which has no in situ surface analysis capability. This result implies that the growing surface of the infinite layer films must be quite smooth, on a near-atomic scale. A direct measure of this surface smoothness is given in the x-ray diffraction pattern of the 1×1 $SrCuO_2/Sr_{0.2}Ca_{0.8}CuO_2$ superlattice. Figure 3 shows an expanded plot about the (001) infinite layer peak, which shows oscillations in the x-ray diffraction intensity due to the finite film thickness. Finite thickness oscillations are observed only if the film surface is flat over a length scale on the order of the coherence length of the x-ray source. For these synchrotron measurements, the coherence length is at least of the order of thousands of angstroms. Over 30 oscillations are present on each side of the (001) peak, giving a fractional

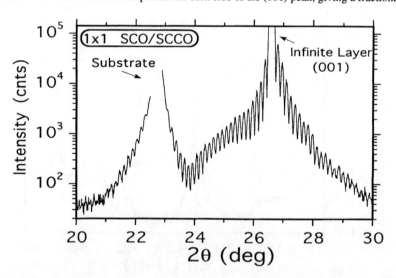

Fig. 3 X-ray diffraction data for a 1×1 $SrCuO_2/Sr_{0.2}Ca_{0.8}CuO_2$ (SCO/SCCO) superlattice about the (001) "infinite layer" peak showing x-ray intensity oscillations due to the finite thickness of the film. The number of resolvable oscillations suggests that the film thickness varies by only 20 Å over a length scale of several thousand angstroms.

variation of film thickness relative to the total film thickness of 1/30, which yields a thickness variation of only 20 Å over thousands of angstroms.[27] This is on the order of the expected substrate surface roughness, and indicates that these infinite layer films are extremely flat under the growth conditions utilized.

SrCuO$_2$/BaCuO$_2$ SUPERLATTICES

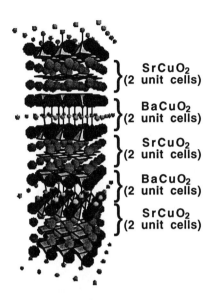

$SrCuO_2$
(2 unit cells)

$BaCuO_2$
(2 unit cells)

$SrCuO_2$
(2 unit cells)

$BaCuO_2$
(2 unit cells)

$SrCuO_2$
(2 unit cells)

Fig. 4 Structural model of a 2×2 SrCuO$_2$/BaCuO$_2$ superlattice compound, also designated as Ba$_2$Sr$_2$Cu$_4$O$_{8+\delta}$. The Ba, Sr, and Cu atoms are represented by the large, medium, and small spheres, respectively. The CuO$_4$ and CuO$_5$ units are shown as shaded polyhedra.

By sequentially depositing from BaCuO$_2$ and SrCuO$_2$ ablation targets in a PLD system, artificially-layered crystalline materials were constructed in which the SrCuO$_2$ and BaCuO$_2$ layers are epitaxially stabilized in the infinite layer structure. The SrCuO$_2$/BaCuO$_2$ superlattices, which can also be described as Ba$_2$Sr$_{n-1}$Cu$_{n+1}$O$_{2n+2+\delta}$, represent a unique HTSc series, with T_c(onset) and T_c(resistance, R=0) as high as 70 K and 50 K, respectively. Note that infinite layer SrCuO$_2$ and BaCuO$_2$, while forming he building blocks for these superconducting compounds, are not stable superconducting materials when grown as single component films. SrCuO$_2$ in the infinite layer structure is an insulator, while BaCuO$_2$ normally does not form the infinite layer structure, even by high-pressure synthesis techniques. The Ba$_2$Sr$_{n-1}$Cu$_{n+1}$O$_{2n+2+\delta}$ series, with $n = 2$, 3, and 4, is structurally analogous to the recently discovered CuBa$_2$Ca$_{n-1}$Cu$_n$O$_{2n+2+\delta}$ high-pressure HTSc phase (4–6), and is obtained by artificially-layering two-unit cells of BaCuO$_2$ and $(n-1)$ unit cells of SrCuO$_2$ in the infinite layer crystal structure. A schematic of the $n = 3$ member is shown in Fig. 4. These 2×2 SrCuO$_2$/BaCuO$_2$ superlattice structures were obtained at temperatures ranging from 500 to 625°C.

Typically, the room temperature resisitivity for these superlattices was 1-10 mΩ-cm. Superconductivity was observed only for superlattice structures grown at T ≥ 525°C. Figure 5 shows the resistivity for 2×2 SrCuO$_2$/BaCuO$_2$ superlattices grown at various temperatures. Although T_c increases with increasing growth temperature up to 600°C, structures grown at higher temperatures contain significant amounts of cubic BaCuO$_2$ as an impurity phase.

The x-ray diffraction pattern for a 1×2 SrCuO$_2$/BaCuO$_2$ superlattice is shown in Fig. 6, along with the corresponding schematic of the ideal atomic arrangement. The solid arrows indicate diffraction peaks from the artificially-layered compounds, while the asterisks designate peaks from the SrCuO$_2$ buffer layer. The vertical dashed lines show the expected locations of the (00ℓ) peaks for the ideal artificially-layered structure. The diffraction pattern clearly indicates the presence of mulitlayer modulation along the c-axis. While the diffraction peaks are close to the ideal (00ℓ) locations, some of the peak intensities are weaker than that predicted from structure calculations with slight deviations or splitting of some of the peaks about the expected peak locations. The most consistent interpretation of these diffraction patterns is to view the peaks as originating either

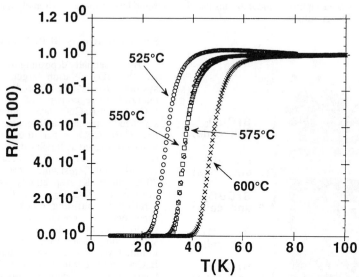

Fig. 5 Normalized resistance plotted as a function of temperature for 2×2 SrCuO$_2$/BaCuO$_2$ superlattices grown at various temperatures.

from the structural modulation or from the superlattice chemical modulation. The structural modulation arises from the crystallinity of the compound. It represents, for instance, the average cation spacing along the c-axis (the growth direction), with some structure peaks present even with no Ba/Sr chemical modulation. For instance, the (003) and (006) peaks for Ba$_2$SrCu$_3$O$_{6+\delta}$ would be present for an infinite layer alloy film with no Ba/Sr ordering along the c-axis. Only as Ba/Sr ordering is realized are other (00ℓ) structure peaks expected. If the Ba/Sr chemical modulation is slightly incommensurate with the structural modulation, superlattice satellite peaks will be present on either side of strong structural peaks at positions which deviate from the ideal (00ℓ) locations. Peaks from the structural and chemical periodicities converge if the deposition process yields exactly integral numbers of SrCuO$_2$ and BaCuO$_2$ infinite layer unit cells per multilayer period. The deviations observed in the peak locations indicate that the chemical modulation is slightly incommensurate with the structural modulation. The fact that the peak intensities are somewhat weaker than expected indicates the presence of Ba/Sr disorder. Since these thin films were formed as artificially-layered superlattices by sequentially depositing SrCuO$_2$ and BaCuO$_2$ layers, it is reasonable to assume that much of this disorder originates from substrate surface roughness and slight inaccuracies in the deposition rates. The x-ray diffraction pattern gives a c-axis lattice constant for the structural periodicity of 12.1 Å, while the chemical periodicity for the same structure is 14.0 Å. Four-circle x-ray diffraction data show these structures to be tetragonal with in-plane lattice constants of 3.9 Å, thus matching the lattice constant for the SrTiO$_3$ substrates.

In these artificially-layered structures, the SrCuO$_2$ and BaCuO$_2$ sub-units could have the ideal infinite layer structure consisting of four-fold coordinated CuO$_2$ planes separated by oxygen-free alkaline earth (Sr or Ba) layers, with no apical oxygen for any of the copper atoms. However, this seems unlikely as all of the hole-doped superconductors have apical oxygen on at least some of the copper atoms, and thermoelectric power measurements indicate that these materials are hole conductors.[19] The more reasonable possibility is for the Ba planes to contain some oxygen, thus creating apical oxygen and increasing the Cu coordination. The schematics of the

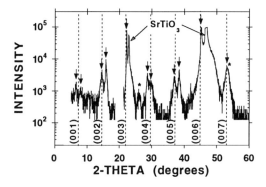 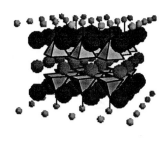

Fig. 6 X-ray diffraction patterns (Cu Kα radiation) and structural model for the $n = 2$ member of the $Ba_2Sr_{n-1}Cu_{n+1}O_{2n+2+\delta}$ series. The dashed lines indicate the nominal locations of the (00ℓ) peaks, signifies $SrCuO_2$ buffer layer, while the solid arrows indicate diffraction peaks due to the artificially-layered structures. In the model, the Ba, Sr, and Cu atoms are represented by the large, medium, and small spheres, respectively, and the CuO_4 and CuO_5 units as shaded polyhedra.

$Ba_2Sr_{n-1}Cu_{1+n}O_{2n+2+\delta}$ structures shown in Figs. 4 and 6 assume this to be true. It is not clear, however, whether oxygen on the Ba planes resides there at the expense of oxygen on specific Cu planes. Because of this uncertainty, we have designated the materials as $Ba_2Sr_{n-1}Cu_{n+1}O_{2n+2+\delta}$ which makes no inference as to whether inequivalent Cu sites are present, although this may occur.

Figure 7 shows the resistivity for the $n = 2$, 3, and 4 members of $Ba_2Sr_{n-1}Cu_{n+1}O_{2n+2+\delta}$. The room-temperature resistivities are 1-5 mΩ-cm. The $n = 2$ member has the highest superconducting transition temperature with T_c(onset) = 70 K and T_c (R=0) = 50 K. The $n = 3$ member has T_c (onset) = 60 K, T_c(R=0) = 40 K, and the $n = 4$ member has T_c (onset) = 40 K and T_c(R=0) = 20 K. The measurement current density was ~50 A/cm^2. Note that for these artificially-layered compounds, T_c decreases as n increases. This behavior differs from that observed for other HTSc series, in which T_c increases as n increases. One possible explanation is that these $Ba_2Sr_{n-1}Cu_{n+1}O_{2n+2+\delta}$ films are not optimally doped, with an increase in T_c possible with an increase or reduction in the hole carrier density. Another possibility is that the use of Sr to separate the CuO_2 planes inhibits the transfer of charge from the $Ba_2Cu_2O_4$ layer into the $SrCuO_2$ layers. This latter scenario would mean that the superconducting transitions observed for all of these structures is due only to the $Ba_2Cu_2O_4$ layers, with little or no coupling into the CuO_2 planes adjacent to the Sr atoms. In fact, this interpretation is consistent with the observed behavior for $YBa_2Cu_3O_7/PrBa_2Cu_3O_7$ superlattices, where T_c decreases as the nonsuperconducting $PrBa_2Cu_3O_7$ layer thickness increases.[27] It also suggests that replacing Sr with Ca may increase charge transfer into adjacent CuO_2 planes and increase T_c. Unfortunately, the formation of high-quality $CaCuO_2/BaCuO_2$ superlattices is more difficult due to the ionic radii differences for Ba and Ca.

$SrTiO_3/BaCuO_2$ SUPERLATTICES

We have also investigated the epitaxial stabilization of infinite layer $BaCuO_2$ using perovskite-like materials other than $SrCuO_2$ in a superlattice structure. Figure 8 shows the x-ray diffraction pattern for a $BaCuO_2/SrTiO_3$ superlattice with a superlattice period of 20.4 Å. The strongest diffraction peaks correspond to structure peaks with an average d-spacing of 4.1 Å, which is consistent with the formation of a $BaCuO_2/SrTiO_3$ superlattice with the $BaCuO_2$ layers

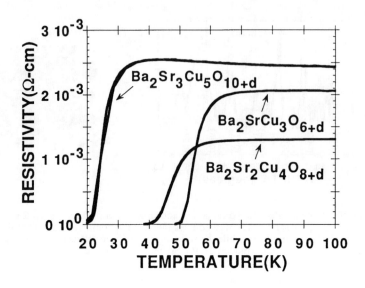

Fig. 7 Resistivity plotted as a function of temperature for the $n = 2$, 3, and 4 members of the $Ba_2Sr_{n-1}Cu_{n+1}O_{2n+2+\delta}$ series.

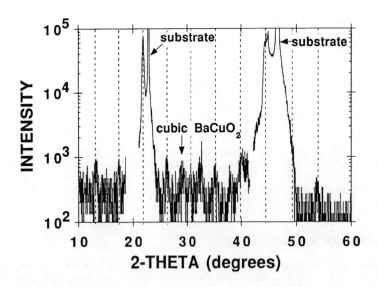

Fig. 8 X-ray diffraction pattern for a $SrTiO_3/BaCuO_2$ superlattice with a superlattice periodicity of 20.4 Å. The dashed lines indicate the expected locations of the superlattice diffraction peaks. Note that a small amount of cubic $BaCuO_2$ is present in this film as an impurity phase.

stabilized in the infinite layer structure. In addition, superlattice satellite peaks are evident as well. While superconductivity has not been observed in these structures to dates, the resistivity is significantly lower than that of $SrTiO_3$ or cubic $BaCuO_2$. The ability to epitaxially stabilize $BaCuO_2$ in the infinite layer structure using $SrTiO_3$ may provide a means of studying the intrinsic transport properties of this material, as $SrTiO_3$ is an excellent insulator.

CONCLUSION

In conclusion, we have grown infinite layer $SrCuO_2/(Ca,Sr)CuO_2$, $SrCuO_2/BaCuO_2$, as well as $SrTiO_3/BaCuO_2$ superlattice structures using conventional pulsed-laser deposition. X-ray diffraction reveals peaks due to the superlattice chemical modulation, even for structures with layers as thin as a single infinite layer unit cell (~3.4 Å). The x-ray diffraction data also show intensity oscillations due to the finite thickness of the film, indicating extremely flat film surfaces with thickness variations of only 20 Å over a length scale of several thousand angstroms. $SrCuO_2/BaCuO_2$ superlattices in the infinite layer structure represent novel artificially-layered superconducting thin-film compounds. These results represent not only the synthesis of new families of superconductors, but also demonstrate that pulsed-laser deposition and epitaxial stabilization can be effectively used to engineer artificially-layered thin-film materials.

REFERENCES

1. J. G. Bednorz and K. A. Muller, *Z. Phys. B, Cond. Matter.* **64**, 189 (1986).
2. T. A. Vanderah, *Chemistry of Superconductor Materials*, (Noyes Publications, Park Ridge, N. J., 1992).
3. A. Schilling, M. Cantoni, J. D. Guo, H. R. Ott, *Nature* **363**, 56 (1993).
4. H. Ihara et al., *Jpn. J. Appl. Phys.* **33**, L503 (1994).
5. C. -Q. Jin, S. Adachi, X.-J. Wu, H. Yamauchi, S. Tanaka, *Physica C* **223**, 238 (1994).
6. M. A. Alario-Franco, C. Chaillout, J. J. Capponi, J.-L. Thoulence, B. Souletie, *Physica C* **222**, 52 (1994).
7. Z. Hiroi, M. Takano, M. Azuma, Y. Takeda, *Nature* **364**, 315 (1993).
8. J. N. Eckstein et al., *Appl. Phys. Lett.* **57**, 931 (1990).
9. T. Terashima et al., *Phys. Rev. Lett.* **60**, 3045 (1992).
10. M. Y. Chern, A. Gupta, B. W. Hussey, *Appl. Phys. Lett.* **60**, 3045 (1992).
11. D. P. Norton, B. C. Chakoumakos, J. D. Budai, D. H. Lowndes, *Appl. Phys. Lett.* **62**, 1679 (1993).
12. C. Niu and C. M. Lieber, *J. Am. Chem. Soc.* **114**, 3570 (1992).
13. M. Yoshimoto, H. Nagata, J. Gong, H. Ohkubo, H. Koinuma, *Physica C* **185**, 2085 (1991).
14. M. Kanai, T. Kawai, S. Kawai, *Appl. Phys. Lett.* **58**, 771 (1991).
15. T. Siegrist, S. M. Zahurak, D. W. Murphy, R. S. Roth, *Nature* **334**, 231 (1988).
16. M. G. Smith, A. Manthiram, J. Zhou, J. B. Goodenough, and J. J. Markert, *Nature* **351**, 549 (1991).
17. G. Er, Y. Miyamoto, F. Kanamaru, and S. Kikkawa, *Physica C* **181**, 206 (1991).
18. M. Takano, M. Azuma, Z. Hiroi, Y. Bando, and Y. Takeda, *Physica C* **176**, 441 (1991).
19. Z. Hiroi, M. Takano, M. Azuma, Y. Takeda, and Y. Bando, *Physica C* **185-189**, 523 (1991).
20. M. Azuma, Z. Hiroi, M. Takano, Y. Bando, and Y. Takeda, *Nature* **356**, 775 (1992).
21. M. Takano, Y. Takeda, H. Okada, M. Miyamoto, and K. Kusaka, *Physica C* **159**, 375 (1989).
22. X. Li, M. Kanai, T. Kawai, and S. Kawai, *Jpn. J. Appl. Phys.* **31**, L217 (1992).
23. X. Li, T. Kawai, and S. Kawai, *Jpn. J. Appl. Phys.* **31**, L934 (1992).
24. D. P. Norton, B. C. Chakoumakos, E. C. Jones, D. K. Christen, and D. H. Lowndes, *Physica C* **217**, 146 (1993).
25. D. P. Norton, J. D. Budai, D. H. Lowndes, and B. C. Chakoumakos, *Appl. Phys. Lett.* **65**, 2869 (1994).

26. D. P. Norton, B. C. Chakoumakos, J. D. Budai, D. H. Lowndes, B. C. Sales, J. R. Thompson, and D. K. Christen, *Science* **265**, 2074 (1994).
27. O. Nakamura, E. Fullerton, J. Guimpel, and I. Schuller, *Appl. Phys. Lett.* **60**, 120 (1992).
28. D. H. Lowndes, D. P. Norton, and J. D. Budai, *Phys. Rev. Lett.* **65**, 1160 (1990).

OBSERVATIONS ON THE GROWTH OF YBa2Cu3O7 THIN FILMS
AT VERY HIGH LASER FLUENCES

RAND R. BIGGERS,* M. GRANT NORTON**, I. MAARTENSE*, T.L. PETERSON*, E. K. MOSER*, D. DEMPSEY*, AND JEFF L. BROWN***
*U. S. Air Force Wright Laboratory, WL/MLPO, Wright-Patterson AFB, OH 45433
**Department of Mechanical and Materials Engineering, Washington State University, Pullman, WA 99164
***U. S. Air Force Wright Laboratory, WL/ELDM, Wright-Patterson AFB, OH 45433

ABSTRACT

The pulsed-laser deposition (PLD) technique utilizes one of the most energetic beams available to form thin films of the superconducting oxide YBa2Cu3O7 (YBCO). In this study we examine the growth of YBCO at very high laser fluences (25 to 40 J/cm^2); a more typical fluence for PLD would be nearer to 3 J/cm^2. The use of high fluences leads to unique film microstructures which, in some cases, appear to be related to the correspondingly higher moveabilities of the adatoms. Films grown on vicinal substrates, using high laser fluences, exhibited well-defined elongated granular morphologies (with excellent transition temperature, T_c, and critical current density, J_c). Films grown on vicinal substrates using off-axis magnetron sputtering, plasma-enhanced metal organic chemical vapor deposition (PE-MOCVD), or PLD at more typical laser fluences showed some similar morphologies, but less well-defined. Under certain growth conditions, using high laser fluences with (001) oriented substrates, the YBCO films can exhibit a mixture of a- and c-axis growth where both crystallographic orientations nucleate on the substrate surface at the same time, and grow in concert. The ratio of a-axis oriented ($a\perp$) to c-axis oriented ($c\perp$) grains is strongly affected by the pulse repetition rate of the laser.

INTRODUCTION

There is a number of very different techniques for producing thin-film high temperature superconductors (HTS). These techniques, which cover quite a large range of deposition conditions, entail significant differences in deposition times, deposition areas, growth rates, growth mechanisms, and film qualities. In pulsed-laser deposition (PLD), we have chosen to utilize one of the most energetic, versatile, and quickest production methods for quality HTS thin films. The laser provides a highly energetic pulse (generally from 10^8 to 10^9 W/cm^2) which is used to ablate material from a stoichiometric target and this material is directed towards a heated substrate. The energy density, footprint, repetition rate, duration, spatial and temporal uniformity, and wavelength of the pulse are the variable laser parameters. These laser parameters coupled with a number of non-laser parameters such as chamber oxygen pressure, substrate temperature, substrate material and orientation, and target material strongly influence the properties of HTS thin films. An understanding of the deposition process coupled with proper adjustment and control of these deposition parameters should allow growth of thin film HTS with film properties and microstructures more closely matched to the intended film applications.

EXPERIMENTAL

The PLD process used in our laboratory has been reported in detail elsewhere [1]. The YBCO films were formed by focusing the output beam from either a Lumonics Hyper EX 400 (L-400) or for the majority of the films, a Lambda Physik LPX 305i (L-305) excimer laser operating on KrF ($\lambda = 248$ nm) onto a stoichiometric YBCO pellet. Nominally an average pulse energy of 200 mJ (L-400) or 400 mJ (L-305) was used which gave average laser fluences of 14 J/cm^2 or 30 J/cm^2 respectively, or equivalent power densities of ~10^9 W/cm^2 with a beam footprint of ~1.3 mm^2 and pulse durations of 13 ns or 25 ns respectively at a pulse repetition rate of 20 Hz. Recently films have been grown utilizing a spatial filter and realigned optics at an average fluence of 7.8 J/cm^2 or 0.3×10^9 W/cm^2 (L-305*).

Vendors provided lanthanum aluminate (LaAlO3) substrates with the following orientations: nominally (001) or "oriented", and "vicinal" (001) substrates with miscuts of 6° or 10.5° about an axis parallel to a <110> direction. All orientations were confirmed by Laue back-reflection.

Mat. Res. Soc. Symp. Proc. Vol. 388 © 1995 Materials Research Society

In a typical film growth experiment, LaAlO₃ substrates (5 x 5 x 0.5 mm and/or 10 x 10 x 0.5 mm) were mounted on a 2.5 cm stainless steel plate. The distance between the YBCO target and the central substrate was 6.8 cm; the angle between the plane of the target surface and the stainless steel plate was either 0° or 26° (L-400) and 26° (L-305) depending on the heater utilized. The laser beam was scanned up to 5 mm across the target, which was rotated at 10 rpm. The substrate position was important in the case of the L-400 because of the thermal gradient intrinsic to the sample holder (used only with the L-400) and for the L-305 because of an asymmetry in the beam spatial energy distribution. The L-305 ablation plume provided regions with fluences of 25, 30, and 40 J/cm². The approximate deposition rates utilized for these lasers are as follows (at 20 Hz): 110 nm/min for the L-400 and 60, 80, and 100 nm/min for the fluence levels of the L-305. (Those of the L-305* have not been determined at this time.)

Our group has made a systematic study on the role of specific deposition parameters on the microstructure and physical properties of YBCO films deposited by PLD. The following deposition parameters were varied (one parameter per experiment) from their nominal values: laser pulse energy, substrate temperature, laser pulse repetition rate, laser plume scanning, deposition time, substrate positioning, substrate orientation [(001) or vicinal], different YBCO targets, spatial filter, different focal lengths, oxygen pressure and oxygen flow rate. The footprint/plume asymmetry was utilized to explore *simultaneously* the effects of the above variations over a range of deposition energies and the effect of mixing different plume components as they would occur in large area depositions. A "**standard**" set of deposition parameters was utilized at intervals during the L-305 studies to determine the overall reproducibility and control. All **standard** films grown in a particular fluence region exhibited essentially the same properties. The L-400 films were utilized to study the effects of substrate temperature on film growth over a wide but continuous range of temperatures on a single film (L-400 had a symmetric spatial energy distribution).

In addition, several films were grown for us at other institutions by PLD, sputtering, and PE-MOCVD on our vicinal and oriented substrates. The PLD processes were similar to ours but they employed lower fluences, ~3 J/cm², with correspondingly lower growth rates (~14 nm/min).

The surface morphology of the films was determined using a Digital Instruments Nanoscope II scanning probe microscope (AFM and STM) and a Hitachi S-900 SEM with a field emission gun operating at an accelerating voltage of 3 kV. J_cs and T_cs were obtained from AC susceptibility and DC transport measurements. Film and substrate axis orientations and lattice spacings were obtained with a Philips Analytical X-Ray Diffractometer.

RESULTS AND DISCUSSION

Effect of Substrate Orientation

Over 250 films, all grown on LaAlO₃ substrates, were examined. For this report a number of selected measured properties of representative YBCO films along with critical deposition parameters have been tabulated. Table I summarizes the strong effects of vicinal LaAlO₃ substrates on the growth of YBCO films over a wide range of growth conditions and processes. Films grown on vicinal substrates possessed good to excellent superconducting properties (T_cs>88 K and $\Delta T\chi$<1.5 K) while under the same conditions films produced on oriented substrates were of a much poorer quality (T_c<80 K and $\Delta T\chi$>2.5 K) and relatively porous (see Fig. 1). In some cases, see Table I, films grown on the vicinal substrates had T_cs that were 13 K higher and the J_cs more than 40 times larger at 77 K. Under deposition conditions that produced the best films for each process, the films grown on the vicinal substrates were of generally equal or better quality, with the single exception of PE-MOCVD films, which were significantly inferior. Again, except for PE-MOCVD, all films grown on vicinal substrates exhibited a characteristic elongated granular structure (so called rope-like morphology: composed of a series of scalloped terraces, a canted series of monolayers) [1]. Of these same films, those grown at the higher fluences (14 to 40 J/cm²) generally had the most well-defined surface morphologies, consisting of elongated and aligned ropes similar to those in Figs. 2a and 2b. Both the height and the width of the ropes increased with increasing fluence in the range of 25 to 40 J/cm².

The rope-like morphology shown in Fig. 2a is typical of that observed in films deposited on vicinal substrates at high fluences. The films seem to consist of overlapping layers one unit cell thick with about 12 to 20 nm of exposed terrace at each level [1]. The films on vicinal substrates also seem to have fewer ($a\perp$) outgrowths. The ($c\perp$) of the film is aligned with the ($c\perp$) of the substrate even though the substrate surface has been rotated 6° or 10.5° about a <110> direction [1].

YBCO films grown on oriented substrates at the three highest fluences (25, 30, and 40 J/cm²), exhibit distinct morphologies and superconducting properties. At the highest fluence, the film surface appeared rather disordered, with essentially no ($a\perp$) and depressed T_c and J_c. This disordered, unstable growth mode is the result of a very high level of supersaturation [6]. At the next highest fluences, 30 to 25 J/cm², the films exhibit a mixture of ($a\perp$) and ($c\perp$) growths, where both crystallographic orientations nucleate on the substrate surface at the same time (see Fig. 4) and grow in concert [2,3]. The ($a\perp$), also referred to as outgrowths, remain quite similar in size and height from film thicknesses ~6 to ~180 nm [2]. By the final film thickness of 360 nm, some ($a\perp$) have significantly increased in size

Table 1: Effects of Using a Vicinal Substrate on YBCO Films Grown Under Various Conditions.

Film	Substrate misorient. to (001)	T_c (K)	ΔT_z (K)	$J_c(.86T_c)$ (10^6 A/cm²)	Rope Height (nm)	Rope Width (nm)	Ougr. per µm² (approx.)	Growth Rate (nm/min) (approx.)	PLD Fluence (J/cm²)	Energy (10^9 W/cm²)	Pulse Rate (Hz)
WL-27416	<1°	75.1	3.7	0.1*	(none)	(none)	.02	100	40	1.6	20
WL-27452	6°	89.0	0.77	4.0*	50	560	.02	90	30	1.2	20
WL-27417	<1°	87.4	0.5	4.4*	(none)	(none)	1.6	-	25	1.0	20
WL-27463	6°	88.9	0.87	-	15	460	1.2				
WL-34578	10.5°	90.1	0.7	4.0*	45	593	.004	110			
WL-27418	<1°	86.7	0.88	-	(none)	(none)	2.7	70	14	.9	20
WL-27464	6°	88.1	1.24	-	10	160	.09	-			
W-26102	<1°	89.2	0.54	-	(none)	(none)	.031	110	8.3	.33	2
W-26109	6°	88.4	0.51	4.4*	47	400	.027				
W-26121	10.5°	89.2	1.15	5.1	150	1100	0				
WL-34606	<1°	82.0	7.4	-	(none)	(none)	.063	88	7.3	.29	20
WL-34608	6°	89.2	0.63	-	75	660	.002	75			
WL-34601	<1°	81.3	3.6	-	(none)	(none)	0	75			
WL-34603	6°	89.9	0.9	-	15	400	.25				
Supercond. Technologies, Inc. (STI)	<1°	89.9	2.7	0.3*	(none)	(none)	.37	14	3	.12	3
	6°	90.7	1.2	2.0*	5	350	.008				
	10.5°	89.5	2.8	0.31	19	281	.04				
McMaster Univ.	<1°	92.0	0.8	-	(none)	(none)	.06	14	2.5	.089	2
	6°	91.7	0.8	4.3*	14	500	.02				
	10.5°	91.6	0.7	4.4	22	320	.008				
Westing-house STC	<1°	87.2	1.0	-	(none)	(none)	0	0.4	Sputtering		cw
	6°	87.6	1.4	2.5*	18	375	0				
EMCORE	<1°	86.7	1.3	-	(none)	(none)	.2	300	MO CVD		cw
	6°	83.5	11.5	-	70	-	.07				
	10.5°	80.5	30.5	-	100	-	.07				

Asterisk (*) indicates J_c values estimated from J_c measurements on similar samples. ΔT_z is the transition width inferred from magnetic susceptibility. Oxygen partial pressures (in mTorr) are as follows: 160 for WL, 100 for W, 400 for STI, 225 for McMaster Univ., ~500 for Westinghouse STC, and 5-10 Torr for EMCORE.

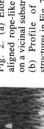
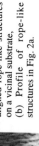

Fig. 1. Unstable growth (disordered) at high fluence, ~40 J/cm².

Fig. 2. (a) Elongated and aligned rope-like structures on a vicinal substrate, (b) Profile of rope-like structures in Fig. 2a.

Fig. 3. Visual differences between areas of unstable (a) and stable (b) island growth, and step-flow growth (c).

Table II
Target Scanning, No Scanning, and Vicinal Effects
at ~40 mJ/cm² (~1.6x10⁹ Wcm²)
with 160 mT of O₂ and at ~780°C

Film#	Rep (Hz)	Target Scan	Tc (K)	ΔTχ (K)	Jc(.86Tc) (x10⁶ W/cm²)	Thickness (nm)	Sub. Orien. ( )	# O.G. (/μm²)	# Voids (/μm²)
27484	20	none	**87.8**	**1.3**	2*	>420*	(001)	~1.0	0
27487	20	none	**87.6**	**1.2**		>420*	(001)	~2.1	0
27490	20	none	**84.9**	**2.3**		>420*	(001)	---	---
27468	2	C	71.8	3.9		>420*	(001)	~1.0	0
27428	5	M	69.7	1.9		>420*	(001)	~0.06	~0.13
27425	10	M	72.8	2.3		420*	(001)	~0	0
27289	20	M	72.2	3.7	~ 0.1	430	(001)	~0.02	0
27416	20	M	75.1	3.7	0.1*	~420	(001)	~0.04	0
27448	40	M	75.5	5.0		<420*	(001)	~0.03	0
27462	20	C	**89.0**	**0.8**	4.0*	420*	6°@<110>	~0.31	0
27480	20	C	**88.3**	**0.7**	4.0*	420*	6°@<110>	~0.37	0

M = manual about 5 x 1 mm, C = computer controlled scans, * = estimation
___ = standard deposition parameters, O.G. = outgrowths (a⊥)

Table III
Frequency Study at ~30 mJ/cm² (~1.2x10⁹ Wcm²)
with 160 mT of O₂ and at ~780°C on (001) LaAlO₃

Film#	Rep. Rate (Hz)	Tc (K)	ΔTχ (K)	Jc(.86Tc) at (x10⁶ W/cm²)	Thick. (nm)	# O.G. (/μm²)	# Voids (/μm²)	Grain Size (μm²)	Target Scan
27469	2	**90.2**	1.0	4.2	~360	~0.1	~0.16	~2.6	C
27429	5	**90.4**	0.8	-----	~360*	~0.6	~0.09	~2.0	M
27426	10	87.4	0.6	-----	~360*	~1.6	0	~1.9	M
27417	20	87.4	0.5	4.5	~360*	~1.6	0	~0.5	M
27448	40	86.7	1.0	2.0	~340	~1.3	0	~0.6	M

M = manual about 5 x 1 mm, C = computer controlled scans, * = estimation
O.G. = outgrowths (a⊥), ___ = standard deposition parameters

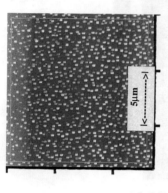

Fig. 4. 100 pulses at 20J/cm². Film about 6 nm thick with ~200 x 300 nm² (a⊥) outgrowths extending ~30 nm above surface.

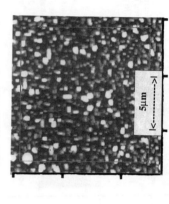

Fig. 5. 5000 pulses at 20J/cm². Film about 360 nm thick with various sizes of (a⊥) outgrowths. Some are merged (a⊥) and many remain the same size as for the 100 pulse film seen above.

while the overall density has slightly decreased due to coalescence, see Fig. 5. Films grown at 25 J/cm^2 had about twice as many ($a\perp$) as did those at 30 J/cm^2. Films grown on oriented substrates consisted of more than one component (areas of different microstructure and macrostructure) when they were positioned in an area spanning ordered and disordered growth regimes. The components can be detected visually, as shown in Fig. 3 where the top three films are on oriented substrates. Note the uniformity of the bottom triangular film which was deposited on a vicinal substrate. AC susceptibility measurements confirm it to be a single component film.

All films (excepting those grown by PE-MOCVD) on the vicinal substrates show, at some scale, a rope-like structure: this is consistent with film growth nucleated at steps/ledges in the substrate. The evolution of the film microstructure from the initial preferred nucleation sites to the final macrostructure is determined by the relative magnitude of the surface diffusion length, L_d, (related to adatom moveability) to the width of the ledges, L_t [4,5]. Having preferred nucleation sites and $L_d \gg L_t$ on the vicinal substrate induces step-flow growth and produces high quality films with distinct rope-like structures.

As the PLD fluence decreases from 40 J/cm^2 through 30 to 25 J/cm^2, the supersaturation (plume density) decreases below the level that causes unstable growth on oriented substrates. Distinct growth regimes, stable and unstable observed on oriented substrates, do not occur when using vicinal substrates. Vicinal substrates force step flow growth over the 25 to 40 J/cm^2 fluence range. The rope heights and widths exhibit a general increase in size over this range. At the very low fluences (3 J/cm^2) the rope structures are rather indistinct but scalloped terraces are found on 300 nm x 300 nm micrographs.

Competition between L_d and L_t determines whether step-flow or island growth occurs, decreasing L_t should favor/enhance the rope-like structures. The 10.5° (narrower steps, smaller L_ts) vicinal substrates all generated more distinct, higher ropes than the 6° under equivalent growth conditions. Vicinal substrates produced films essentially equal in quality to those on oriented substrates only when grown under the conditions optimized to grow the best films on the oriented substrates. Films grown by PE-MOCVD are the exception; they exhibited a macrostructure entirely different than that of any other films in this study.

The vicinal substrate films grown with the L-400 exhibited a multicomponent character. This is due to the large thermal gradient in the substrate heater and the associated range of effective moveabilities. In all cases, the portions of the films with the largest rope-like structures had the higher T_c and J_c characteristics. In several cases, the higher quality regions had a visible arced boundary that matched the curvature of the underlying thermal profiles. The high quality film region was coincident with the highest temperature region of the substrate.

Effect of Target and Plume Scanning at 40 J/cm^2

Table II shows the sensitivity of film growth, under the conditions of the highest fluences, to scanning of the ablation plume across a substrate. Films grown with scanning on oriented substrates produced very poor ($T_c \sim 73$ K and $\Delta T\chi > 2.5$ K) YBCO films. Not scanning the plume produced significantly better films ($T_c \sim 87$ K and $\Delta T\chi < 1.3$ K but by far the best films ($T_c > 88$ K, $\Delta T\chi < 0.8$ K, and reduced numbers of ($a\perp$)), were grown on the vicinal substrates (also with plume scanning).

In PLD, scanning the beam across a rotating target tends to minimize changes in the target surface and reduce undesired effects such as deposition of large particles onto the film surface. Scanning in general also moves the plume across the substrate. The uniformity and degree of movement of the plume determines the uniformity in both density and energy of the arriving adatoms, and thereby can strongly affect the resultant film growth. A highly anisotropic plume where spatially varying fluences ranged from 40 to 25 J/cm^2 was utilized to exaggerate conditions expected for large area film growth. If ($c\perp$) current flow is a critical part of the film J_c (Brick wall model [7]) then the mixing of plume material which produces visibly different surface morphologies, especially on oriented substrates, may further reduce the coupling between grains in the ($c\perp$) direction on the oriented substrates. The scanned areas receiving the highest fluence material exhibited a structure consistent with unstable growth due to very high supersaturations and had very poor T_cs and J_cs. Eliminating plume scanning made no observable differences in the structure of films grown at 25 and 30 J/cm^2 on the oriented substrates; however, the T_cs and J_cs from the region of 40 J/cm^2 fluence adatoms were significantly better, possibly due to better ($c\perp$) current coupling. Our vicinal substrates which induce strong step-flow growth, especially at the highest fluences, are rather insensitive to plume scanning.

Effect of Laser Pulse Rate at ~30 J/cm^2

Table III depicts the strong sensitivity of film growth at 30 J/cm^2 on oriented substrates to the time length between plumes of ablated material (laser pulse rate). Several significant changes and trends were noted. The grain size of these films continuously decreases in average size from 2.6 to 0.6 μm^2 as the pulse rate increases from 2 to 40 Hz. The T_c drops from 90 to 87 K, the number of outgrowths, ($a\perp$), increases from 0.6 to 1.6 /μm^2, and void areas (depressions) disappear, between 5 and 10 Hz. For further increases in pulse rate, T_cs and the number of $(a\perp)$ remain relatively constant with slight decreases at 40 Hz. The J_cs at $.86T_c$ remain at about $4x10^6$ A/cm^2 from 2 to 20 Hz but show a small decrease at 40 Hz.

At the lower pulse rates the effective adatom moveability is increased. The adatoms have more time (until the arrival of the next pulse) to move to the most energetically favorable sites. Under supersaturation conditions favoring stable island growth, this movement is consistent with the formation of larger grains, fewer outgrowths and more voids (the least favorable sites). Since film T_cs and J_cs are determined by grain boundary couplings, larger grains imply less boundary area material and possibly increased T_cs and J_cs. As the time between pulses decreases (higher laser repetition rates) the adatom movements become more limited, and less energetically favorable positions are utilized. The grains become smaller, more outgrowths are nucleated, and the voids disappear. With smaller grains come a larger overall grain coupling area and decreased T_cs and J_cs. In general, the data support these trends but the sharp drop in T_c of 2 K going from 5 to 10 Hz is not understood nor the minimal change in J_c between 2 and 40 Hz. Additional effects like that of plume scanning must also be understood and taken into account to provide a more complete picture of the laser pulse rate effects.

CONCLUSIONS

In our study, PLD and sputtering of YBCO on vicinal substrates (nominally (001) LaAlO$_3$ but with a 6° or 10.5° rotation about a <110> axis) resulted in films with essentially equal or better T_c or J_c than those on (001) oriented substrates. Vicinal substrates induce a step-flow growth mechanism over a wide range of deposition conditions with the structure controlled by the effective moveability of the adatoms. In addition, the films are more likely to be single component with less $(a\perp)$ outgrowths than films on oriented substrates grown under similar conditions. That is, the step-flow growth mechanism is more tolerant of variations in deposition parameters.

Oriented substrates (nominally (001) LaAlO$_3$) produce films with material characteristics that are very sensitive to deposition conditions. Under certain conditions, the laser pulse rate will determine film grain size, T_c, and $(a\perp)$ and surface-void densities with little or no effect on J_c.

ACKNOWLEDGMENT

R.R.B. thanks R. Hughes (McMaster Univ.), M. Eddy (Superconductor Technologies, Inc.), C. Chern (EMCORE Corp.), and J. Talvacchio (Westinghouse Electric Corp.) for their sample films and helpful discussions, M. Capano (WL/MLBM), and P.T. Murray and D. Wolf (Univ. of Dayton) for their measurements and insights, and also B. Lovett, D. Buchanan, S. Murray, R. Bertke, and G. Landis for their support. This research was sponsored by the Wright Laboratory, Aeronautical Systems Center, Air Force Materiel Command, USAF and the Advanced Research Projects Agency (ARPA) under grant number F33615-94-1-5802. E.K.M. gratefully acknowledges the support of a National Research Council-Wright Laboratory Research Associateship. The work at Westinghouse was completed under AFOSR Contract F49620-94-C-0021.

REFERENCES

1. R.R. Biggers, M.G. Norton, I. Maartense, E.K. Moser, D. Dempsey, T.L. Peterson, M.A. Capano, J. Talvacchio, J.L. Brown, and J.D. Wolf, IEEE Trans. Appl. Sup., (1995 in press)
2. M.G. Norton, R.R. Biggers, I. Maartense, E.K. Moser, and J.L. Brown, Phys. C. , **233**, 321 (1994)
3. M.G. Norton, R.R. Biggers, I. Maartense, E.K. Moser, and J.L. Brown, in Epitaxial Oxide Thin films and Heterostructures, edited by D.K. Fork, R. Ramesh, J.M. Phillips, and R. M. Wolf, Mater. Res. Soc. Proc. **341**, 183 (1994)
4. S.J. Pennycook, M.F. Chisholm, D.E. Jesson, R. Feenstra, S. Zhu, X.Y. Zheng, and D.J. Lowndes, Physica C, **202**, 1 (1992)
5. S.A. Kukushkin and A.V. Osipov, Thin Sol. Films, **227**,119 (1993)
6. H.J. Scheel, Adv. in Supercond. VI, Springer, Tokyo (1994)
7. Z.X. Goa, I. Heyvaert, B. Wuyts, E. Osquiguil, C. van Haesendonck, and Y. Bruynseraede, Appl. Phys. Lett. **65**, 6 (1994)

DIELECTRIC SUSCEPTIBILITY AND STRAIN IN $Sr_{1-x}Ba_xTiO_3$ FERROELECTRIC THIN FILMS GROWN BY PULSED LASER DEPOSITION

L.A. KNAUSS*[†], J.S. HORWITZ*, D.B. CHRISEY*, J.M. POND*, K.S. GRABOWSKI*, S.B. QADRI*, E.P. DONOVAN AND C.H. MUELLER**
*Naval Research Laboratory, Code 6670, 4555 Overlook Ave., SW, Washington, DC 20375
**SCT, 720 Corporate Circle, Golden, CO 80401
[†]NRL/NRC Cooperative Research Associate

ABSTRACT

Thin films of $Sr_{1-x}Ba_xTiO_3$ (SBT) with x = 0 to 0.80 have been grown *in situ* by pulsed laser deposition onto single crystals of (001) $LaAlO_3$. The films were grown to thicknesses of 0.6 μm and found to be single phase, highly oriented, and characterized by x-ray ω scan widths of ≤ 0.5°. The temperature dependence of the dielectric constant and the relative dissipation factor were measured at 1 kHz using Au interdigital capacitors deposited on top of the ferroelectric films. The capacitance measurements indicate that the temperature dependence of the dielectric constant of the film is broad and the maximum is shifted relative to the bulk material. The differences between thin film and bulk properties are attributed to strain in the film resulting from film - substrate lattice mismatch. Thick films (~7 μm) gave direct evidence for strain through cracking and delamination. X-ray diffraction measurements have been made to determine the non-uniform strain in the thin films which was approximately 0.1%.

INTRODUCTION

Ferroelectrics are a class of non-linear dielectrics which exhibit an electric field dependent dielectric constant. Recent reports of the electric field effect in bulk materials and thin films at high frequencies (100 MHz - 10 GHz) suggests that these materials could be used to develop active microwave electronics such as phase shifters, tunable filters and tunable high Q resonators [1,2]. Thin film ferroelectrics offer a unique advantage over bulk materials for these applications. Large electric fields (0-200 kV/cm) can be achieved in thin films (~0.5 μm) using low bias voltages (0-10 V) [2]. $Sr_{1-x}Ba_xTiO_3$ (SBT) is currently the material of choice due to its low loss and composition dependent Curie temperature. The Curie temperature of bulk SBT ranges from 30 to 400 K for Ba concentrations ranging from x = 0 to 1, respectively [3]. The ability to control the dielectric properties in a simple way will allow device structures to be easily optimized for maximum tunability and minimum loss at the desired frequency and operating temperature. Also, SBT can be incorporated with high temperature superconductors, further minimizing losses in the devices.

We report an investigation of the temperature dependence of the dielectric susceptibility of oriented SBT thin films for x = 0.35 and 0.80 grown on $LaAlO_3$ substrates. Significant differences are observed in the temperature dependence of the dielectric susceptibility of the thin films compared to the bulk. These differences may be due to strain in the thin films resulting from the large lattice mismatch of ~3% and twinning in the $LaAlO_3$ substrate. A preliminary investigation of non-uniform strain is presented in thin films of SBT for x = 0, 0.35, 0.65 and 0.80.

73

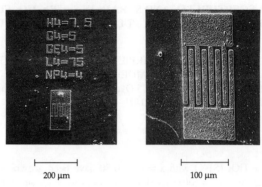

| 200 μm | 100 μm |

Figure 1.   SEM photomicrograph of interdigital electrodes for capacitance measurements. (Chrome/gold)

EXPERIMENTAL

Thin films of SBT on (001) $LaAlO_3$ were grown by pulsed laser deposition for compositions x = 0, 0.35, 0.65 and 0.80. The pulsed laser deposition system used to grow these SBT films has been described previously in more detail [4,5]. A pulsed eximer laser (~30 nsec pulses, ~300 mJ/pulse and λ = 248 nm focused with a 50 cm focal length lens to a fluence of ~2 J/cm^2) was used to ablate the pressed powder targets of various $SrTiO_3$ and $BaTiO_3$ compositions. The vaporized material was deposited onto a heated $LaAlO_3$ substrate approximately 3 cm away from the target. The substrate was heated to 750°C in a 350 mTorr oxygen atmosphere. The films grew at approximately 2 Å/pulse to a total film thickness of 0.6 μm. One 7 μm thick film was grown. The film composition and thickness were determined by Rutherford backscattering of 6.2 MeV He^{2+} ions.

Capacitance measurements were made by e-beam evaporating chrome/gold interdigital electrodes on the surface of the SBT films. The electrodes were patterned by a standard lithography technique. An SEM photomicrograph of these electrodes is shown in Figure 1, where the gap spacing is 5 μm, the finger width is 7.5 μm and the finger length is 75 μm. The electrodes were ~1500 Å thick, and electrical contact was made by wire bonding two gold wires to the large electrode contact pads. One wire was attached to each pad next to the fingers. Measurements of the capacitance and relative dissipation factor as a function of temperature were made at 1 kHz using an HP4284A LCR meter. The temperature control was achieved with an APD Cryogenics closed cycle refrigerator with a Lakeshore 330 temperature controller. A reference measurement was made on an uncoated $LaAlO_3$ substrate which showed a negligible temperature dependence. Thus changes in the capacitance as a function of temperature are due to changes in the dielectric susceptibility of the film. Non-uniform strain was determined from x-ray diffraction using a Rigaku Rotaflex diffractometer with Cu K_α radiation from a rotating anode source.

RESULTS AND DISCUSSION

SBT films grown by pulsed laser deposition were found, by x-ray diffraction, to be single phase, well oriented and nearly epitaxial. Figure 2 shows a θ/2θ scan for x = 0.35 which is typical of all the compositions investigated. The film is exclusively (00ℓ) oriented with a (002) ω scan width of only 0.209°. The ω scan widths for the (002) line of all the compositions were less than

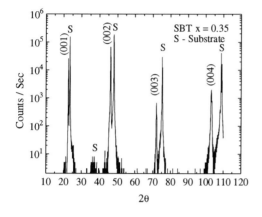

Lattice Parameter	
x	a (Å)
0.00	3.9065
0.35	3.9357
0.65	3.9576
0.80	3.9741

Figure 2.   X-ray diffraction from a thin film of SBT x = 0.35 on (001) oriented LaAlO₃. The lattice parameters listed on the right are calculated from the (004) line position.

0.5°. The lattice constants of the films were determined from the (004) line and are presented in Figure 2.

Capacitance Measurements

Capacitance and dissipation factor measurements were made as a function of temperature for SBT thin films of composition x = 0.35 and 0.80. The data were collected while cooling and warming to detect any thermal hysteresis. In Figure 3 the capacitance and relative dissipation factor are presented for the x = 0.35 composition. The temperature dependence is significantly different from that of the bulk material. The capacitance verses temperature curve is much broader than that of the bulk with a maximum at 153 K, which is about 50 K lower than the bulk maxima. A similar peak shift has been observed previously in a thin film of x = 0.50 composition. The relative dissipation factor shows a large change, ~0.025, between 350 K and the peak at 170 K. Also, the peak in the relative dissipation factor is about 20 K higher than the peak in the capacitance. An unusual feature in the temperature dependence of the capacitance appears close to room temperature. There is a distinct change marked by a cusp at 288 K, and a reproducible thermal hysteresis. The inset in Figure 3 shows the capacitance measured while cooling close to the cusp. The dielectric susceptibility of ferroelectric materials can often be described by a Curie-Weiss law where the susceptibility is proportional to $(T-T_c)^{-1}$. We found that the capacitance could be fit to the Curie-Weiss law in the temperature range above the cusp using the expression,

$$Cap. = \frac{C}{(T - T_c)} + b. \tag{1}$$

This is a three parameter fit of C, T_c and b, where C is the Curie constant for the capacitance, T_c is the transition temperature, and b is a background value to account for parasitic capacitance in the test structure. The results of the fitting are given in the inset where T_c is found to be 214 K which

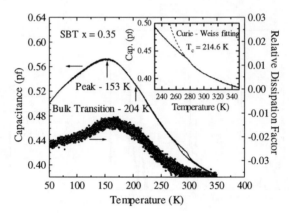

Figure 3.   Capacitance and relative dissipation factor as a function of temperature for SBT
             (x = 0.35).

is within 10 K of the reported bulk transition temperature. The capacitance at temperatures
below the cusp appears to be suppressed relative to the bulk behavior.

Figure 4 shows the capacitance of the x = 0.80 composition. We were unable to measure at
high enough temperatures to see the peak in the capacitance. This was unexpected since the bulk
transition temperature is at 341.6 K Previously we have seen that the peak in the capacitance is
shifted to lower temperature relative to the bulk. There are two changes in the slope of the
capacitance which are also observed in the dissipation factor. These can probably be attributed to
transitions to the two lower symmetry phases, but their temperatures again do not match those of
the bulk [3].

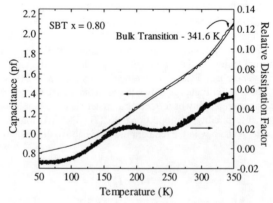

Figure 4.   Capacitance and relative dissipation factor as a function of temperature for SBT
             (x = 0.80).

|_____| 20 μm |_____| 200 μm

Figure 5.   SEM photomicrograph of ~7 μm thick SBT (x = 0.65) film.

Non-uniform Strain Measurements

The presence of strain in SBT films grown on $LaAlO_3$ becomes readily apparent in thick films. Figure 5 shows an SEM photomicrograph of a ~7 μm thick SBT film, x = 0.65. The film exhibits severe cracking and delamination. In particular, the delamination results in curved strips of SBT indicating that the strain is non-uniform perpendicular to the interface. The strain can result from several contributions: lattice mismatch, twinning in the substrate, and differences in the coefficient of thermal expansion between the film and substrate. Differences in the coefficient of thermal expansion can be ruled out as a major contribution, since the film cracked during film growth at which time the temperature was kept constant. To investigate this non-uniform strain in the thin films (~0.6 μm) which typically do not crack, we measured line broadening in the x-ray diffraction peaks. Two major contributions to the broadening are non-uniform strain and crystallite size. Williamson and Hall [6] have shown that these contributions can be combined and expressed as:

$$\frac{\beta \cos\theta}{\lambda} = \frac{1}{D} + 4\varepsilon \frac{\sin\theta}{\lambda} \tag{2}$$

where β represents the line width at half maximum, θ is the Bragg angle and λ is the wavelength of the x-rays. By plotting ($\beta\cos\theta$ / λ) verses ($\sin\theta$ / λ), one expects a straight line where the slope is proportional to strain, ε, and the intercept is proportional to the crystallite size, D. We investigated the non-uniform strain in the [001] direction by measuring the x-ray peaks for the higher order reflections off of (00ℓ) planes out to (004). To obtain the peak width and the Bragg angle, the lines were fit using two lorentzian peaks, one for $K_{\alpha 1}$ and the other for $K_{\alpha 2}$. From the fitting parameters the Bragg angle and the peak width can be obtained. By plotting the fitting results using equation (2), we obtained the straight lines shown in Figure 6. The strain and approximate crystallite size can be determined from the slope and intercept respectively. These values for the four SBT compositions are listed in Figure 6. The strain is ~0.1% in all cases except the x = 0.65 composition. This is a very large strain for a ceramic material considering

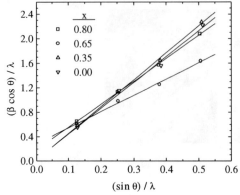

	Strain	Crystallite Size
x	ε	D (μm)
0.80	0.00094	0.5
0.65	0.00067	0.4
0.35	0.00110	10
0.00	0.00106	5

Figure 6.   Plot to determine strain and approximate crystallite size for SBT (x = 0 to 0.80).

that a large strain in metals is approximately 0.2%. The reduction of strain in the x = 0.65 composition can be attributed to the presence of cracks in this film.

CONCLUSIONS

High quality thin films of SBT (x = 0 to 0.80) have been grown by pulsed laser deposition onto (001) LaAlO$_3$ substrates. We have investigated the temperature dependence of the dielectric susceptibility of the thin films by measuring the capacitance as a function of temperature. These measurements have shown that the peak in the dielectric susceptibility as a function of temperature of the film is shifted in temperature relative to the bulk material. The capacitance peak of the film is much broader than the bulk peak and appears to be suppressed as indicated by the deviation from a Curie-Weiss type behavior. We have determined the presence of a significant strain in the SBT films on LaAlO$_3$ with a magnitude of ~0.1%. We propose that this strain accounts for the differences between the thin film and bulk behavior.

ACKNOWLEDGMENTS

Support of this research has been provided by the Office of Naval Research and by Superconducting Core Technologies NCRADA-NRL-94-027.

REFERENCES

1. V.K.Varadan, D.K. Gohdgaonkar, V.V. Varadan, J.F. Kelly and P. Glikerdas, Microwave Journal, Jan **116** (1992).
2. J.S. Horwitz, D.B. Chrisey, J.M. Pond, R.C.Y. Auyeung, C.M. Cotell, K.S. Grabowski, P.C. Dorsey and M.S. Kluskens, Integrated Ferroelectrics **8**, 53 (1995).
3. G.A. Smolenskii and K.I. Rozgachev, Zh. Tekh. Fiz. **24**, 1751 (1954).
4. K.S. Grabowski, J.S. Horwitz and D.B. Chrisey, Ferroelectrics **116**, 19 (1991).
5. J.S. Horwitz, D.B. Chrisey, K.S. Grabowski and R.E. Leuchtner, Surface and Coatings Technology **51**, 290 (1992).
6. W.H. Hall, Proc. Phys. Soc. **62**, 741 (1949).

EPITAXIAL MULTILAYERS OF (Sr,Ba)Nb$_2$O$_6$ AND CONDUCTING FILMS ON (001) MgO SUBSTRATES

T. F. HUANG, K. E. YOUDEN, S. SCHWYN THÖNY, L. HESSELINK, AND
J. S. HARRIS JR., Solid State Laboratory, Stanford University, Stanford, CA 94305.

ABSTRACT

(Sr,Ba)Nb$_2$O$_6$ (SBN) is a very promising material for nonlinear optical applications because of its high electro-optic and nonlinear optical coefficients. For these applications, SBN requires ferroelectric poling along the optical axis. Conducting layers such as Pt or YBCO must therefore be deposited to provide electrodes above and below the SBN film.

We have investigated the epitaxial growth of multilayers of Sr$_{0.61}$Ba$_{0.39}$Nb$_2$O$_6$/Pt and Sr$_{0.61}$Ba$_{0.39}$Nb$_2$O$_6$/YBCO thin films on (001) MgO substrates by pulsed laser deposition. X-ray diffraction 2θ scans indicate epitaxial growth of SBN/Pt/MgO and SBN/YBCO/MgO heterostructures with their c-axes perpendicular to the substrate plane. X-ray phi scans indicate single crystal Pt and YBCO growth with one in plane orientation. Atomic Force Microscopy shows surface roughness of 2.19 nm, and no evidence of particles is observed.

INTRODUCTION

Strontium barium niobate (SBN) possesses extremely high electro-optical coefficients (r_{33} = 1340 pm/V)[1] which makes it desirable for electro-optic modulation and beam steering applications in integrated optics. It also possesses large nonlinear optical coefficients (d_{33} = 12.8 pm/V)[2] making it an attractive material for applications in nonlinear optics such as laser frequency conversion.

SBN is a ferroelectric material at room temperature and in its as-grown state it is presumed to consist of randomly oriented domains. These random domains will not produce an average electro-optic or nonlinear optic effect and for the applications described above the film must therefore be poled. The ability to control the spontaneous polarization direction is also required in order to periodically pole the films for quasi phase matching in optical frequency conversion.

Ferroelectric poling requires the application of an electric field greater than the coercive field (9 KV/cm for SBN) along the optical axis (c axis for SBN). The desired SBN orientation for waveguide applications is with its optical axis orthogonal to the plane of the interface, so that the largest electro-optic and nonlinear optic coefficients can be accessed. This requires the electrodes to be deposited above and below the SBN film. Since the MgO substrate is much thicker and a better insulator than SBN, the bottom electrode must be deposited between the substrate and the SBN film. Otherwise, electrical breakdown will occur before the coercive field strength is reached for the SBN.

The SBN layer will be epitaxially grown on top of the electrode layer and we therefore require a good epitaxial match between the SBN film (tetragonal, a=b=3.935 Å, c= 12.46 Å), electrode material and MgO substrate (cubic, 4.2 Å). Two candidates have been investigated: platinum

79

which is cubic with a lattice constant of 3.92 Å, and yttrium barium copper oxide (YBCO), which is orthorhombic with lattice constants a=3.886 Å, b=3.819 Å and c=11.68 Å.

EXPERIMENTAL

SBN/Pt/MgO and SBN/YBCO/MgO multilayer structures have been deposited using the pulsed laser deposition (PLD) setup described previously[3]. A pure Pt foil and sintered pellets of SBN:61 and YBCO are used as the targets. A KrF excimer laser operating at a wavelength of 248 nm, a pulse duration of 20 ns, and an average pulse energy of 220 mJ was used. The laser repetition rate was 5 Hz and the beam area on the target was 1 x 6 mm^2, resulting in a laser power density of 3.7 J/cm^2. The Pt films were grown at a vacuum pressure of 5x10^{-8} mbar. YBCO and SBN were deposited in an oxygen partial pressure of 10^{-2} mbar, with a flow of molecular oxygen of 15 sccm. A range of substrate temperatures between 600 °C to 775 °C were investigated. After deposition, the substrates were cooled slowly to room temperature at a rate of 5 °C/min under 1 mbar oxygen partial pressure.

RESULTS

SBN / Pt / MgO

Detailed x-ray diffraction (XRD) studies were used to investigate the nature of epitaxial growth and crystalline quality of the multilayer structure. Figure 1 shows a θ-2θ XRD pattern for a SBN/Pt/MgO multilayer structure.

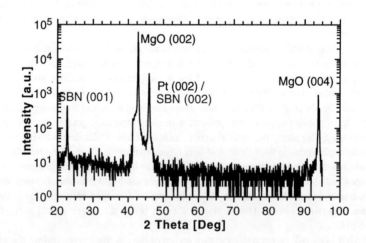

Figure 1: X-ray θ-2θ scan for SBN/Pt/MgO multilayer structure using Cu Kα radiation.

The optimum deposition temperature was determined to be 750°C for the Pt films and 775°C for SBN layers. Figure 1 clearly shows that the films contain only <00c> orientation perpendicular to the substrate. From rocking curve analyses, the full widths at half-maxima (FWHM) were found to be 0.32° and 0.43° for (002) Pt and (001) SBN respectively, indicating good out of plane alignment.

It seems that the crystallinity of Pt films depends greatly on the cleanliness of MgO, which may cause the Pt film to grow with (111) orientation as well as (001). Besides, MgO is slightly hygroscopic, so the substrate deteriorates during exposure to the air. Thus, we clean MgO substrates in acetone and methanol and then etch to create fresh surfaces.

The in-plane epitaxy of the SBN/Pt/MgO heterostructure was determined using x-ray phi scans. The Pt film exhibits four peaks that are characteristic of planar alignment with four-fold symmetry, as shown in Fig. 2. The SBN film appears to be oriented randomly in-plane with respect to Pt.

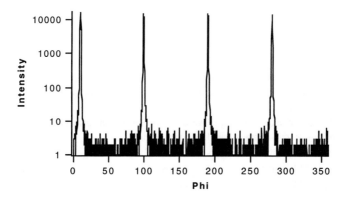

Figure 2 : X-ray phi scans for (111) reflections of Pt on MgO

The chemical composition of the films was analyzed using Rutherford backscattering (RBS) with a 2.2 MeV ^{4}He$^+$ ion beam. In Figure 3, the spectrum of the experimental data and a simulated spectrum are shown for a SBN:61 layer on Pt/MgO substrate. The agreement between the two spectra for the SBN region indicates that the composition of the thin film is SBN:61, the same as the target, and confirms that stoichiometric transfer does indeed occur for pulsed laser deposition. In addition, the simulated thickness of Pt and SBN films are 13 nm and 200 nm which correspond to growth rates of 0.027 Å/pulse and 0.16 Å/pulse, respectively. The flat top of the SBN peaks of the experimental data indicate that the composition of the films is constant with depth. However, diffusion took place between Pt/SBN and Pt/MgO interfaces, which may explain why the experimental data for Pt peak is wider than the simulation.

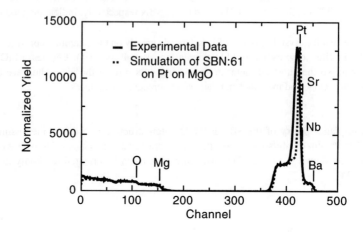

Figure 3: RBS spectrum of a SBN/Pt/ MgO multilayer

For the waveguide applications, the surface morphology of the film is of great importance. Atomic Force Microscope (AFM) images of Pt and SBN films grown sequentially showed smooth surfaces with r.m.s. roughness of 2.44 nm and 2.19 nm for Pt and SBN films respectively. Furthermore, no evidence of particulates were observed for both surfaces, contrary to many other reports on thin film growth using pulsed laser deposition.

The average lateral width of the surface features of the Pt films grown at 750°C for 10 minutes was approximately 140 nm and the height was around 5 nm. The size of features was found to increase with growth time, and therefore thickness, indicating that Pt tends to have a three dimensional island growth mechanism.

SBN / YBCO / MgO

The crystalline structure of the YBCO films deposited at substrate temperatures from 600 to 775°C were analyzed by XRD. At growth temperatures above 675°C the YBCO films are c-axis oriented. At temperatures of 650°C or below, the YBCO films start to develop a 103 peak. Fig. 4 shows the 2 θ scan of the SBN/YBCO/MgO multilayer structure which grew at 775 °C. It indicates crystalline growth of the films with the c axis perpendicular to the substrate.

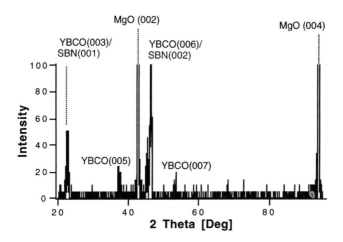

Figure 4: X-ray θ-2θ scan for SBN/YBCO/MgO multilayer structure using Cu Kα radiation.

Phi scans of the (221) SBN, (113) YBCO and (111) MgO peaks are shown in Fig. 5. The study revealed that the YBCO film has excellent in-plane epitaxy with <100>YBCO parallel to <100>MgO. The SBN film has two in-plane orientations, aligned at ±18.2 with respect to the reflections from (113)YBCO. This is the same angle as was observed previously for the growth of SBN on MgO[3].

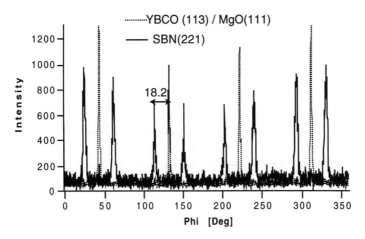

Figure 5: Phi scan of (221) plane of SBN, (113) plane of YBCO and (111) MgO

The orientation of these anti-phase domains of SBN on YBCO layer is similar to that of SBN on MgO substrate which have been discussed in the previous work[3]. It can be explained by considering the lattice mismatch and the electrostatic energy at the interface. The lattice misfit is -7% when the SBN and YBCO films are aligned along x and y axes. It is less than +1% when SBN is rotated by 18.2° with respect to YBCO. In addition to good lattice match, the match of atom positions and charges at the interface minimize the electrostatic energy. Both arguments explain why the SBN films prefer to grow with two in-plane orientations rotated by ±18.2° with respect to the [100] direction of the YBCO films.

CONCLUSION AND FUTURE WORK

We have successfully deposited single crystalline Pt and YBCO on MgO as the bottom electrode for SBN films. Both SBN/Pt/MgO and SBN/YBCO/MgO heterostructures grow epitaxially with c-axis perpendicular to the substrate. The lattice mismatch and the relationship between the interatomic structure of the film and the substrate have been found to greatly influence the orientation of the deposited films. It is interesting that similar in-plane orientations (±18.2) are observed in SBN films grown on YBCO layers as were observed for SBN on MgO substrates. The lattice mismatch and the relationship between the interatomic structure of the film and the substrate have been found to greatly influence the orientation of the deposited films. In addition, the surface is quite smooth with the r.m.s. roughness of 2.19 nm and no particulates have been found, which is crucial for waveguide applications.

We are currently investigating the ferroelectric characteristics of the SBN films. Pt and YBCO electrodes possess quite different electric and optic properties which make it interesting to compare their effect on poling and waveguide propagation. Furthermore, we are simulating the characteristics of optical guided modes in both waveguide structures and optimizing the thickness of electrodes and SBN films.

This work was supported in part by the Center for Materials Research (NSF DMR-9022248-04) and ARPA grant N00014-92-J-1903-P01.

REFERENCES

1. R. A. Vasquez, M. D. Ewbank an R. R. Neurgaokar, Opt. Comm., **80**, 235 (1991).
2. C. R. Jeggo and G. D. Boyd, J. Appl. Phys. **41**, 2741 (1971)
3. S. Schwyn Thöny, K. E. Youden, J. S. Harris Jr. and L. Hesselink, Appl. Phys. Lett., **65**, 2018 (1994).

GROWTH OF HIGHLY DOPED p-TYPE ZnTe FILMS BY PULSED LASER ABLATION IN MOLECULAR NITROGEN

DOUGLAS H. LOWNDES,* C. M. ROULEAU,*† J. W. McCAMY,** J. D. BUDAI,* D. B. POKER,* D. B. GEOHEGAN,* A. PURETZKY,*** and SHEN ZHU*#
* Solid State Division, Oak Ridge National Laboratory,
P. O. Box 2008, Oak Ridge, Tennessee 37831-6056
** Division of Applied Science, Harvard University, Cambridge, MA 02138
*** Institute of Spectroscopy, Troitsk, Russia
  † ORISE postdoctoral researcher
  # Now at Dept. of Physics, U. of Missouri, Columbia, MO 65211

ABSTRACT

Highly p-doped ZnTe films have been grown on semi-insulating GaAs (001) substrates by pulsed-laser ablation (PLA) of a stoichiometric ZnTe target in a high-purity N_2 ambient *without the use of any assisting (DC or AC) plasma source.* Free hole concentrations in the mid-10^{19} cm^{-3} to >10^{20} cm^{-3} range were obtained for a range of nitrogen pressures The maximum hole concentration equals the highest hole doping reported to date for any wide band gap II-VI compound. The highest hole mobilities were attained for nitrogen pressures of 50–100 mTorr (~6.5–13 Pa). Unlike recent experiments in which atomic nitrogen beams, extracted from RF and DC plasma sources, were used to produce p-type doping during molecular beam epitaxy deposition, spectroscopic measurements carried out during PLA of ZnTe in N_2 *do not reveal the presence of atomic nitrogen.* This suggests that the high hole concentrations in laser ablated ZnTe are produced by a new and different mechanism, possibly energetic beam-induced reactions with excited molecular nitrogen adsorbed on the growing film surface, or transient formation of Zn-N complexes in the energetic ablation plume. This appears to be the *first time that any wide band gap ($E_g > 2$ eV) II-VI compound (or other) semiconductor has been impurity-doped from the gas phase by laser ablation.* In combination with the recent discovery that epitaxial $ZnSe_{1-x}S_x$ films and heterostructures with continuously variable composition can be grown by ablation from a single target of fixed composition, these results appear to open the way to explore PLA growth and doping of compound semiconductors as a possible alternative to molecular beam epitaxy.

INTRODUCTION

Wide band gap II-VI semiconductors are recognized as prime candidate materials for blue light-emitting diodes (LEDs) and diode lasers. However, all wide band gap II-VI semiconductors exhibit a pronounced preference for only one type of doping (e.g., ZnSe is normally *n*-type and ZnTe *p*-type) due to self-compensation. Consequently, heteroepitaxial growth seemed to be required for *p-n* junction formation, making necessary careful lattice matching of dissimilar ternary or quaternary materials to avoid strain and rapid degradation of diode performance at high current levels. Recently however, *p*-type doping of ZnSe was achieved by a nonequilibrium method, using MBE growth and a nitrogen RF plasma source,[1,2] and resulted in fabrication of the first blue-green LEDs and diode lasers.[3,4] Formation of ohmic contacts to p-ZnSe remained a problem because the hole concentration was limited to ~10^{18} cm^{-3}. A partial solution was found when a similar nitrogen RF plasma source was used to grow p-ZnTe layers[5] with hole concentrations ~10^{19} cm^{-3} and a graded p-$ZnTe_{1-x}Se_x$ layer was used to contact ZnSe.[6] Infrared spectroscopic measurements were used to verify that the species responsible for efficient p-doping using RF plasma sources is almost certainly atomic nitrogen.[7] Most recently, two groups have shown that still higher hole concentrations, approaching 10^{20} cm^{-3}, can be obtained for ZnTe by using a DC nitrogen plasma and MBE growth.[8,9] The DC plasma source is said to be cheaper and simpler to operate than the RF source.[8]

In this paper we report the growth of highly doped p-type ZnTe films on semi-insulating GaAs (001) substrates by PLA of a stoichiometric ZnTe target through a high purity N_2 ambient *without the use of any assisting (DC or AC) plasma source.* The maximum hole concentration equals the highest obtained for ZnTe by any method thus far.[8] PLA differs from many other film growth methods in that growth occurs from a highly nonequilibrium *pulsed* flux of energetic atoms, ions, and electrons. Spectroscopic measurements carried out during PLA of ZnTe in N_2 do not reveal the presence of atomic nitrogen. This suggests that the high hole concentrations in PLA ZnTe are produced by a new and different mechanism, possibly energetic beam-induced reactions with excited molecular nitrogen adsorbed on the growing film surface, or transient formation of Zn-N complexes in the energetic ablation plume.

All but one of the few previous attempts to deliberately dope laser ablated II–VI films involved substitutions on the column-II site, using ablation targets that were pressed-powder mixtures containing the dopant atom. Shen and Kwok grew (001)-oriented zinc blend CdS films ($E_g = 2.50$ eV) on (001) GaAs and InP substrates by adding Li_3N or In powders, respectively, to CdS targets, and obtained electron and hole concentrations in the 10^{19}-10^{20} cm^{-3} and $\sim 10^{17}$ cm^{-3} ranges, respectively.[10] Similarly, Compaan and co-workers[11,12] grew polycrystalline p-ZnTe films on glass substrates by mixing metallic Cu and ZnTe powders in a pressed target. A hole concentration in the 10^{18}–10^{19} cm^{-3} range was estimated. However, both of these efforts required dopant-atom concentrations in the targets that far exceeded the resulting carrier concentrations, indicating that most dopant atoms either were not incorporated or were electrically inactive. Dubowski and co-workers grew epitaxial CdTe ($E_g = 1.49$ eV) and $Cd_{1-x}Mn_xTe$ films by PLA, as well as dilute magnetic semiconductor superlattice and quantum well structures.[13–16] However, no doping of either material was achieved.[17] Nevertheless, their results are encouraging in demonstrating PLA-grown heterostructures comparable in structural quality to those grown by MBE.[13,14,18]

EXPERIMENTAL

A conventional PLA system was used together with a pulsed KrF (248 nm) excimer laser.[19] Cylindrical lenses brought the laser beam to a horizontal line focus ($\sim$9mm $\times$ $\sim$0.75 mm) at an energy density $E_d \sim 0.66$ J/cm^2 along the radius of a 25.4 mm-diam ZnTe target.[20] The target was rotated at $\sim$8 rpm and the laser beam was switched alternately from one side of the target to the other to prevent cones from forming, thereby maintaining a smooth target surface and a uniform deposition rate.[19] This resulted in ZnTe films that were free of particulates.[21] The ultra-high purity (six nines) N_2 dopant gas was introduced through a mass-flow controller (MFC). In some experiments Ar gas, controlled by a second MFC, was introduced to permit variation of the ablation beam's kinetic energy (via collisions with ambient gas molecules) independent of the N_2 dopant-gas partial pressure. The total chamber pressure was controlled by a capacitance manometer and a throttle valve. Single-crystal semi-insulating (001) GaAs substrates were solvent-cleaned and sulfur-passivated as described elsewhere.[22] The substrate heater face was located 10 cm from the ZnTe target because in situ ion probe measurements had indicated that at this separation the ablation beam's kinetic energy could be reduced by collisions into the range expected to assist epitaxial film growth, while still using N_2 pressures ($\sim$50–100 mTorr) that seemed likely to produce doping.[22,23] No external power supplies or biasing arrangements were used, only the pulsed laser-generated plasma inherent to the ablation process.

The substrate heater was rotated at $\sim$9 rpm which, combined with the ablation beam-switching, produced films with reasonably uniform thickness. The film-growth rate was monitored in situ using a reflectance interferometer consisting of a low-power HeNe laser and large-area Si photodiode. The ZnTe films were grown to a nominal thickness of 676 nm (1 μm in initial experiments) in order to minimize effects of the dense dislocation network propagating from the ZnTe-GaAs interface[24] and make possible meaningful intercomparisons of Hall effect measurements of hole concentration and mobility.[8] All of the results reported here were obtained at a growth temperature of 320°C and an average ZnTe growth rate of 1 Å/sec, corresponding to laser pulse repetition rates of $\sim$2.5 to $\sim$4.5 Hz, depending on the ambient gas pressure.

FILM STOICHIOMETRY, ELECTRICAL, AND STRUCTURAL PROPERTIES

Achievement of the ideal 1:1 stoichiometry ratio is crucial to control the electrical properties of II–VI (and other) compound semiconductors. Relatively small concentrations of vacancies on either sublattice can act as electrically active defects or can form complexes with deliberately introduced dopant atoms, producing compensation. Figure 1 shows results of Rutherford Backscattering Spectrometry (RBS) measurements of the Zn/Te ratio in ~104-nm thick ZnTe films that were deposited on Si substrates in vacuum and in 100 mTorr N_2 at E_d ~ 0.66 J/cm^2. (A Si substrate was used for RBS measurements rather than GaAs to avoid interferences from the Ga or As RBS peaks.) The horizontal lines at 1.00 ± 0.015 represent the bounds of expected results for a series of RBS measurements on a ZnTe film with the ideal 1:1 stoichiometry. The ZnTe film deposited in 100 mT N_2 is stoichiometric over much of its surface, while the film deposited in vacuum is Zn-deficient by a statistically significant amount near the center of the (rotating) substrate heater face. The result that the PLA ZnTe film deposited in 100 mT N_2 is stoichiometric, at least to within the accuracy of RBS, is remarkable because recent MBE growth of ZnTe required Te/Zn flux ratios of 4 to 8 to achieve the highest hole concentration (~7×10^{19} cm^{-3}) and high hole mobility (~25–30 cm^2/V-s) in films ~1 µm thick.[8] The combination of RBS and Hall measurements presented here strongly suggests that ZnTe films with nearly the ideal 1:1 stoichiometry can be grown by PLA of a stoichiometric ZnTe target.

Hall effect measurements were used to determine the free hole concentration, p, and mobility, μ_p, at room temperature and in liquid nitrogen (T = 77 K). As shown in Table I and Fig. 2, a maximum hole concentration >1.1×10^{20} cm^{-3} has been obtained to date, but high hole concentrations, in the mid-10^{19} cm^{-3} range, were obtained for a range of nitrogen pressures, p[N_2], and growth conditions. Figure 2 shows measurements of p and μ_p for a series of ZnTe films grown at different p[N_2] values. All other growth conditions and the final film thickness were held constant. The hole concentration is nearly constant at $2–3 \times 10^{19}$ cm^{-3} for 50 mT < p[N_2] < 200 mT, but falls to ~ 2×10^{18} cm^{-3} at p[N_2] = 25 mT. The hole concentrations are essentially the same at room temperature and 77 K, indicating heavy doping and the formation of an impurity conduction band. ZnTe films grown in pure argon were electrically insulating, while ZnTe films grown in a 25:75 N_2:Ar mixture at 100 mTorr total pressure (using E_d = 2.7 J/cm^2) had hole concentrations of $0.7–1.5 \times 10^{19}$ cm^{-3}, consistent with doping that is controlled by the presence of nitrogen.

Figure 2 shows that the hole mobility peaks sharply at p[N_2] = 50 mT. We attribute this peak, and the less pronounced maximum in the hole concentration, primarily to effects of the ablation beam's kinetic energy, at low p[N_2], and possibly to excessive N incorporation and accompanying lattice strain, at high p[N_2] (see below). At low p[N_2], including vacuum, there is evidence that the kinetic energy of ablated atoms and ions may be sufficient to damage the growing film's crystalline structure, for example by displacing host and nitrogen atoms from substitutional sites and thereby introducing point defects (vacancies and interstitials) that scatter mobile carriers, reducing both the hole mobility and the hole concentration (via compensation). Geohegan et al. have used ion probe measurements to show that for ablation into an ambient gas, the ablation plume quite generally splits into two or more distinct components traveling at different average velocities.[25] They interpret the fastest pulse as consisting of ions that have undergone no collisions with gas molecules before arriving at the substrate. Although this "fast pulse" is exponentially attenuated by increasing the ambient gas pressure or the target-substrate separation, it remains significant for ZnTe ablation into N_2 until the N_2 pressure is increased to ~50 mTorr, consistent with the occurrence of the hole mobility peak in Fig. 2. A second factor that may affect the hole mobility at high p[N_2] values is that the kinetic energy of incident species may actually assist processes of surface diffusion, substitutional incorporation, and/or activation of N atoms on Te sites at moderate pressures, but this energy is reduced by numerous gas-phase collisions at high p[N_2]. These ideas regarding the effects of ablation-beam kinetic energy on the electrical properties of ZnTe films are being evaluated with the help of in situ time-resolved ICCD-camera and ion probe measurements.[23,25]

Table I summarizes the results of high resolution x-ray diffraction (HRXRD) measurements that were performed on ZnTe films grown under various conditions. The film with the highest hole concentration and mobility was fully epitaxial in three dimensions in the "cube on cube" (001) orientation, with no other ZnTe orientations present and with the film and substrate <00ℓ> axes precisely aligned. This orientation was also dominant in films grown in Ar or N_2 ambients at much higher E_d, but very small amounts (< 1%) of the ZnTe (311) and (111) orientations were present under these conditions. In-plane epitaxy was studied via ∅-scans with the other diffractometer angles set for the ZnTe(404) reflection; these revealed sharp peaks every 90° in all

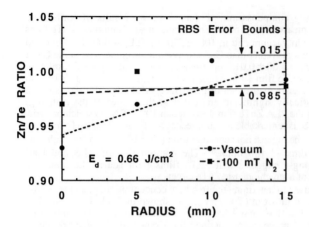

Figure 1. RBS measurements of the Zn/Te ratio in ZnTe films deposited in vacuum and in 100 mT N_2 on a Si (001) substrate at $E_d \sim 0.66$ J/cm^2 and T = 320°C (see text). R = 0 is the common rotation axis passing through the center of the rotating heater face and the center of the rotating ablation target. The dashed straight lines are least squares fits to the data.

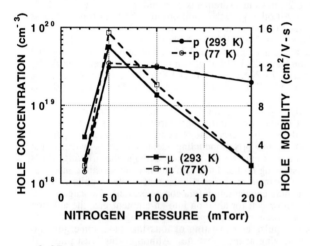

Figure 2. Hole concentration and mobility vs. nitrogen pressure for p-ZnTe films grown at T = 320°C using E_d = 0.66 J/cm^2.

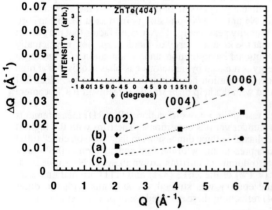

Figure 3. Full width at half maximum (FWHM) peak width ΔQ vs. Q for the (00ℓ) reflections of the ZnTe films whose properties and growth conditions are listed in Table I: (a) E_d = 0.66 J/cm^2, 100 mT N_2; (b) E_d = 2.7 J/cm^2, 100 mT N_2; (c) E_d = 2.7 J/cm^2, 50 mT Ar. The slopes of the plots reveal strain, their intercepts the particle size or coherence length. Inset: $\varnothing$-scan for the ZnTe(404) reflection for film (a), which has the highest hole concentration and mobility.

cases, as expected for a single dominant orientation (see inset, Fig. 3). Most interestingly, for the film with the highest hole concentration and mobility the surface-normal and in-plane lattice parameters were 6.0755 (±0.002) Å and 6.078 Å, respectively, nearly identical but both ~0.4% *smaller* than the bulk ZnTe value of 6.100 Å. A reduced in-plane lattice parameter was also found for another film grown in nitrogen, but not for a film grown in argon (see Table I). These results strongly suggest that substantial nitrogen incorporation on substitutional sites is responsible for the lattice contraction in PLA ZnTe:N films, consistent with their highly p-type conductivity. The decreased hole mobility at high p[N$_2$] (Fig. 2) may be due to the lattice distortion that accompanies very high N concentrations in the solid phase (Table I). Rocking curve (θ-scan) measurements through the ZnTe(004) reflection revealed a significant apparent mosaic spread in all the films (see Table I), but with the greatest line broadening for the N-doped samples, again consistent with heavy doping. (Possible additional effects of different laser E_d values on $\Delta\theta$(004) are also apparent in Table I.) Fig. 3 shows that most of the width of the ZnTe(00ℓ) diffraction peaks is due to strain. From the intercepts in the plot of ΔQ vs. Q, coherence lengths, $\Delta L \sim 5.6 / \Delta Q$, of ~110–190 nm are obtained.

Table I.   HRXRD results and electrical properties for ZnTe films grown in nitrogen and argon.

Ambient	E_d	Thickness	Hole conc.	Mobility	$a_\perp$	$a_\parallel$	$\Delta\theta$(004)
	(J/cm^2)	(nm)	(10^{20} cm^{-3})	(cm^2/V-s)	(Å)	(Å)	(deg)
100 mT N$_2$	0.66	676	1.18 (1.1)	15.8 (22.1)	6.075	6.078	1.06
100 mT N$_2$	2.7	~1000	0.3 (0.6)	5.5 (2.6)	6.102	6.094	1.1
50 mT Ar	2.7	~1000	– – –	– – –	6.100	6.100	0.47

Notes: $a_\perp$ and $a_\parallel$ are the lattice constants (± 0.002 Å) perpendicular and parallel to the film. $\Delta\theta$(004) is the rocking curve FWHM. Hole concentrations and mobilities are measured at room temperature (77 K). The gas flow conditions differ from the data shown in Fig. 1.[26]

IN SITU SPECTROSCOPY AND DOPING MECHANISM

The interaction of the energetic laser-generated plasma "plume" with molecular N$_2$ was studied in a chamber equipped with a gated ($\geq$ 5-ns), intensified-CCD, lens-coupled camera system (200–820 nm response) for time-resolved imaging, and an ion probe for ion/electron current measurements.[23] Optical emission was monitored using a 1.33-m monochromator equipped with an 1800 groove/mm holographic grating, a gated intensified diode array, and a photomultiplier tube.

The principal result of spectroscopic measurements at a distance of 7 cm from the ZnTe target during PLA in molecular N$_2$ was that *no emission was observed from either excited atomic nitrogen or excited molecular nitrogen.* On the other hand, when a nitrogen glow discharge was established between parallel Cu plates in the same chamber, strong near-infrared N$^+$ atomic emission lines and an N$_2^+$ emission band were seen.

The lack of atomic N emission from the PLA plasma in our experiments contrasts with the recent observation by Vaudo et al.[7] of strong near-IR atomic N lines emitted from the nitrogen plasma produced by an RF source that is used for N-doping in MBE growth. They concluded that N atoms, rather than N$_2$ molecules, are the species most likely responsible for p-type doping of ZnSe (and ZnTe) under MBE growth conditions.[7] Although our experiments do not directly rule out the presence of atomic N in the ground state during PLA, this seems very unlikely because the dissociation energy of N$_2$ is nearly 10 eV; if atomic N were produced by dissociation of N$_2$ during PLA then we would expect also to see emission from excited N$_2$, which we did not. Vaudo et al. observed also that the *relative* intensities of the near-IR atomic N and molecular N$_2$ emissions depend sensitively on MBE chamber pressure, with the atomic N emission *nearly disappearing* as the pressure was raised from 2×10^{-6} Torr to 5×10^{-5} Torr.[7] This suggests that even if atomic N were produced in the ablation plume at the higher pressures used here, it would be short-lived due to the increased probability of collisions and rapid recombination, so is unlikely to be involved in doping at the substrate.

However, recent theoretical calculations[27,28] suggest that N_2 molecules in the metastable $A^3\Sigma_u^+$ state may be able to adsorb on a growing film surface and subsequently dissociate. Vaudo et al. ruled out this pathway as being unlikely using a thermal MBE-growth atomic flux. However, the kinetic energies of ablated atoms and ions are more than sufficient, over a wide range of $p[N_2]$ values, to supply the 3.9 eV dissociation energy needed to break the molecular bond of the $A^3\Sigma_u^+$ state. A second possible N-doping mechanism in PLA growth is the direct kinetic energy-enhanced reaction of Zn atoms with N_2 to form (possibly transient) Zn-N compounds in the incoming ablation beam, and their subsequent incorporation in ZnTe:N films. Spectroscopic studies on and very near the growing film surface are needed to resolve these possibilities.

In summary, this appears to be the first time that any wide band gap ($E_g > 2$ eV) II-VI compound (or other) semiconductor has been impurity-doped from the gas phase by laser ablation. In combination with the recent discovery that epitaxial $ZnSe_{1-x}S_x$ films and heterostructures with continuously variable composition can be grown by ablation from a single target of fixed composition,[22] these results appear to open the way to explore PLA growth and doping of compound semiconductors as a possible alternative to molecular beam epitaxy.

This research was sponsored by the Division of Materials Science, U. S. Department of Energy, under contract no. DE-AC05-84OR21400 with Martin Marietta Energy Systems, Inc.

REFERENCES

1. R. M. Park et al., *Appl. Phys. Lett.* **57**, 2127 (1990).
2. K. Ohkawa, T. Karasawa, and T. Mitsuyu, *Jap. J. Appl. Phys.* **30**, L152 (1991).
3. M. Haase, J. Qiu, J. DePuydt, and H. Cheng, *Appl. Phys. Lett.* **59**, 1272 (1991).
4. H. Jeon et al., *Appl. Phys. Lett.* **60**, 2045 (1992).
5. J. Han et al., *Appl. Phys. Lett.* **62**, 840 (1993).
6. Y. Fan et al., *Appl. Phys. Lett.* **61**, 3160 (1992).
7. R. P. Vaudo, J. W. Cook, Jr., and J. F. Schetzina, *J. Cryst. Growth* **138**, 430 (1994).
8. S. O. Ferreira et al., *J. Cryst. Growth* **140**, 282 (1994).
9. T. Baron et al., *Appl. Phys. Lett.* **65**, 1284 (1994).
10. W. P. Shen and H. S. Kwok, *Fall 1993 Annual Mtg. of the Mater. Rres. Soc.*, Boston, MA (to be published, 1994).
11. A. Compaan et al., p. 957 in *Proc. 22nd IEEE Photovoltaic Specialists Conf.*, IEEE, 1991.
12. A. Compaan and A. Bhat, *Int. J. Solar Energy* **12**, 155 (1992).
13. J. J. Dubowski et al., *Superlattices and Microstructures* **9**, 327 (1991).
14. D. Labrie and J. J. Dubowski, *Mat. Res. Soc. Symp. Proc.* **285**, 465 (1993).
15. J. M. Wrobel and J. J. Dubowski, *Appl. Phys. Lett.* **55**, 469 (1989).
16. J. J. Dubowski, *J. Cryst. Growth* **101**, 105 (1990).
17. J. J. Dubowski, private communication.
18. R. L. Harper, Jr., et al., *J. Appl. Phys.* **65**, 624 (1989).
19. D. H. Lowndes, "Growth of Epitaxial Thin Films by Pulsed Laser Ablation", in <u>Modern Topics in Single Crystal Growth</u>, Amer. Inst. of Physics (Eighth Int. Summer School on Crystal Growth, Palm Springs, CA, Aug. 9-15, 1992), in press.
20. The 99.999% pure ZnTe target was prepared by isostatic hot pressing at Plasmetrics, Inc., San Ramon, CA.
21. We find that the areal density of particulates on PLA compound semiconductor films is much lower than for oxide ceramics. By using the techniques that we and others have developed to minimize particulate production due to target "coning" and damage, nearly particulate-free compound semiconductor films can be obtained.
22. J. W. McCamy, D. H. Lowndes, and J. D. Budai, *Appl. Phys. Lett.* **63**, 3008 (1993).
23. D. B. Geohegan, p. 115 in <u>Pulsed Laser Deposition of Thin Films</u>, ed. by D. B. Chrisey and G. K. Hubler, Wiley, New York, 1994.
24. E. Abramof et al., *Semicond. Sci. and Technol.* **6**, A80 (1991).
25. D. B. Geohegan et al., these symposium proceedings.
26. C. M. Rouleau and D. H. Lowndes, unpublished results.
27. T. Nakao and T. Uenoyama, *Jpn. J. Appl. Phys.* **32**, 660 (1993).
28. T. Uenoyama, T. Nakao, and M. Suzuki, *J. Cryst. Growth* **138**, 301 (1994).

P-TYPE II-VI COMPOUND SEMICONDUCTOR THIN FILMS GROWN BY PULSED LASER DEPOSITION

W. P. Shen* and H. S. Kwok**
Applied Laser Laboratory, State University of New York at Buffalo, Amherst, NY 14260
* Current address: R&D Department, Photronics, Inc., 1982 Tarob Ct., Milpitas, CA 95035.
** Current address: Department of Electrical and Electronics Engineering, Hong Kong University of Science and Technology, Clearwater Bay, Hong Kong.

ABSTRACT

In this paper the results on p-type ZnS, ZnSe, CdS and CdSe thin films grown by pulsed laser deposition will be discussed. These films were deposited on GaAs substrates. Li-doping has been shown to be effective in producing p-type II-VI thin films, while In-doping is excellent for n-type CdS and CdSe thin films. No post-annealing process was used. These preliminary results suggest a possible new approach through pulsed laser deposition to solve the doping problem of II-VI compound semiconductors.

INTRODUCTION

II - VI compound semiconductors are well-known for their applications in a wide range of important optoelectronic devices, including detectors, light emitting diodes and solid state lasers operating in the visible spectral range. The major difficulty in fabricating optical devices from II-VI compound semiconductors is that they can only be doped to either p-type or n-type, but not both. Extensive efforts to produce p-type ZnS, ZnSe, CdS and CdSe or n-type ZnTe and CdTe have been ongoing for more than thirty years[1-6]. Several explanations for these difficulties have been advanced: (1) the natural point defects (normally group VI vacancies), which act as donors (S and Se vacancies) or acceptors (Te vacancies) due to their ionic bonding (auto-compensation); (2) the low solid solubility of dopants; (3) dopants at interstitial sites act as compensating dopants (for example, Li and Na act as acceptors at Zn or Cd sites but donors at interstitial positions). Based on thermodynamics, it is believed that the maximum equilibrium carrier concentrations must be very low. A new solution to this problem is obtained by nonequilibrium and low temperature crystal growth techniques - such as molecular beam epitaxy (MBE) and metal-organic chemical vapor deposition (MOCVD) - where the growth is controlled by surface kinetics at a lower temperature. Recently, ZnSe was p-type doped with carrier concentrations in the range of 10^{17} - 10^{18} cm^{-3}, and efficient blue light lasers were produced[7,8]. High quality p-type ZnS, CdS and CdSe thin films have yet to be reported.

Pulsed laser deposition (PLD) of II-VI compounds has been discussed previously[9-13]. The focused pulsed laser beam produces such a rapid temperature rise ($>10^{11}$ Ks^{-1}) on the target that the stoichiometry of the target can be maintained in the growing film[9]. It provides means to reduce the group VI vacancies which generate the native compensation in II-VI compound thin films. High energy atom and ion species in the laser-induced plasma plume along with its pulsed behavior create high surface mobility and supersaturation enables the growth of high quality films at a relatively low substrate temperature[10]. This highly nonequilibrium pulsed process suggests that PLD may be an excellent approach to solve the long-standing II-VI compound doping problem. In this letter our results on p-type ZnS, ZnSe, CdS and CdSe thin films will be discussed. Li-doping has been shown to be effective in producing p-type II-VI thin films, while In-

Mat. Res. Soc. Symp. Proc. Vol. 388 © 1995 Materials Research Society

doping is excellent for n-type CdS and CdSe thin films. No post-annealing process was used. Our experiments were focused on less-known ZnS and CdS systems, although it was found that the doping behavior were similar between ZnS and ZnSe and for CdS and CdSe. These thin films can potentially be applied to optoelectronic devices operating in the visible spectral range.

EXPERIMENTAL

II-VI compound thin films were grown on GaAs and InP substrates by PLD[13]. The deposition conditions were 5 cm target-to-substrate distance, 100 mTorr Ar background gas pressure, and 1.5 J/cm² laser fluence. The target disk was pressed from a ground mixture of pure II-VI compound powder (99.999%) and desired impurity powder (e.g. Li or In powder for p-type or n-type CdS doping, respectively) under 8000 pound/cm². There was no annealing or sintering performed on the target. The same deposition conditions were used for both doped and non-doped samples. It has been observed that the color of the plasma plume was changed due to the luminescence of the dopant species. Plumes with In impurities have a blue-like color while Li impurities produce a pink color. However, the plasma plume dynamics were quite similar to the non-doped case.

The conduction type of doped films can be determined by the sign of the thermal emf generated by a temperature gradient (hotprobe measurement)[14]. The results of our hotprobe measurements are shown in Fig. 1. The samples were cut into the same geometry for comparison and grown on semi-insulating GaAs(100) substrates with a 10^8 cm^{-3} n-type Si-doping. The cold probe was kept at 20 °C. Our pure II-VI thin films are n-type with a very high resistivity (> 10^5 Ω -cm), the n-type behavior being due to native defect centers. However, it was found that Li concentrations as low as 0.5 % in targets converted ZnS, ZnSe, CdS and CdSe films into p-type.

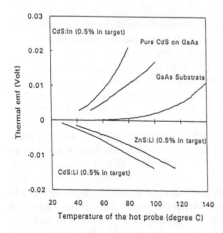

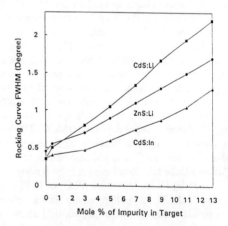

Figure 1 Hotprobe measurement results of ZnS and CdS thin films.

Figure 2 X-ray rocking curve FWHM vs. different impurity concentrations in CdS:Li, CdS:In and ZnS:Li thin films.

Film qualities degrade as doping concentrations were increased. Fig. 2 shows the rocking curve full-width-at-half-maximum (FWHM) around (100) peak vs. impurity concentrations (mole % in the target). The lattice structures apparently were disturbed by dopant atoms. However, x-ray studies showed that these films were polycrystalline films with large nearly in-plane aligned grains. SEM studies showed that surface morphologies were changed for heavily Li-doped films. For ZnS and ZnSe with Li doping concentrations lower than 6% (mole % in the target), the surface was unchanged from the non-doped film. For higher Li-doping concentrations some hillock features appeared, as shown in Fig. 3. For CdS and CdSe, the doping concentration at which the surface morphology degraded was lowered to 3% (mole % in the target), due to the larger lattice stress generated by the larger size difference between Li and Cd atoms.

Resistivities of these films were measured by four-point-probe. Fig. 4 shows the resistivities of p-type Li-doped ZnS and CdS thin films under different impurity concentrations. The lowest resistivities obtained were 102.5 Ω-cm for ZnS:Li (9 mole % in the target) and 5.85 Ω-cm for CdS:Li (11 mole % in the target).

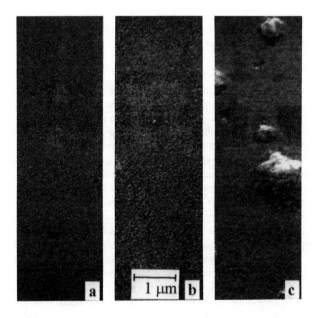

Figure 3   SEM pictures show surface morphology of (a) non-doped ZnS, (b) ZnS deposited by the target with 6 mole % Li, and (c) ZnS deposited by the target with 11 mole % Li.

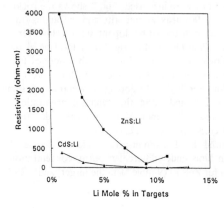

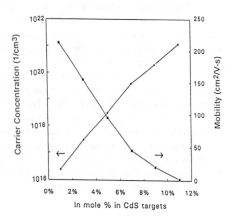

Figure 4 ZnS:Li and CdS:Li resistivities under different Li mole % in targets.

Figure 5 Mobility and carrier concentrations under different In mole % in CdS targets.

To determine carrier concentrations of these films, Hall effect measurements were performed. Due to the difficulties in performing Hall measurements on higher resistivity samples, only CdS:In and some of highly doped CdS:Li films were measured. Other systems, such as ZnS:Li and ZnSe:Li, suffered from difficulties in producing good ohmic contacts[15]. Samples were measured in a 1500 Gauss magnetic field at room temperature. CdS films were patterned into a cross geometry with a line width of 1 mm by Ar RF-plasma etching. Silver contacts were evaporated before the patterning process to ensure good ohmic contacts (confirmed by I-V). Fig. 5 shows the Hall measurement results of CdS:In films with different In concentrations. Carrier concentrations rise dramatically with increasing In content, while the mobility decreased due to defects introduced into the films. The maximum carrier concentration was found to be 1.1×10^{21} cm^{-3}, with 11 % In within the target. However the lowest resistivity, 1.7×10^{-3} Ω-cm, was found with 9% In within the target. This is explained by the high doping concentration reducing the mobility of the samples. CdS:Li films with high Li concentration (7%-11% in targets) were measured also, giving a p-type response. The sample deposited by using a 11% Li target has a mobility 9.79 cm^2/V-s with a carrier concentrations of 1.1×10^{17} cm^{-3}. The mobility of single crystal CdS is 340 cm^2/V-s for electrons and 50 cm^2/V-s for holes. The highest Hall mobility obtained for our PLD films was 230 cm^2/V-s for a 0.5 target mole % CdS:In sample and 22 cm^2/V-s for a 7 target mole % CdS:Li sample. These values are reasonable considering the granular structure of the films.

DISCUSSION

The concept of mole % of impurity "in the target" has to be clarified before further discussions. Since the growth temperatures (typically 300 °C) are much higher than the melting point of the applied dopant (e.g. 180.6 °C for Li and 156.6 °C for In), the sticking coefficients of the dopants must be much lower than ZnS, ZnSe, CdS and CdSe (which have melting points larger than 1200 °C). It can be expected that the mole % of impurity in the target is much higher than the mole % of impurities (or carriers) in the films. For example, a 1 mole % impurity "in the film" corresponds to an impurity level larger than 10^{21} cm^{-3}, which generally produces serious lattice defects, and is often larger than the solid solubility. However, our CdS films deposited with a 9 mole % In-contained target or ZnS films deposited with a 6 mole % Li-contained target showed relatively smooth surface morphologies and only mild degradation on lattice structures. These low-melting-point dopant contained in the plasma plume behave more like a reactive-species background, compared to the normal stoichiometric deposition of PLD. Other effects on this phenomena, such as compensation, are under investigation.

A large variety of potential dopants were tried in our experiments. Two major considerations for doping are: (1) The impurity atom size should be similar to the atom substituted, allowing higher solubility and less lattice strain. (2) The impurity must form a shallow level. C. H. Henry et. al.[4,5] reported that Li and Na, which have covalent radii similar to Cd and Zn, are the only soluble substitutional acceptors in ZnSe, CdS and CdSe with relatively small binding energies (114 meV for Li$^-$ and 90-100 meV for Na$^-$ in ZnSe; 165 meV for Li$^-$ and 169 meV for Na$^-$ in CdS). However, the Li and Na interstitials, which form compensating donors, were found to be very shallow as well (17.67 meV for Li$^+$ and 17.50 meV for Na$^+$ in CdS). The substitutional P, which has a radius similar to S, is a deep acceptor (0.78 eV for CdS). Nevertheless, a shallow acceptor (120 meV) from P-complex was observed for CdS. These observations are consistent with our results. By adding Li and Na impurities, p-type thin films were produced. The CdS:Li thin films had a rougher surface morphology which probably is due to the smaller size of Li. The serious problem of interstitial compensation was apparently improved by the PLD method since very high carrier concentrations have been achieved. Our P-doped thin films had a very smooth surface while the p-type resistivities were normally 1-2 order of magnitude higher than Li- or Na-doped films.

The highest n-type carrier concentration of our CdS:In films was 1.1×10^{21} cm^{-3}. However, an even higher CdS:In carrier concentration (10^{22} cm^{-3}) has been reported by T. Hayashi et. al.[16]. Apparently, the solid solubility of In in CdS is extremely high due to the similar size of Cd and In. These highly doped CdS:In films have a surprisingly smooth surface. The very low resistivity obtained suggests a possible application as a transparent electrodes for wavelength longer then green light.

CONCLUSION

We have demonstrated that p-type Li-doped ZnS, ZnSe, CdS and CdSe epitaxial thin films can be grown on GaAs or InP substrates by pulsed laser deposition. In light of these encouraging preliminary results, further studies to improve the thin film quality for optoelectronic applications are warranted.

ACKNOWLEDGEMENTS

We would like to thank Dr. Wayne A. Anderson, Dr. Ping C. Cheng (SUNY at Buffalo) and Dr. Jim P. Zheng (Army Research Lab.) for their helpful suggestions and discussions.

REFERENCES

1. D. B. Laks, C. G. Van de Walle, G. F. Neumark, and S. T. Pantelides, Appl. Phys. Lett. **63**, 1375 (1993).
2. G. F. Neumark, J. Appl. Phys. **51**, 3383 (1980).
3. R. N. Bhargava, R. J. Seymour, B. J. Fitzpatrick, and S. P. Herko, Phys. Rev. B, **20**, 2407 (1979).
4. C. H. Henry, K. Nassau, and J. W. Shiever, Phys. Rev. B, **4**, 2453 (1971).
5. J. L. Merz, K. Nassau, and J. W. Shiever, Phys. Rev. B, **8**, 1444 (1973).
6. G. Mandel, F. F. Morehead, and P. R. Wagner, Phys. Rev., **136**, A827 (1964).
7. M. A. Haase, H. Cheng, J. M. DePuydt, and J. E. Potts, J. Appl. Phys., **67**, 448 (1990).
8. M. A. Haase, J. Qiu, J. M. DePuydt, and H. Cheng, Appl. Phys. Lett. **59**, 1272 (1991).
9. J. T. Cheung and H. Sankur, CRC Crit. Rev. Solid State Mater. Sci., **15**, 63 (1988).
10. D. B. Chrisey and G. K. Hubler, Pulsed Laser Deposition of Thin Films, John Wiley & Sons, Inc., New York (1994).
11. J. W. McCamy and D. H. Lowndes, Appl. Phys. Lett., **63**, 3008 (1993).
12. H. S. Kwok, J. P. Zheng, S. Witanachchi, L. Shi, and D. T. Shaw, Appl. Phys. Lett., **52**, 1815 (1988).
13. W. P. Shen and H. S. Kwok, Appl. Phys. Lett., **65**, 2162 (1994).
14. D. K. Schroder, Semiconductor Material and Device Characterization, John Wiley & Sons, Inc., New York (1990).
15. Y. Lansari, J. Ren, B. Sneed, K. A. Bowers, J. W. Cook, Jr, and J. F. Schetzina, Appl. Phys. Lett., **61**, 2554 (1992).
16. T. Hayashi, T. Nishikura, T. Suzuki, and Y. Ema, J. Appl. Phys., **64**, 3542 (1988).

INHIBITED SN SURFACE SEGREGATION IN EPITAXIAL SN_xGE_{1-x} ALLOY FILMS GROWN BY PULSED LASER DEPOSITION

M.E. Taylor, G. He, C. Saipetch, and H.A. Atwater
Thomas J. Watson Laboratories of Applied Physics
California Institute of Technology, Pasadena, CA 91125

ABSTRACT

Epitaxial and compositionally homogeneous Sn_xGe_{1-x} alloy films have been grown on Si (001) by pulsed laser deposition using elemental Sn and Ge targets. These results demonstrate that pulsed laser deposition can be used to grow alloys by overcoming the strong tendency for Sn surface segregation seen in growth by other methods such as molecular beam epitaxy.

INTRODUCTION

Electronic structure calculations predict that the binary alloy Sn_xGe_{1-x} is a direct gap semiconductor with a band gap that ranges between 0.55 eV and 0 eV as the composition is varied between x=0.2 and x=0.8, as shown in Figure 1 [1]. Significantly, Sn_xGe_{1-x} also has potential for monolithic integration on Si substrates. As a result, Sn_xGe_{1-x} is of interest for future Si-based infrared optoelectronics applications.

The synthesis of Sn_xGe_{1-x} alloy films has been the subject of many previous investigations, most of which relied on molecular beam epitaxy [2, 3, 4, 5, 6], sputtering [7], or solid phase recrystallization [8, 9, 10] as the growth method. These methods have not as of yet produced crystalline films in the desired composition range. This lack of success is due primarily to the segregation of Sn to the surface of the growing film.

The tendency for Sn surface segregation results from two competing kinetic processes [11, 12]. During steady-state growth, the surface layer is transformed into a bulk layer. Since the bulk equilibrium Sn composition is smaller than the Sn composition of the flux, Sn segregates to the surface. The time available for segregation from the bulk to the surface is limited by the film growth rate. The Sn incorporation coefficient is a strong function of a Peclet number that expresses the ratio of the growth rate to the rate of segregation by diffusion between bulk and surface layers. Deposition methods with high Peclet numbers are predicted to produce films with high incorporation coefficients.

Pulsed laser deposition (PLD) has been used previously to deposit crystalline semiconductor films [13, 14]. PLD differs from more conventional growth methods in that it is characterized by high growth rates and high mean particle energies. In light of the assumptions made concerning the kinetics of Sn surface segregation, PLD is a potentially useful method for synthesizing Sn_xGe_{1-x} films.

SN_xGE_{1-x} FILM GROWTH BY PULSED LASER DEPOSITION

Films were grown in a loadlocked PLD chamber equipped with reflection high energy electron diffraction (RHEED). The target consisted of a polycrystalline Ge (99.999%) wafer overlaid with pie-shaped wedges of Sn (99.9999%) that covered approximately 55% of the surface. The target was rotated at 10 rpm to ensure an average flux composition value of 0.55 and to avoid severe erosion of the target in any particular area. Test-grade Si (001) substrates were cleaned by

Mat. Res. Soc. Symp. Proc. Vol. 388 © 1995 Materials Research Society

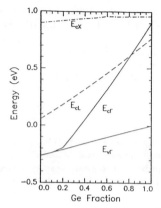

Figure 1: Band structure of Sn_xGe_{1-x} as a function of composition, as predicted by tight-binding calculations.

immersion in 8 H_2O : H_2O_2 : NH_4OH and hydrogen-terminated by immersion in 8 H_2O : HF. The substrate temperature was held at 200 °C for several hours prior to growth to desorb hydrocarbons, ramped to 400 °C shortly before growth to desorb surface hydrogen, and then lowered to the desired growth temperature. At temperatures below 200 °C, deposition led to kinetic roughening to such an extent that the breakdown of epitaxy ensued, placing a lower limit on the growth temperatures used. The pulsed laser is an excimer operating at a wavelength of 248 nm. The pulse repetition rate was varied between 0 and 100 Hz. The beam was attenuated to achieve energy densities on the target of between 1 and 4 Joules/cm^2, as higher energy densities were found to produce amorphous films at a growth temperature of 200 °C. At this level of attenuation, typical deposition rates were on the order of 0.0005 nm/pulse. During growth, a pressure of between 10^{-8} and 10^{-9} Torr was maintained in the PLD chamber, except for some experiments during which the growth chamber was backfilled with Ar to pressures in the 10^{-3} Torr range.

Prior to growth, the RHEED pattern consisted of Bragg rods. After the temperature ramp, additional rods appeared between the original rods, indicating surface reconstruction. During deposition, the Bragg rods were gradually replaced with a pattern of diffuse spots, characteristic of crystalline films with large surface roughnesses. Figure 2 shows a typical post-deposition pattern.

SN_xGE_{1-x} FILM ANALYSIS

After growth, x-ray rocking curve analysis was used to confirm the epitaxial film orientation. Figure 2 shows a x-ray rocking curve spectrum of a PLD film. To the left of the Si substrate peak are two smaller peaks corresponding to the deposited film. The appearance of these peaks indicates that the film is crystalline. The film peaks are broad because the film is thin and is strain-relieved by misfit dislocation formation, resulting in an inhomogeneous strain distribution. The angular offsets between the film peaks and the substrate peak were used to estimate lattice parameters. The lattice parameter estimated from the peak on the left agrees with the lattice parameter expected for a diamond cubic Sn_xGe_{1-x} film if the virtual crystal approximation is assumed. The lattice parameter estimated from the peak in the center matches the lattice

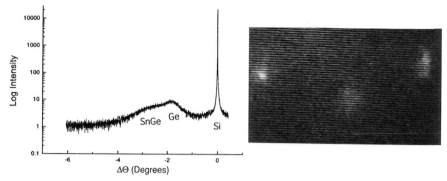

Figure 2: X-ray rocking curve spectrum and RHEED pattern of a 71 nm $Sn_{0.19}Ge_{0.81}$ film grown on Si (001) by pulsed laser deposition at 200 °C with an energy density of 3 J/cm^2.

Magnification 1500

◄─► 10 μm

Figure 3: Scanning electron microscopy photograph of the surface of an 18 nm $Sn_{0.42}Ge_{0.58}$ film grown on Si (001) by pulsed laser deposition at 200 °C with an energy density of 3 J/cm^2, showing the distribution of droplets.

parameter for Ge, indicating that Sn segregation was suppressed rather than eliminated.

Scanning electron microscopy (SEM) and electron microprobe analysis (EMA) were used to examine the film surfaces. Figure 3 is an SEM photograph illustrating the distribution of droplets on the surface. EMA measurements indicated that each droplet was composed either entirely of Sn or entirely of Ge. This is consistent with the droplets being formed by target exfoliation. A majority of the pulses struck either a Sn area or a Ge area on the target, meaning that the plasma corresponding to that pulse consisted entirely of one element. Therefore, the EMA measurements are also consistent with the droplets being formed by condensation in the plume.

Rutherford backscattering spectroscopy (RBS) was used to determine the composition and thickness of the films. The RBS spectrum shown in Figure 4 is representative of the complex spectra generated by the PLD films. The Si edge appears at an energy lower than the energy at which it would appear if it were at the surface, indicating the presence of a continuous film. This energy shift was used to estimate the film thickness. The Ge peak appears at the energy at which Ge films at the surface are expected to appear, indicating that the narrow Sn and Ge peaks represent, for the most part, a homogeneous alloy film. The Sn and Ge droplets appear

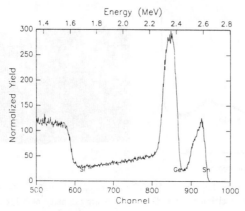

Figure 4: Rutherford backscattering spectrum of a 71 nm $Sn_{0.19}Ge_{0.81}$ film grown on Si (001) by pulsed laser deposition at 200 °C with an energy density of 3 J/cm^2.

as a broad background that increases with increasing channel number. The droplet background was extrapolated across the Sn and Ge peaks and subtracted. The ratio of integrated counts in the Ge peak to integrated counts in the Sn peak was then normalized by the ratio of the Sn and Ge scattering cross-sections to obtain an estimate of the film composition.

CONCLUSIONS

At this point, characterization of the Sn_xGe_{1-x} films grown by PLD leads to four conclusions. Of primary importance, the data confirms that Sn_xGe_{1-x} films with high Sn incorporation coefficients and high Sn compositions can be grown by PLD. For example, the film associated with Figure 4 appears to have a composition value of 0.19 and an incorporation coefficient of 0.35, assuming a flux composition value of 0.55.

Second, a comparison of the Sn incorporation coefficients and the energy densities at the target indicates that higher energy densities lead to larger incorporation coefficients. This is apparent in the films associated with Figure 5, which were grown at different energy densities. The first was grown at an energy density of approximately 2 J/cm^2, the second at approximately 3 J/cm^2, and the third at approximately 4 J/cm^2. The corresponding composition values are 0.25, 0.38, and 0.56. Assuming a flux composition value of 0.55 yields incorporation coefficients of 0.45, 0.69, and 1.0. Increasing the energy density also appears to increase the droplet background, as can be noted in Figure 5.

Third, a second set of growth experiments indicates that for temperatures below 355 °C with an energy density of 4 J/cm^2, the Sn incorporation coefficient is insensitive to temperature. Films were grown with an energy density of 4 J/cm^2 at 214 °C, 238 °C, and 355 °C. The corresponding composition values are 0.44, 0.34, and 0.56. Again assuming a flux composition value of 0.55, the incorporation coefficients are 0.80, 0.62, and 1.0.

Fourth, growth experiments in an Ar atmosphere indicate that an increase in the ambient pressure of inert gas may lead to higher Sn compositions. Monte Carlo simulations of plume expansion in ambient gases predict that the attenuation of the plume due to high gas pressure decreases with increasing mass number when the two elements are ablated sequentially [15]. Thus, the composition is increased due to the increased ratio of Sn atoms to Ge atoms in the

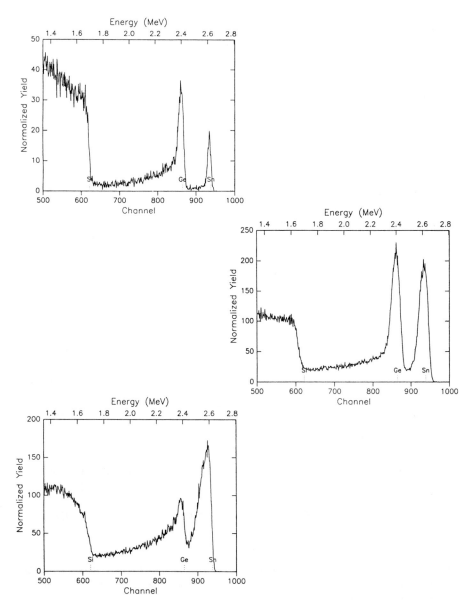

Figure 5: Rutherford backscattering spectra of (a) an 18 nm $Sn_{0.25}Ge_{0.75}$ film grown with an energy density of 2 J/cm^2, (b) a 40 nm $Sn_{0.38}Ge_{0.62}$ film grown with an energy density of 3 J/cm^2, and (c) a 33 nm $Sn_{0.56}Ge_{0.44}$ film grown with an energy density of 4 J/cm^2.

plume that reaches the substrate. In addition, there may be an increase in the incorporation coefficent due to a change in the ratio of Sn and Ge mean energies caused by energy losses related to the ambient gas. Further experiments in conjunction with expanded Monte Carlo simulations have the potential to systematically relate these phenomena.

ACKNOWLEDGEMENTS

This work was supported by the U.S. Department of Energy Contract W-7405-ENG-48, under a subcontract with Lawrence Livermore National Laboratory, and the National Science Foundation. We would like to thank M. Easterbrook for his assistance with the Rutherford backscattering spectroscopy.

REFERENCES

1. H.A. Atwater, G. He, and C. Saipetch, to be published in Mat. Res. Soc. Symp. Proc., **355**, (1995).

2. A. Harwit, P.R. Pukite, J. Angilello and S.S. Iyer, Thin Solid Films, **184**, 395 (1990).

3. J. Piao, R. Beresford, T. Licata, W.I. Wang and H. Homma, J. Vac. Sci. Technol. B, **8**, 221 (1990).

4. H.J. Gossman, J. Appl. Phys., **68**, 2791 (1990).

5. E.A. Fitzgerald, P.E. Freeland, M.T. Asom, W.P. Lowe, R.A. Macharrie, Jr., B.E. Weir, A.R. Kortan, F.A. Thiel, Y.H. Xie, A.M. Sergent, S.L. Cooper, G.A. Thomas, and L.C. Kimerling, J. Elec. Mat., **20**, 489 (1991).

6. W. Wegscheider, J. Olajos, U. Menczigar, W. Dondl and G. Abstreiter, J. Cryst. Growth, **123**, 75 (1992).

7. S.I. Shah, J.E. Greene, L.L. Abels, Q. Yao and P.M. Raccah, J. Cryst. Growth, **83**, 3 (1987).

8. S. Oguz, W. Paul, T.T. Deutsch, B.Y. Tsaur and D.V. Murphy, Appl. Phys. Lett., **43**, 848 (1983).

9. I.T.H. Chang, B. Cantor and A.G. Cullis, J. Non-Cryst. Solids, **117-118**, 263 (1990).

10. S.M. Lee, J. Appl. Phys., **75**, 1987 (1994).

11. J.Y. Tsao, *Materials Fundamentals of Molecular Beam Epitaxy*, Academic Press, Inc., (1993).

12. M.J. Aziz, J. Appl. Phys., **53**, 1158 (1982).

13. J.T. Cheung and H. Sankur, *Laser Ablation of Electronic Materials : Basic Mechanisms and Applications*, edited by E. Fogarassy and S. Lazare, Elsevier Science Publishers, 325 (1992).

14. J.W. McCamy and D.H. Lowndes, Appl. Phys. Lett., **63**, 3008 (1993).

15. D.L. Capewell and D.G. Goodwin, to be published in Mat Res. Soc. Symp. Proc., (this volume), (1995).

PULSED-LASER DEPOSITION OF TITANIUM NITRIDE

WENBIAO JIANG*, M. GRANT NORTON*, J. THOMAS DICKINSON**, AND
N.D. EVANS***
*   Department of Mechanical and Materials Engineering and
**   Department of Physics, Washington State University, Pullman, WA 99164
***   Oak Ridge Institute for Science and Education, Oak Ridge, TN 37831

ABSTRACT

The pulsed-laser deposition technique has been used to form thin films of TiN on (100)-oriented single crystal substrates of silicon and rocksalt. Using atomic force microscopy, it was revealed that TiN films grown on silicon at substrate temperatures ranging from 50°C to 500°C were extremely smooth—the mean roughness being ~ 0.2 nm. Thin TiN films deposited on freshly cleaved NaCl were found to be epitaxial at substrate temperatures as low as 50°C. Epitaxy in this latter system is believed to be due to the structural similarity between film and substrate and the almost exact 4:3 coincident site lattice.

INTRODUCTION

With excellent mechanical, thermal, and electronic properties, such as good thermal stability, high corrosion resistance, and low electrical resistivity, titanium nitride (TiN) thin films have many applications ranging from coatings on cutting tools to diffusion barriers in VLSI microelectronics. This broad range of applications has resulted in the development of a wide variety of deposition techniques to form TiN films. These techniques include both physical and chemical vapor deposition methods. However, most of these techniques require moderate to high substrate temperatures to form oriented crystalline films. The high deposition temperature inhibits the use of TiN films in some applications where the substrate cannot withstand elevated temperatures. It is therefore of interest to study the possibility of TiN deposition at lower substrate temperatures.

The major obstacle to low temperature growth is the difficulty of obtaining the high surface mobility required for the nucleation and growth of crystalline or epitaxial films at low substrate temperatures. This limitation can be overcome by delivering the material to the substrate in unique chemical forms (e.g., clusters or reactive species), in charged or highly excited states, and/or with appreciable kinetic energy. The latter objective may be accomplished by the use of pulsed-laser deposition (PLD). In PLD, a focused high-power excimer laser is directed towards a target in a high-vacuum chamber. The strong laser-target interaction results in the formation of a high temperature, high velocity, and electronically excited plasma consisting of neutral atoms, molecules, ions, and electrons. With little loss of energy in high vacuum, the plasma is transported to and deposited onto a substrate.

Several groups have reported the formation of TiN thin films by PLD using irradiation of either metallic targets in a nitriding atmosphere [1,2] or stoichiometric nitride targets [1,3,4]. Even when the substrate is maintained at room temperature during film deposition, crystalline TiN films have been formed on silicon [3]. Narayan and co-workers have reported the formation of single crystal TiN films on silicon substrates heated during deposition to temperatures in the range 600 - 700°C [4]. This latter result is unusual given the relatively large lattice mismatch (25%) between Si (a_0 = 0.543 nm) and TiN (a_0 = 0.422 nm). The cube-on-cube epitaxy appears to be enhanced by the 4:3 near-coincident site lattice where the mismatch is less than 4%.

Mat. Res. Soc. Symp. Proc. Vol. 388 © 1995 Materials Research Society

EXPERIMENTAL

Titanium nitride films were formed by focusing a Lambda Physik EMG203 excimer laser operating on KrF (λ = 248 nm) onto a hot-pressed TiN target supplied by Target Materials Inc. The laser fluence was 4 J/cm^2, the pulse duration was 20 ns, and the pulse repetition rate was in the range 4 to 8 Hz. The films were deposited in vacuum (base pressure 10^{-6} torr). Two types of substrate material were used in the present study: (100)-oriented single crystal silicon and rocksalt. (No attempt was made to remove the native oxide layer on the silicon substrates.) The substrates were mechanically clamped onto the stainless steel plate of a small resistive heater and were positioned parallel to the target. The distance between the target surface and the substrate was ~ 5 cm. The temperature of the substrate heater was monitored by a thermocouple embedded into the heater block at a position just behind the substrate. All the temperatures reported here are those of the substrate heater as measured by the thermocouple.

The rocksalt substrates were used to facilitate sample preparation for transmission electron microscopy (TEM) and for parallel electron energy loss spectrometry (PEELS). Thin (~10 to 30 nm thick) TiN films were deposited onto freshly cleaved (100)-oriented NaCl substrates. Following deposition, the NaCl was dissolved in deionized water and the film collected on a slotted 400 mesh copper grid and dried in air. This specimen preparation method produces excellent samples for PEELS analysis—they are thin and of a uniform thickness. All the films were analyzed either immediately after deposition or after being stored in a desiccator for a few days. A Hitachi H600 TEM operated at 100 kV was used to examine the films. PEELS was performed with a Philips EM400T/FEG analytical electron microscope operated at 100 kV.

The surface morphology of the films was characterized with a Digital Instruments Nanoscope III Multimode atomic force microscope (AFM) operated in either contact mode or tapping mode. The images shown here are top view images with linear gray scale encoding of the height of features. The films were also examined using a JEOL JSM 6400 scanning electron microscope (SEM) operated at 20 kV. No conductive coating was applied to the samples prior to SEM analysis.

RESULTS AND DISCUSSION

The first obvious feature of all the films grown in this study was their golden-yellow color, which is characteristic of TiN [5]. The surfaces of the films were fairly free of the ubiquitous particulates associated with films formed by PLD. The occurrence of particulates on the film surface is highly dependent on the nature of the target [1]. TiN films formed from the ablation of titanium may have large spherical particles on the surface as a result of the ejection of molten metal from the target [2]. Films made from pre-formed TiN targets often show a lower number of particulates [1].

Examination of the surface morphology of the films by AFM showed that they were extremely smooth for all the substrate temperatures used in this study. Figure 1 shows AFM images of the surface of a silicon substrate and of TiN films grown at three substrate temperatures. Mean roughness values determined by the AFM software for the TiN films varied from 0.18 nm for films deposited at a substrate temperature of 500°C to 0.27 nm for films deposited at 50°C. These values can be compared to the mean roughness of 0.19 nm determined for the uncoated silicon wafer.

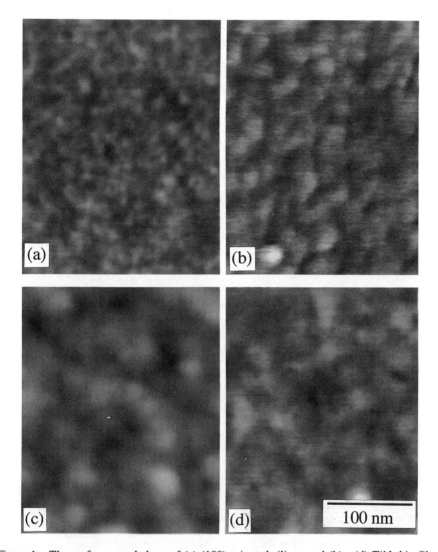

Figure 1.   The surface morphology of (a) (100)-oriented silicon and (b) - (d) TiN thin films
deposited onto silicon substrates at temperatures of (b) 50°C, (c) 300°C, and (d)
500°C. The z-range is 3 nm in all the images.

Figure 2 shows a series of X-ray diffraction patterns obtained from TiN thin films
deposited on (100)-oriented silicon substrates at different temperatures. At a substrate temperature
of 50°C, no strong diffraction maxima are observed. For films deposited at substrate temperatures
of 300°C and 500°C sharp peaks in the patterns corresponding to diffraction from the (200) planes
of TiN are observed indicating that these films are textured. Several groups have reported that TiN

films on silicon are oriented with the [100] of TiN parallel to the [100] of Si [1,4]. Recent results on TiN films prepared by reactive rf magnetron sputtering indicate that the preferred film orientation changes from (200) to (111) with increasing film thickness [6].

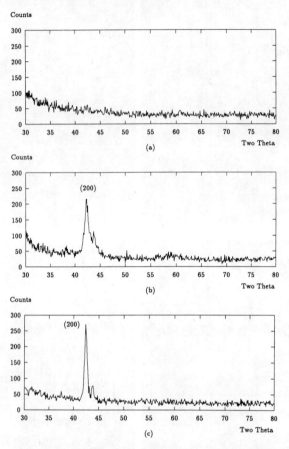

Figure 2.   X-ray diffraction patterns recorded from TiN thin films deposited on (100)-oriented Si substrates at substrate temperatures of (a) 50°C, (b) 300°C, and (c) 500°C.

Electron channeling patterns (ECP), shown in Figure 3, obtained from very thin TiN films deposited on silicon at a substrate temperature of 300°C indicate that the films are highly ordered, at this stage of growth, in the film-substrate interface plane. No clear channelling patterns were acquired from thicker TiN films deposited under equivalent conditions.

Figures 4a and 4b show selected area electron diffraction (SAED) patterns recorded from free-standing thin TiN films deposited on NaCl substrates heated to room temperature and 50°C, respectively. At room temperature the TiN film is crystalline and highly textured. At 50°C the SAED patterns showed only discrete reflections which could be indexed to TiN, indicating that film growth is epitaxial on NaCl at this temperature. SAED patterns obtained from films deposited at higher temperatures all appeared identical to Fig. 4b. This result may at first be surprising because of the large lattice mismatch (29%) between TiN and NaCl ($a_0 = 0.564$ nm). However, there exists an almost exact 4:3 coincident site lattice (where 4 unit cells of TiN match with three unit cells of NaCl) with a mismatch of only 0.2%. Thus as in the case of TiN growth on Si, domain matching appears to be an important feature in determining epitaxial growth in systems with large lattice mismatches [4]. The structural similarity between NaCl and TiN may also favor epitaxy with the TiN molecules occupying cation and anion sites on the NaCl (100) surface. This type of "auto-epitaxial" system can lead to epitaxial growth at low temperatures despite large lattice mismatches [7].

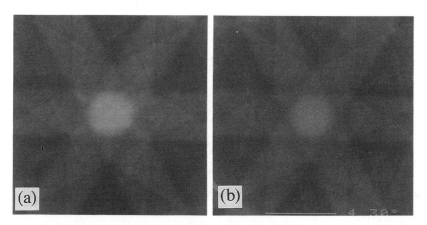

Figure 3. Electron channelling patterns recorded from (a) (100)-oriented Si and (b) a very thin TiN film deposited onto Si at a substrate temperature of 300°C.

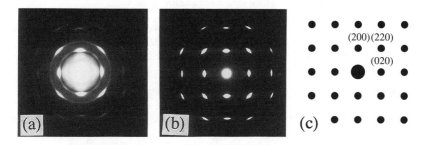

Figure 4. Selected area electron diffraction patterns from TiN thin films deposited onto NaCl substrates heated to (a) room temperature and (b) 50°C. An indexed schematic is shown in (c).

Figure 5 shows a PEELS spectrum obtained from a TiN film deposited at room temperature. The absolute composition of the films could not be determined from these spectra because there is some overlap between the Ti-L_1 edge which occurs at 564 eV and the O-K edge at 532 eV. Absolute composition determination would require multiple-least-squares analysis of second difference spectra using appropriate standards. However, relative changes in film composition were noted from examining the jump ratios of the N-K, Ti-$L_{2,3}$, and O-K edges. The jump ratio is the ratio of the intensity at the edge-onset relative to the background intensity immediately before the edge. These ratios indicate that the amount of oxygen incorporated into the TiN films increases and the amount of nitrogen decreases when the film is grown at room temperature rather than 300°C. Oxygen contamination appears to be a frequent occurrence in TiN films. Several possible sources of this contamination have been identified [1,3].

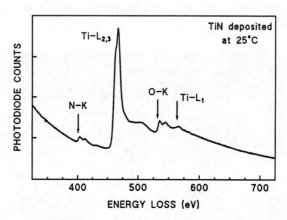

Figure 5.   PEELS spectrum of TiN thin film deposited at 25°C.

CONCLUSION

In conclusion, extremely smooth oriented TiN thin films can be grown on single crystal silicon substrates by pulsed-laser deposition at relatively low substrate temperatures. Epitaxial TiN can be formed on NaCl at a substrate temperature as low as 50°C.

The work at ORNL was sponsored by the Division of Materials Sciences, US Department of Energy, under contract DE-AC05-84OR21400 with Martin Marietta Energy Systems, Inc., and through the SHaRE Program under contract DE-AC05-76OR00033 with Oak Ridge Associated Universities.

REFERENCES

1.   J.C.S. Kools, C.J.C.M. Nillesen, S.H. Brongersma, E. van de Riet, and J. Dieleman, J. Vac. Sci. Technol. A **10**, 1809 (1992)
2.   S. Altshulin, A. Rosen, and J. Zahavi, J. Mater. Sci. **28**, 3749 (1993)
3.   O. Auciello, T. Barnes, S. Chevacharoenkul, A.F. Schreiner, and G.F. McGuire, Thin Solid Films **181**, 73 (1989)
4.   J. Narayan, P. Tiwari, X. Chen, J. Singh, R. Chowdhury, and T. Zheleva, Appl. Phys. Lett. **61**, 1290 (1992)
5.   L.E. Toth, Transition Metal Carbides and Nitrides, Academic Press, New York (1971)
6.   U.C. Oh, J.H. Je, and J.Y. Lee, J. Mater. Res. **10**, 634 (1995)
7.   B. Lewis and D.J. Stirland, J. Cryst. Growth **3,4**, 200 (1968)

GROWTH OF MoN$_x$ BY REACTIVE LASER ABLATION

RANDOLPH E. TREECE,[†] JAMES S. HORWITZ, EDWARD DONOVAN, AND DOUGLAS B. CHRISEY
Code 6674, Naval Research Laboratory, Washington, DC 20375.

ABSTRACT

Control of composition and phase of a series of MoN$_x$ thin films has been accomplished by reactive pulsed laser deposition (PLD). Molybdenum foil targets were ablated in a background gas of N$_2$/H$_2$ (10%) at pressures ranging from 40 to 120 mTorr. The MoN$_x$ films were deposited simultaneously onto (100) MgO and fused silica substrates. The films were characterized by X-ray diffraction (XRD), temperature-dependent resistivity, and Rutherford backscattering spectroscopy (RBS). The composition, phase, and electronic transport properties were found to depend on N$_2$/H$_2$ pressure, substrate temperature, and substrate orientation. The highest superconducting transition temperature (T$_c$) was observed in a hexagonal Mo$_2$N film where T$_c$ (onset) $\approx$ 8 K. In general, T$_c$ was observed to correlate most closely with the N/Mo ratio. As the ratio of N/Mo increased above optimal Mo$_2$N composition, T$_c$ decreased. Films grown on MgO generally had higher N/Mo ratios and hence lower T$_c$ values than films deposited on silica.

INTRODUCTION

While many thin-film deposition techniques are limited in their control over product phase, we are developing pulsed laser deposition (PLD) as a new synthetic technique for the preparation of metastable materials. In PLD, short (30 ns), high-energy laser pulses are focused onto a solid target material. Rapid heating, vaporization, and ionization of the material results in a dense plasma above the target. The vapor is collected on a nearby substrate. While PLD has been used to grow a variety of compounds, it has been used most extensively for the deposition of complex oxides, such as high-temperature superconductors and ferroelectric compounds.[1] Among the advantages of PLD as a synthetic technique for the formation of metastable compounds are the ability to operate in large background pressures ($\leq$ 1 Torr) of reactive gases, the high energies of ablated species ($\leq$ 100 eV) and the very high instantaneous growth rates ($\leq$ 10^5 Å/sec).[2] PLD has been used to deposit films of metastable compounds such as the infinite layer oxides Ca$_{1-x}$Sr$_x$CuO$_2$,[3] cubic boron nitride,[4] and carbon nitride (CN$_x$).[5,6]

This paper will describe the use of PLD for the synthesis of molybdenum nitride materials. The early transition metals usually can form several different, and sometimes metastable, nitride phases as the nitrogen to metal ratios are changed.[7] The non-stoichiometric nature of the transition metal nitrides complicates phase assignment of MoN$_x$ materials. The non-stoichiometries can be present as vacancies or defects on either the nonmetal and/or metal sublattices within a given phase.[11] In addition to the fascinating chemistry seen in the transition metal nitrides, their superconducting properties drive continued research on these materials. The rock-salt (B1) phases of transition metal nitrides are refractory materials with bulk superconducting transition temperatures (T$_c$) as high as 16 K.[7] Thin films of NbN currently are being studied for applications in superconducting electronics.[8]

Previously, we have investigated the growth of the nitrides of niobium (NbN$_x$) by PLD.[9,10,11,12] By varying the nitrogen gas pressure it was possible to prepare Nb, Nb$_2$N, NbN, and Nb$_3$N$_4$.[10] In addition to growing a variety of NbN$_x$ phases, we also isolated a new metastable phase of NbN.[12] Based on our experience with growing the NbN$_x$ materials, we began investigating the growth of MoN$_x$ thin films. Theoretical considerations have been used to predict that if MoN could be grown in cubic B1 phase it should have a T$_c$ $\approx$ 30 K. This paper describes the growth and characterization of MoN$_x$ phases by PLD.

EXPERIMENTAL SECTION

The nitride films were grown in a high-vacuum chamber equipped with a cryopump and a turbopump. A pulsed KrF excimer laser beam (248 nm, 30 ns FWHM) operating at 10 Hz was focused with a 50 cm focal length lens onto a dithering and rotating (30 RPM) Mo target at an angle of 45° to the surface. A laser fluence of 6 J/cm^2 was used. The ambient gas (N$_2$ with 10% H$_2$) input pressure was regulated to a dynamic equilibrium (~10 sccm) by a solenoid-activated leak valve controlled by a capacitance manometer with the chamber under gated or throttled pumping. The substrates were washed with ethanol, attached to the substrate heater with silver paste and maintained ~6 cm from the target. The film compositions and thicknesses were determined by Rutherford Backscattering Spectroscopy (RBS) with 6 MeV He^{2+}. X-Ray diffraction (XRD) patterns were collected using a rotating anode source and a conventional θ–2θ geometry, and indexed using a least squares fit of the data. Rocking curve full width at half maximum of ω scans, reported as Γ (FWHM) for the (200) peak (for the cubic films) or the (006) peak (for the hexagonal films), was measured by fixing 2θ at the peak maximum and scanning through θ. Temperature-dependent resistance (R(T)) measurements were performed by a standard ac four-point probe measurement.

RESULTS AND DISCUSSION

A series of MoN$_x$ films were grown simultaneously on (100) MgO and amorphous fused silica under varied growth conditions. The N$_2$ (+10% H$_2$) reactive gas pressure was varied from 40 to 120 mTorr and substrate temperatures were either 500 or 600 °C. The structures of the deposited films were determined by XRD, the electrical properties were measured by R(T), and the compositions (N/Mo ratios) were found by RBS. The results are summarized in Table I.

XRD Results

The crystal structure adopted by the MoN$_x$ films was dependent on the substrate identity, deposition pressure and deposition temperature. Deposition on amorphous substrates led to the growth of films with poor crystallinity while growth on single crystal MgO promoted the formation of highly oriented material. This is illustrated in Figure 2, where XRD patterns are shown of two MoN$_x$ films grown simultaneously at a pressure of 60 mTorr and substrate temperature of 600 °C on substrates of (a) fused silica and (b) (100) MgO. The XRD pattern of the film grown on fused silica was indexed as a poorly crystalline cubic material with broad weak reflections from (111) and (200) planes consistent with a lattice parameter of a=4.19 Å. The film grown on MgO was indexed as a highly oriented (Γ (FWHM)=0.10 °) hexagonal material with only the (00l) reflections present in the XRD pattern, producing a calculated c lattice parameter equal to 6.00 Å.

The effect of substrate temperature on the structure of the deposited film was demonstrated by growing a film at the same pressure (60 mTorr) but at a lower substrate temperature (500 °C).

Table I. Summary of MoN$_x$ Growth and Characterization Results

P (N$_2$)	T$_{sub}$	Substrate	Crystalline Symmetry	Lattice Parameter(s)	Γ (FWHM)	T$_c$ (onset)	N/Mo Ratio
(mTorr)	(K)			(Å)	(° θ)	(K)	
40	600	MgO	Cub	a=4.24	0.14	4.8	0.65
40	600	SiO$_2$	Amor	NM	NM	4.7	0.85
60	600	MgO	Hex	c=6.00	0.10	6.3	0.65
60	600	MgO	Hex	c=5.97	0.11	7.8	0.20
60	600	SiO$_2$	Cub	a=4.19	NM	5.8	0.45
60	500	MgO	Cub	a=4.23	0.10	NM	0.90
120	600	MgO	Cub	a=4.25	0.12	4.2	0.65
120	600	SiO$_2$	Cub	a=4.17	NM	4.8	0.45

NM=Not Measurable

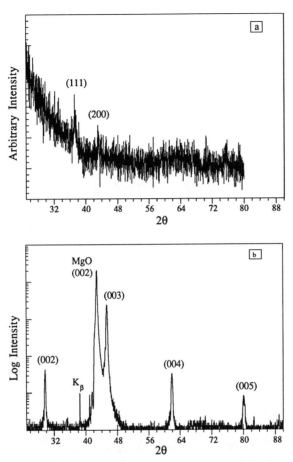

Figure 2. XRD patterns of MoN$_x$ films grown at a pressure of 60 mTorr and substrate temperature of 600 °C on (a) fused silica and (b) (100) MgO substrates.

In contrast to the hexagonal film grown at 600 °C, the highly oriented (Γ (FWHM)=0.10 °) film deposited at 500 °C was cubic with a=4.23 Å.

The effect of gas pressure also was examined. When a pair of films were grown at 600 °C but at the higher deposition pressure of 120 mTorr, both films were cubic, as shown in Figure 3 . While the (111) peak in the film grown on fused silica as shown in Figure 3 (a) is stronger and sharper than the corresponding (111) peak in Figure 2 (a), the cubic lattice parameter is about the same, within experimental error, with a=4.17 Å. The cubic lattice parameter of the highly oriented film (Γ (FWHM)=0.12 °) determined from the (200) and (400) peaks is a=4.25 Å. When the deposition pressure was lowered to 40 mTorr and the substrate temperature was held constant at 600 °C, the film deposited on (100) MgO was also cubic. The highly oriented film (Γ (FWHM)=0.14 °) was found to have a=4.24 Å. The film grown on the fused silica substrate was amorphous.

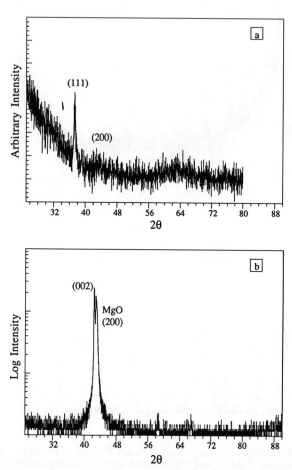

Figure 3. XRD patterns of MoN_x films grown at a pressure of 120 mTorr and
substrate temperature of 600 °C on (a) fused silica and (b) (100) MgO substrates.

R(T) Results

The electrical properties of the MoN_x films were determined by R(T) measurements.
Representative R(T) scans are shown in Figures 4 and 5. As illustrated in the figures, the oriented
films grown on the MgO substrates consistently had lower resistivities (10-15 $\mu\Omega$-cm versus 40-
50 $\mu\Omega$-cm) than the amorphous films grown on the fused silica substrates. The oriented FCC-
MoN_x films displayed lower T_c's than the oriented hexagonal films. The superconducting
transition of the hexagonal film shown in Figure 5 has a higher onset temperature than the
polycrystalline film, but the polycrystalline film reached $R=0$ almost 1 degree higher than the
oriented material. The shape of the transition of the oriented film indicates that there are at least

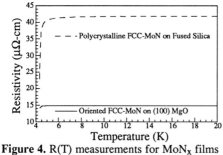
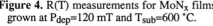

Figure 4. R(T) measurements for MoN$_x$ films grown at P$_{dep}$=120 mT and T$_{sub}$=600 °C.

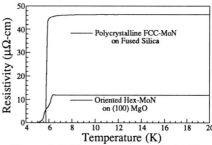

Figure 5. R(T) measurements for MoN$_x$ grown at P$_{dep}$=60 mT and T$_{sub}$=600 °C

two superconducting onsets which can be explained by the presence of more than one superconducting phase in the sample.

<u>Compositional Results</u>

The exact role of the N/Mo ratio in the structural and electrical properties is difficult to determine due to the non-stoichiometric nature of the transition metal nitrides. A plot of the N/Mo ratio versus T$_c$(onset) is shown in Figure 6. Like NbN$_x$, the MoN$_x$ phases are non-stoichiometric and broad ranges of x are allowed for each structural phase.[11] For example, the value of x in NbN$_x$ can range from 0.35<x<0.6.[11] Pure Mo is body centered cubic (T$_c$≈8 K) and can retain its BCC structure while absorbing up to >20 mole % interstitial nitrogen before transforming to the Mo$_2$N phase. Mo$_2$N has been reported to form in cubic[13] (T$_c$(onset)≈7.8 K) and hexagonal[14] phases (T$_c$(onset)≈6 K). MoN has also been reported to form in cubic (T$_c$(onset)≈12.5 K) and hexagonal phases (T$_c$(onset)≈13.5 K).[13]

The films represented in Fig. 6 will be considered in three groups: (1) x <0.3; (2) 0.3<x <0.7; and (3) x >0.7. The film with x <0.3 grew in a hexagonal structure and displayed the highest T$_c$(onset) observed and two superconducting transitions. The hexagonal structure of the film indicates that it cannot be pure Mo, but the low N/Mo ratio together with the presence of two

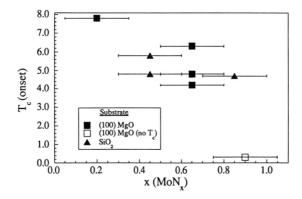

Figure 6. Plot of T$_c$ (onset) versus x in NbN$_x$. The film indicated with the open square did not show a superconducting onset at temperatures greater than 4.2 K.

superconducting transitions can be explained by a mixture of a small amount of Mo (too small to be observed by XRD) present in a film largely composed of Mo_2N. The materials with $0.3 < x < 0.7$ include cubic and hexagonal films with a range of x and T_c values consistent with Mo_2N. The two films with $x > 0.7$ were either amorphous with $T_c(onset) = 4.7$ K or cubic with no $T_c(onset)$ above 4.2 K. Despite the larger N/Mo values, the low T_c's indicate that the films are not MoN. The cubic $MoN_{0.9}$ film is likely Mo_2N with interstitial nitrogen. The interstitial nitrogen does not increase the cubic lattice parameter but does destroy T_c above liquid He temperatures.

CONCLUSIONS

PLD has been used to grow MoN_x films on (100) MgO and fused silica by ablating Mo targets in a reactive atmosphere of N_2 (+10% H_2). The structural and electrical properties of the MoN_x films were shown to depend on gas pressure, substrate temperature and structure. MoN_x films deposited on MgO were highly oriented while those grown on fused silica were nearly amorphous. Films grown on (100) MgO at a gas pressure of 60 mTorr and a substrate temperature of 600 °C displayed the highest superconducting transition temperatures (T_c) and hexagonal structures. Changing the deposition conditions by raising or lower the gas pressure or lowering the substrate temperature led to films with lower T_c's and cubic structures. The hexagonal structure did not grow on the fused silica substrates under any conditions. Normal state resistivities for films grown on fused silica were 3-5 times larger than those observed for films grown on (100) MgO.

In general, T_c was observed to correlate most closely with the N/Mo ratio. As the ratio of N/Mo increased above optimal Mo_2N composition, T_c decreased. Taking into account the low T_c's (4.2 - 7.8 K), crystal structures, and N/Mo ratios ($0.20 \le x \le 0.90$), the superconducting phases in these films are likely Mo_2N in the cubic and hexagonal forms. Nitrogen in excess of x >0.6 is likely located in interstitial defects, and the presence of these nitrogen defects is deleterious to the superconducting properties. Films with $x < 0.3$ are most likely mixtures of Mo_2N and Mo.

REFERENCES

† Present address: Department of Physics, University of Colorado, CO 80309-0390.
1. Growth of electronic ceramics by PLD is reviewed in: J. S. Horwitz, D. B. Chrisey, K. S. Grabowski, and R. E. Leuchtner, *Surf. Coatings Technol.* 51, 290 (1992).
2. "Pulsed Laser Deposition of Thin Films", edited by D. B. Chrisey and G. K. Hubler, (Wiley, New York, 1994).
3. (a) C. Niu and C. M. Lieber, *J. Am. Chem. Soc.* 114, 3570 (1992). (b) D. P. Norton, B. C. Chakoumakos, J. D. Budai and D. H. Lowndes, *Appl. Phys. Lett.* 62, 1679 (1993).
4 . (a) A. K. Ballal, L. Salamanca-Riba, G. L. Doll, C. A. Yaylor II, and R. Clarke, *Mater. Res. Soc. Proc.* 285, 513 (1993). (b) A. K. Ballal, L. Salamanca-Riba, C. A. Yaylor II, and G. L. Doll, *Thin Solid Films* 224, 46 (1993). (c) and F. Qian, V. Nagabushnam and R. K. Singh, *Appl. Phys. Lett.* 63, 317 (1993).
5. (a) F. Xiong and R.P.H. Chang, *Mater. Res. Soc. Proc.* 285, 587 (1993). (b) C. Niu, Y.Z. Lu, and C.M. Lieber, *Science* 261, 334 (1993).
6. R. E. Treece, J. S. Horwitz, and D. B. Chrisey, *Mater. Res. Soc. Proc.* 327, 245 (1994).
7. L. E. Toth in "Transition Metal Carbides and Nitrides"; Margrave, J. L., Ed.; Academic Press: New York, 1971; Refractory Materials Vol. 7.
8. "Superconducting Electronics", edited by H. Weinstock and M. Nisenoff, (Springer-Verlag, New York, 1992).
9. R. E. Treece, J. S. Horwitz, and D. B. Chrisey, *Chem. Mater.* 6, 2205 (1994).
10. R. E. Treece, J. S. Horwitz, J. H. Claassen, and D. B. Chrisey, *Appl. Phys. Lett.* 65, 2860 (1994).
11. R. E. Treece, J. S. Horwitz, and D. B. Chrisey, *Mater. Res. Soc. Proc.* 343, 747 (1994).
12. R. E. Treece, M. Osofsky, E. Skelton, S. Qadri, J. S. Horwitz, and D. B. Chrisey, *Phys. Rev. B.* 51, 9356 (1995)
13. H. Ihara, *et al.*, in "Advances in Cryogenic Engineering Materials", edited by R. P. Reed and A. F. Clark, vol. 32 (Plenum, New York, 1986) pp. 603-616.
14. W. W. Fuller, *et al.*, *J. Vac. Sci. Technol. A*, 1, 517.

ADVANCED PROPERTIES OF SILICON OXYNITRIDE (SiO$_x$N$_y$) THIN FILMS PREPARED BY PULSED LASER DEPOSITION

R.-F. XIAO, L.C. NG, AND H.B. LIAO
Department of Physics, the Hong Kong University of Science and Technology
Clear Water Bay, Kowloon, Hong Kong

Abstract

A pulsed laser deposition technique has been used to grow silicon oxynitride (SiO$_x$N$_y$) thin films at low deposition temperatures (25°C - 300°C). The thin films were found to be quite smooth in surface morphology, extremely inert in chemical solution and highly transparent in the optical range of 0.3 μm to 5 μm. The refractive index was tunable between 1.4 - 2.1 by the addition of oxygen in substitution of nitrogen in the film, and the dielectric constant is much larger than the similar films grown by conventional chemical vapor deposition. The high quality of the SiO$_x$N$_y$ films deposited at such low temperatures was resulted from the large kinetic energy carried by the impinging particles created by the ablation of a high-power pulsed excimer laser. The kinetic energy of the impinged particles on the substrate provides thermal energy for surface diffusion and relaxation.

Introduction

Silicon oxynitride (SiO$_x$N$_y$) can be formed by a partial replacement of nitrogen with oxygen from silicon nitride (Si$_3$O$_4$). It has been found that thin films of SiO$_x$N$_y$ are better for the applications of final passivation, gate dielectric, and functional memory layers in microelectronic and opto-electronic devices [1-5] than that of silicon nitride (Si$_3$N$_4$) and silicon dioxide (SiO$_2$) thin films. This is because the physical properties of SiO$_x$N$_y$ can be tailored between those of SiO$_2$ and Si$_3$N$_4$ to improve thermal and mechanical stability as well as electrical and optical performance.

Traditional techniques for preparing SiO$_x$N$_y$ thin films include high-temperature chemical vapor deposition (HTCVD) [1], rapid thermal processing (RTP), i.e., oxidation of Si [2] or nitridation of SiO$_2$ [3], plasma-enhanced chemical vapor deposition (PECVD) [4], and reactive gas sputtering [5]. However, the specific requirements for each of these techniques limits their applications. For instance, as a high temperature (>600°C) process, HTCVD prohibits deposition of SiO$_x$N$_y$ thin films onto low melting point solid surfaces, in addition to the problems caused by high temperature processing such as enhanced dopant interdiffusion between adjacent layers in devices. RTP technique also requires high processing temperature and works only on the substrates containing silicon. PECVD can be done at relatively low deposition temperatures (~300 °C), but it often incorporates too much hydrogen in the dielectric layers and may cause damage to devices because of the high energy ions.

Recently we have succeeded in growing high quality SiO$_x$N$_y$ thin films by using a pulsed laser deposition (PLD) technique [6]. In contrast to the other deposition techniques mentioned above, in the PLD process thin films are formed by the deposition of highly energetic particles produced by the interaction of a high intensity laser beam with a solid target [7]. The large kinetic energy of the particles impinging on the substrate surface provides certain amount of thermal energy for surface diffusion and relaxation. As the result, high quality thin films can be obtained without the need for high substrate temperature. While early reports have concentrated on the growth process [6,8], in this paper we address some advanced chemical and physical properties of the films relevant to optical and electronic applications.

Experimental setup

Basically, our PLD system consists of a high-power pulsed laser and a multi-window vacuum deposition chamber (Figure 1). The target was made by compressing electronic grade

115

(99.99%) Si_3N_4 powder into a one-inch diameter disk and subsequently sintered at 900°C for 24 hours in a He gas environment. Three different substrates, Si(100), NaCl (100) and electrically-conducting indium-tin-oxide (ITO) coated glass, were used for the purpose of film characterizations for different physical properties. The substrate was mounted on a rotating substrate holder facing the target. The separation between the target and the substrate was about 5.5 cm. The substrate temperature was measured using a thermocouple and controlled by a temperature controller (EUROTHERM, 818p). Irradiation at a wavelength of 193 nm (ArF) was provided by a pulsed (duration width = 23 ns) excimer laser (LAMBDA PHYSIK, LPX210i) at a repetition rate of 5 to 15 Hz, with a laser fluence at the target was about 1- 2 J/cm^2. After loading the substrate and target, the chamber was evacuated to about $2x10^{-6}$ torr, prior to heating the substrate to 300 °C for 30 minutes to reduce outgassing and to clean the substrate. The substrate was then gradually cooled to the deposition temperature. Different pressure levels of oxygen gas were introduced into the chamber by a mass flow controller to alter the oxygen/nitrogen (O/N) ratio in the film. With the above growth conditions, the typical deposition rate was about 0.05 - 0.1 Å per laser pulse depending on the oxygen gas pressure. Deposition was stopped after 20,000 pulses for the Si and NaCl substrates and after 60,000 pulses for the ITO substrates.

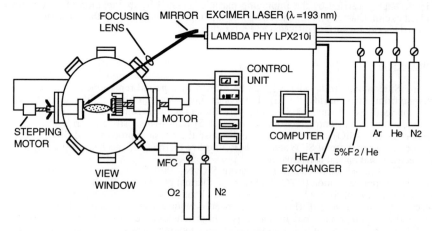

Figure 1. Schematic diagram of pulsed laser deposition system

Results and discussion

The surface morphology of the as-grown SiO_xN_y thin films on Si (100) substrates at different deposition temperatures was examined using an atomic force microscope (TopoMetrix, TMX 2000). Pictures in Figure 2 clearly show that the morphology is a sensitive function of substrate temperature. Islands larger than 500 nm in size are present in the films deposited at 25 °C and 100 °C. The dimensions of the islands become smaller as substrate temperature is increased to 200 °C [Figure 2(c)] and can hardly be seen within an area of 2 µm x 2 µm at T = 300 °C [Figure 2(d)]. This temperature dependence of surface morphology can be understood from a kinetic thin film growth argument. At a low substrate temperature, although the impinged particles (atoms, ions) ablated by the excimer laser carry a certain amount of kinetic energy, they are still not energetic enough to undergo sufficient surface diffusion. Particles hardly have time to find a low energy site to stay before being buried by the incoming particles from the target. Under such circumstances surface becomes kinetically roughened [9]. As the temperature increases, particles gain more kinetic energy once they impinge onto the substrate. They can move farther and have a better chance to find a low energy site, therefore smoothen the surface. Such an effect of surface roughening and smooth-

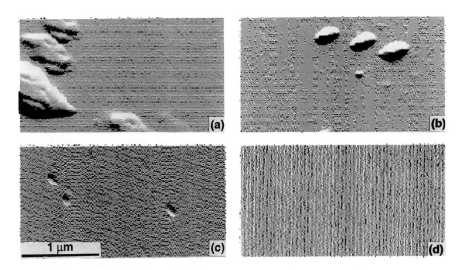

Figure 2. Surface morphologies of SiO$_x$N$_y$ thin films deposited at different substrate temperatures: (a) T = 25 °C; (b) T = 100 °C; (c) T = 200 °C; (d) T = 300 °C.

ening is typical innonequilibrium thin film deposition and has been observed in computer simulations [9]. Hence, to obtain a microscopically smooth thin film in PLD, an elevated temperature is still necessary. However, the deposition temperature in PLD process is much lower than that in the HTCVD process because of the high kinetic energy carried by the impinging particles ablated from the target by the excimer laser.

To evaluate the potential of SiO$_x$N$_y$ thin films for chemical passivation application, we measured the etching rate of the as-grown films in various chemical solutions. The thin films were first put in a KOH solution (25% by weight) for three days. No appreciable change was observed. Similar as-grown films were put in a concentrated HF solution (HF/H$_2$O = 49 % by volume). The etching rates of the SiO$_x$N$_y$ thin films deposited at low temperatures are quite large, i. e., 4500 Å/min for T = 25°C and 850 Å/min for T = 100°C, as one can see from Figure 3. It decreases linearly with increasing deposition temperature up to 200 °C. The etching rates of the films deposited at temperatures higher than 200 °C change slightly (70 Å/min at T = 200 °C and 55 Å/min at T = 300 °C). This indicates the films deposited at above 200 °C have almost reached their maximum chemical passiva-tion strength, more stable than the films grown by HTCVD(80 Å/min) [9]. For comparison, we also measured the etching rate of silicon oxynitride films deposited at 200 °C by a reactive (Ar+O$_2$) sputtering (Denton, DVI, DVB- SJ20C). A very high etching rate of 2000 nm/min was observed in the films deposited by the sputtering method.

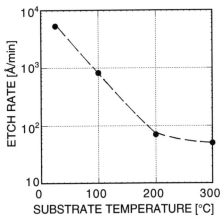

Figure 3. Etching rate of SiO$_x$N$_y$ thin films in HF/H2O=49% solution as function of deposition temperature.

Light transmission of the SiO_xN_y thin films deposited on NaCl substrates was measured by a Fourier Transform (FT) spectrometer (Bruker ISF-66) in the spectral range from 15 μm to 1 μm and by an UV/VIS spectrometer from 900 nm to 190 nm. The thin films deposited at T = 200 °C without the addition of O_2 during deposition are transparent in the spectral range from 0.3 μm to 5 μm with over 80% transmission and the transmission increases as oxygen gas is introduced. The extrapolated optical bandgap of the films from the UV absorption spectrum is about 5.5 eV for the films without any addition of O_2 gas and increases to 8 eV when O_2 gas with a flow rate of 20 sccm was introduced during deposition. The absorption peak due to the Si-N stretching mode at 830 cm^{-1} and the one due to Si-O antisymmetric stretching mode at 1080 cm^{-1} observed in the FT spectra confirm the formation of SiO_xN_y.

Figure 4(a) shows the dependence of the refractive index of SiO_xN_y thin films on substrate temperature measured by an ellipsometer at 6328 Å. The refractive index of these films increases with increasing substrate temperature. At 200 °C, the refractive index reaches a value of 2.1 that has been found for stoichiometric Si_3N_4 films grown by HTCVD [10], with no further improvment for further increases of substrate temperature. Figure 4(b) shows oxygen dependence of the refractive index of SiO_xN_y films at a fixed deposition temperature of 200 °C. The refractive index of the films is strongly dependent on the oxygen content. The refractive index is the highest when no oxygen gas is present, and decreases as oxygen gas is added in the chamber, similar to the results reported in an earlier PLD experiment [8].

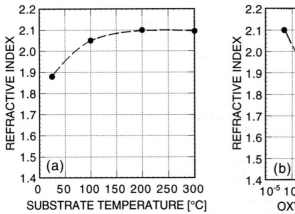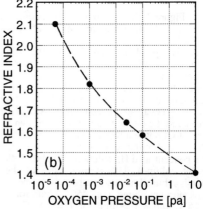

Figure 4. Refractive index of SiO_xN_y thin films: (a) as a function of deposition temperature without oxygen addition; (b) as a function of oxygen gas pressure deposited at T = 200 °C.

The adjustable refractive index of SiO_xN_y thin films can be very useful in the application of protective coatings for optical components. One can control the refractive index of the SiO_xN_y film by adjusting the O/N ratio or substrate temperatures to obtain a refractive index perfectly matched with the optical surface to be protected. The tunable refractive index of SiO_xN_y thin films can also be used in the fabrication of optical wave guides to obtain optimum guiding structure. However, it is important to realize that the mechanism for the reduction of refractive index by decreasing temperature is different from that by increasing oxygen content. The first reduction of refractive index is due to a decrease in film density as deposition temperature is lowered, as is demonstrated by the high etching rate, while the second one results from the formation of SiO_2-like film as more and more oxygen is added and the refractive index approaches that of SiO_2 [11]. X-ray photoelectron spectroscopy also shows that the O/N ratio in the film increased dramatically as oxygen gas was added to the chamber during deposition, while FT spectroscopy shows an absorption peak shifting from the Si-N stretching mode at 830 cm^{-1} to the SiO$_2$ antisymmetric stretching mode at 1080 cm^{-1} as oxygen content is increased. For better chemical stability it is therefore more desirable to tune the refractive index of SiO$_x$N$_y$ using oxygen gas than to decrease the deposition temperature.

Silicon oxynitride is a ceramic material like its counterparts Si$_3$N$_4$ and SiO$_2$. To facilitate the measurement of the dielectric properties of the film, electrically conducting ITO glass was used as a substrate. After deposition, the SiO$_x$N$_y$ film was coated with gold by a DC sputtering machine through a mask with multiple circular openings (diameter = 2mm). The dielectric properties of the film were checked using a C-V measurement system (RT66A). From the C-V scan, we have found that the film exhibits a small but non-negligible hysteresis [Figure 5(a)], a property that ferroelectric materials have. From the measurement of capacitance C we have obtained the dielectric constant κ of the films, using the formulae for a planar capacitor, C = $\varepsilon_0\kappa$A/d, where A and d are the area and thickness of the capacitor formed by ITO/film (0.5 µm thick) /Au. As one can see from Figure 5(b), the dielectric constant depends on both the substrate temperature and the oxygen content. The films deposited at 200 °C or above possess a very large dielectric constant as compared with those SiN$_x$ films deposited by the conventional CVD methods (more than double!) [10], and the dielectric constant decreases monotonically as temperature decreases. When the film is deposited at room temperature, the capacitor formed by ITO/SiO$_x$N$_y$/Au becomes short circuit. This indicates that the Au had penetrated into the holes and vacancies of the film. As for the effect of oxygen, we found that the dielectric constant of the film decreases dramatically as the amount of oxygen increases. Likewise, the electrical resistance of the capacitor decreases rapidly with increasing oxygen content. Therefore, a superior dielectric insulating layer would be the one with less oxygen. The addition of oxygen can reduce both dielectric strength and memory effect of the SiO$_x$N$_y$ film. Hence, one can maximize the performance of the SiO$_x$N$_y$ film by controlling its oxygen content. If a large dielectric strength is needed, less oxygen should be added. And if, in certain circumstances, a quick signal response is desired such as in a high speed, high frequency field-effect-transistor, more oxygen would make the transistor work faster.

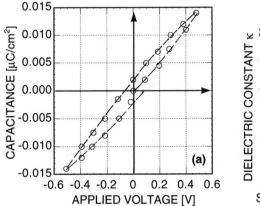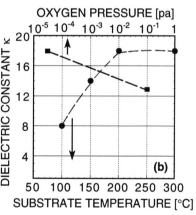

Figure 5. (a) Hysteresis of SiO$_x$N$_y$ thin film deposited at T = 200 °C without oxygen addition; (b) Dielectric constant of SiO$_x$N$_y$ thin films (0.5 mm thickness) as functions of deposition temperature and oxygen pressure.

<u>Summary</u>

We have grown silicon oxynitride (SiO$_x$N$_y$) thin films using a PLD technique. Our results show that these films have promising properties for opto-electronic applications. The novel feature of the experiment is in the lower deposition temperature that is required for producing high quality thin films than that used in the HTCVD technique. A wider range of silicon oxynitride films can be obtained by introducing O_2 gas during the deposition process. Our results show a strong dependence of thin film quality on deposition temperature. The thin films become smooth in surface morphology, extremely inert in HF/H_2O solution, and highly transparent in the optical range from 0.3 μm to 5 μm when the temperature is 200 °C or higher. The refractive index can be tailored by adjusting the deposition temperature and oxygen content. As compared with those SiNx films grown by CVD techniques, the films show a much larger dielectric constant (more than double) and a small but not negligible hysteresis behavior.

Acknowledgment

Financial support from a RGC grant (HKUST640/94P) is acknowledged. The thin film characterizations were done mainly in the Materials Characterisation and Preparation Centre at the Hong Kong University of Science and Technology. Assistance from Mr. Q.X. Su in performing the C-V measurement is appreciated.

References

1. P.K. Mclarty, W.L. Hill, X.-L. Xu, V. Misra, J.J. Wortman, and G.S. Harris, Appl. Phys. Lett. **63**, 3619 (1993).
2. H. Hwang, W. Ting, B. Maiti, D.-L. Kwong, and J. Lee, Appl. Phys. Lett. **57**, 1010 (1990).
3. V. J. Kapoor, and R. S. Bailey, J. Electrochem. Soc., 137 (1990) 3589; D. Xu, and V.J. Kapoor, J. Appl. Phys. **68**, 4127 (1990); D. Xu, and V.J. Kapoor, J. Appl. Phys. **70**, 1570 (1991); N. Novkovski, I. Aizenberg, E. Goin, E. Fullin, and M. Dutoit, Apppl. Phys. Lett. **54**, 2408 (1989); Y. Okada, P.J. Tobin, R.I. Heyde, J. Liao, and P. Rushbrook, Appl. Phys. Lett. **61**, 3163 (1992).
4. J.M. Chovelon, N. Jafrezic-Renault, P. Clechet, Y. Cross, J.J. Fombon, M.I. Baraton, P. Quintard, Sensors and Actuators B**4**, 385 (1991).
5. T. Kanata, H. Takakura, and Y. Hamakawa, Appl. Phys. A**49,** 305 (1989); G.A. Niklasson, T.S. Eriksson, and K. Brantervik, Appl. Phys. Lett. **54**, 965 (1989).
6. R.-F. Xiao, L.C. Ng, C. Jiang, Z.Y. Yang, and G.K.L. Wong, Thin Solid Films, in press.
7. Pulsed Laser Deposition of Thin Films, edited by D.B. Chrisey, and G.K. Hubler (John Wiley & Sons, New York, 1994).
8. E. Fogarassy, C. Fuchs, A. Slaoui, S. De Unamuno, J.-P. Stoquert, W. Marine, and B. Lang, SPIE Vol. **2045**, 148 (1994), and references therein.
9. R.-F. Xiao, and N.B. Ming, Phys. Rev. E**49**, 4720 (1994); R.-F. Xiao, J.I.D. Alexander, and F. Rosenberger, Phys. Rev. A**43**, 2977 (1991).
10. S. Wolf, and R.N. Tauber, Silicon Processing for the VLSI Era, Vol.1- Process Technology (Lattice Press, 1986), pp. 191-195.
11. Handbook of Infrared Optical Materials, edited by P. Klocek (Marcel Dekker, New York, 1991), pp. 346-352.

RAPID GROWTH OF DIAMOND-LIKE-CARBON FILMS
BY COPPER VAPOR LASER ABLATION

W. MCLEAN, B. E. WARNER, M. A. HAVSTAD, M. BALOOCH
University of California, Lawrence Livermore National Laboratory, P.O. Box 808, L-460, Livermore, CA, 94550.

ABSTRACT

Visible light from a copper vapor laser (CVL) operating with 510 and 578 nm radiation (intensity ratio approximately 2:1), an average power of 100 W, a pulse duration of 50 ns, and a repetition frequency of 4.4 kHz has been shown to produce high quality diamond-like-carbon (DLC) films at fluences between $2x10^8$ and $5x10^{10}$ W/cm^2. Maximum deposition rates of 2000 μm$\cdot$cm^2/h were obtained at $5x10^8$ W/cm^2. DLC films with hardness values of approximately 60 GPa were characterized by a variety of techniques to confirm DLC character, hydrogen content, and surface morphology. The presence of C_2 in the vapor plume was confirmed by the presence of the C_2 Swan bands in emission spectra obtained during the process. Economic implications of process scale-up to industrially meaningful component sizes are presented.

INTRODUCTION

PLD of DLC films has been demonstrated with both ultraviolet and infrared light from both excimer and Nd:YAG lasers[1-5] with irradiances on the ablation target of $2x10^8$ to over 10^{11} W/cm^2 respectively. One major impediment to commercialization of PLD processes for DLC (and other materials as well) is the rate at which material can be deposited on a substrate by these commercially available photon sources. This issue has been addressed by employing the visible light output of high power CVLs. By demonstrating peak deposition rates on the order of 1 μm/min, this work lays the foundation for follow-on development of PLD processes by less complex pulsed lasers operating in the visible wavelength regions at high pulse repetition frequencies (PRFs).

EXPERIMENTAL PROCEDURES

The characteristics of the CVLs employed in this work are summarized in Table I. Approximately 190 W (average power) of light from a single CVL oscillator and 500 W amplifier was directed through a fused SiO$_2$ window to a vacuum chamber on a rotating graphite target (POCO AXM-5Q, 5 μm average particle size) moving at a speed of 8 cm/s. Target texturing was minimized by exposing fresh tracks of graphite by indexing the target one beam diameter every revolution. Substrates were biased to - 500 V during the ablation process and to + 40 V between laser pulses.

Table I. CVL Operating Parameters

Wavelength	510, 578 nm (2:1 intensity ratio)
Pulse Width	50 ns (FWHM)
Pulse Energy	25mJ
Pulse Repetition Frequency	4400 Hz

A 500 mm focusing lens outside the vacuum chamber was used to set peak irradiances on the target by adjusting the beam diameter from 50 μm (5×10^{10} W/cm²) to 750 μm (2×10^8 W/cm²). Target-to-substrate spacing was typically 7.6 cm. Film thickness and distribution data on Si(100) substrates were obtained by masking the substrates and measuring the step height from the bare Si to the coating surface with a standard laboratory profilometer. All depositions were performed at ambient temperature in a vacuum of 5×10^{-7} torr or better.

RESULTS AND DISCUSSION

Electron Energy Loss Spectra (EELS) were recorded for primary beam energies of 495 eV and analyzed with a single pass cylindrical mirror analyzer. A typical spectrum of DLC, in this case a 250 nm thick film grown at 6×10^8 W/cm2 [Fig. 1(a)], show the absence of the $\pi \rightarrow \pi*$ loss feature 6 eV below the reflected primary beam. The absence of this feature is characteristic of sp³ bonded material [Fig. 1(b)].[6,7] DLC coatings grown at fluences ranging from 2×10^8 to 5×10^{10} W/cm² exhibited similar characteristics. Raman spectra from these specimens exhibited the asymmetric peak at 1550 cm⁻¹ characteristic of DLC [Fig. 1(c)].[8]

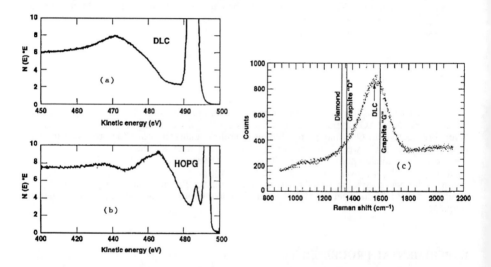

Figure 1. Characterization of DLC and graphite. (a) EELS of DLC grown at 5×10^8 W/cm². (b) EELS of highly oriented pyrolytic graphite. (c) Raman spectrum of DLC from (a).

The deposition profiles on stationary substrates followed a $\cos^n(\phi)$ distribution where ϕ is the angle measured from the surface normal of the target and **n** is approximately equal to 10. A typical deposition profile resulting from 100,000 pulses (i. e., 23 s of laser-on time) at 5×10^8 W/cm² is illustrated in Fig. 2(a). Integration of the volume defined by this curve yields a volume deposition rate of approximately 2000 μm·cm²/h. The variation of peak deposition rate with fluence is illustrated in Fig. 2(b). The rate increases with fluence up to 5×10^8 W/cm². Beyond this level, it is believed that photon interaction with the ablation plume begins to mask the target and the effective coating rate decreases. No differences in the EELS or Raman spectra of material grown under the range of fluences investigated could be detected.

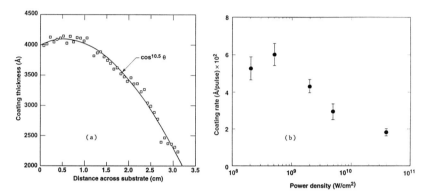

Figure 2. (a) Spatial distribution of DLC grown on a Si(100) substrate after 100,000 laser shots at 190 W average power and 5×10^8 W/cm^2. (b) Variation of peak coating rate with fluence.

In agreement with other investigations at similar laser fluence (but shorter wavelength [248 nm] and higher pulse energy [250 mJ]),[9] the emission spectra obtained from the laser plume viewed parallel to the target surface were found to be dominated by the C_2 vibrational bands.[10] The $\Delta v = + 1$ spectrum of the C_2 Swan bands at 473 nm is shown in Fig. 3. The intensity of the **1,0** line the was seen to roughly correlate with the coating rate measured at the substrate surface. Low and high resolution atomic force microscope images of typical DLC films are shown in Fig. 4. The characteristic grain size is on the order of 150 nm, and the rms surface roughness is less than 10 nm. Although large areas showing no macroscopic inclusions were found, typical 40 μm by 40 μm areas contained 2 - 8 macro particles with an average height of 200 ± 50 nm and width of 50 ± 10 nm.

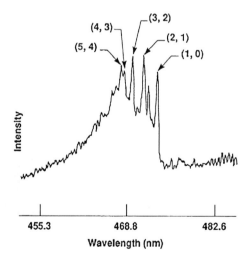

Figure 3. Optical Emission from the C_2 Swan bands during ablation at 5×10^8 W/cm^2.

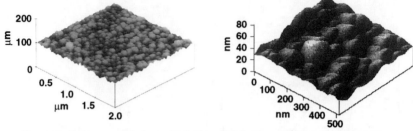

Figure 4. Surface morphology of DLC by AFM at low and high magnification.

A noteworthy companion to most violent PLD processes is the unwanted generation of large particles and their incorporation in the condensed films. It is believed that deep target penetration at the high powers typical of most PLD processes results in macroparticle generation by explosive removal of material around the periphery of the laser spot.[11] By employing relatively low energies per pulse (on the order of 25 mJ), roughly 0.1 μm/pulse (10^{15} atoms) depth was removed from the graphite target surface at maximum deposition rates. Approximately 1.5×10^{-6} coulombs were collected on the biased target per laser pulse, corresponding to a fraction ionized of 1 %. If the target speed is adjusted so that no more than 2 μm of graphite is removed from the target prior to exposing fresh material, the generation of large particles appears to be suppressed. This observation is based on the presence of large incandescent particles in the field of view parallel to the target surface at dwell times that allow drilling to greater depths, and, conversely, their absence at shorter dwell times. Inspections of target surfaces by AFM and optical microscopy provide additional support for this hypothesis.

Examination of DLC by forward recoil spectroscopy (FRS) of hydrogen by a 2.9 MeV He$^+$ beam revealed that the bulk of the material contained little, if any, hydrogen. Monolayer quantities of hydrogen (Fig. 5) were detected at the Si:DLC and DLC:vacuum interfaces. It has been suggested that the hardness of DLC films is inversely proportional to their hydrogen content.[12] Hardness of DLC from this study was characterized by a nanoindentation technique utilizing a diamond tipped AFM probe and found to be approximately 60 GPa. An AFM line trace through an indentation made with an applied force of 1.2×10^{-3} N is shown in Fig. 6.

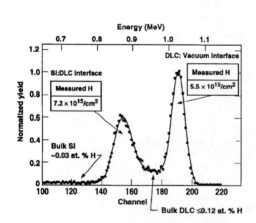

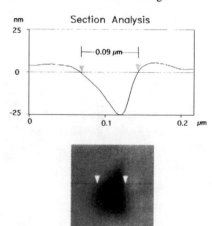

Figure 5. Forward recoil spectroscopy reveals presence of monolayers of H at DLC interfaces.

Figure 6. Nanoindentations in DLC show hardness to be 60 Gpa.

124

CONCLUSIONS

Deposition of DLC with visible light from high repetition rate copper vapor lasers has been demonstrated to be an efficient and effective method for coating specimens. DLC produced by this process at fluences of approximately 5×10^8 W/cm^2 are smooth (10 nm rms roughness), hard (60 GPa) and economical to apply. Coating rates of 2000 μm·cm^2/h at 190 W average laser power set the stage for scaling this technology up to a practical industrial coating tool with the 500 W CVL sources developed at LLNL.

ACKNOWLEDGMENTS

The authors wish to thank L. J. Bernardez of Sandia National Laboratory, Livermore for the Raman spectra and M. A. Schildbach, and R. G. Musket of LLNL for the EELS and forward recoil spectra, respectively. In addition, they wish to thank E. P. Dragon and E. J. Fehring of LLNL for the excellent technical support.

This work was performed under the auspices of the U.S. Department of Energy by the Lawrence Livermore National Laboratory under Contract W-7405-ENG-48.

REFERENCES

1   T. Sato, S. Furuno, S. Iguchi, and M. Hanabusa, Appl. Phys. A **45**, 335 (1988).

2. J. Krishnaswamy, A. Rengan, J. Narayan, K. Vedam, and C. J. McHargue, Appl. Phys. Lett **54**, 2455 (1989).

3. P. T. Murray and D. T. Peeler, Appl. Surface Sci. **69**, 225 (1993).

4. P. T. Murray and D. T. Peeler, J. Elec. Mater., **23**, 855 (1994).

5. C. B. Collins, F. Davanloo, E. M. Juengerman, W. R. Osborn, and D. R. Jander, Appl. Phys. Lett. **54**, 216 (1989).

6. J. Robertson, Adv. in Phys. **4**, 317 (1986).

7. N. Savvides, Mater. Sci. Forum, **52&53**, 407 (1989).

8. H. -C. Tsai and D. B. Bogey, J. Vac. Sci. Technol. A **5**, 3787 (1987).

9. X. Chen, J. Mazumder, and A. Purohit, Appl. Phys. **52**, 328 (1991).

10. L. V. Berzins, Lawrence Livermore National Laboratory (private communication).

11. C. D. Boley and J. T. Early, Lawrence Livermore National Laboratory, UCRL-JC-117078 (1994).

12. F. Jansen, M. Machonkin, S. Kaplan, and S. Hark, J. Vac. Sci. Technol. A **3**, 605 (1985).

MODELING AND SIMULATION OF PULSED LASER ANNEALING AND ABLATION OF SOLID MATERIALS

C. L. Liu and R. F. Wood
Solid State Division, Oak Ridge National Laboratory
Oak Ridge, TN 37831-6032

ABSTRACT

Several different aspects of laser annealing and laser ablation have indicated the need for two-dimensional (2-D) modeling of heat transfer and phase-change effects. We have in mind particularly the laser-induced formation and propagation of buried liquid layers for the case of a-Si on a crystalline silicon substrate, questions related to the early stages of the laser ablation of insulators such as MgO where it is believed that the absorption of the laser radiation occurs at localized but extended regions of high concentrations of defects, and the ejection of particulate material during laser ablation. To deal with these phenomena, a 1-D computational model originally developed for laser annealing has been extended to two-dimensions. The 2-D modeling examines the heat flow and phase changes associated with localized heat sources embedded in a planar material. Concepts such as the state diagram and state array used in the 1-D work have been extended to 2-D and refined. The 2-D program has been rewritten for massively parallel machines such as the Intel Paragons in ORNL's Center for Computational Sciences, thus allowing larger and more accurate calculations for complex systems to be carried out in reasonable times.

I. Introduction

Pulsed laser annealing as a major area in materials processing began about twenty years ago [1]. More recently, interest in pulsed-laser ablation for thin film deposition has grown rapidly [2]. Theoretical treatment and computer modeling of the two processes almost invariably have been carried out in one dimension. However, certain types of phenomena clearly demonstrated the inadequacies of a purely 1-D approach. For example, after the laser annealing of an a-Si layer on a crystalline silicon (c-Si) substrate, transmission electron microscopy [3] reveals microstructure, suggestive of rapid outward growth from point nucleation sites, which clearly can not be explained by a purely 1D approach. In the laser ablation of insulators [4] there is considerable evidence that absorption by isolated, randomly scattered near-surface defects plays an important role. Consequently, there is a need for the development of 2D models to be used in explaining such features. In previous work, a one-dimensional model and a computer program (called LASER8) were developed to deal with various pulsed-laser annealing phenomena [5,6]. This program has now been incorporated into a larger laser ablation [7] computer code. In this paper, we describe recent work in which this 1D approach has been extended to two dimensions. A much more detailed account is given in References 8 and 9.

Crystalline and amorphous silicon have become frequently used model systems for studies of a variety of phase changes under nonequilibrium conditions induced by nanosecond laser pulses. One of the most interesting phenomena is "explosive crystallization" during which the laser pulse may produce a buried liquid layer that propagates within the a-Si. Formation and propagation of the layer is due to the unusual properties, briefly reviewed here, of a-Si compared to c-Si. The melting temperature of a-Si, T_a, is ~225 K lower than that of c-Si (T_c=1685 K) and the thermal conductivity, K_a, is also low, with a temperature-averaged value (room temperature to T_a of 0.01-0.02 W/cm K, between one and two orders of magnitude smaller than the T-dependent conductivity of c-Si, K_c. In addition, the latent heat of a-Si, L_a, is only about 0.73 L_c [10], where L_c is the heat of fusion of c-Si. On melting, Si becomes metallic with the reflectivity R, at λ = 633 nm increasing by 75% and the electrical conductivity by a factor of 20. Therefore, nanosecond

Mat. Res. Soc. Symp. Proc. Vol. 388 © 1995 Materials Research Society

pulsed-laser melting of a-Si at an energy density (Eℓ) just above the threshold for melting results in a highly undercooled liquid whose formation and subsequent behavior can be studied directly by time-resolved electrical [11,12] and optical [13] measurements.

II. Two-dimensional modeling of pulsed-laser interactions with solid materials

We divide the sample into contiguous, semi-infinite hexagons as illustrated in Fig. 1a. An event is assumed to occur at the center of every hexagon more or less simultaneously across the face of the sample. We have in mind specifically here a nucleation event, but it could also be preferential absorption of the laser pulse at defect sites, positions of macromolecules, etc. Under these conditions, the radial outflow of heat from a given hexagon is exactly balanced by the inflow of heat from the neighboring hexagons. Hexagonal boundary conditions, being difficult to deal with, are replaced as follows. Each hexagon is replaced by an inscribed cylinder and attention is focussed on a single cylinder with the boundary condition that the temperature gradient at the outer wall is zero, as indicated in Fig.1b. This is not expected to introduce significant errors since the amount of material in the interstices is small and is still treated approximately. In applications to nucleation phenomena, we may want to simulate the initial laterally uniform melting of the surface which is then followed by nucleation and growth at sites more or less uniformly distributed across the sample, i.e., at the centers of the hexagons or cylinders. Accordingly, the 2D program was set up so that 1D calculations can be carried out until nucleation of a particular phase has occurred and subsequently 2D calculations are initialized to treat the growth of the newly formed phase in both the x and r directions. The finite difference grid is taken to be layers of certain thickness in the x direction (typically $\Delta x = 100$ Å) and rings in the r-direction ($\Delta r = 100 - 200$ Å).

The 2-D heat conduction equation with a source term is given in cylindrical coordinates by

$$\frac{\partial}{\partial t}(\rho h) = \frac{\partial}{\partial x}\left[K_x\frac{\partial T}{\partial x}\right] + \frac{1}{r}\frac{\partial}{\partial r}\left[K_r r\frac{\partial T}{\partial r}\right] + S(x,r,t) \qquad (1)$$

where we have, as in the 1-D case, used Rose's enthalpy formulation [14] of the heat flow problem. Boundary conditions in the x direction are given by

$$\left.\frac{\partial T}{\partial x}\right|_{x=o} = 0 \quad ; \quad T_\infty = T_{in} \qquad (2)$$

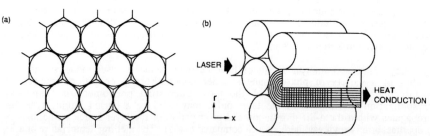

Fig.1 Geometry for 2D calculations showing (a) the hexagons with inscribed cylinders and (b) the FD cells in the x and r directions for one cylinder. After a certain number of cells of constant size in the x-direction is used, additional cells, each successively doubled in size, are added as needed to satisfy the back boundary condition.

and in the r direction by

$$\left.\frac{\partial T}{\partial r}\right|_{r=o} = 0, \quad \left.\frac{\partial T}{\partial r}\right|_{r=r_{max}} = 0 \tag{3}$$

$S(x, r, t)$ is the source term with the form

$$S(x, r, t) = (1 - R) \, P(t) \, \alpha \, \exp(-\alpha x), \tag{4}$$

in which R is the reflectivity of the sample, assumed to be a function of the temperature and phase of the surface, $P(t)$ is the time-dependent intensity of the laser pulse, and α is the absorption coefficient. Discretization of Eq. (1) was carried out using the classical forward time difference scheme. This approach gives an explicit method for updating the enthalpies from time step n to n+1. Further details of the implementation of the discretization of eq.(1) can be found in References 8 and 9.

The LASER8 program incorporates a state array concept which allows a wide variety of physical effects to be simulated as an adjunct to the heat flow and phase change calculations. The state array essentially specifies a number of conditions which the material in a given finite-difference (FD) cell and its neighboring cells must satisfy in order for a phase transition and/or a change in microstructure to occur. Through the introduction of "timers" in a generalized sense these conditions may be made time-dependent. For example, it may be required that the conditions in and around a given FD cell persist for some period of time before a nucleation event can occur. In such a case, we can simulate the nucleation kinetics on a gross scale without going to the level of a molecular dynamics calculation. Space limitation prevents us from discussing this concept here but it is considered at length in References 5 and 6.

II. Illustrative applications

(a) Modeling of Nucleation and Propagation of Buried Liquid Layers in a-Si

We consider a sample composed of an a-Si layer of 415 nm on a c-Si substrate. A 40.0 ns (FWHM) KrF excimer laser pulse with an energy fluence of 0.25 J/cm^2 is incident on the sample. At the KrF wavelength (248 nm), the reflectivity of c-Si and a-Si is 0.58 and that of liquid silicon (L-Si) 0.70. The absorption coefficient was kept constant at $\alpha = 1.8 \times 10^6$ cm^{-1} because of the short wavelength. The temporal and spatial evolution of the various phases is shown in Fig. 2.

As seen in Fig. 2(a), at 27 ns fine-grain silicon (FG-Si) begins to nucleate within the liquid Si in the central surface cell of the cylinder. Before this, a layer of supercooled liquid silicon (L-Si on the figures) was formed on the a-Si layer. The newly solidified material serves as a seed for further growth and the FG-Si phase continues to grow in both x and r directions, as shown in panel (b). However, the growth rates in the x and r directions are not equal. The velocities in the r direction are 12.0 m/s and 13.6 m/s, while in the x direction they are significantly slower, i.e., 8.0 m/s and 8.6 m/s for t=38 ns and t=46 ns, respectively, as shown by the numbers in panels (b) and (c). This phenomenon can be understood as follows. Nucleation of FG-Si releases latent heat into the surrounding liquid and raises its temperature and the temperature of the newly solidified material. Consequently, a temperature or enthalpy gradient is established across the interface. The thermal conductivity of L-Si is much higher than that of a-Si and so heat generated from nucleation and growth of FG-Si is preferentially conducted away in the liquid rather than through the a-Si layer. Thus, the anisotropic growth is initiated.

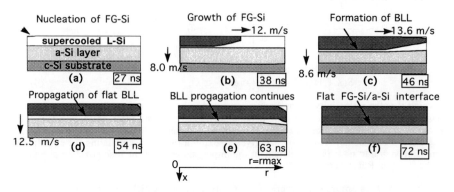

Fig.2 Nucleation and growth of a Si flake showing formation and propagation of the buried liquid layer (BLL), calculated flake growth, and BLL propagation velocities as a function of time.

The fast growth of FG-Si in the r direction reaches a maximum radius at t=54 ns. A small amount of liquid is still left near the surface, and a supercooled liquid layer has been established between the growing FG-Si and the a-Si layer. This buried liquid layer (BLL) is almost flat and propagating at a velocity of 12.5 m/s. The calculated results are consistent with experimental observations [3] in which the BLL was observed to be quite flattened, with no morphological instabilities in the x-direction, to have a thickness of about 10 - 30 nm, and to be moving at 10 m/s. The BLL continues to propagate at t=63 ns as does the growth of FG-Si. Finally, at t=72 ns, the supercooled liquid is exhausted and the whole sample is solidified.

Formation and propagation of the buried liquid layers are related to explosive crystallization [15]. In our case, the FG-Si is grown at the expense of the liquid contained in the buried layer. The latent heat released during this process raises the temperature of the solid phase and the surrounding liquid toward T_c. While crystallization occurs at the back of the buried liquid layer (i.e., the interface between the BLL and FG-Si), previously unmelted a-Si beneath begins to melt since $T > T_a$, resulting in continuous formation of a thin, nearly self-propagating molten layer. The forward motion of the buried liquid layer is driven in part by the difference in the heat of fusion between c-Si and a-Si, $L_c - L_a$, which is positive, and in part by the absorbed laser energy. This means that there is more than enough heat to melt the a-Si beneath. The excess heat will drive the formation and propagation of the buried liquid layer toward the c-Si substrate at an abnormally high velocity, resulting in the explosive crystallization phenomenon. The above simulation results are consistent with experimental observations [16,17] that explosive crystallization is mediated by a BLL and with the previous 1D modeling [5]. A more detailed study using the 2D LASER8 on related subjects will be published in a forthcoming paper [9].

(b) Melting Due to Absorption by Localized Defects

This second illustrative example was intended to roughly simulate the early stages of the laser ablation of ionic crystals such as MgO where absorption is due to localized but extended defects (as opposed to point defects). Still using the geometry of Fig. 2, the absorption process was changed so that t he laser radiation was absorbed only in a region of radius R_a, the radial extent of the defects, lying within the first few radial FD cells. The amount of energy absorbed is proportional to the area so that it scales as $(R_a/R_m)^2$, where R_m

is the maximum radius of the cylinder. Since a nucleation process is not involved, the 2D calculations were initiated at the very beginning of the laser pulse. The radial boundary conditions were kept the same as in the preceding example, to approximate a more or less uniform lateral distribution of the highly absorbing extended defects. Other parameters were chosen to correspond roughly to those of MgO.

Two sets of results are given in Fig. 3 for 10 and 40 ns and pulses of 3 and 5 J/cm^2 as indicated. Because of the low thermal conductivity of MgO, the surface melts very quickly. The diameter of the molten region at 40 ns is ~0.28 μm for the 3 J/cm^2 pulse and ~1.0 μm for the 5 J/cm^2 pulse while the depths are less than half as great. Whether or not MgO ever truly vaporizes seems to be uncertain but apparently it does sublime freely even before it begins to melt. In the laser annealing of GaAs [18,19] it was found that small liquid droplets of Ga formed as gaseous As evolved from the surface of the sample. Similar effects were found in ZnO [19,20] where the evolution of oxygen left the surface covered with a thin layer of Zn after a single pulse from a KrF laser. The coverage was so extensive that good electrical contacts could be soldered on the irradiated spots. In this regard, the results for GaAs and ZnO are so similar to those for MgO that a common mechanism is likely to be involved, namely, the preferential evaporation of the negative ions. However, in ZnO the laser pulse generates a maze of microcracks and this seems to occur in MgO as well, whereas it does not in Si, Ge, and GaAs. Thus, laser ablation of many compounds may not proceed by melting followed by vaporization as is believed to occur in elemental semiconductors and most metals.

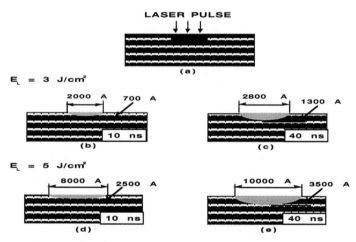

Fig.3 Early stages of the laser ablation of materials like MgO. Laser energy is absorbed only in the central cells where the defects are located, as indicated by the arrows in panel (a).

Because the 2-D calculations described here become computationally intensive, the 2-D LASER8 serial program has been rewritten by one of us (CLL) for massively parallel computers such as ORNL's Intel Paragons. Space decomposition was used for distributing the computational load across the 2-D dimensions of the sample. Each processor takes a certain number of finite difference (FD) cells and calculated all the variables of interest. Communication among the processors is carried out at each time step. The parallel program allows more accurate calculations for complex systems to be carried out in reasonable times.

ACKNOWLEDGMENTS

This research was sponsored in part by the Division of Materials Science, U.S. Department of Energy under contract No. DE-AC05-84OR21400 with Martin Marietta Energy Systems, Inc. This research was also supported in part by an appointment through the Oak Ridge National Laboratory Postdoctoral Research Associates Program administered jointly by the Oak Ridge National Laboratory and the Oak Ridge Institute for Science and Education.

REFERENCES

1.See Laser and Electron beam Processing of Materials, ed.by C. W. White and P. S. Peercy, Academic Press, New York (1980); Laser Annealing of Semiconductors, ed. by J. M. Poate and J. W. Mayer, Academic Press, New York (1982); Pulsed-Laser Processing of Semiconductors, Vol. 23 in the series Semiconductors and Semimetals, ed. by R. F. Wood, C. W. White, and R. T. Young, Academic Press, New York (1984); Martin von Allmen, p.34--39 in Laser-Beam Interactions with Materials, Springer Verlag, Berlin (1987).
2. See Laser Ablation: Mechanisms and Applications, ed. by John C. Miller and R. F. Haglund, Jr., Springer Lecture Notes in Physics 389, Berlin-Heidelberg, Springer (1991); Laser Ablation Mechanisms and Applications-II, ed. by J. C. Miller and D. B. Geohegan, AIP Conference Proceedings 288, New York (1993).
3. D. H. Lowndes, S. J. Pennycook, G. E. Jellison, Jr., S. P. Withrow, and D. N. Mashbrun, J. Mater. Res. 2, 648 (1987).
4. J. T. Dickinson, S. C. Langford, L. C. Jensen, P. A. Eschbach, L. R. Pederson, and D. J. Mashburn, J. Appl. Phys. 68, 1831 (1990); R. L. Webb, S. C. Langford, L. C. Jensen, and J. T. Dickinson, Mat. Res. Soc. Symp. Proc. 236, 21 (1992).
5. R. F. Wood and G. A. Geist, Phys. Rev. Lett. 57, 873 (1986).
6. R. F. Wood and G. A. Geist, Phys. Rev. B34, 2606 (1986) and references therein.
7. C. L. Liu, J. N. Leboeuf, R. F. Wood, D. B. Geohegan, J. M. Donato, K. R. Chen, and A. A. Puretzky, in Proceedings of an Engineering Foundation Conference on Materials Processing and Advanced Application of Lasers, Palm Coast, Florida, May 1–6, 1994.
8. R. F. Wood, Al Geist, and C. L. Liu, submitted to Phys. Rev. B.
9. C. L. Liu, R. F. Wood, and D. H. Lowndes, to be submitted to Phys. Rev. B.
10. E. P. Donovan, F. Spaepen, D. Turnbull, J. M. Poate, and C. C. Jacobson, Appl. Phys. Lett. 42, 698 (1983).
11. G. J. Galvin, M. O. Thompson, J. W. Mayer, R. B. Hammond, N. Paulter, and P. S. Peercy, Phys. Rev. Lett. 48, 33(1982).
12. D. H. Lowndes, R. F. Wood, and J. Narayan, Phys. Rev. Lett. 52, 561 (1984), and references therein.
13. M. von Allmen, Mater. Res. Soc. Symp. Proc. 13, 691 (1983), has also applied an enthalpy equation to laser quenching problems.
14. M. Rose, Math. Comput. 14, 249 (1960).
15. Explosive crystallization, first reported in a-Sb films during the last century [G. Gore, Philos. Mag. 9, 73 (1855)], occurs in a-Si and Ge irradiated by cw lasers [see, e.g., G. Auvert et al, Appl. Phys. Lett. 39, 724 (1981)].
16. G. H. Gilmer and H. J. Leamy, p. 227 in Laser and Electron Beam Processing of Materials, ed. by C. W. White and P. S. Peercy, Academic Press, New York, 1980.
17. H. J. Leamy et al, Appl. Phys. Lett. 38, 137 (1981).
18. D. H. Lowndes, J. W. Cleland, J. Fletcher, J. Narayan, R. D. Westbrook, R. F. Wood, W. H. Christie, and R. E. Eby, in Proceedings of the Fifteenth Photovoltaic Specialists Conference, p. 45, Institute of Electrical and Electronics Engineers, New York (1981).
19. R. F. Wood and D. H. Lowndes, Cryst. Latt. Def. and Amorph. Mat. 12, 475 (1985).
20. D. H. Lowndes et al, in Technical Digest of the 1984 Conference on Lasers and Electro-Optics, p.220, Optical Society of America.

VAPOR BREAKDOWN DURING ABLATION BY NANOSECOND LASER PULSES

C. L. Liu, J. N. Leboeuf,* R. F. Wood, D. B. Geohegan,
J. M. Donato,** K. R. Chen,* and A. A. Puretzky
Solid State Division, Oak Ridge National Laboratory
Oak Ridge, TN 37831-6032

*Fusion Energy Division, ORNL
**Engineering Physics and Mathematics Division, ORNL

ABSTRACT

Plasma generation through vapor breakdown during ablation of a Si target by nanosecond KrF laser pulses is modeled using 0-dimensional rate equations. Although there is some previous work on vapor breakdown by microsecond laser pulses, there have been no successful attempts reported on the same subject by nanosecond laser pulses. This work intends to fill the gap. A kinetic model is developed considering the following factors: (1) temperatures of both electrons and heavy-body particles (ions, neutrals, and excited states of neutrals), (2) absorption mechanisms of the laser energy such as inverse bremsstrahlung (IB) processes and photoionization of excited states, (3) ionization acceleration mechanisms that include electron-impact excitation of ground state neutrals, electron-impact ionization of excited states of neutrals, photoionization of excited states of neutrals, and all necessary reverse processes. The rates of various processes considered are calculated using a second order predictor-corrector numerical scheme. The rate equations are solved for five quantities, namely, densities of electrons, neutrals, and excited states of neutrals, and the temperatures of electrons and heavy-body particles. The total breakdown times (sum of evaporation time and vapor breakdown time) at different energy fluences are then calculated. The results are compared with experimental observations of Si target ablation using a KrF laser.

I. Introduction

One of the most promising techniques for laser materials processing is pulsed laser deposition (PLD) of thin films. The advantages of PLD compared to other techniques include, novel epitaxial and low temperature growth of homogeneous and heterogeneous films by utilizing energetic species, stoichiometric ablation of constituent species of the target, and growth of metastable phases layer-by-layer under nonequilibrium ablation conditions. While experimentalists are trying to find optimal conditions for thin-film growth by PLD, a systematic study in modeling of various physical processes during deposition is needed to assist their effort. For computational modeling of the complicated processes such as occur during PLD, one faces the challenge not only to better understand the fundamentals of the processes, but also to utilize the most appropriate computational techniques. Modeling of vapor breakdown due to the interactions between vapor and incoming laser beam during ablation is such a challenge. Understanding this phenomenon is extremely important in assessing the final state of the plume after the pulse and subsequent plume expansion and transport toward the substrate. There have been no successful attempts reported to model vapor breakdown under irradiation with nanosecond laser pulses, although some work has been done for microsecond laser pulses [1]. In this paper, we present preliminary results from theoretical modeling of vapor breakdown leading to plasma generation (fully ionized gas) through the interactions between the evaporated gas (plume) and the incoming laser beam. A more detailed description of the model and the results will be published later [2].

II. Theory for Vapor Breakdown

2.1 Physical model for vapor breakdown

A reasonable physical model can be set up for the early period of the ablation. During the initial ablation, generated vapor experiences a slow 1-D expansion with a velocity that can be taken as constant (V_o). The vapor can be assumed to be homogeneous during this early stage. In the vapor, ions, electrons, excited states of neutrals, and ground states of neutrals all exist. Evaporation continues to supply these particles $((N_e)_o, (N_o)_o, (N^*)_o)$ into the vapor and thus contributes to the total energy in the vapor. In the meantime, electrons in the vapor absorb energy from photons in the laser beam and exchange this energy with heavy-body particles through elastic collisions, therefore two temperatures, i.e., the electron temperature (T_e) and the heavy-body particle temperature (T_h), are necessary. Some neutrals can be ionized via excited states either by electron-impact or photoionization. A schematic of this physical model is shown in Fig.1 along with a KrF laser pulse used in our model.

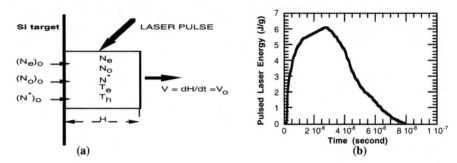

Fig. 1 (a) Schematic of the physical model used for modeling of vapor breakdown (b) laser pulse used in the modeling calculations. H is the thickness of the vapor column perpendicular to the target surface.

2.2 Mechanisms of laser energy absorption

Two absorption mechanisms of laser energy are considered in our model, i.e., inverse bremsstrahlung absorption through electron-atom and electron-ion interactions. The corresponding absorption coefficients, denoted respectively by α_{e-a} and α_{e-i}, are given by [3]

$$\alpha_{e-a} = 2.90\text{x}10^{19} N_e N_o T_e^{1.5} G_{e-a} / v^3 \tag{1}$$

and

$$\alpha_{e-i} = 3.42756\text{x}10^6 N_e^2 G_{e-i} / ((T_h)^{1/2} v^3) , \tag{2}$$

where N_e and N_o are electron and neutral densities, T_e and T_h (in K) are the electron and heavy-body temperatures, G_{e-a} and G_{e-i} are the Gaunt factors for electron-atom IB and electron-ion IB, and v is the frequency of laser pulse. Another absorption mechanism considered is photoionization of an excited state. The cross section σ is obtained according to Zel'dovich and Raizer [4]

$$\sigma = 7.9\text{x}10^{-18} \left(\frac{E_{ion}}{hv}\right)^3 \left(\frac{I_H}{E_{ion}}\right)^{1/2} \tag{3}$$

where E_{ion} (in eV) is the ionization energy of the excited state, $h\nu$ is the photon energy of the laser pulse, and I_H (in eV) is the hydrogen ionization potential. The total energy absorbed (I_{ab}) from the laser pulse of energy (I) is

$$I_{ab} = (\alpha_{e-a} + \alpha_{e-i} + \sigma N^*) I \qquad (4)$$

2.3 Kinetic theory for breakdown

There are two necessary conditions for vapor breakdown, i.e., the presence of "priming" electrons and a mechanism for rapidly accelerating the electron density buildup. One can consider the electron density as an indicator of the ionization level in the plume. In our case, the electron and ion densities are equal since only singly charged Si ions are considered here. There are many sources for "priming" electrons, such as local thermal ionization described by the Saha equation, thermionic emission from the target surface and from heated liquid droplets in the plume, shock heating and ionization, surface defects which are thermally decoupled from the bulk, avalanche and photoionization within the solid target near the surface, etc. Once the "priming" electrons exist, the acceleration mechanism leading to breakdown becomes important. We consider the following forward and reverse processes for electron density acceleration and deceleration [4]:

(1) Electron-impact excitation of ground state neutrals

$$Si + e + E_{ex} = Si^* + e , \qquad (5)$$
where E_{ex} is the excitation energy required for ground state neutrals.

forward rate from ground to excited states, K_{g-e}:

$$K_{g-e} = 6.\times10^{-12} (T_e)^{1/2} (2. + E_{ex} / kT_e) N_e N_o \qquad (5a)$$

backward rate from excited to ground states, K_{e-g}:

$$K_{e-g} = 1.8\times10^{-10} (T_e)^{1/2} (2. + E_{ex} / kT_e) N_e N^* \qquad (5b)$$

(2) Electron-impact ionization of excited states of neutrals

$$Si^* + e + E_{ion} = Si^+ + e + e , \qquad (6)$$

forward rate from excited to ionic states, K_{e-i}:

$$K_{e-i} = 2.2\times10^{-10} (T_h)^{1/2} (\frac{I_H}{E_{ion}})^2 N_e N^* \exp(-E_{ion} / kT_e) \qquad (6a)$$

backward rate from excited to ground states, K_{i-e}:

$$K_{i-e} = 1.\times10^{-25} (T_h)^{-1} (\frac{I_H}{E_{ion}})^2 (N_e)^3 \qquad (6b)$$

(3) Photoionization of excited states of neutrals

$$Si^* + h\nu = Si^+ + e \qquad (7)$$

A schematic of all the kinetic processes considered in the model is shown in Fig.2.

According to Zel'dovich and Raizer [4], we consider heating of heavy-body particles by hot electrons through elastic collisions. The rate of change in electron temperature, τ_e, is

$$\tau_e = \frac{7 \times 10^3 \, (T_e)^{3/2}}{(\ln\Lambda)(N_e + (1/200)(N^* + N_o))} \tag{8}$$

where Λ is the plasma parameter, given by $\Lambda = \dfrac{3(kT_h)^{3/2}}{2(4\pi T_h)^{1/2} \, e^3 \, (N_e)^{1/2}}$

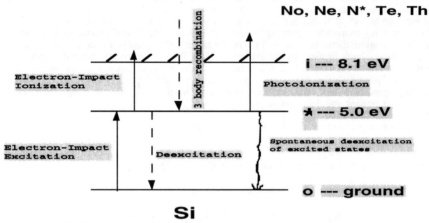

Fig. 2 Schematic of kinetic processes considered in our model.

Similarly the rate of change of the heavy-body temperature, τ_h, is given by

$$\tau_h = \frac{7 \times 10^3 \, (T_e)^{3/2}}{(\ln\Lambda)N_e} \tag{9}$$

III. Numerical calculations for vapor breakdown

From conservation equations, we can write the rate equations for electrons, and for ground and excited states of neutrals as

Electrons:
$$\frac{d}{dt}[N_e] = -N_e \frac{1}{H}\frac{dH}{dt} + (K_{e-i} - K_{i-e}) + \sigma \frac{I_{ab}}{h\nu} N^* + \frac{(N_e)_o}{H} \tag{10}$$

Excited state of neutrals:
$$\frac{d}{dt}[N^*] = -N^* \frac{1}{H}\frac{dH}{dt} + (K_{g-e} - K_{e-g}) + (K_{i-e} - K_{e-i}) - \sigma \frac{I_{ab}}{h\nu} N^* - \frac{N^*}{\tau^*} + \frac{(N^*)_o}{H} \tag{11}$$

Ground state of neutrals:
$$\frac{d}{dt}[N_o] = -N_o \frac{1}{H}\frac{dH}{dt} + (K_{e-g} - K_{g-e}) + \frac{(N_o)_o}{H} \tag{12}$$

where the first terms in eqs. (10,11,12) are due to dilution because of vapor expansion, the last terms are due to resupply of particles from evaporation, and τ^* is the lifetime of excited states. A constant supply of material($(N_e)_o$, $(N_o)_o$, $(N^*)_o$) during ablation is obtained from the Clausius-Clapeyron equation [5]. Conservation equations for the energies are:

Electrons and excited states of neutrals:

$$\frac{d}{dt}\{[N_e(3/2kT_e + \varepsilon_I) + N^* E_{ion}]H\} = (N_e)_o(3/2kT_s + \varepsilon_I) + (N^*)_oE_{ion}$$

$$+ H I_{ab} - 3/2k(T_e - T_h)\frac{N_e}{\tau_e} H \quad (13)$$

Heavy-body particles:

$$\frac{d}{dt}(N_T3/2kT_hH) = (N_T)_o3/2kT_s + 3/2k(T_e - T_h)\frac{N_e}{\tau_e} H \quad (14)$$

where ε_I is the ionization potential of ground states of neutrals, T_s is taken as the evaporation temperature, $N_T = (N_e + N_o + N^*)$, and $(N_T)_o = [(N_e)_o + (N_o)_o + (N^*)_o]$. Substituting eqs.(10,11,12) into eqs.(13,14), we get the rate equations for T_e and T_h:

$$\frac{d}{dt}[T_e] = \sigma\frac{I_{ab}}{hv}\frac{N^*}{N_e}(\frac{hv + E_{ex} - \varepsilon_I}{3/2k} - T_e) + \frac{2}{3kN_e}(K_{e-i} - K_{i-e})(E_{ion} - \varepsilon_I - 3/2kT_e)$$

$$- \frac{2E_{ion}}{3kN_e}(K_{e-g} - K_{g-e}) - \frac{T_e - T_h}{\tau_e} + \frac{2}{3kN_e}(\alpha_{e-a} + \alpha_{e-i})I + \frac{(N_e)_o(T_s - T_h)}{HN_e} \quad (15)$$

$$\frac{d}{dt}[T_h] = \frac{T_e - T_h}{\tau_e}\frac{N_e}{HN_T} + \frac{(N_T)_o(T_s - T_h)}{HN_T} \quad (16)$$

These 0-dimensional rate equations, i.e., eqs.(10,11,12,15,16), are solved for the five quantities of interest using a second-order predictor-corrector algorithm. Starting conditions for the initial distributions of particle densities in the vapor are determined by the Saha equation for N_o and N_e and the Boltzmann distribution for N^* and N_o, i.e.,

Saha $\qquad \frac{(N_e)^2}{N_o} = 2(\frac{2\pi m_e kT_h}{h^2})^{3/2}\frac{g_+}{g_o}\exp(-\varepsilon_I/kT_h) \quad (17)$

Boltzmann $\qquad \frac{N^*}{N_o} = \frac{g_+}{g_o}\exp(-E_{ion}/kT_h) \quad (18)$

where g_+ and g_- are partition function coefficients (both are equal to 6/15 for Si).

IV. Results and discussion

We used a 1-D thermal model (LASER8) [6-9] to calculate the time for a solid to be evaporated under laser irradiation (τ_{vap}) and the kinetic model described here for the time it takes the vapor to be fully ionized (τ_{VB}). Then, the total breakdown time, τ_{TB}, is given by: $\tau_{TB} = \tau_{vap} + \tau_{VB}$. A KrF laser pulse of FWHM=40 ns was used for both models (see Fig.1b). The total breakdown times at two laser energy fluences were calculated.

(1) Laser energy fluence = 6 J/cm^2, $\tau_{vap} = 22$ ns (from the thermal model); $\tau_{VB} = 48$ ns (from the kinetic model); $\tau_{TB} = \tau_{vap} + \tau_{VB} = 22 + 48 = 70$ ns $\approx \tau_{pulse}$

(2) Laser energy fluence = 30 J/cm^2, $\tau_{vap} = 5$ ns; $\tau_{VB} = 16$ ns; $\tau_{TB} = \tau_{vap} + \tau_{VB} = 5 + 16 = 21$ ns $< \tau_{pulse}$

Surprisingly, the results indicated that at an energy fluence of 6 J/cm^2 the total breakdown occurred toward the end of the laser pulse at 70 ns. In this case, one may not see the vapor breakdown visibly by means of light emission. On the other hand, at an energy fluence of 30 J/cm^2, one may readily see the breakdown in the time frame (the total breakdown time ~21 ns) of the laser pulse duration. This is qualitatively consistent with experimental observations.

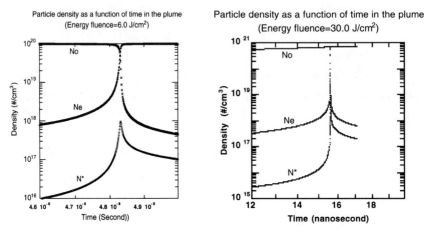

Fig.3 Calculated particle density as a function of time, indicating vapor breakdown times during ablation of a Si target by KrF laser pulses.

ACKNOWLEDGMENTS
This research was sponsored by the Division of Materials Sciences, U.S. Department of Energy under Contract No. DE-AC05-84OR21400 with Martin Marietta Energy Systems, Inc., and in part by an appointment to the Oak Ridge National Laboratory Postdoctoral Research Associates Program administered jointly by the Oak Ridge National Laboratory and the Oak Ridge Institute for Science and Education.

REFERENCES
1. D. I. Rosen, J. Mitteldorf, G. Kothandaraman, A. N. Pirri, and E. R. Pugh, J. Appl. Phys. 53, 3190 (1982).
2. C. L. Liu, J. N. Leboeuf, R. F. Wood, D. B. Geohegan, J. M. Donato, K. R. Chen, and A. A. Puretzky, to be published.
3. Andre Anders, A formulary for plasma physics, Akademie-Verlag Berlin (1990), p.159.
4. Ya. B. Zel'dovich and Yu. P. Raizer, Physics of Shock Waves and High-Temperature Hydrodynamic Phenomena, Academic Press, New York and London, 1966.
5. A. Vertes, P. Juhasz, M. D. Wolf, and R. Gijbels, Scanning Microscopy 2, 1853, 1988.
6. R. F. Wood and G. A. Geist, Phys. Rev. Lett. 57, 873 (1986).
7. R. F. Wood and G. A. Geist, Phys. Rev. B 34, 2606 (1986).
8. C. L. Liu, J. N. Leboeuf, R. F. Wood, D. B. Geohegan, J. M. Donato, K. R. Chen, and A. A. Puretzky, Proceedings of the International Conference on Processing and Advanced Applications of Lasers (ICPAAL), Palm Coast, Florida, May 1-6, 1994.
9. C. L. Liu, J. N. Leboeuf, R. F. Wood, D. B. Geohegan, J. M. Donato, K. R. Chen, and A. A. Puretzky, Mater. Res. Soc. Symp. Proc., Symposium A: Beam-Solid Interactions for Materials Synthesis and Characterization, Nov.28 -Dec.2, 1994, Boston, MA.

MECHANISM OF ArF LASER INDUCED PHOTOLYTIC DEPOSITION OF W FROM WF_6 ON ETCHED Si AND UNETCHED Si

XIAFANG ZHANG, HERBERT J. LEARY, JR. AND SUSAN D. ALLEN
Tulane University, Laser Microfabrication Lab, New Orleans, LA 70118

ABSTRACT

ArF excimer laser-induced photolytic chemical vapor deposition of W on etched Si substrates using tungsten hexafluoride has been studied. Our experimental results show that tungsten film thickness is proportional to the laser irradiation time and fluence, and that the deposition rate initially increases, then decreases with increasing WF_6 pressure. The activation energy obtained from an Arrhenius plot is much less than that for conventional CVD. A deposition mechanism has been proposed which yields results in good agreement with the experimental data. The absorption cross section of WF_6 is determined to be 2.75×10^{-18} cm^2/molecule.

INTRODUCTION

The selective deposition of W by silicon reduction of WF_6 has been used for filling submicrometer contacts in microelectronics manufacturing processes.[1] The most obvious advantage of using an excimer laser is its high output power in the ultraviolet region which allows for photolytic deposition of W films by photo-dissociation of WF_6.[2] Photolytic deposition can be performed with a laser beam aligned parallel or perpendicular to the substrate. In the parallel configuration, absorption occurs exclusively by molecules in the gas phase. In the perpendicular configuration, absorption may occur in the gas phase followed by transport of products to the surface or may occur in an adsorbed layer on the surface. The film deposition only takes place in the laser irradiated area via photochemistry and local heating. The perpendicular configuration was used in the experiments presented in this paper.

ArF excimer laser induced CVD of W by H_2 reduction of WF_6 has been investigated extensively.[2] For non-oxide coated substrates the Si reduction reaction determines the properties of the W/Si interface such as adherence, contact resistance and interdiffusivity. The important reaction mechanisms in the ArF laser photolytic CVD of W by Si reduction of WF_6 are not known at present. The purpose of this research is to study a mechanism of ArF laser induced photolytic deposition of W on Si substrates from WF_6 by correlating the W film thickness and growth conditions with laser fluence, irradiation time, WF_6 pressure and surface temperature.

EXPERIMENTAL

An ArF laser (LAMBDA PHYSIK model LPX 200), which emits a beam of ultraviolet light of wavelength 193 nm passing through an aperture to produce a relatively uniform beam intensity, was focused through a fused quartz window into a chemical vapor deposition cell containing a silicon substrate. One biconvex fused silica lens combined with a cylindrical lens focused the laser beam to an area 2 mm long by 0.4 mm wide. The power of the excimer laser was computer-controlled. We used n-type (111) Si wafers covered with a native oxide (SiO_2) as

139

the substrate. The Si wafers were ultrasonically cleaned 3 times with tetrahydrofuran (THF), isopropyl alcohol (IPA) and acetone to remove all organics from the sample surface, and then rinsed in deionized water to remove the residual solvents. Half of the Si wafer was dipped in dilute hydrofluoric acid (HF:H_2O=1:12) for 2 minutes to partially remove the native oxide from the Si surface, hereafter called "etched Si". The native oxide on the other half of the substrate was left intact, rinsed in alcohol and blown dry to produce "unetched Si". After cleaning, the sample was immediately loaded into the CVD cell that was evacuated to 5×10^{-6} Torr using a turbomolecular pump. WF_6 gas was introduced into the CVD cell at desired pressures, which were monitored by a capacitance-bridge pressure gauge. The cell was then placed on an X-Y translation stage for W deposition at room temperature. The thickness of deposited W films was measured using a Tencor 200 stylus profilometer.

RESULTS AND DISCUSSION

Determination of the laser-induced surface temperature

Since the laser beam was perpendicular to the sample in our experiments, the laser energy is not only absorbed by WF_6 molecules but also heats the surface. Accurate determination of the surface temperature induced by absorption of the laser radiation in the surface layer of the substrate is important for determining the growth mechanism. For a laser beam size much greater than the absorption depth, and for a laser irradiation time much smaller than the thermal diffusion time for the sample thickness and greater than the thermal diffusion time for the optical absorption depth, the peak and average temperature rise of the surface after one pulse by laser heating can be estimated by solving the one dimensional, time-dependent heat-flow equation for surface absorption:[3,4]

$$T(t) = [2\Phi_s(1\text{-}R) / K\pi^{1/2}] (Dt)^{1/2} \qquad (1)$$
$$D = K / \rho C_p \qquad (2)$$

where Φ_s is the peak laser intensity at the surface (W/cm^2), D is the thermal diffusivity (cm^2/s), K is the thermal conductivity (W/cmK), ρ is the mass density (g/cm^3), C_p is the heat capacity (J/g K) and R is the reflectivity of the surface. In our calculation we assumed the parameters in Eq. (1) are temperature independent. K, C_p and R are actually temperature dependent.[5] Therefore, an average value of K in the temperature range of interest was used. The temperature dependence of C_p and R is less than that of K and D in Eq. (1) and can be neglected. In these experiments W atoms initially deposit on a Si surface and after a few monolayers of W have covered the Si surface, W atoms deposit on a W surface. The surface peak temperature and the average temperature of both W and Si during one pulse by laser heating were estimated and listed in Table 1 for a WF_6 pressure of 6.2 Torr. The temperature decay is so fast after the pulse that the average temperature rise between pulses at a repetition rate of 10 Hz is negligible. Table 1 shows that the peak temperature on the silicon surface is higher than on the W surface at the same laser intensity, because the thermal conductivity of W is greater than that of Si at temperatures greater than about 425 K. The real temperature of the W films should be less than the temperature of the Si surfaces predominantly because of the relatively poorer thermal conductivity of the Si underlying the thin W films. Therefore, the actual temperature of our W films should lie between the calculated temperature on the W surfaces and the calculated temperature on the Si surfaces.

Table 1. The values of the estimated temperature for one pulse by eq. (1)

Φ_0 (J/cm^2)	$\nabla T_{Si,peak}$ (°K)	K_{Si} W/cm°K	$\nabla T_{W,peak}$ (°K)	K_W W/cm°K	$\nabla T_{Si,Avg}$ (°K)	$\nabla T_{W,Avg}$ (°K)	Note at 193 nm
0.83	1042.6	0.403	524.4	1.26	711.1	357.6	R_{si}=0.686
0.72	899.0	0.446	452.4	1.28	613.1	308.5	ρ_{Si}=2.33 g/cm^3
0.61	688.8	0.500	380.6	1.30	469.8	259.6	Cp_{Si}=0.702J/g°K
0.41	408.9	0.640	252.5	1.33	278.9	172.2	Rw=0.641
0.35	353.8	0.682	214.7	1.34	241.3	146.4	ρ_w=19.3 g/cm^3
0.22	198.8	0.855	134.5	1.35	135.6	91.7	Cp_w=0.133 J/g°K
0.14	110.1	1.330	116.9	1.38	75.1	79.7	t = 23 ns

Tungsten thickness dependence on growth parameter

The dependence of tungsten film thickness on laser irradiation time was studied for both etched and unetched Si wafers at a WF_6 pressure of 5.5 Torr, a repetition rate of 10 Hz and at a variety of laser energy densities, as shown in Fig. 1. The W deposition rate increases as the laser intensity is increased. The thickness of W films is linearly proportional to laser irradiation time, implying that the overall tungsten deposition reaction in this study is first order with respect to time and is not self-limited in thickness. As shown in Fig. 1, the W film thicknesses on unetched Si are slightly less than on etched Si surfaces for the same laser energy density.

The relationship between the thickness of tungsten films deposited on etched Si and laser energy density was investigated for a WF_6 pressure of 6.2 Torr, a repetition rate of 10 Hz and an irradiation time of 7 minutes. Fig. 2 depicts the results, which show that the thickness of the tungsten films increases linearly with laser intensity, indicating that photolytic deposition of W is linear with laser fluence for this range of experimental parameters. The rate limiting step in ArF excimer laser induced tungsten deposition from WF_6 appears to be a single photon process corresponding to breaking one WF_5 -F bond.

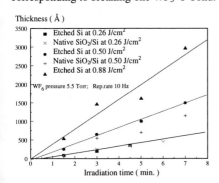

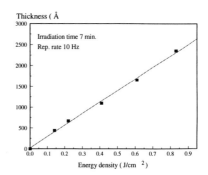

Fig. 1 Thickness of W film as a function of irradiation time at a WF_6 pressure of 5.5 Torr.

Fig. 2 W film thickness on etched Si versus laser intensity at a WF_6 pressure of 6.2 Torr.

The relationship between the deposition rate of tungsten (atoms/cm^2-s) on an etched Si surface as a function of WF_6 pressure is illustrated in Fig. 3 for a laser irradiation time of 7 minutes, a repetition rate of 10 Hz and laser intensities of 0.25 J/cm^2, 0.50 J/cm^2 and 0.88 J/cm^2. The growth rate of W initially increases rapidly with WF$_6$ pressure, peaking at a WF$_6$ pressure of 7.4 Torr, then decrease with pressure. The maximum thickness increases as the laser intensity is increased.

The deposition rates of tungsten films on etched Si as a function of the calculated laser induced surface temperature on a W surface and a Si surface are exhibited in Fig. 4 for a laser irradiation time of 7 minutes and a repetition rate of 10 Hz. The apparent activation energies deduced from the slope of the Arrhenius plots is 11 ± 0.5 kJ/mol and 8 ± 0.5 kJ/mol for the W surface temperature and the Si surface temperature, respectively. Since the real surface temperatures on the W films are between the calculated temperature on the W surface and the calculated temperature on the Si surface, the actual activation energy should lie between 8-11 kJ/mol, which is is much less than that of conventional CVD of W reported by S. L. Lantz (26 kJ/mol)[6] and A. Kuiper (32 kJ/mol).[7] Given the low activation energy this suggests that the deposition rate is dominated by photochemistry in the gas phase.

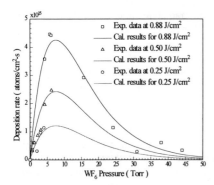

Fig. 3 Deposition rate of W films on etched Si versus WF$_6$ pressure at a laser irradiation time of 7 minutes and a repetition rate of 10 Hz.

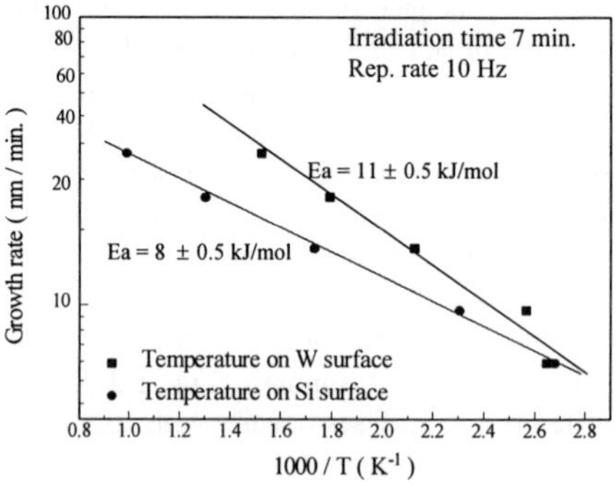

Fig. 4 Arrhenius plot of W film growth rate on etched Si versus surface temperature at a WF$_6$ pressure of 6.2 Torr.

Mechanism of W deposition by ArF laser induced CVD

An analysis of the above experimental results provides evidence for a proposed W deposition mechanism. WF$_6$ gas irradiated by the ArF excimer laser generates F· and WF$_5$· radicals.[2] These radicals travel quickly to the surface, chemically adsorb and react rapidly with surface silicon atoms to produce tungsten atoms under laser heating. The reaction mechanism is given as follows:

$$WF_6(g) + h\nu \Rightarrow WF_5· + F· \qquad\qquad \varphi_3 \qquad (3)$$

$$WF_5· + F· + 3/2Si(s) \Rightarrow W(s) + 1/2SiF_2(g) + SiF_4(g) + 1/2F_2(g) \qquad k_4 \qquad (4)$$

Assuming that Eq. (3) is the rate limiting step, we obtain the W deposition rate from Eq. (4):

$$d[W]/dt = k_4[WF_5\cdot]_s[F\cdot]_s[Si]^{3/2} \qquad (5)$$

where $[WF_5\cdot]_s$ and $[F\cdot]_s$ are concentrations of $WF_5\cdot$ and $F\cdot$ radicals at the surface. As for the Si concentration at the surface, WF_6 gas initially etches native SiO_2 surfaces to produce a Si surface. Kobayashi has found byproduct gases of SiF_4 and WOF_4 during WF_6 gas exposure of the SiO_2 surface at a temperature of 150°C.[8] When W covers the Si surface, Si atoms diffuse through the porous structure of the W films and segregate on top of the W surface.[9] The enhancement of Si diffusion by the porous structure of W films has been observed by Lantz,[6] Kuiper[7] and Green[10] in thermal CVD of W on HF etched Si substrates. Defining [Si] as the concentration of Si atoms at surface and assuming this is constant, and letting $k'_4 = k_4[Si]^{3/2}$, it follows that:

$$d[W]/dt = k'_4[WF_5\cdot]_s[F\cdot]_s \qquad (6)$$

To calculate $F\cdot$ and $WF_5\cdot$ radical concentrations at the surface, we assumed that: 1) half of the $F\cdot$ and $WF_5\cdot$ radicals generated by laser irradiation within a mean free path above the substrate surface do not undergo collision in the gas phase prior to striking the substrate surface, and 2) all $F\cdot$ and $WF_5\cdot$ radicals sticking on the laser irradiated area will participate in reaction (4). Based on the above assumptions, the concentrations of $F\cdot$ and $WF_5\cdot$ radicals on the surface in our experiments can be described by the following formula:

$$[WF_5\cdot]_s[F\cdot]_s = 1/2\lambda\varepsilon[WF_5\cdot][F\cdot] \qquad (7)$$

where $[WF_5\cdot]$ and $[F\cdot]$ are the concentrations of $WF_5\cdot$ and $F\cdot$ radicals in the gas phase, λ is the mean free path of a $WF_5\cdot$ radical (cm), and ε is the sticking coefficient. From Eq. (3), we obtain:

$$[WF_5\cdot][F\cdot] = \varphi_3\alpha I_s[WF_6] \qquad (8)$$

where φ_3 is the quantum yield for Eq. (3), $[WF_6]$ is the concentration of WF_6 molecules in the gas phase, α is the absorption coefficient of WF_6 at 193 nm, and I_s is the laser intensity near the surface. Combining Eqs. (6), (7), and (8), the W deposition rate can be expressed as

$$d[W]/dt = 1/2k'_4\varphi_3\lambda\ \varepsilon\alpha I_s[WF_6] \qquad (9)$$

Setting $[WF_6] = P/(KT)$, $\lambda = KT/(\sqrt{2}\ \pi d^2 P)$, $\alpha = bP$, and $I_s = (1+R)I_o\exp(-bPL)$, Eq. (8) can be expressed as:

$$d[W]/dt = k'_4\varphi_3\varepsilon[(1+R)/(\ 2\sqrt{2}\ \pi d^2)]I_o bP\exp(-bPL\ ) \qquad (10)$$

where R is the surface reflectivity of W at 193 nm, k'_4 is the reaction constant for reaction (4), d is the molecular diameter, P is the partial pressure of WF_6 molecules, I_0 is the incident laser fluence, and b is the molecular absorption coefficient in units of $atm^{-1}\ cm^{-1}$.

For the calculation of deposition rate by Eq. (10), two parameters, $C = k'_4\varphi_3\ \varepsilon$, and b, must be determined. Since the experimental results at a laser intensity of 0.88 J/cm^2 have the most data, a least square fit of this data set with Eq. (10) produced $C = 290 \pm 14$ cm^2/atom-s and $b = 68 \pm 4.7$ $atm^{-1}cm^{-1}$. Using Eq. (10) with these two parameters, we calculate the deposition rates for laser intensities of 0.88 J/cm^2, 0.50 J/cm^2 and 0.25 J/cm^2, which are represented by the solid lines in Fig. 4. The agreement of experimental data with this calculation is remarkably good for all three data sets within experimental error limits. In analyzing the reaction mechanism of W deposition, it is apparent that the activation energy obtained from the Arrhenius plot is the activation energy for Reaction (4).

The absorption cross section σ is determined[11] from σ = 4.06 x10^{-20} b, yielding 2.75x10^{-18} cm^2/molecule. It is a little larger than the value reported by P. Heszler et al for WF$_6$ at 193 nm, σ = 1.7x10^{-18} cm^2/molecule.[2]

CONCLUSION

The W films on both etched Si and unetched Si substrates from WF$_6$ using ArF laser induced CVD have been studied. The activation energy of ArF laser induced CVD of W is much less than thermal CVD. A tungsten deposition mechanism has been proposed. The theoretical calculation of this model fits the experimental results quite well. An absorption cross section of the WF$_6$ obtained from this study agrees with previous measurements within a factor of 1.5.

ACKNOWLEDGMENT

The experimental work was done at the Center for Laser Science & Engineering, University of Iowa. This research was sponsored by NSF (grant number MSS9020825) and DOE/EPSCoR [DOE/LEQSF (1993-95)-01].

REFERENCES

[1] L. F. Tz. Kwakman, W. J. C. Vermeulen, E. H. A. Granneman, and M. L. Hitchman, in Tungsten and other Refractory Metals for VLSL Applications II, edited by Eliot K. Broadbent (Material Research Society, Pittsburgh, PA, 1987) p. 377.

[2] P. Heszler, J. O. Carlsson and P. Mogyorosi, J. Vac. Sci. Technol., A **11**, 2924 (1993).

[3] M. Sparks, J. Appl. Phys. **47**, 837 (1976).

[4] P. K. York, J. G. Eden, J. J. Coleman, G. E. Fernandez and K. J. Beernink, J. Appl. Phys. **66**, 5001 (1989).

[5] K. Imen, J. Y. Lin and S. D. Allan, J. Appl. Phys. **66**, 488 (1989).

[6] S. L. Lantz, W. K. Ford, A. E. Bell and D. Danielson, J. Vac. Sci. Technol. **11**, 911 (1993).

[7] A. E. T. Kuiper, M. F. C. Willemsen and J. E. J. Schmitz, Appl. Surf. Sci. **38**, 338 (1989).

[8] N. Kobayashi, Y. Nakamura, H. Goto and Y. Homma, J. Appl. Phys. **73**, 4637 (1993).

[9] Xiafang Zhang, Ph D thesis, The University of Iowa, 1995.

[10] M. L. Green, Y. S. Ali, T. Boone, B. A. Davidson, L. C. Feldman and S. Nakahara, J. Electrochem. Soc. **134**, 2285 (1987).

[11] Hideo Okabe, in Photochemistry of Small Molecules (John Wiley & Sons, N. Y, 1978) p.373

AMORPHOUS DIAMOND-LIKE CARBON FILM GROWTH BY KrF- AND ArF-EXCIMER LASER PLD: CORRELATION WITH PLUME PROPERTIES

A. A. PURETZKY*, D. B. GEOHEGAN**, G. E. JELLISON Jr.**, AND
M. M. McGIBBON***
*Institute of Spectroscopy, Troitsk, Russia
**Oak Ridge National Laboratory, Oak Ridge, TN
***University of Glasgow, Glasgow, United Kingdom

ABSTRACT

A comparative study of ArF- and KrF-laser deposition of amorphous diamond-like carbon (DLC) films and relevant carbon plasmas has been performed. Spectroscopic ellipsometry and EELS analysis of the DLC films deposited on Si <100> and NaCl substrates were utilized to characterize the high quality ArF- and KrF-laser deposited films (up to 84% of sp^3 bonded carbon in 7 J/cm^2 -ArF-laser DLC film). Gated ICCD imaging, luminescence and ion current probe diagnostics of the carbon plume have revealed quite different properties of carbon plasmas generated by ArF- and KrF- lasers. KrF-laser (6.7 J/cm^2) irradiation produces a less energetic carbon plasma containing larger amounts of C_2 and probably larger clusters compared with ArF-laser irradiation at the same energy fluence. We conclude that the more energetic and highly-atomized ArF-laser carbon plasma results in the better diamond-like properties.

INTRODUCTION

Recently pulsed laser ablation of graphite with ultraviolet excimer laser wavelengths was found to permit the production of high quality amorphous diamond-like carbon (DLC) films.[1,2] Pulsed laser deposition (PLD) of carbon in high vacuum conditions with KrF- (248 nm) and ArF- (193 nm) lasers allows one to produce amorphous diamond like films with smooth (< 20 nm) surface morphology and high degree of diamond like character (large fraction of sp^3 bonded carbon). The important practical feature of the UV excimer PLD process is that much lower laser energy fluences (5–20 J/cm^2) are required compared to visible and IR laser PLD processing.[2]

In this work a comparative study of ArF- (193 nm) and KrF (248 nm) laser deposited diamond-like films has been performed and correlated with gated-ICCD imaging, optical spectroscopic and ion current probe diagnostics of the corresponding laser-generated carbon plasmas. Higher quality DLC films were obtained using ArF-laser irradiation compared to KrF-laser PLD, as determined by EELS and spectroscopic ellipsometry analysis. The plasma diagnostics revealed several key differences between the plasmas generated by these two lasers. ArF-laser (6.7 J/cm^2) generated plasma consisted of a fast propagating ball-shaped component of luminescence containing highly excited C^{++}, C^+, and C species followed by a slower component of cooler, atomized material (C, C^+), followed by a third component which appears to contain C_2, clusters and ultrafine particles. At the same fluence, the KrF-laser does not produce a luminescent plasma ball and appears to produce larger amount of luminescent C_2 and heavier clusters and ultrafine particles at later times after the laser pulse. The higher quality DLC films obtained with 193-nm ArF-laser irradiation appears correlated with both the smaller amount of clusters and ultrafine particles as well as with the higher kinetic energies of carbon ions and atoms generated by this laser.

EXPERIMENTAL

The experimental set up has been described previously.[3,4] It consists of a stainless steel high vacuum chamber (40 cm diameter) equipped with Suprasil quartz windows for optical diagnostics. The chamber is pumped by a turbomolecular pump to a base pressure of 5×10^{-7} Torr. A Questek

145

(Model 2960) excimer laser operating on ArF (22 ns FWHM, 600 mJ) or KrF (28 ns FWHM, 900 mJ) was used. The beam was apertured and focused into the chamber with a spherical lens (500 mm f.l. at 248 nm, 445 mm f.l. at 193nm) to a rectangular beam spot (0.18 cm × 0.11 cm) at an incidence angle of 30° onto 1"-diameter pyrolytic graphite pellets (Specialty Minerals Inc., less than 10 ppm total impurities). The pellets were rotated during the film deposition and plasma plume diagnostics experiments. The maximum laser fluences at the pellet surface were 7 J/cm^2 (ArF) and 20 J/cm^2 (KrF). N-type Si (100) wafers (resistivity 0.2–0.4 ohm-cm) were used as substrates for PLD. The substrates were kept at room temperature and placed at variable distances from the target (d = 4-15 cm). NaCl crystals were also used as substrates for films deposited especially for EELS analysis.

Films were characterized by transmission electron microscopy (STEM), electron energy loss spectroscopy (EELS) and spectroscopic ellipsometry.

Gated imaging was done with an intensified charge-coupled device (ICCD), lens-coupled camera system (Princeton Instruments) with variable gain and gate width (5-ns minimum) and a spectral range from 200-820 nm.

Spectroscopic measurement of the plume luminescence was performed with a 1.33-meter spectrometer (McPherson 209) equipped with an 1800 g/mm holographic grating, an intensified, gated diode array (Princeton Instruments IRY-700RB) and a photomultiplier tube (Hamamatsu R955).

FILMS CHARACTERIZATION

The fraction of sp^3-bonded carbon in the films was estimated using the EELS spectra. The spectra were obtained with a VG HB501 UX dedicated scanning transmission electron microscope (STEM) operated at an accelerating voltage of 100 KeV.[5] The 30 nm-thick DLC films for EELS analysis were deposited on NaCl-substrates using ArF-laser irradiation (11.1 Hz repetition rate, 4660 pulses, 5.6 cm substrate-target distance). The substrate was then dissolved in deionized water and the DLC-film was put on the STEM specimen copper grid. In order to estimate the fraction of sp^3 bonded carbon, the EELS spectrum of crystalline carbon was measured as well. Figure 1(a) and 1(b) shows carbon K-edge spectra of an amorphous diamond-like carbon film (a) and a crystalline graphite film (b) after subtraction of background.

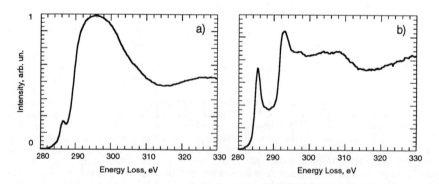

Fig. 1 (a) EELS spectrum of ArF-laser deposited DLC film (NaCl substrate dissolved) in the carbon K-edge core-loss region. (b) EELS spectrum of graphitized carbon, for comparison.

The peak at 285.5 eV [Fig. 1(b)] corresponds to transitions from 1s to π*. The 1s-σ* transition is responsible for the higher energy peaks. In the amorphous DLC film the 1s-π* peak (286.6 eV) is much smaller than that for the graphitized carbon sample due to the small amount of sp^2-bonded carbon in the DLC film. The 1s-σ* peaks (>290 eV) in DLC films are broader

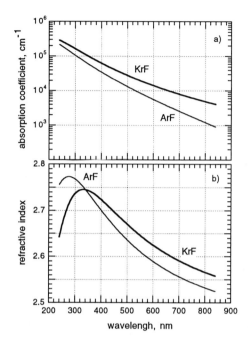

Fig. 2 (a) Absorption coefficient and (b) refractive index versus wavelength for ArF- (7.0 J/cm^2, 11 Hz repetition rate, 3.0x10^4 pulses) and KrF- (16.7 J/cm^2, 11 Hz repetition rate, 2.4 × 10^4 pulses) laser-deposited DLC films. The pellet-substrate distance was 9 cm.

compared to those in crystalline carbon because of the loss of structural order in the amorphous material. The sp^3/(sp^3+sp^2) ratio was estimated by the standard procedure[6] and found to be about 84% for this ArF-laser deposited film (on NaCl-substrate). The low energy spectra also clearly demonstrated the amorphous diamond-like character of the deposited films, i. e., the 26 eV-plasmon peak was shifted to 30 eV and the 6.7 eV- π-π* antibonding transition could not be resolved because of the very low fraction of sp^2 bonded carbon.

The optical properties of both ArF- and KrF-laser deposited films were measured by spectroscopic ellipsometry. These measurements were performed on a two-channel polarization modulation ellipsometer[7] which measures the three associated ellipsometric parameters, N=cos2ψ, S=sin2ψsinΔ and C=sin2ψcosΔ (ψ and Δ are the ellipsometric angles) simultaneously in a single scan. The spectral range investigated was 240–840 nm. The real and imaginary parts of the complex refractive index were obtained by fitting the experimental curves using a five-parameter model developed by Forouhi and Bloomer for amorphous semiconductors[8] (for details see Ref. 9). Two films, one deposited by KrF-laser and one by ArF-laser, were analyzed. For the KrF-laser film (380 nm maximum thickness measured by profilometry), the best fit (χ^2=0.7) to the experimental data was obtained with the following sequence of layers: (1) rough layer (0.74-nm), (2) DLC film (327.1 nm), (3) interface layer (14.5 nm), and finally c-Si. The same layer model applied to the ArF-laser generated film (127 nm maximum thickness measured by profilometry) also gave a good fit to experimental data (χ^2=2.6). The derived thicknesses of the layers are: (1) 0.98 nm, (2) 110.4 nm and (3) 3.9 nm, respectively (see above). Figure 2(a)and 2(b) compare the wavelength dependence of the absorption coefficient and the refractive index for ArF- and KrF-laser deposited DLC films.

The n values measured for this films changed from 2.56 to 2.74 (KrF) and 2.52 to 2.77 (ArF) in the spectral range studied. By comparison, the n value for diamond varies from 2.40 to 2.66 over the same spectral region. The lower absorption coefficient for the ArF-laser DLC (versus the KrF-laser film) corresponds with the trend noted during the entire series of runs, i.e., (a) higher electrical resistance and (b) more transparent diamond-like films are obtained with the ArF-laser compared to those deposited using the KrF-laser. It should be emphasized that the higher-quality ArF-laser DLC of Fig. 2 was achieved with nearly 3-times less laser power compared to that used with the KrF-laser, despite the general trend of higher quality DLC material with increasing laser energy for both wavelengths. The ellipsometry data also indicate substantial differences between the interfacial layer properties for these two cases. The refractive indices of ArF- and KrF-laser produced interfacial layers differ greatly (n=3 and 2.4 at 2 eV, respectively).

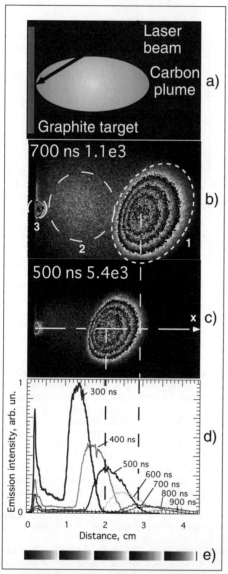

Fig. 3. (a) Irradiation geometry. (b),(c) Gated-ICCD photographs (5 ns gate width) of the total visible carbon plasma emission at (b) $\Delta t = 700$ ns and (c) 500 ns generated by ArF-laser (6.7 J/cm^2). Three distinct spatial regions are indicated in (b) (see text). The 5-grayscale palette is normalized to 1100 counts in (b) and 540 counts in (c). (d) A line-profile of the emission intensity from the irradiated spot along the target normal, **x**, shows the correct relative scaling between the emission intensities at different times (from 300 ns to 900 ns) .

CARBON PLUME DIAGNOSTICS

To understand the difference noted between the ArF- and KrF-laser deposited DLC films, in situ diagnostics of the carbon plasma plume were performed under the respective film growth conditions. Figure 3 show typical gated-ICCD images of the ArF-laser generated plume. Three well-separated regions within the ArF-laser generated carbon plume were discovered.

The first region noted [denoted 1 in Fig. 3 (b)] has the highest emission intensity and a characteristic ellipsoidal-ball shape. This plasma ball propagates with the average velocity v_{pr} about 3.9 cm/μs, estimated from the propagation of the emission maximum. This region expands as well with a characteristic velocity of v_{exp} = 2.6 cm/μs. The second region [2 in Fig. 3(b)] is characteristically round, and it also propagates and expands. The maximum of emission intensity within this region propagates with decreasing average velocity which changes from 1.2 cm/μs (700 ns image) to 0.7 cm/μs (3 μs image, not shown). The third region [3 in Fig.3 (b)] is characterized by relatively intense emission and propagates with a much smaller velocity of only about 0.1 cm/μs.

To clarify what species are responsible for the observed emission, luminescence spectroscopy was performed at different distances from the target surface as well as ion probe current measurements and filtered ICCD-imaging in particular spectral regions corresponding to C$^+$, C, and C$_2$. Figure 4 (a),(b) shows representative luminescence waveforms observed for C$^+$ and C at 2.15 cm (a) and 4.0 cm from the target surface.

A comparison of the total emission measured by imaging with the luminescence profiles [Fig. 4(a)] indicates that predominantly C$^+$ and C species are responsible for the bright emission in the first region. Neutral C luminescence has two characteristic components. As shown in Fig. 4(a), the first occurs during the C^+ luminescence [Fig. 4 (a)] while the second component luminescence (maximum at 0.8 μs) lasts about 2 μs and is correlated with the second imaged region.

At smaller distances from graphite target (d < 1.0 cm) C^{++} luminescence (λ=229.7 nm, from 145 875 cm^{-1}) was also detected. The peak C^{++} emission corresponded with that of C^+ at these distances, and the first component of C emission. One can conclude from these data that the fast propagating ball (Figs. 3(b)–3(d) is a propagating and recombining carbon plasma containing electronically excited species of C^{++}, C^+, and C. Continual recombination results in the appearance of fast neutrals which appears as the first component of the C– luminescence waveform [Fig. 4(a)]. Recombination also maintains the electron temperature within the plasma ball, slowing the drop of T_e during the plasma expansion.

A comparison of the ion current and C^+ emission time profiles in Fig. 4(b) indicate that, in addition to the luminescent C-neutrals, non-emitting ground state ions are also present within the second ball of emission. Region 3 emission was attributed principally to C_2 luminescence from gated-ICCD images of the C_2 Swan bandheads using 515.5 nm and 560.0 nm (10 nm bandwidth, FWHM) filters (not shown).

A comparison between ArF- and KrF-laser generated plasma plumes is given in Fig. 5 at 1μs after the laser pulse. Two notable differences are found. First, the KrF-laser carbon plasma generated at the same fluence (6.7 J/cm^2) does not show a fast-propagating plasma ball [region 1 in Fig. 3 (b)], although the region-2 luminescence is present. To generate the fast plasma ball the KrF-laser fluence needs to be increased to approximately 17 J/cm^2 [see Fig. 5(c)]. However, in this case intense C_2 emission is also observed.

The second main difference between ArF and KrF-generated plasma plumes is the production of many more luminescent C_2 and possibly higher-mass clusters for KrF, as can be seen clearly by comparing the third region luminescence in Figs. 5(a)-5(c).

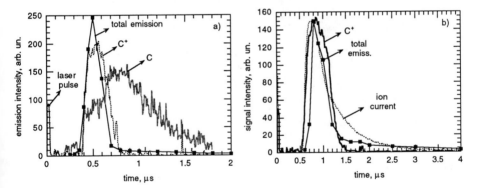

Fig. 4   ArF-laser, (6.7 J/cm^2) generated (a) Luminescence of C^+ (λ=426.7 nm, originating from 168979 cm^{-1} above ground) and C (λ=247.9 nm, 61982 cm^{-1} above ground) and the total emission calculated from the set of ICCD imaging spatial intensity line-profiles (from Fig. 3(d) at d = 2.15 cm from the target surface) . (b) Luminescence, total emission and ion current (peak current 0.3 A) waveforms normalized to their maximum values measured at 4.0 cm from the target surface . The ion probe was biased to -70 V (floating with respect to its shield).

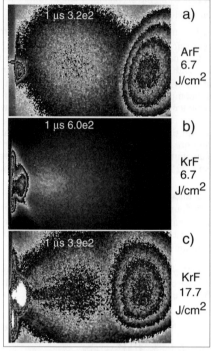

a)
ArF
6.7
J/cm^2

b)
KrF
6.7
J/cm^2

c)
KrF
17.7
J/cm^2

Fig. 5 Gated ICCD images of ArF- (a) (6.7 J/cm^2) and (b,c) KrF- (6.7 J/cm^2 and 17.7 J/cm^2, respectively) laser generated plasmas at 1 µs after the laser pulse. The 5-grayscale palette [Fig.3(e)] is normalized to (a) 320 counts , (b) 600 counts and (c) 390 counts (equal exposures, 50 ns gates).

CONCLUSION

A comparative study of ArF- and KrF-laser deposited amorphous diamond like carbon (DLC) films has been performed in conjunction with relevant plasma diagnostics of the plasma plumes. ArF-laser irradiation produces the highest quality amorphous DLC films, with higher optical transmission, higher index of refraction, better electrical insulation, and higher sp^3: sp^2 ratios than KrF-laser deposited DLC films. ArF-laser deposited films on NaCl substrates were estimated to contain 84% sp^3-bonded carbon. Unfiltered and species-resolved gated-ICCD imaging, spectroscopic, and ion probe current measurements show that the KrF-laser generates substantially larger amounts of ultrafine particles and C$_2$ (and probably larger clusters) compared to that generated with the ArF-laser. In addition, the KrF-laser generated plume at comparable fluences lacks the fast-propagating plasma ball noted using ArF irradiation. The KrF fluence must be increased to at least 17 J/cm^2 to generate the similar fast component. However, spectroscopic ellipsometric analysis shows that the ArF (7 J/cm^2) DLC films are still superior to the high-fluence KrF (17 J/cm^2) DLC films

The authors gratefully acknowledge many helpful discussions with S.J. Pennycook and D. H. Lowndes. This work was supported by the Division of Material Sciences, U.S. Department of Energy under contract DE-AC05-84OR21400 with Martin Marietta Energy Systems, Inc., and NIS/ IPP program sponsored by Division of Defense Programs, U.S. Departement of Energy under contract DP-15.

REFERENCES

1. D. L. Pappas, K. L. Saenger, J. Bruley, W. Krakov, J. J. Cuomo and R. W. Collins, *J. Appl. Phys.* **71**, 5675 (1992).
2. F.Xiong, Y. Y. wang, V. Leppert, R. P. H. Chang, *J. Mater. Res.* **8**, 2265 (1993).
3. D. B. Geohegan, *Thin Solid Films* **220**,138 (1992).
4. D. B. Geohegan, *Appl. Phys. Lett.* **60**, 2732 (1992).
5. N. D. Browning and S. J. Pennycook, *Microbeam Analysis* **2**, 81 (1993).
6. S. D. Berger, D. R. McKenzie, P. J. Martin, *Philosophical Magazine Letters* **57**, 285 (1988).
7. G. E. Jellison, Jr. and F. A. Modine, *Appl. Optics* **29**, 959 (1990).
8. A. R. Forouhi and I. Bloomer, *Phys. Rev. B* **34**, 7018 (1986).
9. G. E. Jellison, Jr., *Thin Solid Films* **234**, 416 (1993).
10. D. F. Edwards and H. R. Philipp, *Handbook of Optical Constants of Solids*, ed. by E. D. Palic, Academic Press, San-Diego, 1985, p.665.

On- AND *Off-*AXIS LARGE-AREA PULSED LASER DEPOSITION

James A. Greer and M. D. Tabat
Electronics Systems Division, Raytheon Company, Lexington, MA.

INTRODUCTION

Over the past few years Pulsed Laser Deposition (PLD) has become a popular technique for the deposition of a wide variety of thin films, and PLD systems are currently found in numerous industrial, government, university, and military laboratories. At present, it is estimated that well over 200 different materials have been deposited by PLD and the list keeps growing. However, even with all the interest in laser deposition the technique has not yet emerged as an industrial process. At the moment, industry still prefers standard thin film growth techniques such as magnetron and ion beam sputtering, chemical vapor deposition, and electron beam evaporation for production applications. These processes have been in use for decades and have demonstrated the ability to deposit films of most materials over large areas with excellent uniformity at reasonable cost and deposition rates. Furthermore, an entire infrastructure has been built up to support these processes including standardization of deposition rate monitors, power sources, target and crucible sizes, etc. On the other hand, laser-deposition is still an emerging technology, and relatively little infrastructure exists to adequately support either research or industrial applications. Since there are several materials which are difficult if not impossible to grow in thin-film form by more conventional techniques, it is expected that as pulsed laser-deposition matures this unique process will take its rightful place on the manufacturing line.

This paper will focus on some of the scale-up issues relating to PLD, and the results presented will demonstrate that large-area laser deposition utilizing laser-beam rastering over large diameter targets is capable of depositing films with uniform properties on substrate sizes which are compatible with today's mainstream semiconductor process lines. Thus, the most relevant issue for the PLD process at present is the elimination of the particulates typically found in laser-deposited films. These particles are characterized by the so-called Normalized Particle Density (NPD) expressed as the number of particles per square centimeter per angstrom of film, and can vary by several orders of magnitude depending on target material, target density, grain size, on-target laser fluence, and other process parameters [1]. Recently *"off-*axis" PLD has become a popular way to deposit films with reported low NPD's [2,3]. Issues relevant to the scaling-up of *off-*axis PLD process will be examined, and the properties of films grown by both *on-* and *off-*axis PLD will be compared.

MATERIALS AND METHODS

A variety of PLD techniques have been used to deposit films over substrates of at least 75 mm (3-inch) in diameter including Off-Set [4,5] and Rotational/Translational PLD [6,7]. The alternative large-area approach, which utilizes laser beam rastering over large diameter targets has been scaled to 125 mm (5-inch) [8] and more recently 150 mm (6-inch) diameter substrates [9]. Laser beam rastering offers advantages over these other "fixed beam" techniques, and has been extended to deposit uniform films of yttria onto 200 mm (8-inch) diameter substrates. Yttria (Y_2O_3) has a variety of uses in thin film form including insulators for transistor gates in

151

electrolumenescent displays, as MIS diodes, and as capacitors [10,11,12,13]. Furthermore, due to the fact that PLD is inherently a low temperature process, yttria can be deposited onto temperature sensitive substrates such as CdTe or HgCdTe [8]. In this case the oxide films act as anti-reflective coatings for IR detector applications. Yttria is also a useful material to use when trying to optimize laser beam raster-scan programs in an attempt to obtain uniform film properties over large substrates. This is due to the colorful interference fringes which are produced on the substrate by thickness variations in the laser deposited film. Yttria has been deposited onto both 150 mm and 200 mm diameter rotating silicon substrates using either a 75 mm (3-inch) or 100 mm (4-inch) diameter offset target as previously described. Thickness profiles of the laser-deposited films were readily obtained in two mutually perpendicular directions ("horizontal" and "vertical") across large diameter substrates using a Nanospec Ellipsometer.

Furthermore, in order to map out compositional profiles of large-area PLD films YBCO was deposited onto 150 mm (6-inch) diameter silicon substrates. In this case a 90 mm (3.5-inch) diameter stoichiometric target was used and depositions were carried out on room temperature substrates in an oxygen ambient. The composition of the film was first evaluated by Rutherford Backscattering (RBS) at three locations. This data then served as a standard for Energy Dispersive X-ray Analysis (EDXA) which was used to map out the film composition in two mutually perpendicular directions across the 150 mm diameter substrate. EDXA data was collected for 700 seconds at each of 50 points across the surface in order to obtain sufficient statistics. While EDXA alone may not be the best technique to determine the precise film composition, it does provide the ability to obtain accurate *variations* in film composition at different substrate locations.

Large-area *off*-axis PLD experiments have also been conducted using laser-beam rastering over large diameter targets. Figure 1 displays a schematic of the *off*-axis set-up used to deposit films over 75 mm (3-inch) diameter silicon substrates. In this case, the center of the substrates were located 75 mm above the ablation target, and off-set from the center of the plume by d = 6 mm as shown in Figures 1a and 1b. A 100 mm diameter target of Y_2O_3 was used for film thickness profiles, and a 90 mm diameter YBCO target was used for compositional profiles.

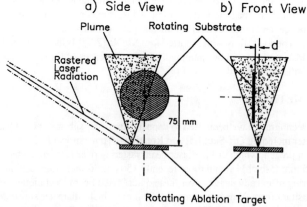

Figure 1. a) Side and b) Front View of the Large-Area *Off*-Axis PLD Apparatus.

The laser beam raster program was adjusted to improve film uniformity, and both the substrate and target were rotated during deposition. Thickness and composition profiles were obtained from yttria films using ellipsometery, and from YBCO films using EDXA, respectively.

SEM micrographs were made of both *on-*, and *off*-axis laser-deposited films and the normalized particle density was determined [1]. The NPD is defined here as the total number of particulates larger than ~ 0.50 μm which were found in a field of view of ~2.6 X 10^{-5}/cm^2 (that area obtained from a 2,000X SEM, taking into account the tilt angle) divided by the nominal film thickness. The NPD is meaningful as long as the film is not more than a few times thicker than the smallest particle size of interest. Furthermore, the intrinsic stress of the laser deposited films was evaluated using a Flexus Thin Film Stress Measurement System.

Deposition parameters for the laser deposited films included 248 nm (KrF) radiation with a pulse length of 18 ns, and an on-target laser fluence of between 1 and 4.0 J/cm^2, 100 Hz repetition rate, and background gas (Ar or O_2) pressures ranging from 0.1 to 100 mTorr. The target to substrate separation for the normal *on*-axis PLD films was set at 12.7 cm (5-inches).

RESULTS AND DISCUSSION

On-axis PLD Results:

Figure 2 shows the normalized film thickness profiles obtained from Y_2O_3 films grown by *on*-axis PLD on 150 mm and 200 mm diameter substrates. In both cases the film thickness was measured in two mutually perpendicular directions (horizontal and vertical) across the complete substrate. The *maximum* variation in film thickness (defined as the difference between the maximum and minimum values obtained across the substrate divided by the average film thickness) for the 7,800 Å thick yttria film deposited onto the 150 mm diameter substrate was found to be 7.1% (± 3.55%). Neglecting the four points located at a radial position of ±74.5 mm (0.5 mm from the substrate edge), the *maximum* variation in film thickness was found to be only 4.6% (± 2.3%). It should be realized that for most semiconductor applications the outer ~3 mm edge of the substrate is not useable. Furthermore, the *maximum* variation in film thickness measured from a 5,700 Å yttria film deposited over a 200 mm diameter silicon substrate was found to be only 6.7% (± 3.4%). Using an on-target laser fluence of ~3.0 J/cm^2 and laser repetition rates of 100 Hz (30 Watts) with an Ar background pressure of 7.5 mTorr, uniform average film growth rates for Y_2O_3 of 0.4 and 0.3 μm/hour, or 1.1 and 0.83 Å/sec were obtained over the 150 mm (182 cm^2) and 200 mm (324 cm^2) diameter substrates, respectively.

RBS data indicated that the nominal composition of the large-area 8,000 Å thick YBCO film was stoichiometric, and furthermore, there was no discernible difference in film composition from center to edge. Figure 3 shows the composition profiles obtained at 25 locations in each of two mutually perpendicular directions from this film grown on the 150 mm diameter silicon substrate. Here the *maximum* variation in composition (defined as the difference between the maximum and minimum atomic concentrations obtained over the whole substrate as determined by EDXA) for the Y, Ba, and Cu cations were 1.4, 1.2, and 0.8 atomic percent (± 0.7, ± 0.6, and ± 0.4 atomic percent), respectively. These values compare to the standard errors calculated by the EDXA software which were ± 1.48, ± 0.17, and ± 0.36 atomic percent for the Y, Ba, and Cu

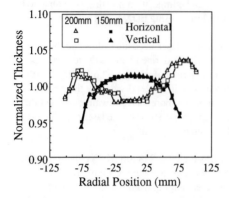

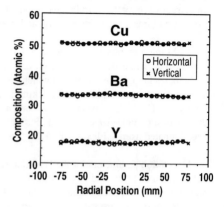

Figure 2. Horizontal and Vertical Thickness Profiles of Y_2O_3 Films Deposited Onto 150 mm (6-inch) and 200 mm (8-inch) Diameter Silicon Substrates Using *On*-Axis PLD.

Figure 3. Horizontal and Vertical Composition Profile of a YBCO Film Deposited on a 150 mm (6-inch) Diameter Silicon Substrate Using *On*-Axis PLD.

species, respectively. Similar profiles were not obtained for 200 mm diameter substrates as a larger YBCO target was not available. The average deposition rate for YBCO obtained over 150 mm diameter substrates (182 cm²) using *on*-axis PLD with only 30 Watts on-target, was found to be 1.45 μm/hour (~4.0 Å/sec). High power industrial lasers with 100 Watt outputs are currently available, and the use of these tools could increase deposition rates by a factor of three. The uniformity of both film thickness and composition displayed in Figures 2 and 3 define the present *state-of-the-art* for the PLD process, and such results would be difficult to reproduce by standard deposition techniques for complex chemical compounds such as YBCO. Furthermore, other emerging deposition technology, such as off-axis magnetron sputtering, has extremely low deposition rates (~0.1 to 0.3 Å/s) in comparison to those already demonstrated by large-area PLD.

Off-axis PLD Results

Figure 4 displays the thickness profile obtained from a 1.55 μm thick yttria film deposited by *off*-axis PLD using an on-target laser fluence of 3.0 J/cm², a repetition rate of 100 Hz, in an Ar gas pressure of 25 mTorr. The *maximum* variation in film thickness was found to be 4.1 % (± 2.05%) with an average growth rate of 1.0 μm/hour over a 46 cm² area. Typical deposition rates for on-axis laser-deposition of Y_2O_3 were about 1.5 μm/hour using similar conditions.

Figure 5 shows the Y, Ba, and Cu composition profiles as determined by EDXA obtained from a YBCO film deposited by *off*-axis PLD onto a room temperature 75 mm diameter silicon substrate using a fluence of 3 J/cm², 100 Hz repetition rate, in 25 mTorr of oxygen. The *maximum* variations in composition for Y, Ba, and Cu elements was found to be 2.09 (± 1.04), 2.45 (± 1.23), and 1.49 (±1.25) atomic percent, respectively. For comparison, the standard statistical errors generated by the EDXA software were found to be ±1.25, ± 0.27, and ± 0.24 atomic percent, for Y, Ba, and Cu elements, respectively. Furthermore, EDXA data was taken

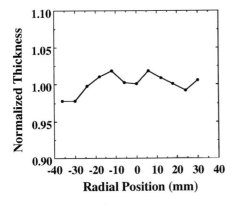

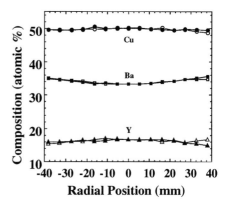

Figure 4. Thickness Profile of a Y_2O_3 film Deposited onto a 75 mm diameter Silicon Substrate Using Large-Area *Off*-Axis PLD.

Figure 5. Composition Profile of a YBCO Film Grown On a 75 mm diameter Silicon Substrate Using Large-Area *Off*-Axis PLD.

at the center of this substrate three different times. The *maximum* variations in the elemental composition measured from this location were found to be only 0.00, 0.03, and 0.03 atomic percent for the Y, Ba, and Cu elements respectively. This latter result demonstrates that EDXA is an excellent tool which provides reproducible data which can be used to accurately determine *variations* in a films composition over large substrate sizes. The average growth rate obtained over the 75 mm diameter substrate (46 cm²) for YBCO using the conditions stated above for the *off*-axis process was found to be 1.3 μm/hour (3.7 Å/sec). Comparing the data in Figures 3 and 5 clearly indicate that the *on*-axis PLD technique provides significantly better uniformity of film composition, even over a much larger substrate size. Furthermore, it is likely that as *off*-axis PLD is scaled-up further the non-uniformity in film composition as noted in Figure 5 will likely increase.

<u>Stress, Crystallinity, and Morphology of *On*- and *Off*-axis PLD Films</u>

The intrinsic film stresses of all of the *off*-axis PLD yttria films deposited onto silicon substrates were found to be extremely low (σ < 5 MPa) and compressive in nature. In contrast, *on*-axis PLD films grown under similar conditions (pressure and fluence) typically display significantly larger intrinsic compressive stress values, ranging from over -1 GPa at low deposition pressures, dropping to about -100 MPa at pressures above about 7 mTorr [8]. SEM micrographs of *off*-axis deposited films indicated a less dense, more granular film structure as shown in Figure 6. Furthermore, X-ray powder diffraction scans for both *on*- and *off*-axis Y_2O_3 films laser deposited on silicon substrates are shown in Figure 7 along with the constructs for both Y and Y_2O_3. Very weak diffraction was obtained from the *off*-axis PLD film as noted, with the dominant peak at an angle of 2Θ = 33° referenced to the (101) line of metallic Y, indicating that the oxide component of this film is completely amorphous. X-ray diffraction obtained for *on*-axis PLD films yields strong diffraction at 2Θ = 29° indicating a highly oriented (222) oxide film. Similar results have been obtained when depositing Y_2O_3 onto CdTe substrates

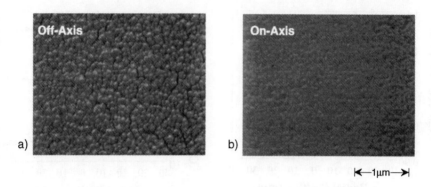

Figure 6. SEM Micrographs of Y_2O_3 Grown on (100) Silicon Substrates at Room Temperature By a) *Off*-Axis and b) *On*-Axis PLD. The Scale is the Same for Both Micrographs.

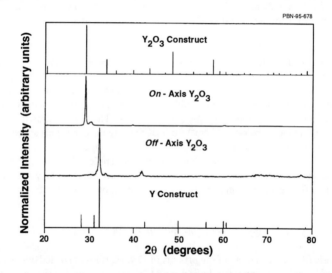

Figure 7. X-ray Powder Diffraction Scans of Y_2O_3 Films Deposited at Room-Temperature Onto (100) Si Substrates Using *On*- and *Off*-Axis PLD. The Constructs for Both Y_2O_3 and Metallic Y are Also Shown for Reference.

using *on*-axis PLD [8]. EDXA data obtained from both *on*- and *off*-axis samples indicated that both films were slightly oxygen deficient, but comparable in oxygen content. It would be expected that the total kinetic energy of the condensing species is likely to be higher for *off*-axis films due to the smaller average target-substrate distance, (75 mm versus 127 mm for the *off*- and *on*-axis cases respectively) since the ablated species would make fewer collisions with background gas. However, the results obtained above are consistent with that of a very low

156

kinetic energy in a direction parallel to that of the substrate normal for the *off*-axis PLD condensing species. Thus, the kinetic energy is not efficiently utilized in the condensation process to form fully dense and highly organized films. This also brings into question the role that Rydberg atoms (formed by the neutralization of an ion) may play in the condensation process. It is believed that these long-lived Rydberg atoms can add significant amounts of energy to the growing film upon condensation [14]. However, the Rydberg atoms usually have relatively high kinetic energy, and probably do not obtain a sufficient amount of transverse energy to be directed to, and condense on the substrate in *off*-axis PLD.

Using *off*-axis PLD to grow Y_2O_3 films on room temperature silicon substrates at deposition pressures below ~ 2 mTorr yielded films which were very porous and soft, and could be easily removed from the substrate with a Q-tip. As the pressure was increased to ~ 7 mTorr, the films became more dense and adhered to the substrate surface. When the pressure was increased to over 90 mTorr, the films remained hard and well adhered, but displayed very non-uniform film thickness (± 22%) with increased NPD's. Thus, at low pressure where very few collisions with the background gas occur, only material with some initial transverse energy will condense on the substrate. As the pressure is increased an increasing fraction of the ablated atoms and molecules are deflected providing more species with higher transverse energy which condense on the substrate. At even higher pressures, most of the ablated species make numerous collisions with the background gas, and loose a significant amount of their initial kinetic energy yielding poor quality films. It is also interesting to note that *on*-axis laser-deposited Y_2O_3 films display a significant reduction in the intrinsic compressive film stress as the growth pressure is increased above ~7 mTorr for a variety of background gases indicative of multiple collisions yielding reduced kinetic energy of the condensing species [8]. Similar curves of film stress versus pressure are obtained for standard sputtering processes [15]. *Off*-axis YBCO films deposited at ~ 7 mTorr in oxygen were shiny, hard and adhered to the substrate surface. Increasing the pressure to 25 mTorr increased the amount of particulates in the *off*-axis YBCO films. Growing YBCO films at room temperature in 90 mTorr of O_2 using *off*-axis PLD yielded films which were soft and poorly adhered at the substrate edge, but well adhered in the center of the substrate.

Figures 8 displays scanning electron micrographs of YBCO films deposited onto 75 mm diameter silicon substrates using *off*-axis PLD with a fluence of 3 J/cm^2 with an oxygen pressure of 25 mTorr. The film shown in Figures 8a, 8b, and 8c were deposited onto the center, middle, and edge, respectively, of the substrate which was off-set from the plume center by d = 6 mm as shown in Figure 1b. The SEM's shown in Figures 8d, 8e, and 8f were again obtained from the center, middle, and edge, respectively, of a film deposited using identical deposition parameters as above, except that the off-set was increased to 12 mm. The NPD's for the *off*-axis Y_2O_3 films seen in Figures 8a, 8b, and 8c were found to be below 10 particles/cm^2/Å, which is about six times lower than that obtained for the *on*-axis Y_2O_3 films using the same fluence and pressure. The NPD's of the films deposited with the larger substrate off-set were found to be 96, 148, and 215 particulates/cm^2/Å at the center, middle, and edge of the substrate, respectively. As noted, more particulates were found at the outer edge of the substrate. The gradient in NPD appears to be due to a decrease in the amounts of small particulates (diameters less than ~ 1μm) at the substrate center, more so than any significant drop in the number density of larger particles. This effect may be due to differences in the angular distribution of various particle sizes, or more likely, due to the possibility that the smaller, lighter particles are deflected more quickly by the background gas. Figure 9 shows a higher magnification SEM obtained near the edge of the

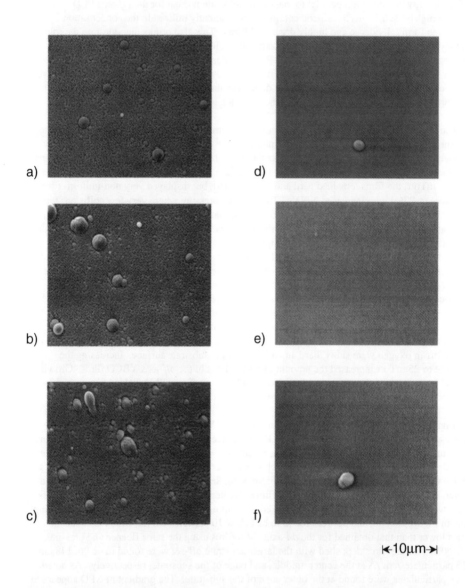

⊢10μm⊣

Figure 8. SEM Micrographs of *Off*-Axis deposited Y₂O₃ films grown with identical deposition conditions taken at: a) the center, b) the middle, and c) the edge of a 75 mm diameter substrate using a substrate off-set of d = 6 mm; and d) the center, e) the middle, and f) the edge of a 75 mm diameter substrate with a substrate off-set of d = 12 mm. The Scale is the Same in All of the Micrographs.

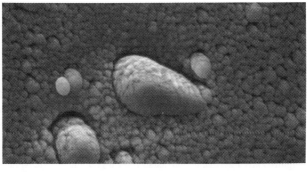

|←→|
1 μm

Figure 9. SEM Micrograph of Y_2O_3 Particles Deposited Near the Edge of a 75 mm Diameter Substrate Using *Off*-Axis PLD. Note the Shading Effect Produced by the Particle Due to the Preferred Directionality of the *Off*-Axis Deposition Process.

substrate of a the Y_2O_3 film deposited using the larger substrate off-set. The effect of the directionality of the *off*-axis deposition process can clearly be seen. Particles which are deposited early in the deposition tend to shade the substrate directly behind the particle when they are closer to the target. As the large substrate spins 180 degrees the flux at this location is greatly reduced. Thus, voids are formed as noted in Figure 9. These voids may be deleterious to some of the intended device applications for the films. Finally, looking at the SEMs obtained from the center, middle, and edges of a substrate, as seen in Figures 8, indicate that the voiding effect is strongest at the substrates edge, and nearly non-existent at the substrate center.

It is important to remember that these *off*-axis films were all deposited on unheated substrates. Depositing films onto substrates at elevated temperatures would likely lead to films with increased density and improved crystalline quality. However, any arguments about the energetics of the laser-ejected species providing higher quality films at lower substrate temperatures using *on*-axis PLD probably do not hold for the *off*-axis PLD process. It is also unclear whether the voids noted in Figure 9 will form when depositing films onto substrates held at elevated temperatures where thermally activated surface mobility of the arriving species may be sufficient to fill in the voids produced by the shading effect.

The NPD's of *off*-axis YBCO films were also evaluated. Typical values for films deposited at room-temperature on 75 mm diameter Si substrates in 25 mTorr of oxygen, using a fluence of 3.0 J/cm^2 were found to be 100, 150, and 440 particles/cm^2/Å the center, middle, and edge of the substrate, respectively. Typical values obtained for *on*-axis YBCO films using fluences of 1.5 J/cm^2 are typically ~140 particles/cm^2/Å [1]. The relatively high values obtained for the *off*-axis case may be due to the higher fluence used. However, high fluence is needed in *off*-axis PLD in order to maintain a reasonable deposition rate.

CONCLUSIONS

Standard *on*-axis pulsed laser deposition based on laser beam rastering over large diameter targets has been used to deposit films over both 150 mm (6-inch) and 200 mm (8-inch) diameter substrates. These films display excellent uniformity in film thickness and composition, thus demonstrating that the laser-deposition process is compatible with mainstream semiconductor manufacturing equipment. The composition profiles of YBCO films deposited over 150 mm diameter substrates using *on*-axis PLD were significantly better than that obtained from films deposited over 75 mm diameter substrates using the *off*-axis PLD approach. Furthermore, the deposition rates of *on*-axis PLD are higher than those obtained using *off*-axis PLD. *Off*-axis laser-deposited Y_2O_3 films display very low intrinsic compressive stress and are amorphous when grown on room-temperature substrates indicating that the condensing species have very low transverse kinetic energies over a wide pressure range. SEM micrographs indicated that the *off*-axis PLD films were more porous than *on*-axis PLD films. However, substrate heating during deposition may significantly improve the *off*-axis PLD film qualities. The NPD's of *off*-axis PLD films tended to increase from the substrate center to edge. Furthermore, particulates which are deposited near the edges of substrates tended to shadow the local substrate surface leaving voids which may cause significant problems for many *off*-axis PLD film applications. Also, the number of particles which condense on the substrate is a strong function of substrate off-set with respect to the plume center. This may make it difficult to reliably obtain films with low NPD's when using *off*-axis PLD. Coupling these facts together with the difficulty in heating large-area substrates in an *off*-axis mode with black-body type heaters [16, 17] makes it unlikely that *off*-axis PLD will be become a useful production tool in the near future. Other techniques to minimize the NPD such as rastering the laser beam over the complete target diameter, or varying the rotation rates with the fixed-beam approach may both lower the amount of particulates in laser-deposited films [18]. However, standard *on*-axis large-area PLD provides films with excellent uniformity at useful growth rates over relevant substrate sizes, and it is therefore expected that this approach will move to the production floor in the near future.

ACKNOWLEDGMENTS

Thanks to Mr. S. Hein and Dr. L. Kupferberg for their assistance with the SEM and EDXA data. Thanks to Dr. Ross E. Muenchausen for providing the RBS data. Finaly, special thanks to Dr. Vilnis Guntis Kriesmanis for the use of his Apple Computer and persistent source of entertainment.

REFERENCES

1) J. A. Greer and H. J. Van Hook, Mat. Res. Soc. Symp. Proc. Vol. 191. pp. 171-176, 1990.
2) S. Lee, D. Hwang, Y. K. Park, and J. Park, Appl. Phys. Lett., **Vol. 65**, pp. 764-766, 1994.
3) B. Holzapfel, B. Roas, L. Schultz, P. Bauer, and G. Saemann-Ischenko, Appl. Phys. Lett. **Vol. 61**, pp. 3178-3180 (1992).
4) H. Buhay, S. Sinharou, M. H. Francombe, W. H. Kasner, J. Tavlacchio, B. K. Park, N. J. Doyle, D. R. Lampe, and M. Polinsky, Proc. Integrated Ferroelectrics, **Vol. 1**, 213 (1992).
5) M. Lorenz, H. Hochmuth, H. Borner, and D. Natusch, Mat. Res. Soc. Symp. Proc. **Vol. 341**, pp. 87-94 (1994).

6) E. J. Smith, M. T. Smith, M. M. Eddy, G. G. Firpo, B. J. L. Nilsson, B. F. Zuck, and D.D. Strother, presented at The Fall Meeting of the materials Research Society, Boston, MA, 1992.

7) M. D. Robinson, M. M. Eddy, B. J. L. Nilsson, W. L. Olson, E. J. Smith, K. H. Young, G. V. Negrete, and R. B. Hammond, presented at The International Conference on Metallurgical Coatings and Thin Films, San Diego, 1991.

8) J. A. Greer and M. D. Tabat, Mater. Res. Soc. Symp. **Vol. 341**, pp. 87-94, 1994.

9) J. A. Greer and M. D. Tabat, to be published in J. Vac. Sci. and Technol. **A 13(3),** May/June 1995.

10) W. M. Cranton, D. M. Spink, R. Stevens, and C. B. Thomas, Thin Solid Films, **Vol. 226**, pp. 156-160, 1993.

11) H. Fukumoto, T. Imura, and Y. Osaka, Appl. Phys. Lett., Vol. **55**(4), pp. 360-361, 1989.

12) T. S. Kalkur, R. Y. Kwor, and C. A. De Araujo, Thin Solid Films, Vol. **170**, pp. 185-188, 1989.

13) J. C. Vyas, G. P. Kothiayal, K. P. Muthe, d. P. Gandhi, A. K. Debnath, S. C. Sabharawal, and M. K. Gupta, Journal of Crystal Growth, Vol. **130**, pp. 59-66, 1993.

14) R. Leuchtner, To be Published in Mater. Res. Soc. Symp. Proc. Vol. **354** (1995).

15) D. W. Hoffman and J. A Thornton, J. Vac. Sci. Technol., **16(2)**, pp. 134-137, 1979.

16) J. A. Greer, J. Vac. Sci. Technol. A **10(4)**, pp. 1821-1826, (1992).

17) *Pulsed Laser Deposition of Thin Films*, ed. by D. B. Chrisey and G. K. Hubler, John Wiley and Sons, New York, 1994.

18) C. Doughty, A. T. Findikoglu, and T. Venkatesan, Appl Phys. Lett., **Vol. 66**, pp. 1276-1278, 1995.

Part II

Ion-Assisted Deposition, Hyperthermal Beams, and Surface Dynamics

ION-ASSISTED PULSED LASER DEPOSITION OF BN FILMS

G. L. DOLL[a], D. C. CHANCE[b], AND L. SALAMANCA-RIBA[c]
[a]General Motors Research & Development Center, Warren, Michigan 48090
[b]Physics Department, Wayne State University, Detroit, Michigan 48201
[c]University of Maryland, College Park, Maryland 20742 US

ABSTRACT

Boron nitride films grown by ion-assisted pulsed laser deposition have been characterized by infrared absorption, Auger electron spectroscopy, and transmission electron microscopy. Elemental bonding and the crystallinity of BN films grown in three nitrogen ion energy regimes: high (2500 eV), low (700 eV), and without ions (0 eV) are examined, and the results interpreted within the framework of a compressive stress mechanism for cBN film growth.

INTRODUCTION

Boron nitride crystallizes in four major structures -- hexagonal (hBN), rhombohedral (rBN), wurtzitic (wBN), and cubic (cBN), all of which are isostructual to specific phases of carbon. The crystal structures of these four phases are illustrated in Figure 1 [1]. The soft hBN and rBN phases are comprised of sp^2 bonds, while the hard wBN and cBN phases have sp^3 bonds. The four BN phases are useful in tribological applications, but because its properties closely resemble those of diamond, cBN is highly valued as a high temperature electronic material as well. The physical properties of the different phases of BN, as well as most other III-N compounds have been recently compiled in a data review by Edgar [2].

Pouch and Alterovitz [3] have compiled a review of deposition techniques used to grow boron nitride. Generally, only those deposition techniques that employ ions in the process are successful in growing cBN. In this work, we review some of our research into the growth and characterization of BN films by ion-assisted pulsed laser deposition undertaken over the past several years.

EXPERIMENTAL DETAILS

BN films examined in our studies were grown by ion-assisted pulsed laser deposition. In this technique, a pulsed KrF laser (λ=248 nm) is focused onto a pyrolytic hexagonal BN target. The BN from the target evaporates and subsequently condenses onto a substrate (n-type Si (001) wafers) heated to temperatures ranging from 400 to 600 °C. When cBN films are desired, nitrogen ions from a Kaufman-type broad beam ion source are utilized. Our cBN films are grown in two distinct ion energy regimes, high and low. We define the low ion energy regime as nitrogen (N_2^+/N^+) ions with energies of 500 to 1000 eV, and the high ion energy regime as nitrogen ion energies with greater than 2000 eV. Whereas a Kaufman source is the sole source of the low energy regime ions, substrate bias is added to obtain the high energy ion regime. Films that are grown under conditions with no ions or with ion energies less than 500 eV, are not cBN, but are nano-crystalline sp^2 bonded materials. Deposition parameters for the films discussed in this work are given in greater detail in earlier publications [4-6]. In this work, we examine BN films grown by pulsed laser deposition using 0, 700, and 2500 eV nitrogen ions.

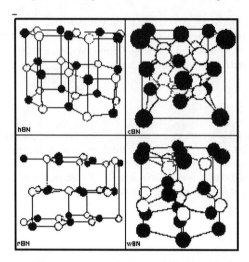

Figure 1: The four phases of BN, from Reference 1.

Film characterization techniques using infrared absorption and Auger electron spectroscopy (AES) were performed in the laboratories of the Physics and Analytical Chemistry departments at the GM R&D Center. The transmission electron microscopy measurements were performed at the University of Maryland. BN powders that were used as standards were obtained from Advanced Ceramics and GE Superabrasives.

RESULTS AND DISCUSSION

Infrared absorption spectroscopy is well suited to the characterization of the chemical bonding (sp^2 or sp^3) in BN films [7]. Whereas the sp^3-bonded materials (cBN and wBN) have one strong absorption peak near 1060 cm^{-1}, the sp^2-bonded hBN has two features near 1360 and 780 cm^{-1} corresponding to energies of the A_{2u} and E_{1u} infrared-active phonons, respectively. While the E_{1u} mode of hBN is an in-plane vibration excited by light polarized perpendicular to the $\hat{c}$-axis ($\bar{E} \perp \hat{c}$), the A_{2u} mode is an out-of-plane vibration excited by light polarized parallel to the $\hat{c}$-axis ($\bar{E} \| \hat{c}$). Rhombohedral BN has the $C_3^4, R3$ space group, so there are again two infrared active modes, (A and E symmetry) which are excited by $\bar{E} \| \hat{c}$ and $\bar{E} \perp \hat{c}$ light, respectively. Because of the similarities in crystal structures and chemical bondings between rBN and hBN, we expect that the energies of the rBN A and E modes to be similar to the hBN A_{2u} and E_{1u} modes, respectively.

The infrared absorption spectra of the BN films grown with 2500, 700, and 0 eV ions are shown in Figure 2. The spectrum of a BN film grown with 2500 eV ions (top) shows a strong absorption peak near 1080 cm^{-1} characteristic of sp^3-bonding and two weak shoulders near 1400 and 800 cm^{-1} that arise from sp^2-bonds. The 20 cm^{-1} difference between the absorption peak in the spectrum of this film and the spectra of cBN and wBN powders is observed with most cBN films [2]. The origin of the upshift has been speculated by Ikeda *et al.* to arise from compressive stress in the cBN film [8]. The spectrum of the BN film grown with 700 eV ions (middle) has similar features to the spectrum of the 2500 eV sample, but the sp^2 absorption features dominate those characteristic of sp^3 bonding. Finally, the spectrum of the sample grown without ions (bottom) shows only sp^2 bonding. We note that since the absorption spectra of the films were obtained from transmittance measurements, the infrared measurements are characterizing the bonding in the entire thickness of the films.

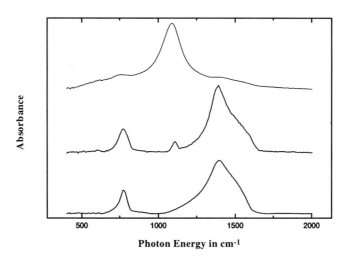

Figure 2: Infrared absorption spectra of BN films grown with 2500 eV (top), 700 eV (middle), and without (0 eV) nitrogen ions (bottom).

Auger Electron Spectroscopy (AES) has been shown to be a sensitive probe of the chemical bonding in the surface (40 Å) of BN films [9]. High resolution window scans of the B KVV Auger transitions for hBN and cBN powders are shown in Figure 3, along with a scan taken from the BN film grown with 2500 eV ions. Similar scans of the N KVV Auger transitions were also performed, but are not shown here. The scans obtained from the hBN and cBN powders are in excellent agreement with previously published AES scans [9]. Whereas scans obtained from films grown with 700 eV ions and without ions (not shown) display the general characteristics of only the hBN standard, features from both cBN and hBN powder standards appear in the scan of the high ion energy BN film. The film data have been modeled as the sum of contributions from hBN and cBN components, and the result of a least squares fit of that model, (shown as the dotted line) indicates that within the sampling depth of the surface, the film is actually composed of approximately equal parts of sp^2 and sp^3 bonded BN. A similar result was obtained by fitting the N KVV scan of the 2500 eV film. The absence of any sp^3-bonded features in the AES scan of the 700 eV film indicates that the sp^3-bonding in that sample (as observed by infrared absorption) is not near the films surface. A detailed study of the AES spectra for the four BN phases is underway [10].

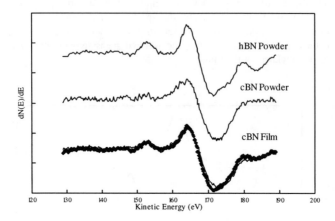

Figure 3: High-resolution window scans of the B-KVV Auger electron transitions in cBN and hBN powders, and a BN film grown with 2500 eV nitrogen ions. The scan of the film is shown fit with equal contributions of hBN and cBN components. From Reference 6.

The bonding in a BN film grown with 700 eV ions was examined by Terminello *et al.* [1] using near-edge x-ray absorption fine structure (NEXAFS) techniques. They reported that the 700 eV BN film was comprised predominantly of a sp^2-bonded phase plus a smaller amount of sp^3-bonded material. By comparing the core level photoabsorption data of the four phases of BN with that of the film, they reported that the characteristics of the sp^2 bonding in the film more closely resembled rBN than hBN. We do not necessarily think that this means that rBN is present in that film. The ABCA ĉ-axis stacking of the hexagons in the rBN lattice means that some atoms do not reside directly above <u>and</u> below atoms in adjacent planes (see Figure 1). Thus, the bonding of these atoms would differ from those in hBN which has an ABA ĉ-axis stacking of the hexagons (i.e., every atom of the BN hexagons has atoms directly above and below it in adjacent layers). A turbostratic arrangement of hBN would also generate a large number of atoms not bonded to atoms directly above and below. Hence, the NEXAFS measurements which probe empty electronic states of the B and N atoms, may not be able to distinguish rBN from a turbostratic arrangement of hBN. In an earlier work, Chaiken *et al.* [11] examined the 700 eV BN film with a polarization-dependent NEXAFS measurement and observed a preferenetial alignment of the sp^2-bonded BN planes normal to the substrate surface.

As shown above, infrared absorption, NEXAFS, and AES provide sensitive information about the bonding in BN films, but not necessarily the crystalline structure. The crystallinity of BN films is usually studied by means of electron microscopy since BN is reasonably transparent to x-rays. Cross sectional and plan-view transmission electron microscopy (TEM) studies of the cBN film have been performed by Ballal *et al.* [4,5], and show that the film grown with 2500 eV has three distinct layers. An "amorphous" region ~50 Å thick resides at the Si/BN interface. Upon closer inspection with high-resolution TEM imaging, this "amorphous" region actually consists of randomly oriented, nano-crystalline hBN. The middle layer is ~150 Å thick, and consists of hBN crystallites oriented such that their [001] axes are perpendicular to the growth direction (i.e., the Si [001] direction). No evidence for the rBN phase was obseved in either of these sp²-bonded layers.ions is cubic with a lattice constant of 3.62 Å. The top layer, which extends from ~200 Å to the total film thickness (~1800 Å) is cBN. The cBN layer was found to have a columnar structure with average diameters and lengths of the columns equal to 500 and 750 Å, respectively. Notably, the columns were observed to be preferentially oriented with the cBN [110] axis parallel to the Si [001] direction. No in-plane orientation was observed.

In a later study [6], we reported TEM diffraction patterns for the pulsed laser deposited BN film grown without ions on a (001) silicon wafer. Ring patterns were observed that index to lattice planes of hBN. The measured interplanar spacings of the polycrystalline hBN film agree with those of bulk hBN (a=2.50 Å, and c=6.66 Å). By tilting the sample to obtain the Si [221] plane, the electron diffraction patterns did not show the same continuous ring pattern from hBN, but rather sections of the (002) hBN ring were observed. The observation of ring sections clearly indicates that there is a preferred orientation of the hBN film with the [001] hBN axis perpendicular to the Si [001] direction (i.e., the growth direction). No evidence for the rBN phase was observed in this film. We have not performed TEM experiments on our BN films that were grown by pulsed laser depotion with low energy nitrogen ions.

CONCLUSIONS

We first summarize the results of the characterization experiments performed on the ion-assisted, pulsed laser deposited BN films. From the interface to the first ~50 Å, the BN film grown with the 2500 eV nitrogen ions has a randomly oriented, nano-crystalline hBN phase. In the middle layer, an oriented hBN layer appears and extends to ~200 Å from the interface. Beyond that, columnar cBN grains oriented with their [110] axes parallel to the growth direction emerge, and extend to the surface of the film. Within ~40 Å of the surface, sp²- and sp³-bonded regions coexist. The BN film grown without ions also has an amorphous region adjacent to the interface. Beyond that, a somewhat less oriented hBN phase (hBN [001] ∥ Si [001]) extends to the film surface. McKenzie et al. [12,13] have formulated a thermodynamic argument in which a biaxial stress term explains the preferred orientation of the hBN. That is, since the [001] direction of hBN is more easily compressed than the in-plane directions, a substantial reduction in the free energy occurs when the [001] hBN direction lies in the plane of the stress field.

Davis et al. have performed TEM experiments on BN films that were grown by ion-assited e-beam deposition [14]. The ion energies used in that study are comparable to those used to grow the 700 eV BN film studied here. Using cross-sectional high-resolution TEM scans, Davis et al. observed a three-layer structure in their BN film: an initial layer (~20 Å) of amorphous BN, a preferentially oriented layer of hBN (hBN [001] ∥ Si [001]) ~20 - 60 Å thick, and a randomly oriented, polycrystalline layer of cBN. This type of layered structure has also been reported for BN films grown by low energy ion-assisted pulsed laser deposition [15].

Davis et al. [14] have offered the following explanation to account for the layered structure observed in their BN films. A thin, randomly oriented, nano-crystalline hBN layer (which appears amorphous in the TEM experiments) possessing a small amount of residual stress is produced during the initial stages of deposition. As the deposition continues, the increasing stress is partially relieved by the formation of the preferentially oriented crystalline hBN phase, consistent with the thermodynamic arguments of McKenzie *et al.* [12,13] discussed above. The stress continues to build through the hBN layer, eventually attaining a large enough value to cross the P-T threshold where cBN becomes the stable phase. This explanation is sometimes referred to as the compressive stress mechanism for cBN growth.

We find that the results of the characterizations on our ion-assisted pulsed laser deposited BN films are consistent with the compressive stress mechanism for cBN growth discussed above. The tri-layered structure observed in the BN film grown with 2500 eV ions is similar to that observed in BN films grown with lower ion energies, although the thicknesses of the individual layers differ. What has not been addressed is the origin of the [110] preferred orientation of the cBN phase during the high ion energy growths. Recently, Johansson *et al.* [16] have also observed this same preferred orientation of cBN which was grown by unbalanced magnetron sputtering of boron carbide targets in an Ar/N_2 environment with a high voltage rf substrate bias.

REFERENCES
1. L. J. Terminello, A. Chaiken, D. A. Lapiano-Smith, G. L. Doll, and T. Sato, J. Vac. Sci. Technol. A **12**, 2462 (1994).
2. J. H. Edgar (ed.), Properties of Group III Nitrides, EMIS Datareview Series No. 11, (INSPEC, London, UK, 1994).
3. J. J. Pouch and S. A. Alterovitz (eds.), Synthesis and Properties of Boron Nitride, (Trans Tech, Brookfield, Vermont, 1990).
4. A. K. Ballal, L. Salamanca-Riba, G. L. Doll, C. A. Taylor II, and R. Clarke, J. Mater. Res. **7**, 1618 (1992).
5. A. K. Ballal, L. Salamanca-Riba, C. A. Taylor II, and G. L. Doll, Thin Solid Films **224**, 46 (1993).
6. G. L. Doll, M. C. Militello, S. J. Simko, A. K. Ballal, and L. Salamanca-Riba, in Silicon Carbide and Related Materials, edited by M. G. Spencer, R. P. Devaty, J. A. Edmond, M. Asif Khan, R. kaplan, and M. Rahman (Institute of Physics Conference Series Number **137**, Bristol, UK, 1994) pp. 449-452.
7. G. L. Doll in Reference 2, pp. 241-248.
8. T. Ikeda, T. Satou, and H. Satoh, Surface Coat. Technol. **50**, 33 (1991).
9. R. Trehan, Y. Lifshitz, and J. W. Rabalais, J. Vac. Sci. Technol. A **8**, 4026 (1990).
10. G. L. Doll and S. J. Simko, (unpublished).
11. A. Chaiken, L. J. Terminello, J. Wong, G. L. Doll, and C. A. Taylor II, Appl. Phys. Lett. **63**, 2112 (1993).
12. D. R. McKenzie, W. D. McFall, W. G. Sainty, C. A. Davis, and R. E. Collins, Dia. Rel. Mater. **2**, 970 (1993).
13. D. R. McKenzie, J. Vac. Sci. Technol. B **11**, 1928 (1993).
14. R. F. Davis, K. S. Ailey, R. S. Kern, D. J. Kester, Z. Sitar, L. Smith, S. Tanaka, and C. Wang in Diamond, SiC and Nitride Wide Bandgap Semiconductors, edited by C. H. Carter Jr., G. Gildenblat, S. Nakamura, and R. J. Nemanich (Mater. Res. Soc. Proc. **339**, Pittsburgh, PA, 1994) pp. 351-362.
15. T. A. Friedman, W. M. Clift, H. A. Johnsen, E. J. Klaus, K. F. McCarty, D. L. Medlin, M. J. Mills, and D. K. Ottesen, in Laser Ablation in Materials Processing: Fundamentals and Applications, edited by B. Braren, J. J. Dubowski, and D. P. Norton (Mater. Res. Soc. Proc. **285**, Pittsburgh, PA, 1993) pp. 507-512.
16. M. Johansson, L. Hultman, H. Luthje, K. Bewilogua, S. Daaud, P. Willich, and S. Jager, presented at the 1994 Intl. Conf. on Met. Coatings and Thin Films, San Diego, CA, 1994 (unpublished).

SUBSTRATE HEATING MEASUREMENTS IN PULSED ION BEAM FILM DEPOSITION

J. C. Olson*, M. O. Thompson**, H. A. Davis*, D. J. Rej*, W. J. Waganaar*, and D. R. Tallant***
*Los Alamos National Laboratory, MS-E526, Los Alamos, NM 87544
**Materials Science and Engineering Department, Cornell University, Ithaca, NY 14853
***Sandia National Laboratories, P. O. Box 5800, Albuquerque, NM 87185-0343

ABSTRACT

Diamond-like Carbon(DLC) films have been deposited at Los Alamos National Laboratory by pulsed ion beam ablation of graphite targets. The targets are illuminated by an intense beam of hydrogen, carbon, and oxygen ions at a fluence of 15-45 J/cm^2. Ion energies are on the order of 350 keV, with beam current rising to 35 kA over a 400 ns ion current pulse.

Raman spectra of the deposited films indicate an increasing ratio of sp^3 to sp^2 bonding as the substrate is moved further away from the target and further off the target normal. Using a thin film platinum resistor at various positions, we have measured the heating of the substrate surface due to the kinetic energy and heat of condensation of the ablated material. Plume power density and energy input are inferred from the temperature measurements. This information is used to determine if substrate heating is responsible for the lack of DLC in positions close to the target and near the target normal.

INTRODUCTION

Pulsed ion beam deposition has the potential to become a low cost, high throughput (a few \$/m^2 at 50 m^2/hr for 1 μm coatings) film production process for flat panel displays, photovoltaic cells, and other applications. This process is similar to pulsed laser deposition. An intense beam of ions (100-1000 keV, 10-100 kA, 0.1-1.0 μs) strikes a target, ablating target material which condenses on a substrate as a thin film. Experiments at Los Alamos National Laboratory are examining both science and technology issues related to pulsed ion beam deposition. The present work is concerned with heating of the substrate by the hot, dense ablation plume from a graphite target and the effect of this heating on the properties of the deposited material. Substrate heating by the ablated material was examined using thin film metal thermometers.

EXPERIMENTAL SETUP

A schematic of the experimental configuration is shown in Figure 1. The intense ion beams used in these experiments is produced by the Anaconda generator at Los Alamos. The machine is coupled to a focusing, magnetically insulated diode, out of which a 400 keV, 30 kA, 700 ns ion beam is extracted. Ablation targets are placed 35 cm from the anode, at the beam focus, where a peak beam energy density of approximately 30($\pm$15) J/cm^2 is

7 ms duration, and was necessary in order to acquire reasonable signals from the thermometers without heating them appreciably. The voltages above and below the thin film thermometer layer, $V_{in}(t)$ and $V_{out}(t)$ respectively, were monitored using HP54111D digitizing oscilloscopes. The quantity $V_{in}(t) + V_{out}(t)$ should remain equal to V_0 and provided a measure of the noise. A signal to noise ratio, S/N, was computed from

$$S/N(t) = \frac{[V_{in}(t) - V_{out}(t)] - [V_{in}(0) - V_{out}(0)]}{V_0 - (V_{in}(t) + V_{out}(t))}. \tag{1}$$

Shots with a peak S/N of less than 5 were rejected. Out of forty shots taken at the thermometer positions discussed in this report, 34 produced data with acceptable S/N. All of the rejected shots were at the $d = 20$ cm, $\theta = 30°$ position where signals were very low.

$R_s(t)$ was calculated from

$$R_s(t) = \frac{100}{\dfrac{V_0}{V_{in}(t) - V_{out}(t)} - 1}. \tag{2}$$

Using (2), the time dependent temperature at the surface of the thermometer was computed from

$$\Delta T(t) = \frac{1}{\gamma} \frac{R_s(t) - R_s(0)}{R_s(0)}. \tag{3}$$

Figure 2 shows the signal, noise and time dependent resistance measured on shot 3246, with the thermometer located at $d = 15$ cm, $\theta = 0°$. The spike near $t = 0$ is due to the firing of Anaconda and can be ignored.

Heating power per unit area delivered to the substrate, $P(t)$, was found by deconvolving the solution to the heat equation

$$\Delta T(t) = \int_0^t \frac{P(t')}{\sqrt{\pi \rho c_p \kappa (t - t')}} dt' \tag{4}$$

where ρ, c_p and κ are the density, heat capacity, and thermal conductivity of the thermometer. For the time scales of interest in our experiment, it is sufficient to use ρ, c_p and κ for the bulk material, namely silicon. Diffusion time through the thin isolation, thermometer, and SiO_2 layers is about 8 ns. From the form of (4) it is clear that the experimentally measured $T(t)$ can be used to infer a temperature rise for other substrate materials simply by scaling by the appropriate ρ, c_p and κ. Figure 3 shows calculated peak power density on the substrate for many shots, to illustrate the magnitude of shot to shot variations.

RESULTS

Figure 4 shows the substrate temperature rise during carbon deposition at three different positions: $d = 15$ cm, $\theta = 0°$; $d = 20$ cm, $\theta = 0°$; and $d = 20$ cm, $\theta = 30°$. The right vertical axis shows the inferred temperature rise of a glass substrate at those positions. Figure 5 shows the heating power due to the ablation plume at the same three positions. We also took data with the conditions $d = 10$ cm, $\theta = 0°$.

Substrate heating decreased with increasing angle and with increasing target substrate separation, from 70 kW/cm^2 for approximately 10 μs duration at $d = 10$ cm, $\theta = 0°$ to

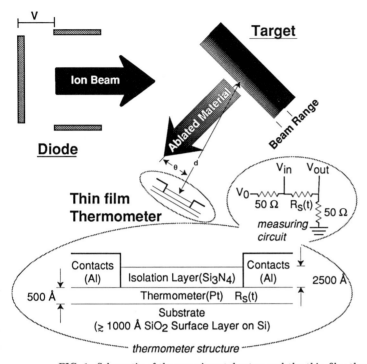

FIG. 1. Schematic of the experimental setup and the thin film thermometer.

delivered [1,2]. A single machine pulse typically ablates 10 mg of carbon from a POCO graphite target.

Substrate heating for various target-substrate separations and orientations was examined by the use of thin film metal thermometers. This technique relies on the temperature dependent resistance of a thin metal film to provide heating information with nanosecond time resolution. Previously, the technique has been used to examine ablation in pulsed laser deposition experiments and melt dynamics in pulsed laser induced melting experiments [3-7]. In our experiments, the target-thermometer separation, d, was varied from 10 to 20 cm and the thermometer position relative to the target normal was varied from $\theta=0$ to $\theta=30°$.

Thermometer construction and a circuit description are shown in Figure 1. The relatively thick (2500 Å) Si_3N_4 isolation layer served two purposes: to diminish capacitive coupling between the thermometer and beam created plasmas; and, more importantly, to prevent plasmas from shorting out the resistance of the thermometer. Details about the fabrication and calibration of this type of thermometer may be found elsewhere [3]. Thermometers in our experiments had room temperature resistances around 150 Ω with temperature coefficients of resistance ($\gamma = 1/R \ dR/dT$) of 0.0032 K^{-1}.

Temperature measurement was accomplished by measuring the time-dependent thermometer resistance, $R_s(t)$. This was done by applying a pulsed bias, V_0, of 10 V to the measuring circuit. The pulsed bias was applied 5.3 ms prior to the ion beam pulse, had

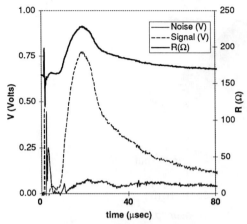

FIG. 2. Signal (thin dashed line), noise (thin solid line) and $R_s(t)$ (thick solid line - right vertical axis) from shot 3246, thermometer position $d = 15$ cm, $\theta = 0°$.

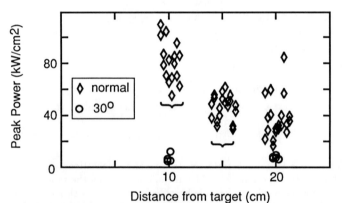

FIG. 3. Peak power density data for shots taken at all positions, showing shot to shot reproducibility.

5 kW/cm^2 for approximately 20 μs duration at $d = 20$ cm, $\theta = 30°$. Arrival time of the leading edge of the ablation plume was used to compute a plume expansion velocity. The measured value was 1 cm/μs normal to the target, decreasing to 0.4 cm/μs at $\theta = 30°$. These values are necessarily crude, due to low statistics and large shot to shot irreproducibilities in the pulsed ion beam, but nevertheless are in reasonable agreement with the plume expansion velocity of 2 cm/μs measured by visible light framing pictures. Another measure of the plume energy content can be made by integrating the power to find energy, and dividing the energy by the number of atoms deposited. 25 nm of film at a density of 1.7($\pm$0.3) g/cm^2 is deposited in a typical pulse [8], and the energy found by integrating the power curve for either position on the normal is 0.7 J/cm^2, giving a plume energy of 19 eV/particle. This calculation assumes that all the heat absorbed by the substrate is contributed by particles

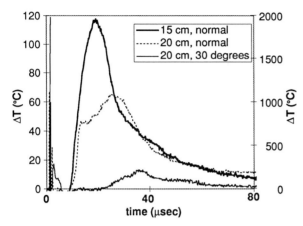

FIG. 4. Temperature rise as measured by thin film thermometers located: $d = 15$ cm, $\theta = 0°$ (thick solid line); $d = 20$ cm, $\theta = 0°$ (thin dashed line); and $d = 20$ cm, $\theta = 30°$(thin solid line). Right vertical axis shows inferred temperature rise for glass substrates at the same positions. All curves are for single shots.

which condense on the surface, and is therefore likely to be an overestimate of the plume temperature.

DISCUSSION

The Raman spectra of carbon films deposited on glass have been analyzed. Films deposited at $d = 22.5$ cm, $\theta = 0°$ and $d = 15$ cm, $\theta = 50°$ showed diamond like character, but the film obtained at $d = 15$ cm, $\theta = 0°$ was composed of nanocrystalline graphite, which is also known as "glassy" carbon [8]. This appears to be due to the high substrate temperature reached during deposition at this position, which promotes formation of thermodynamically stable sp^2 bonds. Plume energies appear well-suited to the formation of DLC [9], and deposition onto a more thermally conductive substrate should allow the production of high quality DLC films at high rates even on the target normal at low target-substrate separations.

ACKNOWLEDGMENTS

This work was supported by the U. S. Department of Energy, under contract number W-7405-ENG-36, through Los Alamos National Laboratory, Laboratory Directed Research and Development. One of us (JCO) was supported by the Director's Office of Los Alamos National Laboratory through a Director's Funded Postdoctoral Fellowship. Production of the thin film thermometers and participation of one of us (MOT) was supported by the

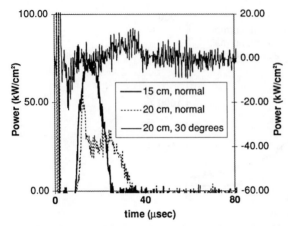

FIG. 5. Calculated heating power density due to the ablation plume at positions: $d = 15$ cm, $\theta = 0°$ (thick solid line - left vertical axis); $d = 20$ cm, $\theta = 0°$ (thin dashed line - left vertical axis); and $d = 20$ cm, $\theta = 0°$ (thin solid line - right vertical axis). All curves are for single shots.

National Science Foundation. The authors gratefully acknowledge the assistance of Jeff Espinoza and James Lopez in carrying out these experiments.

REFERENCES

[1] H. A. Davis, G. P. Johnston, D. J. Rej, W. J. Waganaar, C. L. Ruiz, and F. A. Schmidlapp, in *Proceedings of the Tenth International Conference on High-Power Particle Beams(NTIS PB95-144317)*, edited by W. Rix and R. White (National Technical Information Service, Springfield, VA, 1994), pp. 226–231.

[2] D. J. Rej, R. R. Bartsch, H. A. Davis, R. J. Faehl, J. B. Greenly, and W. J. Waganaar, The Review of Scientific Instruments **64**, 2753 (1993).

[3] D. P. Brunco, J. A. Kittl, C. E. Otis, P. M. Goodwin, M. O. Thompson, and M. J. Aziz, The Review of Scientific Instruments **64**, 2615 (1993).

[4] J. A. Kittl, R. Reitano, M. J. Aziz, D. P. Brunco, and M. O. Thompson, Journal of Applied Physics **73**, 3725 (1993).

[5] D. P. Brunco, M. O. Thompson, C. E. Otis, and P. M. Goodwin, Journal of Applied Physics **72**, 4344 (1992).

[6] J. Y. Tsao, S. T. Picraux, P. S. Peercy, and M. O. Thompson, Applied Physics Letters **48**, 278 (1986).

[7] M. O. Thompson, G. J. Galvin, J. W. Mayer, P. S. Peercy, and R. B. Hammond, Applied Physics Letters **42**, 445 (1983).

[8] G. P. Johnston, P. Tiwari, D. J. Rej, H. A. Davis, W. J. Waganaar, R. E. Muenchausen, K. C. Walter, M. Nastasi, H. K. Schmidt, N. Kumar, B. Lin, D. R. Tallant, R. L. Simpson, D. B. Williams, and X. Qiu, Journal of Applied Physics **76**, 5949 (1994).

[9] J. J. Cuomo, D. L. Pappas, J. Bruley, J. P. Doyle, and K. L. Saenger, Journal of Applied Physics **70**, 1706 (1991).

ROLE OF IONS IN BIAS ENHANCED NUCLEATION OF DIAMOND

J.M. LANNON, JR.*, J.S. GOLD*, AND C.D. STINESPRING**
*Department of Physics, West Virginia University, Morgantown, WV 26506
**Department of Chemical Engineering, West Virginia University, Morgantown, WV 26506

ABSTRACT

Ion-surface interactions are thought to play a role in bias enhanced nucleation of diamond. To explore this hypothesis and understand the mechanisms, surface studies of hydrogen and hydrocarbon ion interactions with silicon and silicon carbide have been performed. The experiments were carried out at room temperature and used *in-situ* Auger analyses to monitor the surface composition of thin films produced or modified by the ions. Ion energies ranged from 10 to 2000 eV. Hydrogen ions were found to modify silicon carbide thin films by removing silicon and converting the resulting carbon-rich layers to a mixture of sp^2- and sp^3-C. The interaction of hydrocarbon ions with silicon was shown to produce a thin film containing SiC-, sp^2-, and sp^3-C species. In general, the relative amount of each species formed was dependent upon ion energy, fluence, and mass. The results of these studies, interpreted in terms of chemical and energy transfer processes, provide key insights into the mechanisms of bias enhanced nucleation.

INTRODUCTION

Control of the nucleation density is a critical issue during plasma-assisted chemical vapor deposition (PACVD) of diamond. For growth on clean silicon, nucleation densities are quite low, and values higher than 10^7 cm^{-2} are typically achieved only by resorting to *ex-situ* substrate abrasion[1-3] or other "seeding" techniques.[4-6] Recently, however, Glass and coworkers,[7-9] reported enhancement of the nucleation density by five orders of magnitude using an *in-situ* process called bias enhanced nucleation. This technique involves biasing the silicon substrate at -250 eV while it is immersed in the plasma. Their results suggest that C-C species produced early in the growth process are the precursors to diamond nucleation, and they speculate that the increased flux of hydrogen and hydrocarbon ions in the plasma plays a role in the formation of these precursors.

Motivated by these results, we have initiated ultrahigh vacuum (UHV) surface studies to investigate the interactions of hydrogen and hydrocarbon ions with silicon and silicon carbide. This paper briefly reviews our previously reported results for hydrogen ion interactions with silicon carbide[10] and presents new data concerning hydrocarbon ion interactions with Si(100).

EXPERIMENTAL

The UHV growth and analysis system used in these studies included facilities for Auger electron spectroscopy (AES) and low energy electron diffraction (LEED).[10,11] The hydrocarbon species used for thin film growth and surface modification were introduced using an effusive

Mat. Res. Soc. Symp. Proc. Vol. 388 © 1995 Materials Research Society

beam gas source. An ion gun was used to supply argon ions for substrate cleaning and hydrogen or hydrocarbon ions for growth and surface modification.

Si(100) 2x1 surfaces were prepared from n-type electronic grade substrates with a resistivity of 5 ohm-cm. These were cleaned by argon ion bombardment and annealed to produce the 2x1 surface as verified by LEED. Silicon carbide films (~100 nm thick) were grown *in-situ* at 1200 K by exposing Si(100) 2x1 surfaces to an ethylene flux.[10,11]

Using a back pressure of hydrogen, methane, or ethylene, ions were produced in the ion gun by electron impact ionization. Consequently, the relative concentrations of the dominant ion species produced could be estimated using mass spectrometer cracking patterns.[12] For hydrogen, H^+ and H_2^+ ions were produced in the ratio of 2:100, respectively. For methane, CH_3^+ and CH_4^+ were formed in the ratio of 85:100, respectively. For ethylene, $C_2H_2^+$, $C_2H_3^+$, and $C_2H_4^+$ ions were formed in the ratio of 61:62:100, respectively. In the remaining discussion, H_x^+, CH_x^+, and $C_2H_x^+$ will be used to refer to these dominant ionic species.

The effects of ions on the composition of the silicon and silicon carbide substrates were followed using AES to monitor the Si-LMM and C-KLL Auger peak positions, line shapes, and intensities as functions of ion fluence. As described elsewhere,[10,11] quantitative determinations of the sp^2-C, sp^3-C, and SiC-C contribution to the C-KLL Auger line shape were obtained by factor analysis using graphite, diamond, and silicon carbide standards.[13]

RESULTS

For all ion energies studied, exposure of silicon carbide films to H_x^+ ions resulted in a reduction in the Si-LMM AES peak intensity, while the C-KLL intensity increased slightly and then remained relatively constant.[10] This is illustrated in Figure 1 using the data for 500 eV H_x^+ ions. Although the C-KLL intensity remained nearly constant, the line shape was found to vary with both ion fluence and energy. These line shape variations represent changes in carbon bonding which were quantitatively described in terms of SiC-C, sp^2-C, and sp^3-C bonding using factor analysis.[10,11] The results of

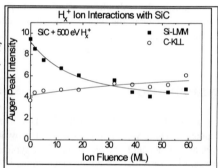

Figure 1. Si-LMM and C-KLL intensity versus carbon fluence for 500 eV H_x^+.

these analyses indicated that the SiC-C species simply decreased with ion fluence for all ion energies studied. As shown by Figures 2 and 3, however, the sp^2-C and sp^3-C components were dependent on both ion energy and fluence.

The sp^2-C component for 10 eV ions (Figure 2) increased steadily (albeit slowly) from the start of the exposure. In contrast, sp^2-C formation for higher ion energies exhibited an energy dependent induction period. On increasing the ion energy from 100 to 500 eV, the induction period for sp^2-C formation increased by a factor of eight. For 2000 eV ions, although the data are somewhat limited, the induction period for this energy is at its greatest comparable to that for 500 eV ions, and in fact, the induction period for 2000 eV ions may be shorter than that for either 100 or 500 eV ions. As seen in Figure 3, sp^3-C formation occurred promptly (i.e. without an induction period) for 100, 500, and 2000 eV ions; while little, if any, sp^3-C was produced by 10 eV ions. For low fluences, 500 eV ions were the most efficient at producing sp^3-C species.

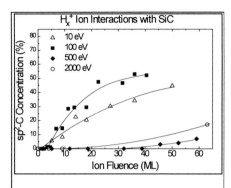

Figure 2. Concentration of sp^2-C species versus H$_x^+$ ion fluence.

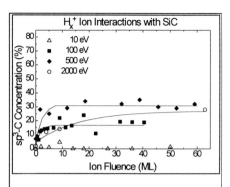

Figure 3. Concentration of sp^3-C species versus H$_x^+$ ion fluence.

Corresponding studies of hydrocarbon ion interactions on both silicon and silicon carbide are in progress. The data discussed here, however, are limited to the interaction of 500 eV CH$_x^+$ and C$_2$H$_x^+$ ions with the Si(100) 2x1 surface. Using AES to follow this interaction, it was found that a thin film was formed which consisted of a distribution of C-species. Figure 4 compares the increase in C-KLL intensity (due to all C-species) versus carbon fluence for CH$_x^+$ and C$_2$H$_x^+$ ions. The key point illustrated by these data is that the initial rate of increase and the saturation level for C$_2$H$_x^+$ ions are approximately two to three times as great as those for CH$_x^+$ ions.

The line shapes of the C-KLL peak produced by hydrocarbon ion interactions were also found to depend on fluence and ion species identity. These results are illustrated in Figures 5, 6, and 7 which show the distribution of carbon species produced versus ion fluence. For both hydrocarbon ions, sp^2-C and SiC-C were the dominant initial interaction product with the rate of sp^2-C formation being slightly greater than that for SiC-C. The sp^2-C saturation value produced by C$_2$H$_x^+$ ions was approximately four times that produced by CH$_x^+$ ions. The saturation values for SiC-C were essentially the same. The formation of sp^3-C species occurred less rapidly than either sp^2- and SiC-C.

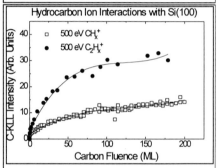

Figure 4. C-KLL intensity versus carbon fluence for 500 eV CH$_x^+$ and C$_2$H$_x^+$ ions.

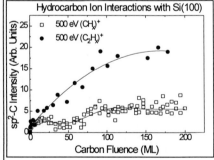

Figure 5. sp^2-C intensity versus carbon fluence for 500 eV CH$_x^+$ and C$_2$H$_x^+$ ions.

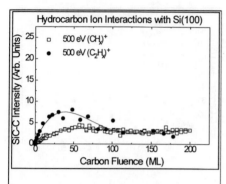

Figure 6. SiC-C intensity versus carbon fluence for 500 eV CH_x^+ and $C_2H_x^+$ ions.

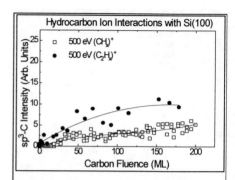

Figure 7. sp^3-C intensity versus carbon fluence for 500 eV CH_x^+ and $C_2H_x^+$ ions.

DISCUSSION

a) Ion Interaction Mechanisms

These results show that ion interactions produce compositional changes in silicon and silicon carbide thin films. The question arises as to the mechanisms by which these changes occur and, in particular, the relative importance of chemical and energy transfer effects. Insight into these issues is provided by Seitz,[15] who theoretically described the transfer of ion energy to the lattice.

As ions travel through the solid, they dissipate energy either by elastic or inelastic collisions. Elastic collisions cause lattice vibrations and may displace atoms from their lattice sites. Inelastic collisions result in electronic excitation or ionization of atoms in the lattice. Atoms can be displaced from the lattice when the ion energy is greater than a threshold value (~60 eV for the hydrogen ions / 20 - 30 eV for hydrocarbon ions). As the ion energy increases above this value, the rate of energy loss due to atom displacements increases rapidly, reaches a maximum, and then decreases rapidly to a low level. A similar threshold exists for electronic excitation (>1000 eV for ions used in this study). The projected range or normal penetration depth of the ions depends on their mass and energy. For 10, 100, 500, and 2000 eV H_x^+ ions, the ranges are 0.1, 1, 6, and 22 nm, respectively. For 500 eV CH_x^+ and $C_2H_x^+$ ions, the ranges are 4 and 2 nm, respectively.

Based on these considerations, 10 eV H_x^+ ions should produce chemical effects. In contrast, 100, 500, and 2000 eV H_x^+ ions and 500 eV hydrocarbon ions should produce both chemical and energy transfer effects. The 2000 eV H_x^+ ions should also produce electronic excitation effects (e.g. decomposition of Si-C or C-C species).

The results for 10 eV H_x^+ ions suggest that the chemical effects include the removal of SiC-Si from the near surface layers of the silicon carbide lattice. This most likely involves the formation and desorption of SiH_x species. Changes in the C-KLL line shape for 10 eV H_x^+ ions indicate that the carbon rich surface produced by the removal of SiC-Si has a high degree of sp^2 character which increases with ion fluence without an induction period. This could result from reconstruction of the surface to minimize the energy of the remaining carbon-network. Although 10 eV ions do not have sufficient energy to displace atoms from their bulk lattice sites, they may enhance the surface mobility of the carbon atoms and aid in minimizing the energy. For 100, 500, and 2000 eV ions, the observed decrease in SiC-Si is also presumably a chemical effect.

Results for 100, and 500 eV H_x^+ ions suggest the prompt formation of sp^3-C and the induction period for sp^2-C formation are energy transfer effects. The ability of these ions to produce sp^3-C species can be most simply explained by the displacement of carbon atoms into vacant silicon sites to produce sp^3 bonded C-C species. The induction period for sp^2-C can be explained by either of two mechanisms. The first involves the formation of a carbon network at the projected range of the H_x^+ ions. The second involves displacement of carbon atoms (damage) in the promptly formed sp^3-C layers.[10]

Energy transfer processes dominate the effects of 500 eV hydrocarbon ions. The initial rate of increase of the C-KLL peak for $C_2H_x^+$ ions are about two to three times those for CH_x^+ ions. This is most likely due to the fact that the projected range for CH_x^+ ions is twice that for $C_2H_x^+$ ions. The fact that sp^2-C species are a dominant initial interaction product suggests these species are associated with a high probability event such as the formation (implantation) of a carbon network at the projected range of the hydrocarbon ions. The slightly slower formation of SiC-C may then result from a lower probability event such as the displacement of silicon atoms from their lattice sites and their replacement by carbon atoms. The still slower formation of sp^3-C represents a continuation of this process.

The formation and desorption of SiH_x species by hydrogen ions has been discussed. It is, in fact, reasonable to expect that both silicon and carbons species are desorbed either as volatile hydrides or by sputtering when ion energies are above the threshold value. Since the observed trend is always toward the formation of carbon rich surfaces, the results of the present study suggest that the desorption or sputtering rate is lower for carbon species than it is for silicon species.

b) **Implications for Bias Enhanced Nucleation**

Using x-ray photoelectron spectroscopy (XPS) to follow the evolution of the diamond thin film during bias enhanced nucleation, Glass and coworkers[7] observed the formation of silicon carbide after the first minute and a C-C species after 5 minutes. For the next hour and twenty-five minutes, the relative amounts of SiC-C and C-C species remained unchanged. Then, a rapid increase in the C-C species intensity was observed that corresponded to the nucleation and growth of the diamond thin film.

Our results suggest that the situation prior to the appearance of the diamond thin film is a quasi-steady-state in which the growth of silicon carbide, using silicon provided by the substrate and hydrocarbon ions (and neutrals) provided by the plasma, is nearly offset by the removal of silicon and carbon by the formation of volatile hydrides and sputtering. Our results suggest that the silicon removal rate is greater than that for carbon which accounts for the appearance of the C-C species early in the nucleation process. Our results further suggest that the C-C species are a mixture of predominantly sp^3-C (based on the 250 eV ion energy which is above the threshold for displacements) and some sp^2-C. This can not be discerned, however, from the XPS data of Glass and coworkers.[7]

After ~1.5 hours, rapid growth of the diamond film occurs as the slowly growing silicon carbide layer reaches ~9 nm in thickness. Previous studies[15,16] indicate that at this thickness the availability of silicon becomes transport limited for the growth temperature used by Glass and coworkers. This additional depletion of silicon shifts the reaction from one in which silicon carbide is slowly formed to one in which a carbon rich diamond film is formed. The enhanced nucleation density of this film results from the fact that the initial C-C species is predominantly

sp^3-C (due to the 250 eV ion energy). That is, in our view, the initial C-C species observed by Glass and coworkers are in fact the diamond nuclei.

CONCLUSIONS

Hydrogen ions with energies of 10, 100, 500, and 2000 eV modify the silicon carbide surface and near surface layers by chemical and energy transfer mediated processes to produce sp^2- and sp^3-C species. The relative amounts of sp^2- and sp^3-C produced depend strongly on ion energy and fluence. The production of significant quantities of sp^3-C requires the displacement of carbon atoms from their lattice sites. The process leading to an induction period for sp^2-C formation is less obvious but may involve chemical effects due to ions thermalized in the sub-surface layers of the silicon carbide lattice. Alternatively, the induction period may represent slowly accumulated ion damage in the sp^3-C layers. The interaction of 500 eV hydrocarbon ions with silicon appears to be dominated by energy transfer effects and produces SiC-, sp^2-, and sp^3-C. These results provide a number of insights with regard to the mechanisms of bias enhanced nucleation. Specifically, hydrocarbon ions are responsible for the deposition of SiC and C-rich species, while hydrogen ions enhance the levels of sp^3-C.

ACKNOWLEDGMENTS

This research was supported by the Ballistic Missile Defense Organization/IST through the Office of Naval Research and the Advanced Research Programs Administration.

REFERENCES

1. S. Iijima, Y. Aikawa, and K. Baba, Appl. Phys. Lett. **57**, 2646 (1990).
2. S. Iijima, Y. Aikawa, and K. Baba, J. Mater. Res. **6**, 1491 (1991).
3. C.-P. Chang, D.L. Flamm, D.E. Ibbotson, and J.A. Mucha, J. Appl. Phys. **63**, 1744 (1988).
4. K.V. Ravi and C.A. Koch, Appl. Phys. Lett. **57**, 348 (1990).
5. K.V. Ravi, C.A. Koch, H.S. Hu, and A. Joshi, J. Mater. Res. **5**, 2356 (1990).
6. A.A. Morrish and P.E. Pehrsson, Appl. Phys. Lett. **59**, 417 (1991).
7. B.R. Stoner, G.-H.M. Ma, S.D. Wolter, and J.T. Glass, Phys. Rev. B **45**, 11067 (1992).
8. B.R. Stoner , B.E. Williams, S.D. Wolter, K. Nishimura, and J.T. Glass, J. Mater. Res. **7**, 257 (1992).
9. S.D. Wolter, B.R. Stoner, J.T. Glass, P.J. Ellis, D.S. Buhaenko, C.E. Jenkins, and P. Southworth, Appl. Phys. Lett. **62**, 1215 (1993).
10. J.M. Lannon, Jr., J.S. Gold, and C.D. Stinespring, J. Appl. Phys. **77**, 3823 (1995).
11. J.M. Lannon, Jr., J.S. Gold, and C.D. Stinespring, Proc. Mat. Res. Soc. **339**, 63 (1994).
12. *Index of Mass Spectral Data*, ASTM-STP #356, (Published by ASTM, 1963), p 1.
13. E.R. Malinowski, Analytical Chem. **49**, 606 (1977) and ibid, **49**, 612 (1977).
14. F. Seitz, Disc. Faraday Soc. **5**, 271 (1947).
15. C.D. Stinespring and J.C. Wormhoudt, J. Appl. Phys. **65**, 1377 (1989).
16. F. Bozso, J.T. Yates, Jr., W.J. Choyke, and L.Muehlhoff, J. Appl. Phys. **57**, 2771 (1985).

ION BEAM DEPOSITION OF BORON-ALUMINUM NITRIDE THIN FILMS

J.H. EDGAR[+], C.R. EDDY, JR.[*], J.A. SPRAGUE[*] and B.D. SARTWELL[**]
[+]Department of Chemical Engineering, Kansas State University, Manhattan, KS 66506-5102
[*]Code 6671, Naval Research Laboratory, Washington, DC 20375-5345
[**]Code 6170, Navel Research Laboratory, Washington, DC 20375-5342

ABSTRACT

Analysis of the phase behavior, structure, and composition of aluminum nitride thin films with up to 22% boron prepared by ion-beam assisted deposition is presented. The c-lattice constant of the film decreased with increasing boron content as expected from the formation of an AlN - wurtzite BN alloy. There was no evidence for separate boron nitride precipitation from either X-ray diffraction or FTIR. In contrast, Auger electron spectroscopy of the boron present in the films suggested that two types of boron bonding was present.

INTRODUCTION

The group III nitrides have come under increasing scrutiny in the past seven years for their potential applications as wide band gap semiconductors for high power-high frequency electronics and short wavelength optoelectronics.[1-3] A practical advantage of the group III nitrides is their formation of a continuous range of psuedobinary alloys, at least among AlN, GaN, and InN. Even greater control of the lattice constants and energy band gap of the group III nitrides could be achieved if the alloying capability also included solid solutions with boron nitride (BN).

There are considerable structural and bonding differences between boron nitride and the other group III alloys suggesting that the boron solubility would be limited. Previous studies of BN-AlN and BN-GaN solubility have been ambiguous. Lyutaya and Bartnitskaya[4] prepared nonstoichiometric (mole fraction B, Al, Ga greater than the N mole fraction) compounds, speculating that the boron substitutes on both Al or Ga and N positions. The B-Al-N thin films prepared by Noreika and Francombe[5] by reactive sputtering had smaller lattice constants than pure AlN, which the authors suggest might be caused by B substitution on Al sites or from the formation of aluminum-boron compounds.[5]

In this paper, the results on the solubility of boron in aluminum nitride thin films prepared by ion-beam assisted deposition (IBAD) are reported. The film's composition and structure were analyzed by Auger electron spectroscopy (AES) and x-ray diffraction, respectively. Fourier transform infrared (FTIR) spectroscopy of the films was taken to provide insights into the localized structure of the bonds, and to determine if BN precipitates form.

EXPERIMENTAL

The BAlN films were prepared by IBAD from electron beam evaporated boron and aluminum, and nitrogen ions supplied from a Kaufman ion source. Silicon (111), both on-axis and 1 to 2 degrees off-axis, were used as substrates. The Al evaporation rate, as detected by quartz crystal monitors, was held constant at 2.2 Å/second, and the B evaporation rate was adjusted to achieve the desired composition. Two nitrogen beam energies were employed, 100 and 200 V, with depositions at two substrate temperatures, 600 and 675 °C. The nitrogen current flux, as measured by Faraday cups positioned in the same plane as the substrates, was maintained such that the arrival rate of N/(B+Al) was always greater than 1.1. Substrates were transferred into the deposition chamber via a vacuum load lock. The system base pressure was typically 1.0 x 10^{-8} Torr and during deposition the pressure was maintained at 9.0 x 10^{-5} Torr by adjusting the flow of nitrogen to the ion gun. All films were approximately 8000 Å thick.

Auger electron spectroscopic (AES) analyses were performed on the alloy films using a Physical Electronics Industries Model 10-155 cylindrical mirror analyzer with a co-axial electron gun. For all analyses, the electron beam energy and current were 3.0 keV and 25 µA (35 mA/cm^2), respectively and the CMA modulation was 3.7 eV. Although depth profiles of the films were not obtained because of their thickness, sputtering using 3.1 keV Ar$^+$ (current density 20 µA/cm^2) ions was employed to remove surface impurities and a sufficient thickness of the film to achieve a steady-state composition. Charging of the films during analysis was a problem which was overcome by rotating the substrate such that the angles of incidence of the electron beam and argon ion beam were 50° and 40°, respectively, with respect to the substrate normal. This resulted in a balance between the incident electrons and argon ions, and the emitted secondary electrons such that Auger spectra could be acquired.

All spectra were acquired manually in the dN(E)/dE mode after sputtering for approximately 15 minutes. In order to determine the Al, B, and N concentrations in the films, calibration samples were produced by hot pressing cubic BN, hexagonal BN, and AlN powders into pellets. Auger spectra were obtained for the BN samples; however, significant oxygen contamination which could not be removed by sputtering precluded obtaining a satisfactory AlN standard spectrum. Therefore, an assumption was made that the Auger spectrum obtained for the Al-N films deposited with no boron were, in fact, stoichiometric AlN. With this information it was concluded that the sensitivity factors of 0.15, 0.24, and 0.35 given in the Handbook of Auger Electron Spectroscopy for B-KLL, Al-LVV, and N-KLL, respectively, were inaccurate for determining concentrations in the Al-B-N films. For example, using identical analysis parameters, the B peak height in the BN calibration sample was 50% greater than the Al peak height in the AlN film. Also, the N peak height in the AlN films was twice that of the N in the BN sample. From these analyses, a sensitivity factor, S, of 0.15 was assumed for the Al peak, which resulted in values of 0.23, 0.44, and 0.23 for B in BN, N in AlN, and N in BN, respectively. These sensitivity factors were used to calculate atomic fractions X(i) from the AES peak heights I(i) using the expression:

$$X(i) = \left[\frac{I(i)}{S(i)}\right]\bigg/\left[\sum_j \frac{I(j)}{S(j)}\right] \qquad (1)$$

Since the B concentrations in all of the films was less than 10% and the Al/N peak height ratios remained essentially constant, the value of 0.44 for the N sensitivity factor was used for all of the films.

The lattice constants of the films were measured by x-ray diffraction with Cu-K$_\alpha$ radiation. Symmetric reflections of the (0002) and (0004) planes were taken to determine the c-lattice constant and asymmetric reflections were taken from the (10$\bar{1}$5) and the (21$\bar{3}$3) planes in order to calculate the in plane a lattice constants. The infrared absorption of the films was measured by Fourier transform infrared spectroscopy from films deposited on double side polished silicon.

RESULTS

The AES spectra for AlN, BN, and one alloy film is presented in Figure 1. The line shape of the boron in the alloy had a doublet and was clearly different from the signal for the pure boron nitride film. The characteristic Auger signal for boron is known to depend its bonding, with the signal for hBN and cBN peaking at two different energies[6]. In the alloy film the doublet in the peak shape for the alloy film indicates the boron is present in two bonding states. The imprecision in the energy scale due to charging effects made identification of these states impossible. The calculated atomic fraction for B+Al was equal to the nitrogen fraction (50%), within experimental error, for all films.

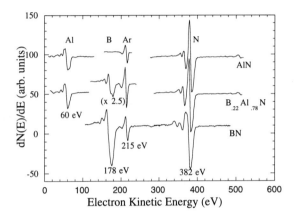

Figure 1. Auger electron spectra of an AlN, BN, and BN-AlN alloy thin film.

For pure AlN and alloy films deposited with a nitrogen beam voltage of 100 V, the films were highly aligned with the c-axis perpendicular to the substrate surface. Only the (0002) and (0004) diffraction peaks were present in symmetric theta-2theta x-ray diffraction patterns (Fig 2.a). With a nitrogen beam voltage of 200 V, the peak widths increased, the intensities decreased, and additional peaks were present (Fig2.b).

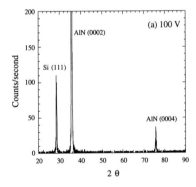 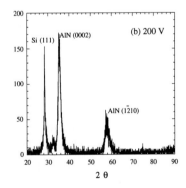

Figure 2.   X-ray diffraction patterns from AlN deposited with a nitrogen beam voltage of (a) 100 V and (b) 200 V.

For pure AlN films the c-lattice constant was higher than the reported bulk value and the a-lattice constants was smaller indicating the films were under biaxial stress: in compression in the plane of the film and in tension perpendicular to the surface. The presence of compressive stresses in IBAD AlN films was also observed by Windischmann[7] and Martin et al[8]. The stress increased with the voltage of the nitrogen beam as evidenced by the increase in the c-lattice constant. From electron diffraction taken in a transmission electron microscope, the films were determined to be polycrystalline with random a-axis orientation in the plane of the film.

For pure BN, no x-ray diffraction peaks other than the Si(111) and Si(222) substrate peaks were observed. This is not too surprising because of the low atomic numbers of the constituent elements in BN. Electron diffraction produced in a transmission electron microscopy revealed that the films were h-BN, with a very fine grain size (10-20 Å).

Adding B to the AlN produced two effects: a broadening of the diffraction peak width and a shift in the peak positions to greater angles corresponding to a reduction in the lattice constants. This effect can be seen in detailed XRD from the (0004) peak in Fig. 3. No additional peaks attributable to BN were observed either due to the absence of precipitates or their weak interaction with x-rays.

The c-lattice constant is plotted as a function of the boron content as measured by AES in Fig. 4. Also included in Fig. 4 is the predicted lattice constant for the alloy based on a linear change from the reported values for AlN (c= 4.98 Å)[9] and wurtzitic BN (wBN, c= 4.213 Å)[10]. The measured c-lattice constants were all higher than expected based on the predictions from bulk materials due to stress induced in the films from ion bombardment, and the values for films deposited with the 200 V nitrogen beam were higher than those deposited at 100 V. Nevertheless, the c-lattice constant decreases at the same rate as the predicted values. This suggests that AlN-wBN alloys were formed.

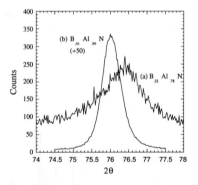

Figure 3. The x-ray diffraction pattern from the (0004) plane for (a) $B_{.22}Al_{.78}N$ and (b) $B_{.01}Al_{.99}N$ alloy films.

Unfortunately, the values of the measured a-lattice constant had a great deal of scatter, and no clear change with boron concentration could be discerned. This is possibly due to difficulties in accurately measuring the peak positions from the asymmetric x-ray diffraction as the intensities from the $(10\bar{1}5)$ and the $(21\bar{3}3)$ planes were much weaker than the diffraction from the (0002) and (0004) planes. In fact, due to the relatively poor crystallinity of the films deposited with a nitrogen

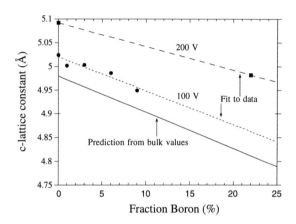

Figure 4.   The c-lattice constant for films as a function of boron content at 100 V (●) and 200 V (■). The solid line is the predicted value based on bulk values of AlN and wBN, and the dashed curves are linear regression fits to the data at 100 V and 200 V.

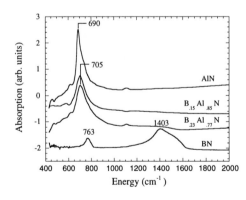

Figure 5. FTIR spectra for AlN, BN, and two alloy films.

The characteristic infrared absorption for AlN is reported to occur at 666.7 cm^{-1} (TO mode)[11] but for our films, the peak absorption occurred at slightly higher values; 690 cm^{-1} for films deposited at a nitrogen beam voltage of 100 V and 698 cm^{-1} for films deposited at 200 V (Fig. 5). This shift in peak position was due to stress in the films caused by the nitrogen ion bombardment, as was also observed by Windischmann.[7] The reported characteristic infrared absorption peaks for hexagonal BN (hBN) are positioned at 780 and 1400 cm^{-1} [12]; for our films the peak positions were 763 and 1403 cm^{-1} (Fig. 5). No evidence for either wurtzite BN (1090, 1120, and 1230 cm^{-1}) or cubic BN (1065 and 1340 cm^{-1})[12] was observed.

The infrared absorption of the alloy films was similar to that of pure AlN, except for a shift in the peak position to higher wavenumbers and a broadening of the peak width. Such a shift would be expected from alloy formation. No additional absorption peaks characteristic of any type of BN were observed.

CONCLUSIONS

Characterization of the thin films of AlN-BN prepared by IBAD presented contradictory evidence about the extent of solid solubility. A doublet in the boron signal detected by Auger electron spectroscopy of the alloy films indicates two types of boron bonding, However, no additional absorption peaks were present in the FTIR spectra as would be expected from precipitated BN. The c-lattice constants of the alloys had the same dependence on boron content as would be expected from a AlN-wBN alloy. It seems likely that at least some B solubility in AlN occurs resulting in the stabilization of sp^3 bonded alloy.

REFERENCES

1. R.F. Davis, Proc. IEEE **7**, 702 (1991).

2. J.H. Edgar, J. Mater. Sci. **7**, 235 (1992).

3. S. Strite and H. Morkoc, J. Vac. Sci. Technol. B **10**, 1237 (1992).

4. M.D. Lyutaya and T.S. Barnitskaya, Inorg. Mater. **9**, 1052 (1973).

5. A.J. Noreika and M.H. Francombe, J. Vac. Sci. Technol. **6**, 722 (1967).

6. G. Hanke, M. Kramer and K. Muller, Mater. Sci. Forum **54&55**, 207 (1990).

7. H. Windischmann, Thin Solid Films **154**, 159 (1987).

8. P. Martin, R. Netterfield, T. Kinder, and A. Bendavid, Appl. Optics **31**, 6734 (1992).

9. W.J. Meng in Properties of Group III Nitrides, J.H. Edgar ed. (INSPEC, London, 1994) p. 22.

10. J.H. Edgar, ibid p. 7.

11. L.E. McNeil, ibid p. 249.

12. G.L. Doll, ibid p. 241.

ION-BEAM-INDUCED EPITAXY AND SOLID PHASE EPITAXY OF SiGeC ON Si FORMED BY Ge AND C ION IMPLANTATION AND THEIR STRUCTURAL AND OPTICAL PROPERTIES

N. KOBAYASHI*, H. KATSUMATA**, Y. MAKITA*, M. HASEGAWA*, N. HAYASHI*, H. SHIBATA*, S. UEKUSA** and S. HISHITA***

* Electrotechnical Laboratory, 1-1-4 Umezono, Tsukuba, Ibaraki 305 Japan
** Meiji University, 1-1-1 Higashimita, Tama, Kawasaki, Kanagawa 214, Japan
*** National Institute for Research in Inorganic Materials, 1-1 Namiki, Tsukuba, Ibaraki 305 Japan

ABSTRACT

Epitaxial layers of $Si_{1-x-y}Ge_xC_y$ on Si(100) (x=0.13 and y=0.014 at peak concentration) were formed by ion implantation of Ge ions and C ions at room temperature (RT) and by subsequent IBIEC (ion-beam-induced epitaxial crystallization) process with 400keV Ge and Ar ions at 300-400°C and SPEG (solid phase epitaxial growth) process up to 840°C. Crystallization up to the surface both by IBIEC and SPEG processes has been confirmed with RBS-channeling analysis. X-ray diffraction experiments have demonstrated strain compensation by incorporation of C atoms for IBIEC-grown $Si_{1-x-y}Ge_xC_y$/Si samples, whereas strain accommodation due to C precipitation has been observed for SPEG-grown $Si_{1-x-y}Ge_xC_y$/Si samples. Photoluminescence (PL) observed at 2K from IBIEC-grown samples has shown intense I_1 peak with/without I_1 related (Ar) peak and that from SPEG-grown samples has shown G line emission. These optical properties could suggest that small vacancy agglomeration is dominant in IBIEC-grown samples and C agglomeration is dominant in SPEG-grown samples, respectively.

INTRODUCTION

SiGeC on Si has recently attracted interest because of the structural advantage in strain compensation and of bandgap modification by C atom incorporation in SiGe [1-3]. The optical properties, however, have not been well understood so far for SiGeC on Si. Gupta et al. have investigated donor complex formation in high-dose Ge implanted Si with and without C implantation [4]. Boucaud et al. have reported the band-edge luminescence at 77K from $Si_{1-x-y}Ge_xC_y$ alloys formed by the rapid thermal chemical vapor deposition [5].
On the other hand, ion-beam-induced epitaxial crystallization (IBIEC) in semiconductors has an attractive feature that it induces crystallization at lower temperatures than in the process of solid phase epitaxial growth (SPEG) [6-8]. Moreover, IBIEC can provide non-thermal equilibrium incorporation of foreign atoms far in excess of the solubility limit. C is known to be highly insoluble in Si and C precipitation is reported to be a severe problem for SiGeC in the SPEG process [9]. Comparison of IBIEC process to SPEG process in $Si_{1-x-y}Ge_xC_y$/Si growth is, therefore, an investigation of interest from the material synthesis point of view.
In this paper we present the structural and optical properties of $Si_{1-x-y}Ge_xC_y$ layers on Si(100) formed by high-dose Ge and C ion implantation and subsequently grown both by IBIEC and SPEG processes. Since part of the structural properties for IBIEC growth for $Si_{1-x-y}Ge_xC_y$/Si has been already reported elsewhere [10], optical properties that are relevant to structural properties are focused in the following.

EXPERIMENTAL

Non-doped (100)-oriented Si wafers (float-zone, $\rho > 0.1$ Ωcm) were implanted at room temperature with 80keV ^{74}Ge ions to 10^{16} cm^{-2} with a current density of about 0.7μA/cm^2 and

189

with 17keV ^{12}C ions to 5.5×10^{15} cm^{-2} with a current density of about 0.1 μA/cm^2. Energies of Ge and C ions were selected so as to give almost the same depth profiles in the implanted layer. An amorphous layer of Si$_{1-x-y}$Ge$_x$C$_y$ with about 90nm thickness and a deeper amorphous Si layer with about 50nm thickness were formed by these processes. The peak concentration of Ge atoms is about 13 at.%, which was observed by Rutherford backscattering spectrometry (RBS) analysis and that of C atoms is calculated to be about 1.4 at.%, which was confirmed by secondary ion mass spectrometry (SIMS) analysis. These samples were then subjected to bombardments of 400keV Ge or Ar ions at 300-400°C for the IBIEC process. The ion current density for IBIEC was maintained to be 0.6 μA/cm^2 for Ge ions and 2 μA/cm^2 for Ar ions. SPEG experiments on Si$_{1-x-y}$Ge$_x$C$_y$ layers on Si were performed in a vacuum up to 840°C using an infrared gold image furnace.

Crystalline growth and structural properties of the grown layers were analyzed by the RBS-channeling technique using 2MeV He$^+$ ions. X-ray rocking curve measurements with 4 Ge crystals (220 reflection) (quadruple crystal diffraction x-ray diffraction (QCD-XRD)) using Cu Kα_1 radiation were performed to evaluate the properties of the strain in crystallized layers.

Photoluminescence (PL) measurements at 2K were performed using 514.5 nm line of an Ar$^+$ ion laser with average excitation power of 50 mW for exciting light source. The emission was dispersed by 1m grating monochromator and was detected by using a liquid nitrogen-cooled Ge p-i-n photodiode.

RESULTS AND DISCUSSION

Structural Properties

Figure 1 shows RBS-channeling spectra (random and ‹100› channeled direction) at high resolution glancing angle (105° to the beam incidence direction) representing the crystalline growth up to the surface for Si$_{1-x-y}$Ge$_x$C$_y$/Si by the IBIEC process with 400keV Ge ion bombardments to 6×10^{15} cm^{-2} at 350°C. Observed value of χ_{min} just below the surface in the scattering yield by Si amounts to 10% and the ratio of scattering yields by whole implanted Ge atoms in aligned direction to those in the random direction amounts to 0.13. An increase in dechanneling yield near the depth of peak Ge concentration was observed both in Si and Ge yields. The increase of dechanneling yield deeper than 340 channel reflects defects around the end-of-range (EOR) created by 400keV Ge bombardments.

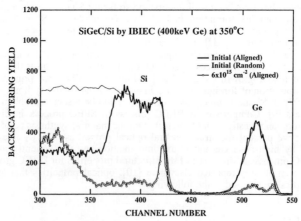

Fig.1 RBS-channeling spectra representing the crystalline growth of SiGeC/Si by IBIEC with 400keV Ge at 350°C.

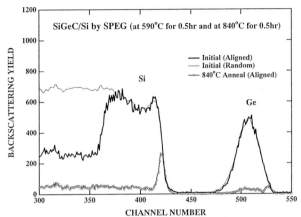

Fig.2 RBS-channeling spectra representing the
crystalline growth of SiGeC/Si by SPEG up to 840°C.

Cross-sectional TEM observation has indicated that no dislocation has been seen in the surface $Si_{1-x-y}Ge_xC_y$ layers but feeble small defects are observed in the region from the surface to a depth of about 200nm. Dense dislocation networks were observed in the region between depths from 200nm to 500nm, which corresponds to the distorted region around EOR. $Si_{1-x-y}Ge_xC_y$/Si samples bombarded with 400keV Ar ions to 2.2×10^{16} cm^{-2} at 350°C have shown similar RBS-channeling spectra representing the crystallization up to the surface except that the regions of defects at EOR are slightly deeper than in the Ge-bombarded samples. SIMS measurements for Ar-bombarded samples have shown the same depth profiles for Ge and C atoms with peak concentration around 50-60 nm in depth and the atomic concentration of Ar atoms is around 0.1% in the surface layer [10].

Figure 2 shows random and ‹100› channeled RBS spectra observed at 105° to beam incidence, which showing crystalline growth for the $Si_{1-x-y}Ge_xC_y$/Si sample by SPEG experiments performed at 590°C for 0.5hour and at 840°C for 0.5hour. The value of χ_{min} observed at the surface in Si yield amounts to 6.2% and the ratio of scattering yields by whole implanted Ge atoms in aligned direction to those in the random direction amounts to 0.13. These values reflect better crystalline quality for SPEG-grown samples than for IBIEC-grown samples.

Figure 3 shows QCD-XRD rocking curve diffraction scans along (004) reflection planes for samples of $Si_{1-x-y}Ge_xC_y$/Si that were crystallized up to the surface by the IBIEC process with 400keV Ar to 2×10^{16} cm^{-2} at 400°C and by the SPEG process at 590°C for 0.5hour and 840°C for 0.5hour, which is the same sample shown in fig.2. Main peaks observed both in the spectra at $\Delta\Theta=0$ are the (004) diffraction peaks from the Si substrates. The IBIEC-grown $Si_{1-x-y}Ge_xC_y$/Si sample shows no distinctive subpeaks both in negative and positive angles, which suggesting the strain compensation by the incorporation of C atoms in the lattice sites. The thin solid curve denotes a simulation for reflection from the $Si_{1-x-y}Ge_xC_y$/Si with a polygonal depth profile of Ge and C atoms. The polygonal shape was composed of a linear increase in concentration of incorporated atoms from a depth of x=0 to x=48nm, a constant concentration between depths of x=48nm and x=60nm and a linear decrease from x=60 to x=116nm [10]. The SPEG-grown $Si_{1-x-y}Ge_xC_y$/Si sample shows subpeaks mainly at negative angles. The simulation (dotted line) was obtained by the same procedure indicated above but the whole C concentration was decreased down to 75% of implanted atoms. This fact reflects the incomplete C incorporation into lattice sites probably due to C precipitation in the SPEG-

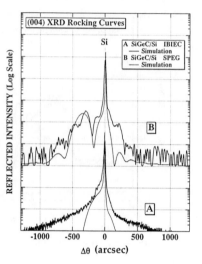

Fig.3 X-ray rocking curves and simulations for (004) reflections from (A) SiGeC/Si grown by IBIEC with 400keV Ar at 400°C and (B) SiGeC/Si grown by SPEG up to 840°C.

grown $Si_{1-x-y}Ge_xC_y/Si$ sample, which was also observed in the $Si_{1-x-y}Ge_xC_y/Si$ and $Si_{1-y}C_y/Si$ samples grown by SPEG around 850°C [9].

<u>Optical Properties</u>

Figure 4 demonstrates PL spectra observed at 2K for the $Si_{1-x-y}Ge_xC_y/Si$ sample grown by the IBIEC process with 400keV Ge bombardments to $6x10^{15}$ cm^{-2} at 350°C (SiGeC/Si (IBIEC-Ge)) which is the same sample shown in fig. 1, and that grown with 400keV Ar bombardments to $3.2x10^{16}$ cm^{-2} at 300°C (SiGeC/Si (IBIEC-Ar)). A sharp peak at 1.019eV observed both in the samples is identical to I_1 (or W) non-phonon peak, which is frequently observed in ion-implanted and annealed Si samples [11, 12]. Although the origin of this peak has not yet been completely identified, it was found that the I_1 emission spectrum didn't depend on the doping species of starting materials nor on the persistent impurities like C or O atoms in Si. Small vacancy clusters are thought to be one of the possible candidates for this optical center [11, 13]. The vibronic sideband peaks are observed between 0.975eV and 1.005eV. The feeble peak at 0.968eV corresponds to G peak [14], and the peak at 1.040eV is identical to I_3 peak [11, 12]. The SiGeC/Si (IBIEC-Ar) sample shows, in addition, another sharp peak at 1.0095eV and its sideband emissions. This I_1-related peak at 1.0095eV was observed in Ar-implanted and annealed Si sample and interpreted to be owing to incorporation of noble gases in the intrinsic defects which constituting related deep optical defects [15]. The overall feature of the present PL spectra is, therefore, found to be similar to those observed in ion-implanted and annealed Si.

Our recent observations by positron lifetime experiments on defects for IBIEC-grown Si layer pre-amorphized by 130keV Ge implantation to $7x10^{14}$ cm^{-2} have revealed the formation of small vacancy clusters (trivacancies and quadrivacancies) in the grown layer [16]. This is consistent with the fact that IBIEC-grown layer shows strong I_1 emission, when I_1 peak is assumed to be due to vacancy clusters (a trigonal tetravacancy cluster [13]) in Si. Northrop *et al.* have reported the effect of the strained $Si_{1-x}Ge_x/Si$ structure or $Si_{1-x}Ge_x/Si$ multiquantum-well (MQW) structure on the I_1 peak emission in the Si-implanted and annealed samples [17]. They have argued that I_1 becomes broadened when the emission originates within the alloy layers. Our results didn't show any broadening nor shift of the I1 peak in the $Si_{1-x-y}Ge_xC_y$ alloy layers. One possible reason is that electronic structure doesn't undergo any big difference in the surface $Si_{1-x-y}Ge_xC_y$ layer [2]. Nevertheless, we cannot neglect the possibility that I_1 peak emission originates mainly from the underlying Si substrate and not

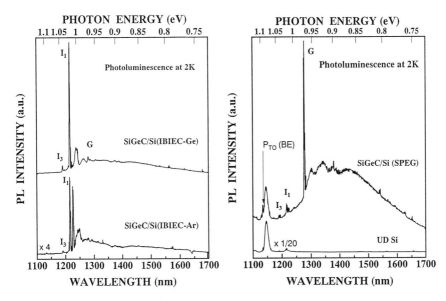

Fig.4 PL spectra at 2K for two SiGeC/Si samples, grown by IBIEC with 400keV Ge at 350°C (IBIEC-Ge) and with 400keV Ar at 300°C (IBIEC-Ar).

Fig.5 PL spectra at 2K for the SiGeC/Si sample, grown by SPEG up to 840°C and for undoped Si.

from the surface $Si_{1-x-y}Ge_xC_y$ layer. Although further investigation is necessary to reveal more precise nature of PL emission, it is noted that we observe the similar PL emissions in $Si_{1-x-y}Ge_xC_y$/Si samples grown by IBIEC to those observed in ion-implanted and annealed Si samples.

Figure 5 shows the PL spectra for the $Si_{1-x-y}Ge_xC_y$/Si grown by SPEG (annealed at 590°C for 0.5hour and 840°C for 0.5hour, (SiGeC/Si (SPEG))), which is the same sample shown in figs. 2 and 3, and for the undoped Si crystal layer. The spectrum for SiGeC/Si (SPEG) shows a different feature from those observed in IBIEC-grown $Si_{1-x-y}Ge_xC_y$/Si samples. The narrow peak at 0.969eV corresponds to G peak, which has been identified to be an emission due to C_S-Si_i-C_S carbon pair [14]. The broad sideband peaks around 0.952eV and 0.924eV are identical to perturbed emission form of G peak [18]. The I_1 peak is weak in this sample, which is constant with the fact that I_1 peak is normally annealed out by the annealings above 400°C in Si [11, 15]. G line is, on the other hand, still survives in the present $Si_{1-x-y}Ge_xC_y$/Si (SPEG) sample, although it is also normally annealed out by the annealings above 400°C in Si [11]. Moreover, we have observed no G peak emission from the SPEG-grown $Si_{1-x-y}Ge_xC_y$/Si sample annealed up to 700°C. By taking into account the origin of the G peak, therefore, C agglomeration in the sample grown by SPEG up to 840°C has occurred in the annealing process. Although the RBS spectrum for the SPEG-grown sample shown in fig.2 has demonstrated better crystalline quality than the IBIEC-grown sample, the C precipitation could result in the optical G center formation in the SPEG-grown samples This is consistent with the above-mentioned XRD observations, where the C precipitation has revealed the incomplete strain compensation.

Although overall features of PL emissions in $Si_{1-x-y}Ge_xC_y$/Si (IBIEC) and $Si_{1-x-y}Ge_xC_y$/Si (SPEG) are similar to those in ion-implanted and annealed Si, the main process producing the optical centers for non-phonon peaks I_1 and G are thought to be different. The intense I_1 and I_1-related (Ar) emissions in the IBIEC-grown samples could be due to IBIEC-induced small vacancy agglomeration and the G peak emission in the SPEG-grown samples could be due to C precipitation in the high-temperature annealing process for the highly C implanted samples.

SUMMARY AND CONCLUSIONS

Epitaxial layers of $Si_{1-x-y}Ge_xC_y$ on Si(100) were formed by ion implantation of Ge ions and C ions at room temperature (RT) and subsequent IBIEC process with 400keV Ge and Ar ions at 300-400°C and SPEG process up to 840°C. RBS measurements have revealed crystallization up to the surface both in IBIEC and SPEG processes. Although XRD experiments have demonstrated good strain compensation by incorporation of C atoms for $Si_{1-x-y}Ge_xC_y/Si$ (IBIEC), incomplete strain compensation due to C precipitation has been observed in $Si_{1-x-y}Ge_xC_y/Si$ (SPEG) samples. PL emissions from both IBIEC-grown and SPEG-grown samples have shown similar peaks to those by ion-implanted and annealed Si samples with intense I_1 and I_1-related (Ar) peaks in the IBIEC-grown samples and with intense G peak emission due to C_S-Si_i-C_S defect formation in the SPEG-grown samples. These features are thought to reflect the fact that small vacancy agglomeration is dominant in IBIEC-grown samples and C precipitation is dominant in SPEG-grown samples.

REFERENCES

1. S. Furukawa, H. Etoh, A. Ishizawa and T. Shimada, U.S. Patent No. **4885614** (1989).
2. R.A. Soref, J. Appl. Phys. **70**, 2470 (1991).
3. K. Eberl, S.S. Iyer, S. Zollner, J.C. Tsang and F. K. LeGoues, Appl. Phys. Lett. **60**, 3033 (1992).
4. A. Gupta, C. Cook, M.M. Rahman, J. Qiao, C.Y. Yang, S. Im, N.W. Cheung and P.K.L. Yu, J. Appl. Phys. **75**, 4252 (1994).
5. P. Boucaud, C. Trancis, F.H. Julien, J.-M. Lourtioz, D. Bouchier, S. Bodnar, B. Lambert and J.L. Regolini, Appl. Phys. Lett. **64**, 875 (1994).
6. J. Linnros, G. Holmen and B. Svensson, Phys. Rev. B **32**, 2770 (1985).
7. J.S. Williams, R.G. Elliman, W.L. Brown and T.E. Seidel, Phys. Rev. Lett. **55**, 1482 (1985).
8. J. Nakata, Phys. Rev. B **43**, 14643 (1991).
9. J.W. Strane, W.J. Edwards, J.W. Mayer, H.J. Stein, S.R. Lee, B.L. Doyle and S.T. Picraux, Mat. Res Soc. Symp. Proc. **280**, 609 (1993).
10. N. Kobayashi, M. Hasegawa, N. Hayashi, H. Tanoue, Y. Makita and H. Shibata, Nucl. Instrum. Meth. (in press).
11. C.G. Kirkpatrick, J. R. Noonan and B.G. Streetman, Rad. Effects **30**, 97 (1976).
12. G. Davies, Phys. Rep. **176**, 83 (1989).
13. Y.H. Lee and J.W. Corbett, Phys. Rev. B **9**, 4351 (1974).
14. K. P. O'Donnell, K.M. Lee and G.D. Watkins, Physica **116B**, 258 (1983).
15 N. Burger, K. Thonke and R. Sauer, Phys. Rev. Lett. **52**, 1645 (1984).
16. N. Hayashi, R. Suzuki, M. Hasegawa, N. Kobayashi, S. Tanigawa and T. Mikado, Phys. Rev. Lett. **70**, 45 (1993) .
17. G.A. Northlop, D.J. Wolford and S.S. Iyer, Appl. Phys, Lett. **60**, 865 (1992).
18. G.Davies, E.C. Lightowlers and M.C. do Carmo, J. Phys. C **16**, 5503 (1983).

SOLID PHASE EPITAXY OF IMPLANTED SI-GE-C ALLOYS

Xiang Lu, Applied Science and Technology Graduate Group,
Nathan W. Cheung, Department of Electrical Engineering and Computer Sciences,
University of California, Berkeley, CA 94720

ABSTRACT

$Si_{1-x-y}Ge_xC_y$/Si heterostuctures were formed on Si (100) surface by Ge and C implantation with a high dose rate MEtal - Vapor Vacuum Arc (MEVVA) ion source and subsequent Solid Phase Epitaxy (SPE). After thermal annealing in the temperature range from 600 °C to 1200 °C, the implanted layer was studied using Rutherford Back-scattering Spectrometry (RBS), cross-sectional High Resolution Transmission Electron Microscopy (HRTEM) and four-bounce X-ray Diffraction (XRD) measurement. Due to the small lattice constant and wide bandgap of SiC, the incorporation of C into Si-Ge can provide a complementary material to Si-Ge for bandgap engineering of Si-based heterojunction structure. Polycrystals are formed at temperature at and below 1000 °C thermal growth, while single crystal epitaxial layer is formed at 1100 °C and beyond. XRD measurements near Si (004) peak confirm the compensation of the $Si_{1-x} Ge_x$ lattice mismatch strain by substitutional C. C implantation is also found to suppress the End Of Range (EOR) defect growth.

INTRODUCTION

The Si-Ge system has been extensively synthesized by researchers using various synthesis techniques, such as Molecular Beam Epitaxy (MBE), Ultra-High Vacuum Chemical Vapor Deposition (UHV-CVD), and ion implantation with subsequent Solid Phase Epitaxy (SPE) [1-3]. The high-dose Ge implantation process is advantageous over the other two direct epitaxial methods because it is compatible with conventional Si processing technology, and can readily be integrated into standard IC process sequence. However, crystal defects introduced by the implantation process, such as the End Of Range (EOR) dislocation loops formed near the amorphous/crystalline (a/c) interface after SPE, have to be eliminated for any device applications.

Recently, interest in the group IV Si-Ge system has extended to the Si-Ge-C ternary system [4, 5]. One of the fundamental difficulty to form defect-free $Si_{1-x} Ge_x$/Si heterojunction is due to the over 4% lattice constant mismatch between Si ($a_{Si} = 0.543$ nm) and Ge ($a_{Ge} = 0.566$ nm). C is the only element in group IV with a single crystal (diamond) lattice constant less than Si, and has the potential to compensate the built-in strain in the Si-Ge system. Recent work by Im et al. [5] has demonstrated that $Si_{1-x-y}Ge_xC_y$ shows less strain induced dislocation formation than $Si_{1-x}Ge_x$ alloy. Being isoelectronic with Si and Ge, C is not expected to be a dopant. Nevertheless, it is also known that C inhibits the kinetics of Si SPE such that a low growth rate is expected.

This paper presents the material synthesis results of $Si_{1-x-y}Ge_xC_y$ alloy formed by high-dose, implantation using a MEtal-Vapor Vacuum Arc (MEVVA) source [6, 7]. Different from conventional implanters, the MEVVA source provides a high current density implantation using solid Ge and C (graphite) sources. A relatively low voltage (i.e. 20 kV) is needed in our work to synthesis surface layer of Si-Ge and Si-Ge-C alloys. Issues on the solid phase epitaxy crystalline quality of the $Si_{1-x-y}Ge_xC_y$ layer, as well as the recovery of the implantation damage are reported in this paper. Pfiester et al. [8] has demonstrated that Ge implantation can retard

Mat. Res. Soc. Symp. Proc. Vol. 388 © 1995 Materials Research Society

the Boron diffusion at the field oxide area. Oztwk et al. [9] has shown shallower source and drain junction depth using surface Ge pre-amorphization technique. Besides that, we expect the surface Ge-Si-C alloys to be potential candidates for shallow junction technology for the following reason: the heterojunction formation between the source/drain area and the channel region can provide the extra potential barrier for P-MOSFET off-state leakage current. It will minimize the Drain Induced Barrier Lowering (DIBL) effect, and improve MOSFET performance when the transistor channel length scales down to the deep sub-micron regime.

EXPERIMENTAL DETAILS

CZ grow p-type (20 ohm-cm) Si wafers were implanted with Ge ions to a dose of 8.0×10^{15} cm^{-2} at 20 kV using the MEVVA source [6,7]. Both single charged Ge^+ and double charged Ge^{++} were produced by the ion source with the composition ratio of 60% and 40% respectively. Without mass/charge selective devices, both Ge^+ and Ge^{++} were implanted at 20 kV simultaneously. We used 0.3 ms beam pulses at a maximum beam current of 3 mA/sec. The beam pulse repetition rate was 5 pulses/second. The C ions produced in the MEVVA source were single charged C^+, and they were implanted at 10 kV following the Ge implantation. A dose of 2.0×10^{15} cm^{-2} C^+ was implanted with a current condition similar to the Ge implantation. Since the regrowth annealing temperature was up to 1200 °C, and the implanted layer was quite shallow, an extra oxidation barrier coating was necessary for surface passivation. The samples were coated with 30 nm Si_3N_4 film using Plasma Enhanced Chemical Vapor Deposition (PECVD) at 300 °C with a deposition rate of 15 nm/min. With the surface protected by the Si_3N_4 coating, the implanted samples were annealed in the temperature range from 600 to 1200 °C in a N_2 ambient for 30 minutes. After the high temperature SPE, the Si_3N_4 coating were etched away in phosphoric acid at 160 °C.

The Ge concentration profiles were measured using Rutherford Back-scattering Spectrometry (RBS) with a 2.0 MeV He ion beam. The RBS experiments were performed at a scattering angle $\Theta = 165°$. The solid phase epitaxial growth kinetics, the epitaxial layer crystallinity and the implantation damage were measured by ion channeling experiment in both <110> and <111> axial directions. Cross-sectional High Resolution Transmission Electron Microscopy (HRTEM) was performed using a JEOL JEM 200CX microscope operating at 200 kV. The epitaxial layer strain was examined using x-ray rocking scan techniques near the Si (004) peak with a Siemans D5000 X-ray Diffractometer.

RESULTS AND DISCUSSION

Based on TRIM-90 Monte Carlo simulation, only a single Ge peak concentration will form even though both the Ge^+ and Ge^{++} are implanted at the same acceleration voltage (20 kV). According to the simulation, the Ge peak concentration should be at $R_p = 22$ nm, agreeing with the RBS results shown in Figure 1. However, the calculated longitudinal straggling ΔR_p is only 9 nm, about half of the value we find in Figure 1. In other words, the Ge profile is much broader than expected. One possible cause for the broader Ge distribution is the Si surface sputtering during implantation. Our calculation results show that a 5 to 10 nm Si surface layer sputtering will form a similar Ge profile as shown in Figure 1. The peak Ge concentration is 3.7% in the sample. No detectable Ge profile redistribution is found after annealing at temperature up to 1100 °C. Heavy metals, possibly Pb or Au, with dose less than 1.5 x $10^{13}/cm^2$ is found in the RBS spectrum. They may have been co-implanted into Si during the Ge and/or C implantation process. Figure 1 also shows that a calculated 10 kV C implantation results in a profile with a projected range R_p = 33 nm, and a straggling ΔR_p = 14 nm. We expect that the C peak (1.2%) is located 11 nm deeper than the Ge peak.

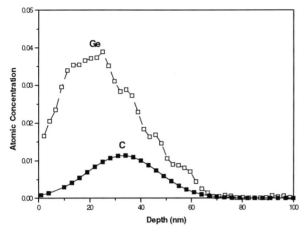

Figure 1. Ge and C profiles of $Si_{0.95}Ge_{0.04}C_{0.01}$. Ge profile is measured by RBS, while C is extracted using the nominal dose and implantation energy.

The regrowth of the amorphous layer during thermal annealing is shown to occur in a layer-by-layer fashion, not by the nucleation of crystals inside the amorphous layers. Figure 2 shows the <111> channeling spectrum of the samples just after implantation (as-implanted), and after annealed at different temperatures for 30 minutes. The original amorphous layer thickness is about 95 nm based on RBS energy loss analysis. The crystal epitaxial growth starts at less than 600 °C (not shown in Figure 2). However, even at 900 °C, the epitaxial growth stops at about 62 nm, near the Ge and C-rich area. At this temperature, no detectable substitutional Ge is found. At 1000 °C, the epitaxial layer grows further into the Ge and C high concentration area. However, the growth terminates at about 44 nm from the surface. Partial Ge substitution in the epitaxial layer is also found in the Ge channeling profile. The exact physical nature of the

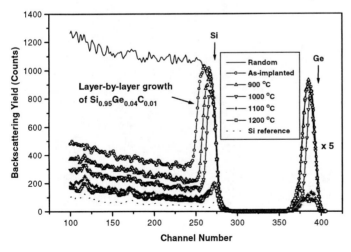

Figure 2. RBS <111> channeling spectra of $Si_{0.95}Ge_{0.04}C_{0.01}$. The RBS experiments are performed using 2 MeV He ion at a scattering angle 165°. The Channel resolution is 3.57 kV/Channel, while the cross-section is at 40 kV.

alloy structure that causes the strong dechanneling beyond 44 nm can not be determined from the RBS spectrum. However, HRTEM results show that polycrystals are formed in the alloy. At 1100 °C, the epitaxial growth finally reaches the surface as shown in Figure 2. From the Ge signal of the spectrum, we can tell that a majority part of Ge reached substitutional site, and the integrated Ge channeling yield is 14 % at 1100 °C, and 20 % at 1200 °C. Ge diffusion is detected at 1200 °C annealing, and higher Ge channeling yield (compared with 1100 °C) is shown at this temperature. On both 1100 °C and 1200 °C curves, the channeling yield is significantly higher than that from a pure Si crystal, implicating the existence of defects in the alloy. However, EOR defects that are normally located at the a/c interface are annealed out, invisible in the RBS spectra.

Samples underwent 1000 °C and 1100 °C epitaxial regrowth are taken for cross-sectional HRTEM examination. As shown in Figure 3, the epitaxial regrowth at 1000 °C reaches 46 nm from the surface. Above that, the crystallites lost its orientation, and polycrystals with nominal size 5 - 10 nm are formed. Figure 4 shows the micrograph of $Si_{0.95}Ge_{0.04}C_{0.01}$ layer from 1100 °C epitaxial regrowth. Apparently, single crystal $Si_{0.95}Ge_{0.04}C_{0.01}$ layer is grown. Some defects are found along the <111> directions, most likely to be localized stacking faults. Those defects shown in this image correspond the high channeling yield observed in the RBS spectra. EOR loops are not visible in the HRTEM observation either.

X-ray rocking scan can effectively resolve the crystal strain by measuring the lattice constant from the satellite diffraction peaks. In this experiment, x-ray rocking scans are performed near the Si (004) peak. In Figure 5, rocking curves from three different samples are given:

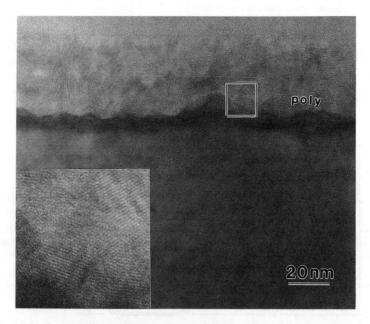

Figure 3. Cross-sectional TEM image of $Si_{0.95}Ge_{0.04}C_{0.01}$ after 1000 °C epitaxial growth shows polycrystalline formation at 46 nm underneath the Si surface.

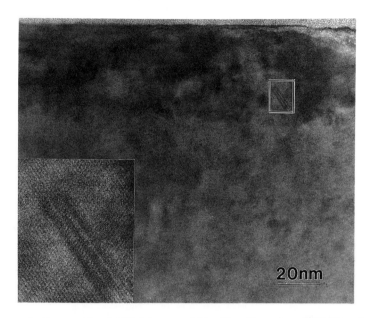

Figure 4. Cross-sectional TEM image of $Si_{0.95}Ge_{0.04}C_{0.01}$ after 1100 °C epitaxial growth shows single crystal formation in the alloy.

$Si_{0.96}Ge_{0.04}$, $Si_{0.95}Ge_{0.04}C_{0.01}$ and virgin Si reference. The $Si_{0.96}Ge_{0.04}$ sample is also formed by implantation of Ge and subsequent SPE. But the Ge implantation energy is 46 kV in stead of

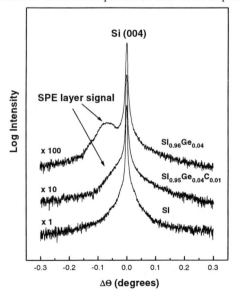

Figure 5. X-ray rocking curves scanned near the Si (004) peak. The satellite peaks are from the epitaxial layers.

20 kV, so the Ge profile is deeper and thicker than $Si_{0.95}Ge_{0.04}C_{0.01}$. Nevertheless it has the same Ge peak concentration. Both alloy samples are measured after $1100\,^{\circ}C$ epitaxial growth. The satellite peaks shown in the Figure are interpreted as the (004) signal from the top epitaxial layers. As shown in the Figure, the satellite peak from the $Si_{0.95}Ge_{0.04}C_{0.01}$ is much weaker than that from the $Si_{0.96}Ge_{0.04}$. More importantly, it is much closer to the Si substrate (004) peak, indicating a lattice constant very close to pure Si crystal. This result confirms that the C implantation can relieve the strain formation in the Si-Ge alloy. The satellite peak in $Si_{0.96}Ge_{0.04}$ locates at $\Delta\Theta = -0.0650^{\circ}$, suggesting a 0.171% lattice expansion comparing with pure Si (a_{Si} = 0.543 nm). Ge lattice constant is a_{Ge} = 0.566 nm, 4.18 % larger than that of pure Si. Based on first order linear approximation, the 0.171% lattice expansion counts for 4.09% Ge concentration in the epitaxial layer, which is close to the results seen in the RBS spectra shown in Figure 1.

SUMMARY

$Si_{1-x-y}Ge_xC_y$ epitaxial layer is formed by Ge and C implantation followed by solid phase epitaxy. Localized defects along <111> directions are found in the surface alloy after thermal annealing at $1100\,^{\circ}C$ or higher temperature. Polycrystalline layer starts to form in the alloy when the annealing temperature is $1000\,^{\circ}C$ or lower. The polycrystalline formation may have been caused by the high dose, deeper (comparing with Ge) C implantation. C implantation is found to retard the epitaxial growth, while its compensation of the lattice strain at substitutional site is confirmed by XRD measurement. Deep C implantation also limits the growth of EOR defects.

ACKNOWLEDGMENT

The authors thank Dr. I. Brown and Dr. S. Anders for providing the MEVVA ion implantation. We are also grateful to Dr. K. M. Yu for performing the RBS analysis. This work is sponsored by Joint Services of Electronics Program under Contract F49620-94-C-0038 and by National Science Foundation under Contract 442427.

REFERENCES

[1] H. Presting, H. Kibbel, M. Jaros, R. M. Turton, U. Menczigar, G. Abstreiter and H. G. Grimmeiss, Semicond. Sci. Tech., 1(1992), 1127 - 1148.

[2] S. Verdonekt-Vandebroek, E. Crabbe, B. S. Meyerson, D. L. Harame, P. J. Restle, J. C. M. Stork, and J. B. Johnson, IEEE Tran. Electron Devices, vol. 41 (1994), No. 1, 90 - 100.

[3] G. L. Patton, S. S. Iyer, S. L. Delage, S. Tiwari, and J. M. C. Stork, IEEE Elec. Dev. Lett., vol. 9 (1988), No.4, 165 - 167.

[4] J. W. Strane, H. J. Stein, S. R. Lee, B. L. Doyle, S. T. Picraux and J. W. Mayer, Appl. Phys. Lett., 63 (20) 1993, 2786 - 2788.

[5] S. Im, J. Washburn, R. Gronsky, N. W. Cheung, K. M. Yu, and J. W. Ager, Appl. Phys. Lett., 63 (19) 1993, 2682 - 2684.

[6] I. G. Brown, Rev. Sci. Instrum., 63 (4) 1992, 2351 - 2356.

[7] I. G. Brown, B. Feinberg, and J. E. Galvin, J. Appl. Phys. 63 (10) 1988, 4889 - 4898.

[8] J. R. Pfiester and J. R. Alvis, IEEE Elect. Dev. Lett. vol. 9. No. 8 (1988), 391-393.

[9] M. C. Ozturk, J. J. Wortman, C. M. Osburn, A. Ajmera, G. A. Rozgonyi, E. Frey, W. Chu, and C. Lee, IEEE Tran. Elect. Dev., vol. 35, No. 5 (1988), 659 - 668.

A NEW ION BEAM DEPOSITION TECHNIQUE FOR LOW TEMPERATURE SILICON EPITAXY

S. Mohajerzadeh*, C.R. Selvakumar*, D.E. Brodie**, M.D. Robertson** and J.M. Corbett**

*Department of Electrical and Computer Engineering, **Department of Physics, University of Waterloo, Waterloo, Ontario, Canada, N2L 3G1

Abstract

We report the results of an investigation to grow thin Si films on Si substrates at low substrate temperatures using ionized SiH_4 gas generated with a Kaufman type ion gun. This investigation shows island-growth at higher substrate temperatures (500-700°C) in the form of square-based pyramids. By lowering the substrate temperature to 300°C, we were able to achieve a planar growth. The growth rate can be enhanced by introducing elemental Si from a thermal evaporation source. Scanning electron microscopy, transmission electron microscopy and electron diffraction analysis were used to study the crystalline quality of the samples prepared at different temperatures.

Introduction

Ion beam deposition techniques are of great interest to many researchers in view of their potential capabilities for the growth of thin film semiconductors[1-3]. The kinetic energy of the ions can enhance the likelihood of epitaxial growth at lower substrate temperatures. Most of the reports based on the use of energetic ions to grow Si thin films have concentrated on sputtering techniques[2]. The use of a Si^+ beam for direct deposition onto a Si substrate is another alternative for Si homoepitaxy at low substrate temperatures[3]. The plasma enhanced chemical vapor deposition, although mostly a chemical reaction technique, uses energetic reactive ions to enhance the growth rate of the films.

In this paper, we report the preliminary results of the use of a new ion beam deposition technique to grow thin Si films on Si substrates. Silicon films were grown at different substrate temperatures and their atomic structures studied using electron microscopy.

Mat. Res. Soc. Symp. Proc. Vol. 388 © 1995 Materials Research Society

Experimental Set-up

The equipment used for this study is illustrated in Figure 1. A cylindrical vacuum chamber pumped by a diffusion pump and backed by a rotary mechanical pump was used. A base pressure of 1×10^{-7} Torr (1.3×10^{-5} Pa) is typically attained. A commercially available IonTek Kaufman type ion source was used to ionize the gas and to direct the ion beam towards the substrate. During the film growth

the substrate is grounded to reduce any chance of charge pilage on the substrate surface. The ion gun bombards the surface of the substrate at an angle of 20° with respect to the normal of its surface. The ion beam energy and current can be adjusted between 30-1000 eV and 0-30 mA, respectively. Ultra high purity argon and silane (SiH_4) gasses are used as sources for the ion beam.

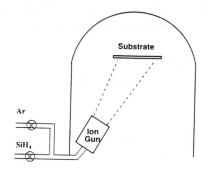

Figure 1: Schematic diagram of the deposition chamber.

For the gun to operate, a chamber pressure of about 1×10^{-4} Torr (1.3×10^{-2} Pa) is required.

For this study, $1\Omega - cm$ n-type <100> Si substrates were used. Each Si substrate was cleaned for 15 mins. in a standard RCA solution. Before loading the Si wafer into the chamber, it was given an HF deglaze followed by a rinse with D.I. water and blow-dried with N_2. After loading the sample and pump-down cycle, the sample was heated to 300°C.

In-situ cleaning of the substrate surface was done by argon ion bombardment prior to the start of deposition. This is a crucial step which significantly influences the quality of the grown film. A 200 eV argon ion beam was used to sputter-clean the substrate. This low energy ion cleaning was found sufficient to remove the native oxide grown during the handling.

After 5 mins. of Ar ion bombardment, the Ar line was closed and the ion gun was shut down. The substrate temperature was rapidly raised to 800°C (in about one minute) and held at that temperature for about 1 min. to anneal surface damage caused by the energetic ion bombardment[4]. The substrate temperature was then lowered to its desired value (300-700°C) to start the deposition. SiH_4 gas was introduced to the ion source and the deposition took place at a chamber pressure of 8×10^{-5} Torr. This pressure is necessary to maintain the plasma in the ion source. During the film growth, an ion beam energy of 50eV and a current of 13mA were used. The ion beam intensity at the substrate surface was $30\mu A$ cm^{-2}. Following the deposition cycle, the substrate temperature was allowed to cool down. It was observed that at this pressure (8×10^{-5} Torr), no deposition was made on the parts of the substrate which were not directly exposed to the ion gun.

Physical Characteristics

The *as-grown* samples were used for the electron microscopy study. Figure 2.a shows a scanning electron microscopy (SEM) image of the sample grown at 700°C. Rectangle-based pyramids as large as $1\mu m$ in length, oriented with the (100) directions parallel to the substrate, are observed. As shown in this figure, the substrate surface is covered by pyramids with (111) and (113) faces exposed. Some of the adjacent pyramid have merged to form a larger structure. Figure 2.b shows a micrograph of the sample grown at 500°C evidencing 3-D growth. It seems that the substrate surface is more completely covered at this lower temperature. The size of the pyramids is also smaller.

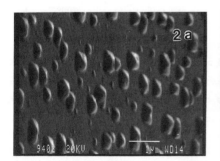 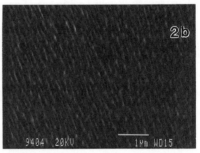

Figure 2: SEM micrograph of (a) the sample prepared at 700°C, (b) the sample prepared at 500°C.

The difference in the coverage of the substrate surface can be due to differences in the Si adatom mobilities. Part of the adatom surface mobility comes from thermal energy gained from the substrate and part from the kinetic energy of the ion beam. At higher substrate temperatures the surface mobility of adatoms is high and they can travel further before they bond with the underlayer. This could result in the formation of islands which are elongated in a preferred direction influenced by the ion beam. By lowering the temperature of the growth, the total surface mobility is reduced and the probability for planar growth is increased. The sample prepared using a 300°C substrate temperature shows a planar growth.

Transmission electron microscopy (TEM) was used to study the Si film grown at this temperature and to compare it with the sample grown at 700°C. Figure 3 shows a bright field [110] zone axis image of the sample grown at 700°C. The exposed atomic planes, (111) and (113), are clearly observed in this figure. High resolution TEM study also indicates the epitaxial growth.

Figure 3: Bright field image of one of the pyramids prepared at 700°C.

Figure 4.a displays the dark field [110] zone axis image for the sample grown at 300°C indicating planar growth. Figure 4.b displays a high resolution micrograph of the same sample clearly showing the epitaxial growth. The high resolution image also shows the presence of twins and microtwins, originating from the interface. The damage created due to 200 eV Ar ion bombardment might be too large to be adequately removed by the short annealing time.

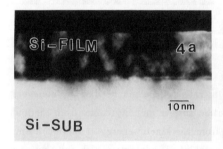

Figure 4: (a) Dark field image of the sample prepared at 300°C indicating a planar growth, (b) high resolution image of the film-substrate interface.

Growth Rate Enhancement

The growth rate can be enhanced by introducing elemental Si atoms from a thermally evaporated source. Figure 5 shows the schematic diagram of the chamber modified to include a $B - N$ source for Si evaporation. The elemental Si adatoms do not have much kinetic energy except for their thermal energy. Part of the energy of the Si adatoms generated by the ion beam is transferred to the thermally evaporated Si adatoms, imparting mobility to them.

Depending on the ratio of the arrival rate of the thermally evaporated adatoms to the arrival rate of the ion beam adatoms, the growth mechanism can be ion beam assisted deposition or ion beam vapor deposition. In either case the energy of the adatoms is mainly supplied through the ion beam bombardment.

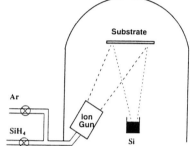

Figure 5: Elemental Si incorporation from a $B - N$ crucible source.

Using this modification, a growth rate of $300\mathring{A}$ per minute has been obtained. In this experiment, it was also found that the films are B-doped with a concentration of about 5×10^{18} cm^{-3}.

We found that by using the ion gun alone, a planar growth can be obtained only if the substrate temperature is lower than 300°C. However, by adding more Si atoms from an elemental evaporation source (like a $B - N$ crucible) one can achieve a planar growth at higher substrate temperatures. This technique can be extended to grow Si-based alloys such as SiGe layers at low substrate temperatures.

Conclusion

In conclusion, a new, simple and inexpensive technique for growing silicon thin films has been investigated. Depending on the substrate temperature, the films appeared to attain island or planar growth. With the substrate at a higher temperature, 3-D growth with pronounced pyramids was observed. By lowering the substrate temperature, the chance of achieving planar growth appeared to be enhanced. At the low substrate temperature of 300°C, planar growth was achieved. In-situ cleaning of the sample is an important step for this technique and it was achieved by Ar ion bombardment at low temperatures followed by an in-situ annealing. HRXTEM studies show twins and microtwins in the film grown at 300°C. This work has been partially supported by

MICRONET, a University of Waterloo Interdisciplinary Grant, and by Natural Science and Engineering Research Council of Canada.

References

[1] T. Ohmi, T. Ichikawa, H. Iwabuchi, and T. Shibata, J. Appl. Physics, **66**, 4756 (1989).

[2] I. Yamada, F.W. Saris, T. Takagi, K. Matsubara, H. Takaoka, and S. Ishiyama, Jpn. J. Appl. Phys. **19**, L181 (1980).

[3] K.J. Orrman-Rossiter, A.H. Al-Bayati, D. G. Armour, S.E. Donnely, and J.A. van den Berg, Nucl. Instrum. Methods B **59/60**, 197 (1991).

[4] J.C. Bean, G.E. Becker, P.M. Petroff, and T.E. Seidell, J. Appl. Phys., **48**, 907 (1977).

A NEW LOW TEMPERATURE THIN FILM DEPOSITION PROCESS: ENERGETIC CLUSTER IMPACT (ECI)

H.HABERLAND, M. LEBER, M.MOSELER, Y.QIANG, O.RATTUNDE, T.REINERS, AND Y.THURNER
Freiburger Materialforschungszentrum, Stefan Meierstrasse 21, 79104 Freiburg, Germany

ABSTRACT

A beam of metal cluster ions of variable size is deposited with variable kinetic energy on a substrate. Mirror–like and strongly adhering films are produced on unheated substrates for sufficiently high cluster impact energies. Numerical simulations provide the physical insight why this novel technique gives different, and sometimes superior results compared to conventional methods. Several examples are presented.

I. INTRODUCTION

Two parameters influence most thin film properties: the kinetic energy of the arriving atoms and the temperature of the substrate. Both parameters determine the surface mobility, if one is reduced, the other one has to be increased. We have developed a new physical vapor deposition method [1, 2] where a very high temperature (several thousand K) is generated at each impact, thus reducing the influence of the substrate temperature considerably. An intense beam of ionized large metal clusters (like $Al_{1000}^-, Co_{2000}^+, Cu_{5000}^-, Ti_{7000}^+$, or $(TiN)_{9000}^-$ is produced, electrically accelerated, and deposited with variable kinetic energy on a surface.

A numerical simulation [3, 4] of a single impact is shown in Fig. 1. The top figure shows a Cu_{1000} cluster hovering over the surface. Its kinetic energy is 10 eV/atom, or 10 keV. Both cluster and surface have a temperature of 300 K. The time after the first contact (in units of 10^{-12} s) is indicated in each frame. At 0.2 ps, about half of the cluster has dived into the surface, which had little time to react. When the first atoms in the cluster get into contact with the surface they are scattered backwards. The next line of atoms in the cluster hit these recoiling atoms, etc. This produces a "nano" shock wave [1] which moves into the surface. Due to this strong compression, a pressure of about 10^{11} Pa and a temperature of 3.000 K is generated locally. At 0.5 ps, the nano shock wave has moved into the substrate, leaving behind a small temporary crater. At 1 and 5 ps, the compressed material expands, while about 50 ps later the area has cooled and a well annealed deposit is left.

These extreme conditions lead to a solid anchoring of the deposited material. Because of the high local temperature caused by the impact of such a massive object, there is only little influence of the substrate temperature on the outcome of a single collision event. In the experiment, each spot is hit by a cluster about every second, which is about 11 orders of magnitude longer than a typical deposition event, so that diffusion processes between the cluster impacts can of course be influenced by the substrate temperature.

Deposition by clusters was pioneered by Takagi and Yamada of the university of Kyoto, who called their process ICB, short for ionized cluster beam deposition [5]. The two methods are quite different as discussed below, so that we choose to call our method ECI, short for energetic cluster impact.

Mat. Res. Soc. Symp. Proc. Vol. 388 © 1995 Materials Research Society

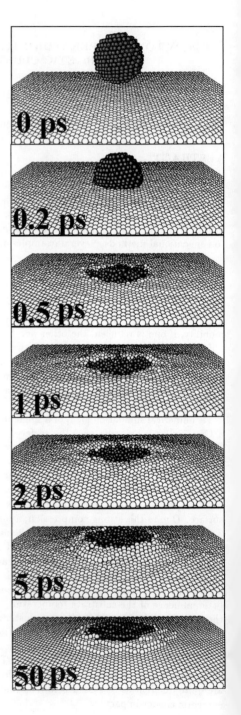

Fig. 1: Langevin Molecular Dynamics Simulation of an impact of a Cu_{1000} cluster on a Cu surface with a kinetic energy of 10.000 eV. The time after the cluster first hit the surface is given in units of 10^{-12}s. After about 50 ps a well annealed deposit remains from the violent encounter of the massive cluster with the surface. In the experiment 10^{12} clusters/s hit a cm^2 giving a compact thin film on a room temperature substrate.

II. EXPERIMENT

The principle of the present set–up is as described earlier [1, 2], but another chamber for separating the charged and uncharged components of the beam has been added, and the time–of–flight (TOF) spectrometer was completely redesigned. Fig. 2 shows an overview of the apparatus, which consists of five chambers: 1.) the chamber for cluster generation containing a liquid nitrogen cooled aggregation tube with a magnetron sputter discharge K, 2.) a chamber with the ion optic to separate neutral and charged clusters, 3.) the deposition chamber with a substrate holder S which can be floated up to ±30 kV, 4.) a load–lock chamber for rapid substrate change containing a broad Ar^+ beam from a Kaufman type gun for substrate cleaning, and 5.) a chamber which is not shown containing the time of flight (TOF) mass spectrometer. Briefly, the cluster ions are generated in the aggregation tube, they go through two diaphragms B1 and B2, are bent and focused on B3, and finally accelerated onto the substrate S. Their size can be monitored by the TOF spectrometer.

The cluster source, a 5 cm diameter magnetron sputter discharge K, is operated at a pressure of about 100 Pa in a mixture of argon and helium. The argon is floated directly over the sputter target, while the helium is introduced from the back. Increasing the He to Ar flux ratio decreases the cluster size. The typical discharge power is 50 to 200 W at about 200 V. The target and the magnetron magnet are water cooled. The number of clusters coming from the target is vanishingly small under these conditions [1, 6]. The clusters are formed in the rare gas surrounding the discharge. The condensation process is similar to the well known gas–aggregation process, but strongly modified due to the many charges present [1, 7, 8].

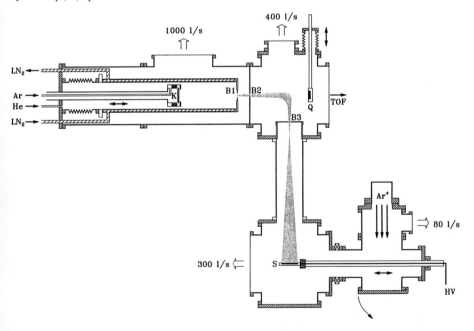

Fig. 2 Schematic of the experimental set–up. See text for details.

The clusters are ionized by the many charged or energetic particles in the discharge, so that no additional electron impact is necessary [1]. The degree of ionization of the cluster beam is unusually high and depends on the material used. Typical values are 60 to 80% for Al, 20 to 60% for Mo, and 20 to 50% for Cu. These extremely high values are easy to understand: The clusters are formed in a region where the electron and ion density is high. This leads to many charge transfer processes, as discussed in detail elsewhere [1]. Note, that no additional electron impact is used as in the ICB method [3]. One problem with gas aggregation sources is to stop the aggregation process, before the clusters become too large. This is done by letting the rare gas sweep the clusters through the diaphragm B1, where the cluster growth is effectively stopped. Most of the gas is pumped by a 1000 l/s turbo pump. The heavy clusters continue to fly through a second diaphragm B2. Charged clusters are deflected by an ion optic, which is not shown in Fig. 2, and are focused through B3 onto a substrate holder S which can be floated to ±30 kV. The thermal loading of the substrate by the deposition process is below 0.01 Watt/cm^2 and thus completely negligible. The substrate holder S can be retracted into a load–lock system for substrate cleaning and changing.

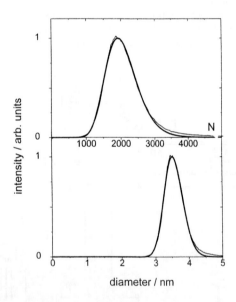

Fig. 3: Mass spectrum of negatively charged copper clusters. The horizontal axis has been converted from flight time to cluster size. The maximum of the size distribution lies around 800 Cu–atoms, with a full width at half maximum of about 560 atoms. Note, that no small clusters are detected. The lognormal distribution used earlier for heavily clustered material gives a very good fit to the data [7].

The total beam intensity can be monitored by the micro balance Q, which can be retracted so that the charged beam component can be analyzed in a time–of–flight mass spectrometer. It uses a ± 30 kV post acceleration detector, so that clusters containing several thousand atoms can be detected with sufficient probability [1, 5].

Metal clusters (Al, Co, Cu, Mo, and Ti) have been studied mostly so far. Reactive ECI is also possible. Good TiN layers have been made, by mixing about 0.5% N_2 into the aggregation gas. For the production of clusters from non–conducting materials one can operate the source with a radio frequency power supply. A gas aggregation cluster source can also be operated with a thermal evaporator [5]. This is presently used in this laboratory for the production of sodium cluster ions, whose optical absorption is measured [8], but can similarly be used for deposition purposes. Thus it is possible to produce many kinds of thin films with this new technology. All materials which can be sputtered or evaporated can be used.

Many Ar^+ ions exist in the sputter discharge. They can escape from the source under unfavorable conditions and impinge on the substrate, if it is negatively charged. This has indeed sometimes been observed when working with positively charged clusters. In this case both the clusters and the Ar^+ are accelerated onto the substrate. This leads to sputtering of the substrate and to an undesirable high argon content in the thin film. This problem can be avoided by working with negatively charged clusters, as argon and helium do not have stable negative ion states. The magnetic field of a small permanent magnet deflects any electrons which might be in the beam, so that one can work with a pure, negatively charged cluster beam of known composition.

Fig. 3 shows a size distribution for negatively charged Cu clusters. Note that no atomic or small cluster ions are present in the beam. The thin line gives the experimental result, while the thick solid line is the lognormal distribution often found for heavily clustered material [7]. The measured flight time has been converted to cluster mass and cluster diameter. The size distribution was checked by AFM and TEM [9]. The intensity is typically 10^{12} cluster ions/s·cm^2, which corresponds to a film growth rate of about 3 Å/s.

The pressure in the deposition chamber is below 10^{-8} Pa, and does not rise significantly when the beam is going. In principle, a UHV deposition chamber can be attached now, as the beam is not accompanied by any gas load. The beam is highly collimated, and has a very high mass to charge ratio (number of atoms per electric charge 10^3 to 10^4). Thus the charging problem which often arises when depositing ions on a non–conducting surface is much reduced. In fact, it is possible to make a durable Mo or Cu–coating on untreated paper or even Teflon [9].

III. Simulations

The cluster impact was simulated by Langevin Molecular Dynamics [10, 11]. Fig. 1 shows a result. A n–body Embedded Atom Method Potential (EAM) was used, which is the best potential one can use today for this size of simulation. For the calculation shown in Fig. 1 up to 50.000 mobile atoms had to be taken into account. The figure shows a result for the collision of copper clusters with a copper surface. The electron–phonon coupling, or in other words the electronic part of the heat conductivity, had to be taken into account for simulations of metals with high electric conductivity; otherwise the experimental results could not be reproduced [12].

IV. Results

The ECI method has some unique advantages which result from the high local temperature and pressure generated at each cluster impact, the ensuing low substrate temperature, the directionality of the beam, the fact that the substrate is not in contact with a plasma, and the low thermal loading of the substrate. A list of achievements is given below, which cannot be reached by conventional methods of thin film formation, like evaporation or sputtering. The substrate was at room temperature ($\sim 20°C$) for all examples, save for the last one. The mean cluster size varied between 1500 and 7000 atoms, and the kinetic energy between 12 and 30 kV. The film quality is not very dependent on the last two parameters, provided that the kinetic energy is higher than the total cohesive energy.

1. ECI molybdenum films (thickness up to 5μm) can be mechanically polished, which cannot be done with sputtered films. These are too brittle and will splinter away upon polishing. The slowly decaying high temperature at each impact leads to a local annealing, which releases the mechanical stress and thus reduces the brittleness. The optical reflectivity at 10.6 μm (wave length of the CO_2 laser) did not change significantly upon polishing, indicating that the films had already been rather smooth before the polishing. The ability to clean a Mo–mirror by mechanical polishing is important in industrial applications of the CO_2 laser where the last mirror in a beam handling system can easily become dirty and needs frequent cleaning.

2. Untreated Teflon, paper, and ceramics can be coated with metal at room temperature. Due to the extremely large mass to charge ratio of the clusters no charging problem was noticed. No other method exists which can produce a well adhering thin film on untreated Teflon.

3. TiN clusters can be produced by sputtering Ti and adding a small amount of N_2 ($\sim 0.5\%$) into the argon gas of the cluster ion source. Due to the high reactivity of titanium TiN clusters form probably already in the cluster ion source. TiN films are very hard and have a golden appearance. Normally they are deposited on substrates heated to about 300°C. With ECI one arrives to make good, golden looking films at room temperature. The smoothness of the TiN films deposited at room temperature onto a quartz substrate was below 1 nm as measured by an interferometric optical technique, and 1.8 nm as measured by atomic force microscopy used in the tapping mode [9].

4. It was possible to fill small holes of 0.8 μm diameter with Cu at a substrate temperature of 200°C [2]. This step–covering problem arises in the semi–conductor industry during the manufacture of VLSI circuits [2, 13].

These are four examples where this new method has some distinct advantages. As the method has been developed only very recently [1, 2], many more can be expected. Several are under active investigation [9]. Another feature which distinguishes ECI produced films from those produced by standard sputter techniques, is that no columnar growth is observed. The reason is probably again the high pressure and temperature generated at each impact.

Table 1: Comparison of the two methods, which use charged clusters for thin film formation. ICB stands for Ionized Cluster Beam, while ECI stands for Energetic Cluster Impact.

METHOD	CLUSTER CONTENT	IONIZATION
ICB/Kyoto	10^{-5} or smaller	up to 5%
ECI/Freiburg	100 %	100 %

V. Discussion

The idea of using clusters and cluster ions for deposition purposes was pioneered by Takagi and Yamada [3]. Although conceptually similar, the actual experimental implementation is very different. As shown in table 1, the degree of ionization is low and the cluster content is very low in the ICB method [3]. While in the ECI method both parameters are 100% if the neutral and charged component are separated as shown in Fig. 2. Good thin films can be made by both methods, but due to the vastly different beam parameters, we assume that the thin film formation mechanisms are different, too.

The morphology of thin films produced by evaporation depends mainly on the substrate temperature. This leads to the one dimensional nature of the Movchan, Demchishin zone structure model [4]. For a typical sputter deposition one has an additional control parameter: the mean kinetic energy of the impinging atoms, which allowed Thornton the two–dimensional extension of the zone structure model [4]. Deposition by Energetic Cluster Impact has a third control parameter: the number of atoms in the cluster. If a three–dimensional extension of the zone–structure model is feasible, remains to be seen. In any way, the added flexibility may lead to a variety of better films or novel applications.

Acknowledgements

This research is supported by Commission of the European Communities, Program Brite–Euram II, contract CT93.0534 and by the Deutsche Forschungsgemeinschaft through SPP "Ionen– und Plasmaoberflächentechnik".

References

[1] H. Haberland, M. Karrais, M. Mall, and Y. Thurner, J. Vac. Sci. Technol. A10, 3266 (1992)

[2] H. Haberland, M. Mall, M. Moseler, Y.Qiang, Th. Reiners, and Y. Thurner, J. Vac. Sci. Technol. A12, 2925 (1994)

[3] H. Haberland, Z. Insepov, and M. Moseler, Z. Phys. D26, 229 (1993)

[4] H. Haberland, Z. Insepov, and M. Moseler, Phys. Rev. B., 51, 11061 (1995)

[5] R. L. McEachern, W. L. Brown, M. F. Jarrold, M. Sosnowski, G. Takaoka, H. Usui, and I. Yamada, J. Vac. Sci. Technol. A9, 3105 (1991) and references therein

[6] B. Chapman, *Glow Discharge Processes*, Wiley, New York 1982

[7] The operational principles of gas aggregation sources and the high voltage post acceleration detector is described in detail in: *Experimental Techniques*, chapter 3 of: Clusters of Atoms and Molecules I (H. Haberland, ed.) Springer Berlin 1994.

[8] S.G. Choi and M. Kushner, Mat. Res. Soc. Symp. 206, 283 (1991)

[9] C. Granqvist and R. Buhrmann, J. Appl. Phys. 47, 2200 (1976)

[10] Th. Reiners, C.Ellert, M.Schmidt, and H.Haberland, Phys.Rev.Lett. 74, 1558 (1995)

[11] H. Haberland, M. Leber, M.Moseler, Y.Qiang, O. Rattunde, Th. Reiners, and Y. Thurner, unpublished results

[12] M. Moseler and H. Haberland, manuscript in preparation

[13] S. M. Rossnagel, D. Mikalsen, H. Kinoshita, and J. J. Cuomo, J. Vac. Sci. Technol. A9, 261 (1991), and references therein

FILM SYNTHESIS ON POWDERS BY CATHODIC ARC PLASMA DEPOSITION

ANDRÉ ANDERS*, SIMONE ANDERS*, IAN G. BROWN*, AND IGOR C. IVANOV**

* Lawrence Berkeley Laboratory, University of California, Berkeley, CA 94720
** Charles Evans & Associates, 301 Chesapeake Drive, Redwood City, CA 94063

ABSTRACT

Cathodic arc plasma deposition was used to coat Al_2O_3 powder (mesh size 60) with platinum. The power particles were moved during deposition using a mechanical system operating at a resonance frequency of 20 Hz. Scanning electron microscopy and Auger electron microscopy show that all particles are completely coated with a platinum film having a thickness of about 100 nm. The actual deposition time was only 20 s, thus the deposition rate was very high (5 nm/s).

INTRODUCTION

Coating of powders is of interest for a number applications such as catalysis and powder metallurgy and ceramic powder sintering. In particular, coating of ceramic powder with platinum is of interest in catalysis due to the large platinum surface area per mass unit. Complete coverage of the particles is sometimes difficult to achieve, and adhesion of the film to powder grains can be a problem.

Powder coatings are generally done by fluidized-bed chemical vapor deposition [1], rotary-bed chemical vapor deposition [2], plasma spraying, rotary-bed physical vapor deposition [3] or in-situ precipitation in aqueous solvents [4] or non-aqueous solvents [5].

In the present paper we describe a new approach to film synthesis on power particles using cathodic arc plasma deposition. In section II we outline the features and capabilities of cathodic arc plasma deposition in general, and focus in section III on the application of coatings on power particles. As an example we deposited platinum on alumina powder.

CATHODIC ARC PLASMA DEPOSITION

A high current discharge between a cold cathode and an anode in vacuum or in a gas is called a "cathodic arc". The term "vacuum arc" is frequently used if no gas is present. Cathodic arcs are characterized by plasma production at small, non-stationary cathode spots (see, for instance, [6-8]). The arc current of typically 100 A or more is concentrated in a few micron-sized spots which move quickly over the cathode surface. The number of simultaneously active cathode spots depends on current, cathode material and temperature, and cathode surface conditions. Since the current density can be as high as 10^{12} A/m^2 [9], the solid cathode material is transformed into a completely ionized plasma which leaves the cathode spot region with a velocity of $1\text{-}2 \times 10^4$ m/s (this number is valid for all cathode materials). This velocity corresponds to kinetic ion energies in the range from 20 eV (light ions such as carbon) to 300 eV (heavy ions such as platinum). Thus the cathode material is eroded and forms an energetic plasma jet. A current-carrying part of the plasma streams to the anode, and another part of the plasma can be used to synthesize metal films on arbitrary substrates. Sometimes the term "cathodic arc evaporation" is used. This is term suggests misleadingly that neutral vapor is involved. However, it is fully ionized plasma that is actually used in depositing a film.

Cathodic arc plasma deposition can operate in the complete absence of any gases ("vacuum arc plasma deposition"), and thus pure metal films are be obtained. If the cathodic arc is operated in reactive gases (e.g., nitrogen, oxygen), compound films can be formed (e.g., nitrides, oxides). Due to the high ion energy, the films formed have very interesting properties such as high density, smooth surface, and good adhesion. It is believed that a broad intermixed region which forms between the substrate and the film is responsible for the observed good adhesion properties. Cathodic arc plasma deposition of TiN films [10, 11] and (hydrogen-free) hard amorphous carbon films [12, 13] have been up to now the most successful applications of the method.

Cathodic arc plasma deposition can be modified by filtering the plasma prior deposition to remove micron-size "macroparticles" and a small neutral component [14, 15]. Some applications such as optical coatings require filtering to obtain acceptable film quality. Other applications such as metallurgical coatings can be done without filtering but with very high deposition rate [16]. Another modification of cathodic arc plasma deposition results when a negative bias is applied to the (conductive) substrate. An electric sheath forms around the substrate, and ions are accelerated in the sheath and arrive at the surface with an energy given by the sheath voltage and ion charge state. This represents an alternative way of doing ion implantation [17]. Because we used (non-conductive!) alumina powders in the present experiments, no substrate bias was applied.

EXPERIMENT

The metal plasma in our experiments was produced using a small cathodic arc plasma gun (Fig. 1). The cathode is a rod made of the material to be transformed into a plasma (here platinum). The cathode rod is insulated from an annular anode cylinder by an alumina ceramic tube. Each arc pulse is triggered by a surface flashover from a trigger electrode to the cathode surface (trigger duration about 10 µs, trigger current 4 A, initial voltage 12 kV). The voltage between cathode and anode breaks down to about 20 V after the arc is triggered. Cathode spots burn only on the front face of the cathode rod because the rest of the cathode is shielded by the ceramic tube. More details and photos of our cathodic arc sources have been published previously [17].

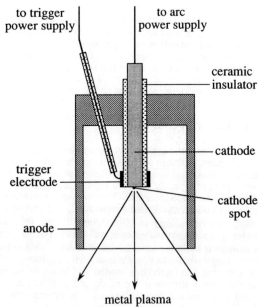

Fig. 1 Construction of cathodic arc plasma source

Cathodic arc plasma deposition has been used up to now for deposition of metal or compound films on more-or-less regularly formed, relatively large substrates. There is no report - to our knowledge - about coatings on powder particles. Since the flow of plasma from a cathodic arc plasma source is directional, not all sides of a substrate or sample are coated but only the side facing the source. In order to obtain a film covering all sides, either the substrate or the source must be moved, and we have done this in the past (substrate movement [18], source movement [19]). A powder, however, consists of many "individual miniature substrates" so that this conventional approach will not work.

The usual approach is to move many powder particles simultaneously in a rotary bed (see, for instance, [2-5]). Another approach is to use a mechanical resonance system. We modified a large loudspeaker system to make it vacuum compatible and mounted it on a frame (Fig. 2). A Petri dish of 5 cm diameter was attached to the moving central magnet. Powder was put in the Petri dish and remained there during evacuation held by gravity. The pressure was 3×10^{-7} Torr (4×10^{-5} Pa). For most experiments we used Al_2O_3 powder (activated, neutral, gamma, 96% purity of metal basis, from Johnson Matthey Electronics) of mesh size 60, corresponding to grain sizes of 250 μm and smaller.

Above this mechanical arrangement we mounted a vacuum arc plasma source which was equipped with a platinum cathode (rod diameter 6.2 mm). Platinum was chosen due to its importance in catalysis. The distance between source exit and Petri dish was 5 cm.

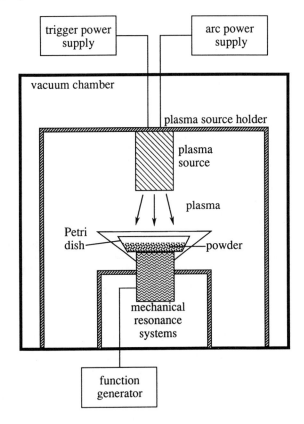

Fig. 2 Experimental arrangement for cathodic arc plasma deposition on powder particles

The Petri dish was set in motion after evacuation using the vibration system which was powered by a function generator. The frequency was tuned to the resonance frequency of the mechanical system (20 Hz). The motion of the dish including the powder could be observed through a window and easily controlled by choosing a suitable power amplitude (about 5 V into 8 Ω impedance, i.e. about 0.6 W) and frequency. The motion of powder particles depends very much on these parameters; they were chosen such as to make the particles "dance" up to a few millimeters above the bottom of the Petri dish. Experiments with 1.5 micron Al_2O_3 powder (alpha, 99.99% purity of metal basis) showed that these very small powder particles tend to agglomerate, and therefore we restricted our experiments to powder of mesh size 60.

After the motion of powder particles was initiated, the platinum plasma source was switched on. It operated in a pulsed mode with an arc current of 150 A, arc duration 1 ms, and arc repetition rate of 5 pulses per second (thus the arc duty cycle was 0.5%). No macroparticle filter was used because platinum macroparticles can be tolerated in catalytic application. The deposition experiments were stopped after 4000 arc pulses (about 13 min.).

RESULTS AND DISCUSSION

The powder particles were investigated using scanning electron microscopy and Auger electron spectroscopy. Both methods show clearly that all particles were completely covered with platinum. An example of an Auger electron spectroscopy survey is shown in Fig. 3; no oxygen or aluminum signals were detected, hence platinum coverage is complete. A platinum film thickness of about 100 nm was derived.

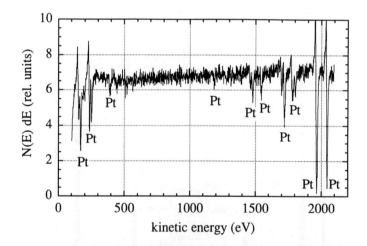

Fig. 3 Auger electron spectroscopy survey of alumina powder (mesh 60) coated with platinum. Note that only Pt signals are observed.

The typical kinetic ion energy of platinum ions in the cathodic arc plasma jet is about 300 eV, and thus the impact onto a substrate surface causes intermixing. We could not find any indication that the platinum film peels off the powder particles, and so we assume that adhesion is very good. Previous experiments in the field of cathodic arc plasma deposition show that intermixing of substrate and film occurs, and it seems reasonable to attribute good adhesion to intermixing also under the present conditions. Fig. 4 shows deposition with intermixing calculated by the Monte Carlo code T-DYN [20].

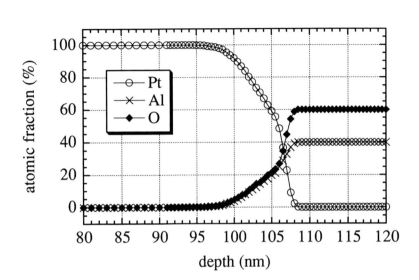

Fig. 4 Deposition of 300 eV platinum ions on a Al_2O_3 substrate (T-DYN simulation).

The pulse operating mode allows working with a small average arc power (in the present experiment 15 W), and thus it requires only small, relatively low-cost equipment and no cooling of plasma source and substrate. For industrial application, however, it seems inevitable to operate a dc arc facility. The actual deposition time was only 20 seconds in our experiment, and thus the actual deposition rate was 5 nm/s. For comparison, sputter coating of $YBa_2Cu_3O_7$ powder (diameter 80 μm and smaller) with silver took several hours with a rate smaller than 0.1 nm/s [3]. Therefore, cathodic arc plasma deposition can be considered as a high-rate (and therefore low-cost) method providing superior film properties such as good adhesion. Another advantage is that deposition (including reactive deposition) can be done with the substrate at room temperature. For comparison, deposition of TiN on iron powder (mesh 150-200) was done at a temperature of about 1000°C with a deposition rate of 0.2 nm/s [2]. Cathodic arc plasma deposition is an environmentally friendly method since no hazardous waste is generated. In contrast, reactive gases and solvents are used in chemical vapor deposition on powders[1, 2].

In conclusion, we have shown that cathodic arc plasma deposition is a high-rate, low-cost, environmentally friendly deposition method which can be extended to coatings of powders. As an example we have demonstrated film synthesis of platinum (100 nm thickness) on Al_2O_3 powder (mesh 60) at a rate of 5 nm/s.

ACKNOWLEDGMENT

This work was supported by the Electric Power Research Institute under RP 8042-03 and the U.S. Department of Energy, Division of Advanced Energy Projects, under contract No. DE-AC03-76SF00098.

REFERENCES

[1] C. F. Powell, J. H. Oxley, and J. M. Blocher Jr., Vapour Deposition, (Wiley, New York, 1966).

[2] H. Itoh, K. Hattoria and S. Naka, J. Mater. Sci. 24, 3643 (1989).

[3] H. R. Khan and H. Frey, J. of the Less-Common Metals 175, 177 (1991).

[4] A. K. Garg and L. C. De Jonghe, J. Mater. Res. 5, 136 (1990).

[5] W.-J. Kim, Y. T. Moon, C. H. Kim, D. K. Kim, H.-W. Lee, J. Mater. Sci. Lett. 13, 1349 (1994).

[6] J. M. Lafferty (Editor), Vacuum Arcs, (Wiley Interscience, New York, 1980).

[7] B. Jüttner, IEEE Trans. Plasma Sci. 15, 474 (1987).

[8] G. A. Mesyats and D. I. Proskurovsky, Pulsed Electrical Discharge in Vacuum, (Springer, Berlin, 1989).

[9] A. Anders, S. Anders, B. Jüttner, W. Bötticher, H. Lück, and G. Schröder, IEEE Trans. Plasma Sci. 20, 466 (1992).

[10] P. J. Martin, R. P. Netterfield, and T. J. Kinder, Thin Solid Films 193, 77 (1990).

[11] A. Bendavid, P. J. Martin, R. P. Netterfield, and T. J. Kinder, Surface and Coatings Technology 70, 97 (1994).

[12] D. R. McKenzie, D. Muller, B. A. Pailthorpe, Z. H. Wang, E. Kravtchinskaya, D. Segal, P. B. Lukins, P. D. Swift, P. J. Martin, G. Amaratunga, P. H. Gaskell, and A. Saeed, Diamond and Related Materials 1, 51 (1991).

[13] R. Lossy, D. L. Pappas, R. A. Roy, J. J. Cuomo, and V. M. Sura, Appl. Phys. Lett. 61, 171 (1992).

[14] I. I. Aksenov, S. I. Vakula, V. G. Padalka, V. E. Strel'nitskii, and V. M. Khoroshikh, Zhurn. Tekh. Fiz. 50, 2000 (1980); Sov. Phys.-Tech. Phys. 25, 1164 (1980).

[15] A. Anders, S. Anders, and I. G. Brown, Plasma Sources Sci. & Technol. 4, 1 (1995).

[16] R. L. Boxman and S. Goldsmith, Surface Coatings Technol. 52, 39 (1992).

[17] A. Anders, S. Anders, I. G. Brown, Mat. Res. Soc. Symp. Proc. 316, 833 (1994).

[18] I. G. Brown, A. Anders, S. Anders, R. A. Castro, M. R. Dickinson, R. A. MacGill, and Z. Wang, presented at the Ninth International Conference on Ion Beam Modification of Materials, Canberra, Australia, 1995 (unpublished).

[19] R. A. MacGill, S. Anders, A. Anders, R. A. Castro, M. R. Dickinson, K. M. Yu, and I. G. Brown, Surface Coatings Technol. (1995) accepted for publication.

[20] J. P. Biersack, S. Berg, and C. Nender, Nucl. Instrum. Meth. Phys. Res. B59/60, 21 (1991).

DISSOCIATIVE CHEMISORPTION AT HYPERTHERMAL ENERGIES: BENCHMARK STUDIES IN GROUP IV SYSTEMS

L.-Q. Xia, M. E. Jones, N. Maity, S. E. Roadman, and J. R. Engstrom[†]
School of Chemical Engineering, Cornell University, Ithaca, NY 14853 USA

ABSTRACT

We present a review of our recent work concerning supersonic molecular beam scattering of thin film precursors from the Si(100) and Si(111) surfaces. Both SiH_4 and Si_2H_6 exhibit translationally activated dissociation channels at sufficiently high incident kinetic energies, $\langle E_\perp \rangle >$ 0.5 eV. The dominant variables under our reaction conditions are the incident kinetic energy and the angle of incidence, whereas mean internal energy and substrate temperature play relatively minor roles. The former two variables couple to produce a universal relationship between the reaction probability and a scaled kinetic energy given by $\langle E_\perp \rangle = E_i[(1-\Delta)\cos^2\theta_i + 3\Delta\sin^2\theta_i]$, where θ_i is the angle of incidence, Δ is a corrugation parameter, and $0 \leq \Delta \leq 1$. In addition to the reaction probability, the reaction mechanism for Si_2H_6 is also dependent upon incident kinetic energy and surface structure, where a $SiH_4(g)$ production channel is observed on the Si(111)-(7x7) surface at low to moderate incident kinetic energies. The reactions of SiH_3CH_3 and PH_3 provide convenient comparative examples. Methylsilane, reacting on a β-SiC surface, exhibits a translationally activated dissociation channel, similar to what is observed for SiH_4 and Si_2H_6. Phosphine, on the other hand, exhibits the characteristics of trapping, precursor-mediated dissociative chemisorption. These results act to underscore the important role played by the frontier orbital topology, even at hyperthermal incident kinetic energies.

INTRODUCTION

Recently a number of research groups have launched efforts directed at making use of supersonic molecular beams as sources for thin film deposition. The work conducted by Eres and co-workers on Ge thin film deposition from Ge_2H_6 represents perhaps the most significant study conducted to date [1]. There are a number of reasons as to why such molecular beams may be useful for thin film growth. The most important feature of the supersonic beam is its narrow velocity distribution, the first moment of which (i.e., the mean velocity) can be varied over a wide range employing "seeding" techniques [2]. If the reactant gas possesses a moderately high molecular weight (> 30), kinetic energies ≥ 1 eV can be routinely achieved employing H_2 as the carrier gas. At these hyperthermal kinetic energies three effects may come into play that could prove useful for thin film deposition: (i) enhanced reaction (dissociative chemisorption) probabilities; (ii) enhanced mobility of adsorbed molecular fragments; and (iii) "self-annealing" of near-surface layers due to energetic, yet non-reactive molecular impacts. Our work to date has focussed exclusively on the first of these— the effect of incident kinetic energy on both the kinetics and mechanisms of gas-surface reactions relevant to thin film deposition. Here we shall review the major results we have obtained concerning the reactions of SiH_4 and Si_2H_6 on the Si(100) and Si(111) surfaces [3–8]. Convenient comparative examples will be provided by results we have obtained concerning the reactions of SiH_3CH_3 and PH_3. Rather than focus in detail on any one experimental result, we shall identify the most important trends we have observed in our work to date, which should provide insight into the behavior of a variety of gas-surface systems of importance to thin film deposition.

Mat. Res. Soc. Symp. Proc. Vol. 388 ©1995 Materials Research Society

EXPERIMENTAL PROCEDURES

The experiments were conducted in a custom-designed multiple-stage stainless steel ultrahigh vacuum (UHV) chamber (base pressure $< 10^{-10}$ Torr) that is coupled to a triply differentially pumped supersonic beam line, which is described in detail elsewhere [9]. The supersonic beam is generated via a heatable nozzle source (orifice dia. = 150-250 μm), and passes through two intermediate stages before striking the sample. The main UHV chamber houses a differentially pumped, rotatable quadrupole mass spectrometer (QMS) that may be employed to measure angular distributions of species scattered from the surface, or in direct (e.g., time-of-flight) analysis of the incoming molecular beam. Both the Si(100) and Si(111) samples (cut to 1.9 x 2.3 cm from 0.04 cm. thick, 1-10 Ω-cm Si wafers) were prepared by standard RCA clean, inserted into vacuum, and annealed *in situ* to 1000°C [1150°C for Si(111)] for 4 minutes to remove the surface oxide. Surface cleanliness and order is verified *in situ* via x-ray photoelectron spectroscopy (XPS) and low energy electron diffraction. The substrates are mounted on a precision manipulator, the axis of which is coincident with the molecular beam, that permits both polar rotation and tilt, the latter providing variation of the angle of incidence (±1°) of the molecular beam with respect to the surface normal. Substrate temperatures are measured by a calibrated infrared pyrometer, and the sample is heated via radiation by a hot filament on the back side of the wafer.

Primarily due to safety concerns the reactants were provided in pre-diluted mixtures (1% reactant in ultrahigh purity H_2 or He) directly from the vendor. The translational energies of the reactant gases were varied by employing a combination of seeding techniques (i.e., H_2 or He carrier) and heating the nozzle, and were measured directly via time-of-flight (TOF) techniques employing the rotatable QMS. Reaction probabilities were measured by the beam reflectivity technique pioneered by King and Wells [10]. Briefly, the reaction probability is calculated from the change in partial pressure of the reactant produced by the scattered flux when the molecular beam is either impinging upon the Si substrate or a non-reactive beam flag (Macor). In our approach the beam is additionally modulated prior to entering the scattering chamber to clearly distinguish between the partial pressure rise produced by the collimated supersonic beam and that produced from the small effusive component (<1% total gas load) emanating from the last differential stage. This permits, under steady-state conditions, summing and averaging of the partial pressure waveforms detected by the QMS, thus increasing the signal-to-noise ratio— measurements of S_R as small as 0.005 are possible employing this approach [6, 9].

RESULTS

Direct dissociative adsorption of SiH_4 and Si_2H_6 on Si(100) and Si(111)

The focus of most of our work has been on an explicit examination of the dynamics of dissociative adsorption of SiH_4 and Si_2H_6 on the Si(100) and Si(111) surfaces [3–8]. At sufficiently high substrate temperatures ($T_s \geq 600$°C), and for the incident reactant fluxes considered here (< 1 ML-s^{-1}), under steady-state conditions the coverage of adsorbed hydrogen atoms is small (< 0.01 ML). Thus, the reaction probabilities reported here represent those for the limit of zero-coverage, $S_{R,0}$. Displayed in Fig. 1 is a summary of our measurements of the reaction probability of SiH_4 and Si_2H_6 on the Si(100) and Si(111) surfaces. The data presented in Fig. 1 represent 160 unique combinations of reactant gas, substrate orientation and temperature, nozzle temperature, and angle of incidence. Here the reaction probability is plotted as a function of scaled incident kinetic energy, $\langle E_\perp \rangle$:

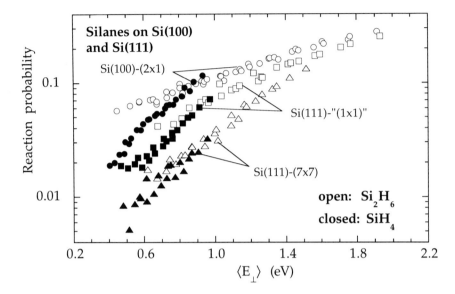

Fig. 1 Reaction probability of SiH₄ and Si₂H₆ as a function of scaled incident kinetic energy. The substrate temperature was 900 °C [Si(100) and Si(111)-"(1x1)"] and 700 °C [Si(111)-(7x7)].

$$\langle E_\perp \rangle = E_i \left[(1 - \Delta)\cos^2\theta_i + 3\Delta\sin^2\theta_i \right] \qquad (1)$$

where Δ is a corrugation parameter [5]. The parameter $\Delta = 1 - \langle n_z^2 \rangle$, where n_z is the z-component of the local surface normal, and the brackets indicate an average over the unit cell. For a gas-surface potential exhibiting a finite corrugation, the local surface normal will vary across the unit cell. Thus $\Delta = 0$ for a gas-surface potential exhibiting zero corrugation. For the data shown in Fig. 1 values of Δ were found to lie in the range 0.12–0.19, and exhibit the expected trends with surface microstructure and molecular "size" (SiH₄ vs. Si₂H₆) [5,6,9]. Physically, the quantity $\langle E_\perp \rangle$ represents the fraction of the incident kinetic energy that is directed along the local surface normal, averaged over the unit cell [5]. As may be seen, for all 6 combinations of gas, substrate orientation and temperature considered in this figure, this scaling analysis describes the data well.

Several points can be made from the data shown in Fig. 1. First, in all cases, the reaction probability is observed to increase nearly exponentially with increasing (scaled) energy. As such, these data underscore the dominant roles played by E_i and θ_i for these reaction conditions. For example, experiments involving measurements of $S_{R,0}$ at fixed E_i, yet differing gas (nozzle) temperatures suggest that internal energy does not contribute significantly under our reaction conditions [3,6]. The exponential dependence upon energy, on the other hand, could be explained by models based upon quantum mechanical tunneling, or a statistical, RRKM dissociation process such as that proposed by Harrison [11]. We find that the latter provides an excellent description of our data set for SiH₄ and SiD₄ [9]. Structural sensitivity for both the reaction of SiH₄ and Si₂H₆ is clearly observed under our reaction conditions. As may be seen in Fig. 1, for both molecules, the absolute reactivity follows the order Si(100)-(2x1) > Si(111)-"(1x1)" > Si(111)-(7x7). The simplest, and perhaps best interpretation for this involves the dangling bond density: 6.78×10^{14} cm^{-2} for Si(100)-(2x1), unknown for Si(111)-"(1x1)", and 3.04×10^{14} cm^{-2} for Si(111)-(7x7). The structural sensitivity of $S_{R,0}$ for Si_2H_6, however, does appear to be decreasing with increasing kinetic energy.

Dissociation and pyrolysis of Si_2H_6 on Si(100) and Si(111)

In addition to examining the effect of incident kinetic energy and angle of incidence on the reaction probability, or "sticking" coefficient, we have also investigated their influence on the dissociation mechanism of Si_2H_6 [8]. Two reaction channels are thought to exist for this molecule, given by:

$$Si_2H_6(g) \rightarrow 2Si(s) + 3H_2(g) \tag{2}$$

and

$$Si_2H_6(g) \rightarrow Si(s) + SiH_4(g) + H_2(g) \tag{3}$$

Making use of published electron impact cross-sections for SiH_2^+ formation from SiH_4 and Si_2H_6 one can directly measure the fraction of Si_2H_6 that reacts on the Si surface that produces SiH_4 as a gas phase reaction product [8]. This quantity is plotted in Fig. 2 as a function of scaled kinetic energy for Si_2H_6 reacting on both the Si(100) and Si(111)-(7x7) surface. As may be seen, the $SiH_4(g)$ production channel is completely absent on the (clean) Si(100) surface for these reaction conditions. On the Si(111)-(7x7) surface, however, the $SiH_4(g)$ production channel is

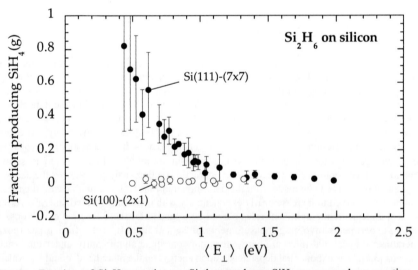

Fig. 2   Fraction of Si_2H_6 reacting on Si that produces SiH_4 as a gas phase reaction product as a function of scaled incident kinetic energy. The substrate temperature was 700 °C [Si(111)-(7x7)] and 900 °C [Si(100)].

clearly evident, becoming decreasingly important as the scaled incident kinetic energy increases. The implication is clear— not only is the absolute reactivity of this molecule sensitive to incident kinetic energy, but the dissociation *mechanism* is also, with complete pyrolysis being favored at high incident kinetic energies.

<u>Universality of translational activation of dissociative chemisorption</u>

Ligand substitution: SiH₃-X

Two issues arise if we consider the following question: how general is the effect of incident kinetic energy on gas-surface reactivity? First, we might consider the effect of ligand substitution. This issue has already been addressed by the data shown in Fig. 1, where the relevant difference in ligands was H vs. SiH₃. As may be seen, for these reaction conditions we observe remarkable similarities between the reactivity of these two molecules, especially on the Si(111) surface. This similarity has lead us to propose that these molecules react via a similar mechanism, namely, Si-H bond activation, for the reaction conditions considered here. To consider further the effects of ligand substitution we have examined the reaction of SiH_3CH_3 on a β-SiC surface. These data are displayed in Fig. 3 where we plot the reaction probability as a function of scaled incident kinetic energy. As may be seen, as with SiH_4 and Si_2H_6, the reactivity of this molecule increases exponentially with increasing kinetic energy. The absolute reactivity is considerably less than the corresponding hydrides at similar energies, however, which could reflect an intrinsically decreased reactivity for this molecule and/or a decreased reactivity of the β-SiC surface. What is clear, however, is that SiH_3CH_3 exhibits the behavior expected for dissociative chemisorption via a direct translationally activated reaction channel.

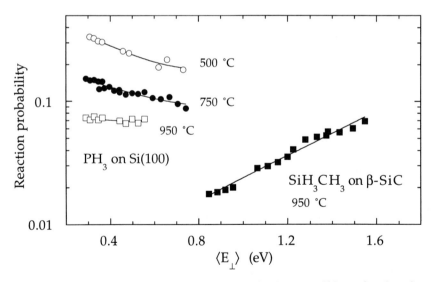

Fig. 3   Reaction probabilities of PH₃ on Si(100) and SiH₃CH₃ on β-SiC as a function of scaled incident kinetic energy. Substrate temperatures as indicated.

Central atom substitution: Group V vs. Group IV

A second issue that can be addressed concerns the role of the central atom. In particular, we have examined the reaction of PH_3 on a $Si(100)$ surface employing supersonic molecular beam techniques. Representative data are shown in Fig. 3, where again we plot the reaction probability as a function of scaled incident kinetic energy. In contrast to the silanes, the reaction probability of PH_3 *decreases* with increasing incident kinetic energy. Also, unlike the silanes, the reaction probability is a strong function of substrate temperature, where S_R is observed to decrease with increasing substrate temperature. Both of these observations point to a dissociation mechanism involving a reaction intermediate that is adsorbed on the surface. Such a precursor-mediated dissociation channel involves two steps: (i) trapping into to precursor state; and (ii) reaction on the surface, where the latter is controlled by the surface temperature. Thus, the dependence of S_R on scaled kinetic energy reflects the dynamics of *trapping*, whereas the dependence on substrate temperature represents primarily the competition between thermal desorption of PH_3 vs. thermal decomposition on the surface to $PH_{3-x}(a) + xH(a)$. The behavior observed for PH_3 is a direct consequence of its electronic structure, where the lone pair (HOMO) is capable of sharing electron density with the Si substrate forming a strongly bound adsorbed molecular species. The silanes, on the other hand, cannot strongly interact with the substrate due to their coordinative saturation, and thus react primarily through a direct dissociation channel.

Acknowledgements

This research was supported by the NSF through a Presidential Young Investigator Award (NSF/CTS-9157892), the Cornell Materials Science Center (NSF/MRL-DMR-9121654), and the Semiconductor Research Corporation through the Cornell Center of Excellence in Microscience and Technology (94-SC-069).

† Address correspondence to this author. Electronic mail: jre@cheme.cornell.edu

REFERENCES

1. D. Eres, D. H. Lowndes, and J. Z. Tischler, Appl. Phys. Lett. **55**, 1008 (1989).
2. See, e.g., *Atomic and Molecular Beam Methods, Volume 1*, G. Scoles, Ed. (Oxford, New York, 1988) and references therein.
3. J. R. Engstrom, D. A. Hansen, M. J. Furjanic and L.-Q. Xia, J. Chem Phys. **99**, 4051 (1993).
4. J. R. Engstrom, L.-Q. Xia, M. J. Furjanic and D. A. Hansen, Appl. Phys. Lett. **63**, 1821 (1993).
5. L.-Q. Xia, and J. R. Engstrom, J. Chem. Phys. **101**, 5329 (1994).
6. M. E. Jones, L.-Q. Xia, N. Maity, and J. R. Engstrom, Chem. Phys. Lett. **229**, 401 (1994).
7. N. Maity, L.-Q. Xia and J. R. Engstrom, Appl. Phys. Lett. **66**, 1909 (1995).
8. L.-Q. Xia, M. E. Jones, N. Maity and J. R. Engstrom, J. Chem. Phys. **103** (in press).
9. L.-Q. Xia, M. E. Jones, N. Maity and J. R. Engstrom (submitted for publication).
10. D. A. King and M. G. Wells, Surface Sci. **29**, 454 (1972).
11. V. A. Ukraintsev and I. Harrison, J. Chem. Phys. **101**, 1564 (1994).

HYPERTHERMAL H ATOM REACTIONS ON D/Si(100)

STEVEN A. BUNTIN
National Institute of Standards and Technology
Gaithersburg, MD 20899

ABSTRACT

A photolytic source of H atoms is used to probe abstraction and adsorption reactions on the D/Si(100) monodeuteride surface. Surface H and D coverages are determined for H atom exposures, with incident average kinetic energies of 1.0 and 2.8 eV. The D atom depletion probability per incident H atom is 0.3±0.2 for both kinetic energies, and is likely due to abstraction. These results, together with previous studies, indicate that the rates of depletion of surface D and the uptake of H are nominally independent of H atom kinetic energy over the range of about 0.3 to nearly 3 eV.

INTRODUCTION

Radicals play an important role in the chemistry of materials processing. Even for these highly reactive species, however, kinetic energy can serve to enhance greatly surface reaction rates, particularly for activated processes [1]. Studies with atomic hydrogen have demonstrated the potential of this species in UHV-compatible surface cleaning [2] and atomic layer epitaxy [3]. While aspects of the kinetics of hydrogen atom reactions with halide [4,5] and hydride [6] adlayers on silicon have been characterized, the effects of H atom kinetic energy in silicon surface chemistry remain to be examined. Characterization of the incident kinetic energy and other dynamical parameters is vital to both the understanding of the mechanisms and the full exploitation of radical/surface chemistry.

EXPERIMENTAL

The experiments were conducted in an ultrahigh vacuum (UHV) chamber that was evacuated by a liquid nitrogen-trapped diffusion pump, and equipped with facilities for Auger electron spectroscopy (AES) and temperature programmed desorption (TPD). A 10x10 mm Si(100) crystal (P-doped, 0.2-0.8 ohm·cm) was cleaned initially by repeated sputter (2 keV Kr^+, 7μA, 900s)/anneal (1050K, 300s) cycles until C and O contaminants were below the AES detection limit. Iodine was found to be present on the sample following each HI exposure run, and one sputter/anneal cycle reduced the I concentration to a few atomic percent or less.

Surface hydrogen adlayers were prepared by exposing the clean Si(100) surface held at a temperature T_s=690 K to a flux of atomic species generated from H_2 or D_2 in the presence of a hot tungsten filament ($\approx$ 1950 K). TPD spectra obtained following atomic D dosing are in good agreement with previous results [7,8], indicating that the surface is dominated by the monohydride phase having a saturation coverage of 1.0 ML (1 ML=6.8x10^{14} cm^{-2}). For the data below, the initial adlayer was prepared in the saturation regime. By comparing TPD integrated intensities for saturated H- and D-terminated/Si(100) surfaces, the D_2/H_2 and D_2/HD relative detection sensitivity ratios were determined.

A pulsed valve was mounted directly in the UHV chamber with no differential pumping and operated at 10 Hz. HI was transferred from the lecture bottle to a reservoir that was held at room temperature. The reservoir contained only 0.21±0.06 HI (mole fraction), with the balance being H_2 (originally present in the lecture bottle). The depletion from this reservoir, as measured with a capacitance manometer, gave a valve throughput of $1.3(\pm 0.1) \times 10^{16}$ molecules (H_2 and HI)/pulse; valve/laser photolysis exposures were done using from 1200 to 9000 pulses. The speed, angular and temporal distributions of the valve output were characterized with a fast ionization gauge (FIG) [these measurements were preformed in an unbaked chamber after the completion of all other experiments].

HI was photolyzed with the polarized, focused output of an excimer laser operating at 193 or 248 nm. Beam intensity profiles at the center of the photolytic volume were measured. The pulse energy in the photolytic volume was 210 mJ (90 mJ)/pulse at 248 nm (193 nm). Time-of-flight (TOF) spectra of H atoms resulting from HI photolysis were measured with quadrupole mass spectrometer (TOF-QMS) housed in a doubly-differentially pumped chamber and equipped with a multichannel scaler having a channel width of 640 ns. The geometry was such that the surface normal at the center of the crystal, the axial centerline of the valve and the axial centerline of the TOF-QMS were all nominally coplanar and perpendicular to the propagation axis of the excimer laser beam. The center of the excimer laser beam (i.e., the center of the photolytic volume) was 3.5 cm, 1.0 cm and 32.0 cm from the Si surface, the valve nozzle and the center of the TOF-QMS ion source, respectively.

The initial preparation of the D/Si(100) surface (i.e., the "as-prepared" surface) and adlayer modifications following valve/laser photolysis exposures with the sample held at T_s=650 K were probed by TPD and AES. Thermal desorption was multiplexed to acquire spectra for H_2, HD and D_2. Auger spectra were recorded following the acquisition of TPD spectra.

RESULTS

Characterization of the free-jet expansion/HI photolysis

Angular distributions of the output from the valve were fit well by a $\cos^n\phi$ distribution, with n=12.2±1.5 and ϕ being the angle between the axial centerline of the valve and the FIG. The temporal distribution was nearly Gaussian and had a FWHM of about 330 μs. From the FIG measurements [9], the most probable velocity $v_{mp}=8.0(\pm 1.2) \times 10^4$ cm/s and the velocity dispersion $\Delta v/v=0.25\pm0.10$ are determined.

TOF spectra of H atoms generated in the photolysis of HI were recorded simultaneously during D/Si(100) adlayer exposures, and the energy distributions are representative of the H atoms that strike the crystal. Based solely on the energetics of photofragmentation and the wavelength of the incident light [10], a single peak due to H atoms with a kinetic energy of 3.29 eV (0.94 eV) at 193 nm (248 nm) is expected. Simulated TOF spectra for a delta function energy distribution at each wavelength indicate that the measured TOF spectra show more intensity in a long-time (low energy) tail. This is not unexpected since photolysis occurred close to the nozzle in a high density region of the free jet. H atom energy distributions, obtained by direct inversion of the experimental TOF spectra, have average kinetic energies and standard deviations of 2.9 and 0.4 eV (0.85 and 0.08 eV), respectively, at 193 nm (248 nm).

Evaluation of the reactive species

Relative (193 vs 248 nm) and absolute exposures at the crystal of the photofragments (H and I atoms) are determined by integration of the valve/laser photolysis over time, coordinate and velocity spaces. In this analysis, which is described in detail elsewhere [11], *all* of the relevant parameters of the photolysis (i.e., the free-jet expansion and photofragmentation dynamics) are either characterized by direct measurement or are well-established in the literature [10].

Results from the integration are summarized as follows: 1) For both photolysis wavelengths, *no* I photofragment was found to strike the crystal, 2) The average energies for H atoms that strike the crystal are 1.0 eV and 2.8 eV for 248 and 193 nm, respectively and 3) The H atom exposure/pulse at the crystal for 248 nm is 2.3±0.6 times that for 193 nm; this *relative* H atom exposure factor S_H is used to scale results at 248 nm to those at 193 nm. In terms of valve throughput, *absolute* H atom exposures are 0.0014±0.0005 and 0.0032±0.0012 ML/Pa·L at 193 and 248 nm, respectively.

Hydrogen adlayer modifications

Following each exposure run, H and D atom coverages were obtained as follows,

$$\theta(H) = [I'(H_2)+0.5I'(HD)]/[I(D_2)+I'(HD)+I'(H_2)]_0 \qquad (1a)$$
$$\theta(D) = [I(D_2)+0.5I'(HD)]/[I(D_2)+I'(HD)+I'(H_2)]_0 \qquad (1b)$$

where $I(x)$ is the integrated TPD intensity of species x, with the prime indicating scaling by the x/D_2 relative detection sensitivity factor. The denominator in these expressions is the average result of typically two as-prepared surfaces (subscript 0 indicates no valve/laser photolysis exposure) that were taken on the same day as the data in the numerator.

For valve only exposures, both $\theta(H)$ and $\theta(D)$ change slightly ($\approx$ 10% of ML) and monotonically over the examined throughput range, with these changes being due presumably to the interactions of HI. In addition, the net surface coverage of hydrogen does not change with valve only exposure; i.e., $[\theta(H)+\theta(D)]=1.0$ ML over the examined throughput range. The exposure of the Si surface to H_2 and HI from the valve is dominated by the pump-out time of the pulse from the chamber. From measured H_2 and HI pumping speeds, it is estimated that the crystal is exposed to 0.2 ML HI and 4.0 ML H_2 per Pa·L of valve throughput. Exposures done with no gas output from the valve gave $\theta(H)$ and $\theta(D)$ results that were indistinguishable from the as-prepared surface.

Adlayer coverage changes resulting from valve/laser photolysis as a function of valve throughput are shown in Fig. 1. For both 193 and 248 nm, $\theta(H)$ and $\theta(D)$ change monotonically, and the extent of the coverage changes are *much more pronounced* than when compared with the results for valve only exposures. That is, the presence of the photolysis laser leads a marked decrease [increase] in the $\theta(D)$ [$\theta(H)$]. Also, the total surface hydrogen coverage [$\theta(H)+\theta(D)$] increases by 10-15%, independent of throughput over the throughput range examined.

AES signal ratios for I to Si as a function of valve throughput were computed; while the precision is poor (relative uncertainties are ≈ 50%), qualitative results can be drawn from these data. I uptake for valve only and valve with laser photolysis for both 193 and 248 nm are essentially indistinguishable; that is, the I/Si AES signal ratios are effectively independent of valve throughput, over the throughput range examined. While the I/Si AES ratios were not calibrated, the nominal surface I uptake was about 3-4 times the background level.

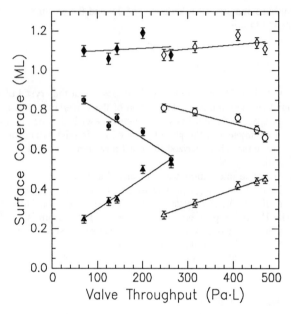

Fig 1.: θ(H) and θ(D) for valve/laser photolysis exposures are given by the triangles and circles, respectively. Total hydrogen coverages are given by the diamonds. Open (solid) symbols are for 193 nm (248 nm). Solid lines are to guide the eye.

<u>Kinetic energy dependence of H atom-induced adlayer modifications</u>

While adlayer changes with the valve only are not fully understood, the photolysis of HI and subsequent H atom interactions do not appear to alter significantly the HI/H$_2$ exposure interactions. This is supported by integration results which indicate that laser photolysis depletes only a very small fraction of the HI throughput per pulse (≤ 0.3%) and the similarity of the AES I/Si ratios for valve only and valve with laser photolysis. It is apparent that the large enhancement in adlayer modification with laser photolysis of HI is due to the interaction of energetic H atoms with the surface. To quantify the effects of H atom kinetic energy, the data in Fig. 1 must be further evaluated to account for two factors. First, adlayer changes resulting from valve only exposures are treated as unperturbed and hence separable from the effects due to H atom exposure. The changes in θ(D) and θ(H) as a function of exposure over the relevant exposure range are fit to linear functions, and subtracted from the data in Fig. 1. After having corrected the results for background HI/H$_2$ exposure, the valve throughput for the data at 248 nm is then scaled by S$_H$, thereby casting the results in terms of relative H atom exposures. The final results are given in Fig. 2, where changes in θ(D) and θ(H) due to H atom exposure are plotted for H atoms having nominal kinetic energies of 1.0 and 2.8 eV. The absolute H atom exposures shown in Fig. 2 range from about 0.2 to 0.8 ML.

DISCUSSION

To within the experimental uncertainties, the D atom depletion rate is identical for 1.0 and 2.8 eV H atoms, and the surface D atom depletion probability per incident H atom is 0.3±0.2. While these results do not

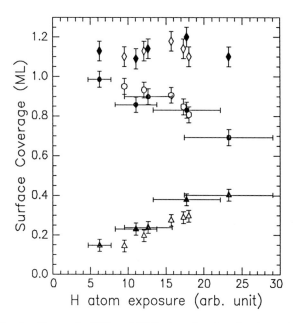

Fig. 2: Change in θ(H) and θ(D) given by the triangles and circles, respectively. The total hydrogen coverages are given by the diamonds (abscissa error bars are omitted). The open (solid) symbols are for 2.8 eV (1.0 eV) kinetic energy H atoms.

specify the depletion mechanism, it is likely that abstraction, rather than exchange, is occurring. Given the Si-H bond energy of 3.4 eV [12], abstraction is about 1 eV exothermic, while the exchange process is effectively thermoneutral. While there are no calculations of the barriers for abstraction or exchange on the Si(100) monohydride surface, electronic structure calculations give E_a=2.4 and 4.59 kcal/mol for H abstraction from Si_2H_6 [13] and SiH_4 [14], respectively. Two channels are found for exchange [14]; back-side attack with inversion (E_a=12.93 kcal/mol) and front-side attack with retention (E_a=13.46 kcal/mol). Projecting these results for SiH_4 to the surface system, the inversion mechanism for exchange is expected to be highly disfavored based on steric constraints, and the front-side attack with retention would be nominally equivalent to formation of the dihydride species. Indeed, the apparent activation energy for the removal of surface hydrogen by incident hydrogen atoms generated with a hot-filament (average energy about 0.3 eV) is 0.8±0.6 kcal/mol, which is consistent with an abstraction mechanism [6]. While the barrier for abstraction for the surface system is similar to calculations for the small silicon hydride species, the exchange barrier(s), given the nature of the transition state(s), are likely to be higher for the surface.

For hot filament-generated atomic H, Koleske *et al.* derived an absolute surface D atom depletion probability of 0.06-0.36 by assuming a "sticking probability" between 0.2 and 1. (the absolute atomic flux for the hot-filament source is not known) [6]. The surface D atom depletion probability by hyperthermal H from the present work is quite similar to the value derived for hot-filament species, suggesting that the reaction probability (abstraction) is *independent* of incident kinetic energy over the range of 0.3 to nearly 3 eV.

From the absolute H atom exposures and mass balance considerations, Fig. 2 indicates that most of the incident H atoms react (abstract D or adsorb). The H atom uptake characteristics are identical for both H atom kinetic energies. While these experiments do not identify the bonding nature of the adsorbed H, the uptake of H can occur by either dihydride formation or addition of H to dangling bonds liberated in abstraction. Similar to results obtained for hot-filament generated atomic hydrogen [6,7], the adsorption probability of H under the conditions of these experiments is greater than the depletion probability.

SUMMARY

D/Si(100) adlayer modifications due to interactions with H atoms having average kinetic energies of 1.0 and 2.8 eV are investigated. The D atom depletion probability per incident H atom is 0.3±0.2 for both kinetic energies, and is likely due to abstraction. These results, together with previous studies for hot-filament generated atomic species, indicate that the rates of depletion of surface D and the uptake of H are nominally independent of H atom kinetic energy over the range of about 0.3 to nearly 3 eV.

REFERENCES

1. M.A. Golub and T. Wydeven, Polymer Degrad. and Stab. **22**, 325 (1988).
2. E. Miyauchi, Crit. Rev. Sol. State. Matl. Sci. **17**, 107 (1991).
3. S. Imai, T. Iizuka, O. Sugiura and M. Matsumura, Thin Solid Films **225**, 168 (1993); S.M. Bedair, J. Vac. Sci. Technol. B **12**, 179 (1994).
4. J.T. Yates, Jr., C.C. Cheng, Q. Gao, M.L. Colaianni and W.J. Choyke, Thin Solid Films **225**, 150 (1993) and references therein.
5. D.D. Koleske and S.M. Gates, J. Chem. Phys. **99**, 8218 (1993).
6. D.D. Koleske, S.M. Gates and J.A. Schultz, J. Chem. Phys. **99**, 5619 (1993); D.D. Koleske, S.M. Gates and B. Jackson, *ibid.* **101**, 3301 (1994).
7. K. Sinniah, M.G. Sherman, L.B. Lewis, W.H. Weinberg, J.T. Yates, Jr. and K.C. Janda, J. Chem. Phys. **92**, 5700 (1990).
8. M.L. Wise, B.G. Koehler, P. Gupta, P.A. Coon and S.M. George, Surf. Sci. **258**, 166 (1991); C.C. Cheng and J.T. Yates, Jr., Phys. Rev. B **43**, 4041 (1991).
9. W.R. Gentry and C.F. Giese, Rev. Sci. Instrum. **49**, 595 (1978).
10. G.N.A. van Veen, K.A. Mohamed, T. Baller and A.E. de Vries, Chem. Phys. **80**, 113 (1983).
11. S.A. Buntin (to be submitted).
12. K.W. Kolasinski, Int. J. Mod. Phys. B (in press) and references therein.
13. K.D. Dobbs and D.J. Doren, J. Amer. Chem. Soc. **115**, 3731 (1993).
14. K.D. Dobbs and D.A. Dixon, J. Phys. Chem. **98**, 5290 (1994).

LOW TEMPERATURE FILM GROWTH
BY SUPERSONIC JETS OF CBr$_4$

DOUGLAS A. A. OHLBERG, GARRY ROSE, JAMES REN, AND R. STANLEY
WILLIAMS
Department of Chemistry and Biochemistry, University of California
Los Angeles, Los Angeles, CA 90024-1569

ABSTRACT

Pulsed, supersonic jets of CBr$_4$ seeded in a hydrogen bath gas have
been used to deposit films on silicon at low temperatures (ca. 100° C)
in a high vacuum chamber. In situ analysis of the films using x-ray
photoelectron spectroscopy (XPS) and depth profiling indicate a
surface composition of 34% Br and 66 % C and a bulk composition of
88% C and 12% Br. The deposition efficiency of CBr$_4$ was found to drop
dramatically when seeded in bath gases of deuterium, helium, and
argon, suggesting that the film growth is an activated process,
requiring precursor energies of at least 3.6 eV.

I. INTRODUCTION

A persistant problem with diamond growth through convential
CVD methods has been the poor incorporation efficiency of carbon in
the growing film. It has been suggested that the efficiency may be
improved through the use of supersonic beams.[1] At first glance, the
method promises many advantages: the ability to achieve high kinetic
energies which can surmount the activation barriers to dissociative
chemisorption, increased flux, and the ability to precisely control the
amount of material hitting the substrate. Indeed, the method has been
attempted for diamond growth[2], but thus far, results have been
discouraging. One possibility for this may be that precursors used
were not massive enough to retain the initial kinetic energy they
gained during expansion, losing it, instead, through subsequent
collisions with bath gas molecules rebounding from the growth
substrate to form a boundary layer.
Evidence for the efficacy of translational activation in film
growth has been seen, however, in the encouraging work of Connolly
et al. with chlorinated hydrocarbons.[3] This study, along with others[4,5]
indicates that the full advantages of supersonic expansion for film
growth are best realized by using precursors with large molecular
weight and low bond strengths. Thus, CBr$_4$ is of special interest: the

233

molecular weight is 331.65 amu and the relatively weak CBr_3-Br bonds have been calculated to be 2.44 eV.[6] Although CBr_4 is a solid up to 95 °C, it has a significant vapor pressure over a temperature range easily accessible to a conventional bubbler.

II. EXPERIMENTAL PROCEDURE

Figure 1 is a schematic of the apparatus used for the deposition experiments. CBr_4 of 99% purity from Aldrich[7] was used without further purification and packed between glass wool plugs into a small stainless steel container. The quarter inch stainless steel tubing which served as the outlet line for the container led to a pulsed solenoid valve[8], which was mounted inside a small vacuum chamber (ca. 6 liters) about 1 cm from the substrate.

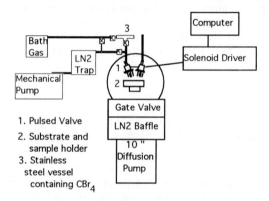

1. Pulsed Valve
2. Substrate and sample holder
3. Stainless steel vessel containing CBr_4

Figure 1. Pulsed supersonic jet deposition apparatus.

From various reports of the vapor pressure of CBr_4 in the literature[9,10], the following Clausius-Clapeyron relation for the sublimation pressure was derived:

$$\ln P = 27.1 - 8616/T \tag{1}$$

where P is the pressure in Torr and T the temperature in Kelvin. The stainless vessel containing the CBr_4 was heated to temperatures ranging from 80° to 100° C, which corresponded to partial vapor pressures of 20 to 50 Torr, respectively. This in turn corresponded to seeding ratios of one to two percent in a bath gas at stagnation pressures around 24 psig.

The kinetic energy of the precursors was estimated using the following formula for the terminal velocity (V_T) reached by a gas after undergoing a supersonic expansion:

$$V_T = (2RT\gamma/(\gamma-1)M)^{0.5} \qquad (2)$$

where R is the gas constant, T the temperature, γ the average ratio of C_p/C_v for the gas mixture, and M the average molecular weight of the gas mixture.[11] Although we were unable to measure the velocities of the precursor directly, such measurements have been performed on a system similar to ours and good agreement was found with the predicted velocities.[12]

Typical base pressures of the chamber before deposition were around 5×10^{-10} Torr. These pressures were only realized, however, when the lines behind the valves were evacuated with a mechanical pump. As soon as the valves were fully backed with the normal operating stagnation pressure of 24 psig, the base pressure of the chamber rose to 10^{-6} Torr due to leakage from the poppets in the pulsed valves.

In order to ensure that the precursor didn't recrystallize on the inside of the valves the valves were heated with a quartz IR lamp mounted inside the vacuum chamber. Consequently, the growth temperature of the otherwise unheated substrates ranged typically from 70° C to 120° C. Temperature measurements were made from a thermocouple probe spot-welded to the sample support.

Various substrate surfaces were used in the experiments including etched, unetched, and sputter-cleaned Si(100) and Si(111). No pre-annealing of the substrates was done prior to deposition.

The chemical composition of the deposits was deduced from x-ray photoelectron spectroscopy (XPS) performed in a UHV chamber which communicated with the growth chamber and allowed for sample analysis without removal of the sample from vacuum. The sensitivity factors of carbon and bromine relative to our electrostatic energy analyzer were calibrated with a dibromoanthracene standard that was purified by sublimation onto a cold finger in the growth chamber before subsequent examination in the analysis chamber.

Two methods were adopted to determine the fluxes seen by the substrate. The first involved measurement of a gas pulse with a fast ionization gauge[13] by which an instantaneous pulse pressure, and hence a number density, was obtained. Fluxes were then calculated by multiplying the resulting density by the calculated velocity from Eq. 2. This method did not consider the correction for the trenndusen effect, i.e. the segregation of the heavier species along the beam centerline. The second method was able to account for this by measuring the actual area of the substrate exposed to the precursor flux from the spot size of the deposition. The number densities of molecules seen by this area were then determined by isolating the chamber from the diffusion pump, and allowing the pulsed valve to fill the chamber to a certain pressure at the same flow rate that was used during deposition. Pressures were measured with a Granville-Phillips convectron gauge.[14] Corrections for each gas measured were made using the conversion graphs supplied by the manufacturer. Good agreement was found between these two calibration methods and precursor fluxes on the order of $10^{17} cm^{-2} sec^{-1}$ were measured with a seeded pulsed hydrogen beam.

III. RESULTS AND DISCUSSION

After an exposure of an hour and a half to a pulsed beam of CBr_4 in hydrogen, non-volatile deposits that were soluble in acetone were observed with spot areas ranging from 1 to 5 mm^2. Film thicknesses of half a micron were measured with an AFM on films deposited through a mask. Typically, the film was thickest in the middle and tapered away toward the edges. Figure 2 shows a representative XPS spectrum of such a deposit.

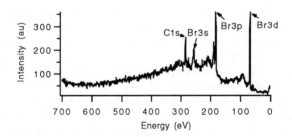

Figure 2.   XPS spectrum of a film deposited with a CBr_4/H_2 beam.

The surface composition of the deposit as determined by XPS indicated a Br to C ratio of 1:2 as deposited and 1:8 after sputtering.
Deposition was also attempted with a variety of other bath gases to determine whether the efficiency of deposition varied with precursor kinetic energy. Because the flux seen by the substrate depends on the terminal velocity of the expansion, longer exposure times were necessary for the slower moving, heavier bath gases to insure that samples experienced the same net fluences. Figure 3 shows the XPS spectrum of an unetched Si(111) wafer which has experienced twice the total fluence of CBr_4 in a seeded helium beam

as the sample represented in Fig. 2. The striking feature about Fig. 3 is that in spite of an increase in fluence, negligible amounts of C and Br were observed on the substrate exposed to CBr_4 in He, whereas a one half micron film was deposited when the bath gas was H_2.

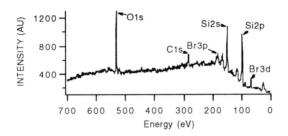

Figure 3. XPS of unetched Si wafer exposed to pulsed CBr_4/He beam

The same trend was seen with exposure to other bath gases, this time on etched Si(111) substrates. Table I summarizes the estimated kinetic energies and fluences involved in experiments with other bath gases.

Table I. Deposition conditions for CBr_4 in various bath gases.

Bath Gas	Estimated Velocity (m/s)	Kinetic Energy(eV)	Total Fluence (molecule/cm^2)
hydrogen	1.6×10^3	4.5	1.1×10^{21}
helium	1.2×10^3	2.6	3.0×10^{21}
deuterium	1.4×10^3	3.7	4.6×10^{21}
argon	590	0.6	3.3×10^{21}

Although the fluences of precursor in the experiments involving helium, deuterium and argon were several times that of the precursor in hydrogen, the deposition efficiencies were so low that once again, insignificant amounts of carbon and bromine were seen with XPS. It should be mentioned that before and after each deposition with a bath gas other than hydrogen, an experiment with hydrogen as the bath gas was run, to ensure that any observed deviations in the efficiency of deposition were not attributable to a change in the beam position due to deformation of the poppet in the pulsed valve.

V. SUMMARY

The deposition efficiency for carbon atoms onto various Si surfaces was determined to be somewhat less than 1% for CBr_4 seeded in an H_2 expansion, but completely negligible when other bath gases were used. The non-volatile nature and chemical composition of the film indicate that it is polymeric. The fact that film growth was not possible with a seeded deuterium beam suggests that the increase in carbon incorporation efficiency with hydrogen beams is not a chemical effect, but rather due to the higher kinetic energy gained with an expansion involving hydrogen gas. Thus, activation energies greater than 3.7 eV are necessary to initiate film growth with CBr_4.

VI. ACKNOWLEDGMENTS

This research was supported by the BMDO with a grant monitored by the ONR.

VII. REFERENCES

1. Max N. Yoder, Thin Solid Films **225**, 145 (1993).
2. H. Sprang, H. Busmann, S. Lauer, and I.V. Hertel, Appl. Phys. **A55**, 347 (1992).
3. M. S. Connolly, E. F. Greene, C. Gupta, P. Marzuk, T.H. Morton, C. Parks, and G. Staker, J. Phys. Chem. **85**, 235 (1981).
4   J. Fernandez de la Mora, B.L Halpern, and J.A. Wilson, J. Fluid Mech. **149**, 217 (1984).
5   J. Fernandez de la Mora, J. Chem. Phys. **82**, 3453 (1985).
6. D. F. McMillen and D. M. Golden, Ann. Rev. Phys. Chem., **33**, (1982).
7. Aldrich Chemical Co., Milwaukee, Wisconsin.
8. General Valve Corporation, Fairfield, New Jersey.
9. D. R. Lide, ed. CRC Handbook of Chemistry and Physics. 73th ed. (CRC Press, Boca Raton, Florida, 1980), p3-320.
10. Beilstein's Handbuch der Organischen Chemie EIII/IV (Beilstein Institute, Frankfurt-am-Main, Germany, 1959), B1, 85.
11. R. B. Bernstein, Chemical Dynamics via Molecular Beam and Laser Techniques (Oxford Univ. Press, New York, 1982), Sec. 3.3.
12. T. Curtiss, PhD Thesis, University of California Los Angeles 1990.
13. Beam Dynamics Inc., Minneapolis, Minnesota.
14. Granville-Phillips, Boulder, Colorado.

Part III

Opto-Electronic Materials, Nitrides, and Carbon Films

EFFECTS OF CARBON-ION IRRADIATION-ENERGIES ON THE MOLECULAR BEAM EPITAXY OF GAAS AND INGAAS

Y. Makita, T. Iida, T.Shima [a], S.Kimura, A.Obara and K. Harada [b]
Electrotechnical Laboratory, 1-1-4 Umezono,Tsukuba, Ibaraki 305, Japan
and
C.W. Tu
University of California, San Diego, La Jolla, CA 92093-0407, USA
and
S. Uekusa
Meiji University, 1-1-1 Higashi-mita, Tama, Kawasaki 214, Japan
and
T. Matsumori
Tokai University, 1117 Kitakaname, Hiratsuka 259-12, Japan.
and
K. Kudo and K. Tanaka
Chiba University, 1-33 Yayoi-cho, Inage, Chiba, Chiba 263, Japan

Abstract

Carbon ion (C^+) irradiation during molecular beam epitaxy (MBE) growth of GaAs/GaAs and $In_{0.53}Ga_{0.47}As/InP$ layers was carried out using CIBMBE (combined ion beam and molecular beam epitaxy) method as a function of wide acceleration energy (E_a=30 eV-30 keV) at a constant ion beam current density. Judging from the monitored current density and the net hole concentration ($|N_A-N_D|$) obtained from Hall effect measurements, activation rate as high as 88% was achieved for as-grown GaAs layers by C^+ ion irradiation of E_a=~170eV. It was revealed that by annealing at 800°C, a slight enhancement (~10%) of $|N_A-N_D|$ is practiced for E_a<~130eV but a significant increase of $|N_A-N_D|$ is realized for E_a>~1keV. In $In_{0.53}Ga_{0.47}As/InP$ layers with increasing E_a, a type conversion of electric conduction from n to p was found to occur at E_a= ~70~100eV. These observations describe that E_a plays a vital role to determine the location of incorporated electrical and optical active impurities in GaAs and InGaAs. Further for comparison, C^+-implanted GaAs layers were prepared by high-energy (400 keV) ion-implantation as a function of substrate temperature (T_i=RT-600 °C). For C dose concentration of $1x10^{18}cm^{-3}$, the highest activation rate of ~20 % was obtained at T_i=~150 °C. This result states that CIBMBE method is a superior doping method in view of activation rate of introduced dopants and the formation of damage-free ion-irradiated layers.

1. Introduction

The advantage of ion-beam doping technique is the high controllability of important doping parameters such as dopant species (high mass purity) and concentration. Since this method is a thermally non-equilibrium technique, the above parameters are independent of growth conditions of host materials such as growth temperature for the case of molecular beam epitaxy (MBE). In

---

a) On leave from Chiba University, 1-33 Yayoi-cho, Inage, Chiba, Chiba 263, Japan.
b) On leave from Tokai University, 1117 Kitakaname, Hiratsuka 259-12, Japan.

Mat. Res. Soc. Symp. Proc. Vol. 388 © 1995 Materials Research Society

solid-source MBE system, evaporation of dopant material using effusion cell is a most commonly used doping method. If the growth chamber is properly designed, the accidental incorporation of undesired impurities is normally governed by the purity of source materials. The decomposition and evaporation of crucible and heater materials are also other sources of contamination. In solid-source MBE, the doping efficiency and the dynamic range of dopant concentration are mainly determined by growth temperature, sticking coefficient and available vapor pressure. Accordingly except Si and Be, doping with a wide dynamic range of impurity concentration (10^{15} - 10^{20}cm^{-3}) is usually very difficult to be achieved.

High-energy ion implantation (HE-I^2) is one of the most frequently used ion-beam doping technologies. Two main drawbacks about HE-I^2 are the unelimeated radiation damage and the low activation rate even after suitable high-temperature annealing. The use of low-energy ion-beam (LE-IB) is one of the practical methods to minimize radiation damage [1-5]. To fabricate thick doped-layers, LE-IB technique should be combined with material-growth technology associated with high-vacuum technique, i.e., MBE method. In this case, LE-IB can alter the kinetics of arriving atoms and/or molecules at the surface and induce ion-assisted surface processing, which leads to an enhanced activation rate of dopants and an establishment of low-temperature MBE growth.

We here report the doping features of C atoms that are irradiated in form of accelerated ion-beam onto the MBE-growing GaAs surface. This is defined as combined ion beam and molecular beam epitaxy (CIBMBE) method [6-12]. The energy of C$^+$ ions (E_c^+) was varied between 30 eV and 30 keV which covers the three major beam-solid interaction regions; epiplantation, sputtering and ion implantation.

2. Experimental

Two key growth-parameters for CIBMBE, E_c^+ and the substrate temperature (T_g) during ion-beam irradiation are directly related with the creation and annihilation of damage. The high T_g is supposed to bring about in-growth annealing. Since the threshold displacement energy (E_{ds}) of GaAs impinged by C$^+$ ions is ~14 -23 eV [13], the irradiation of ion beam with E_a=~30 eV is expected to produce damage-free layers.

To see the in-growth annealing, the incorporation of C$^+$ ions by HE-I^2 was made for GaAs as a function of substrate temperature (T_i)[10]. T_i was varied from room temperature (RT) to 600 °C. HE-I^2 was made for undoped GaAs grown by MBE with C concentration ([C]) of 1×10^{18} and 1×10^{19} cm^{-3}. Three energies of 100, 200 and 400 keV were used to obtain a flat [C] distribution of 0.8 μm below the surface. Furnace annealing was made at 850 °C for 20 min in a purified-H$_2$ ambient. Samples were characterized by Raman scattering and Hall effect measurements at RT. Photoluminescence (PL) spectra were obtained at 2 K using a 1-m monochromator and a 514.5 nm line of Ar$^+$ laser. Same characterization methods were applied for GaAs and In$_{0.53}$Ga$_{0.47}$As layers fabricated by CIBMBE.

Figure 1 shows Raman spectra for unimplanted (UI) and C$^+$-implanted GaAs prepared by HE-I^2 with [C]=10^{19} cm^{-3} as a function of T_i. A peak at ~292 cm^{-1} is the longitudinal optical (LO) phonon mode of GaAs. Since lattice destruction caused by radiation damage is greatly introduced, LO phonon intensity for T_i= RT is extremely weak. For low T_i, LO phonon line broadens asymmetrically and is shifted towards lower frequency. For T_i< 400 °C, LO phonon intensities gradually increase with growing T_i and their central energy shifts towards the original position of LO-phonon for UI-GaAs. LO phonon intensities for samples with T_i > 500 °C are nearly the same

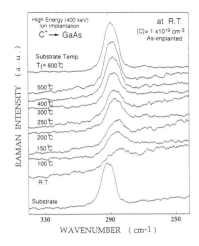

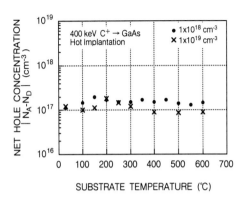

Fig.1 Raman scattering spectra from unimplanted GaAs substrate and 400 keV C$^+$-implanted GaAs as a function of ion-implantation substrate temperature (T_I). C concentration ([C]) is 1×10^{19} cm^{-3}.

Fig.2 Variation of net hole concentration ($|N_A - N_D|$) in 400 keV C$^+$-implanted GaAs as a function of ion-implantation substrate temperature (T_I). C concentration ([C]) is 10^{18} and 10^{19} cm^{-3} for closed circles and crosses, respectively.

to that for UI-GaAs. The results in Fig.1 imply that radiation damage introduced by HE-I^2 of C$^+$ ions can be annealed out in-situ for $T_I > \sim 500$ °C.

Hall effects measurements for as-implanted GaAs present no in-situ activation of C atoms during HE-I^2. These samples were furnace-annealed at 850°C for 20min. The variation of net hole concentration ($|N_A - N_D|$) with increasing T_I is shown in Fig.2 for [C]=1×10^{18} and 1×10^{19}cm^{-3}. For all samples, $|N_A - N_D|$ indicates nearly identical value of $\sim 1 \times 10^{17}$ cm^{-3}. The highest $|N_A - N_D|$ was obtained at T_I=150 °C with activation rate of ~ 10 % and at 200 °C with merely ~ 1 % for [C]=10^{18} and 10^{19} cm^{-3}, respectively. One should note that highly improved activation rate by hot implantation is not achieved even for annealed samples. The extremely low activation rate is principally attributed to the formation of serious radiation damage, deep acceptor levels [14], C_{Ga} donors [15] and point- or complex- defects behaving as donors [16].

The 2 K PL measurements for as-implanted samples, no emission peak was observed, stating again that in-situ activation of radiation recombination centers does not take place during hot ion-implantation by means of HE-I^2. The 2 K PL spectra from annealed samples as a function of T_I exhibit that a broad emission denoted by $[g-g]_\beta$, attributed to acceptor-acceptor pairs is observed for the entire T_I region. This emission energy corresponds to $|N_A - N_D|$= $\sim 1 \times 10^{17}$ cm^{-3} [6,17]. The above results of PL and electrical measurements tell that C atoms introduced by HE-I^2 are not activated in-situ even at T_I as high as 600 °C which corresponds to the highest T_g of C$^+$-irradiated GaAs prepared by CIBMBE.

To study the effects of E_a on the physical properties of C$^+$-impinged GaAs, C-doped GaAs layers were fabricated by CIBMBE system shown in Fig.3. CIBMBE system consists of a solid-

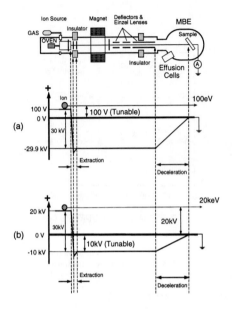

Fig.3 Schematic diagram of CIBMBE system and its voltage configuration.

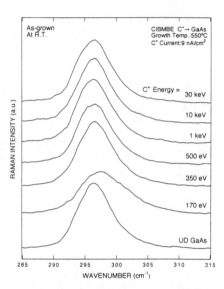

Fig.4 Raman scattering spectra from C$^+$-irradiated GaAs as a function of C$^+$ energy (E$_C$$^+$). All samples were grown at 550°C using CIBMBE system with the same C$^+$ ion beam current density of 9 nA/cm^2 .

source MBE and a LE-IB accelerator [6-11]. This accelerator produces ions with E$_a$ = 30 eV - 30 keV with mass-purity of ΔM/M=1/100. The voltage-configuration is shown for E$_a$ = 30 eV-3 keV and E$_a$ = 3 keV~30 keV in Fig.3-(a) and -(b), respectively. Samples were grown on semi-insulating (100) GaAs at As$_4$/Ga flux ratio of 2-3 with growth rate of 1μm/h. Mass-separated ^{12}C$^+$ ion irradiation was made during MBE growth.

3. Results and discussion for C$^+$-irradiated GaAs prepared by CIBMBE

3.1. Raman scattering

Figure 4 shows Raman spectra for as-irradiated CIBMBE-GaAs as a function of E$_C$$^+$. Layers were grown at T$_g$= 550 °C with constant ion beam current density (I$_C$) of 9nA/cm^2. In all samples, a single peak at ~292 cm^{-1} is observed. In low E$_C$$^+$ region, LO phonon line broadens asymmetrically and shifts slightly from ~292 cm^{-1} towards higher frequency. As are discussed later, as-irradiated GaAs layer with E$_C$$^+$=170 eV presents highest electrical activation rate (~88 %). The majority of C atoms are accordingly located at As sites, working as acceptors. Since the atomic radius of C is smaller than those of host Ga and As atoms, lattice compression is automatically introduced. The broadening and blue-energy shift of LO phonon observed for E$_C$$^+$=170 eV reflect this feature about crystal lattice [18].

Raman spectra also indicate that LO phonon signals shift towards higher frequency side with increasing |N$_A$-N$_D$| and they broaden asymmetrically. The reducing LO phonon intensity with growing

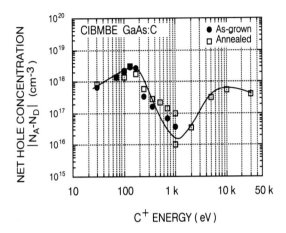

$|N_A-N_D|$ is partly attributed to the destruction of lattice symmetry due to increased [C]. This situation leads to the observation of forbidden transverse optical (TO) Raman signals from (100) surface. These features imply the presence of compressive stress and the hardening of the layer [18-20].

In contrast, HE-I² GaAs indicate that LO signals shift towards lower frequency side with growing $|N_A-N_D|$. This feature is opposite to that observed for CIBMBE GaAs and it is attributed to the softening of the lattice [21]. In HE-I² GaAs, TO phonon is the second dominant signal for high $|N_A-N_D|$, stating that unelimiated damage is existing even after high temperature annealing [14]. Raman signals accordingly reveal that the lattice destruction produced by C⁺ ion-irradiation with E_C^+=170 eV is negligible and damage-free ion-beam doping is realized in CIBMBE.

Fig.5 Variation of net hole concentration ($|N_A-N_D|$) obtained from room temperature Hall effects measurements in C⁺-irradiated GaAs as a function of C⁺ energy (E_C^+). All samples were grown at 550°C using CIBMBE with the same C⁺ ion beam current density of 9 nA/cm².

3.2. Electrical properties

3.2.1 Dependence on C⁺ ion beam current density (I_C^+)

For C⁺-irradiated GaAs, the dependence of $|N_A-N_D|$ on C⁺ ion beam current density, I_C was investigated for E_C^+=100 eV at T_g=500 ~ 590°C. Results reveal that a linear increase of $|N_A-N_D|$ is established with increasing I_C [6,9].

3.2.2. Dependence on C⁺ ion beam acceleration energy (E_C^+)

Figure 5 shows the results of Hall effect measurements as a function of E_C^+. Closed circles are for as-irradiated samples and open squares for annealed ones, respectively. E_C^+ was varied between 30 eV and 30 k eV. For E_C^+<130 eV, $|N_A-N_D|$ of as-irradiated layers gradually increases with elevating E_C^+ up to the highest concentration of 2.82x10¹⁸cm⁻³ at E_C^+ =170 eV [7,11]. Annealing of as-irradiated layers does not result in the increase of $|N_A-N_D|$ and some samples present a decrease of $|N_A-N_D|$. For E_C^+>170 eV, $|N_A-N_D|$ of as-irradiated layers steeply decreases down to one order of magnitude smaller than that obtained for E_C^+=170 eV. By high-temperature annealing, $|N_A-N_D|$ of as-irradiated layers is increased and the rate of $|N_A-N_D|$ enhancement seems to be gradually growing with increasing E_C^+. One should note that these values of $|N_A-N_D|$ for E_C^+>170 eV are still low as compared with the highest $|N_A-N_D|$ obtained for as-irradiated layers with E_C^+=170 eV. When E_C^+ exceeds 1 keV, $|N_A-N_D|$ of as-irradiated layers becomes smaller than 1x10¹⁵cm⁻³. $|N_A-N_D|$ for annealed

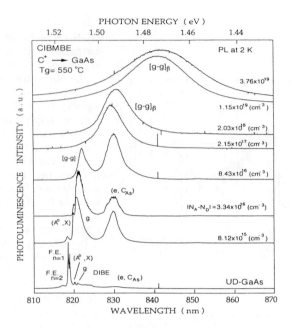

Fig.6 Low temperature (2 K) photoluminescence (PL) spectra from C^+-irradiated GaAs as a function of net hole concentration ($|N_A-N_D|$). The grown layers were made at 550°C using CIBMBE system with E_C^+=100 eV.

layers gradually increases with increasing E_C^+ and for E_C^+=~10 keV it eventually reaches a saturated value.

The above results indicate that the optimum E_C^+ to obtain a highest activation rate is located at ~170 eV. Judging from I_C^+ and $|N_A-N_D|$, ~88 % of C atoms was found to be activated for E_C^+= 170 eV. The growing $|N_A-N_D|$ with increasing E_C^+ for E_C^+<130 eV is ascribed to the accelerated migration of arriving C^+ ions that is enhanced by the kinetic energy of LE-IB. For obtaining damage-free layers, E_C^+ has long been considered to be reduced as low as the displacement energy (E_{ds}) of target materials which are impinged by ions with specific mass. The results in Fig. 5 tell that this enhanced migration effects should be taken into account to raise the activation rate of incorporated dopant ions. The undetectable $|N_A-N_D|$ in as-irradiated GaAs layers for E_C^+ > 1 keV is attributed to the inactivation of introduced C atoms even at T_g=550°C. The saturated activation rate obtained for E_C^+=30 keV is merely ~10% and this is the usual maximum activation rate for GaAs layers prepared by HE-I^2. One should note that the highest activation rate of CIBMBE-layers is by one order of magnitude greater than that of HE-I^2 layers.

3.3. Optical properties

3.3.1 Dependence on C^+ ion beam current density (I_C^+)

The 2 K PL spectra of undoped (UD) GaAs layer grown by MBE and C^+-irradiated GaAs layers prepared by CIBMBE are shown in Fig.6 as a function of $|N_A-N_D|$ [6,9]. All layers in Fig.6

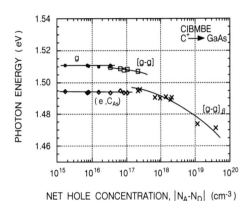

NET HOLE CONCENTRATION, $|N_A-N_D|$ (cm^{-3})

Fig.7 Variation of emission energies in the low temperature (2 K) photoluminescence (PL) spectra from C$^+$-irradiated GaAs as a function of net hole concentration ($|N_A-N_D|$). The grown layers were made at 550°C using CIBMBE system with E_c^+=100 eV.

were grown at 550 °C. In UD-GaAs, FE(n=1 and n=2) is a dominant emission with a small contribution of (A°,X) and (e, C). Three donor-related emissions such as (D°, X), (D$^+$, X) and (D, A) are not appreciably observed. Here each emission has the usual meaning and DIBE stands for defect-induce bound excitons emissions [22]. These observations tell that the accidental contamination can be strictly suppressed in this growth system, which is a prerequisite condition for the study of intentional doping of impurities. [6,9,17]

In CIBMBE layers, two emissions denoted by 'g' and [g-g] are observed. They are commonly obtained in GaAs doped with shallow acceptor impurities such as C, Be, Ge, Mg, Zn, and Cd [14,17,23-26]. 'g' lies just below (A°, X) and [g-g] is located slightly lower than 'g'. [g-g] is attributed to acceptor-acceptor pairs and exhibits a strong energy shift towards the lower energy side (red shift) with increasing $|N_A-N_D|$, reflecting the dependence of binding energy on pair-distance [27]. The results in Fig.6 reveal that C atoms introduced by CIBMBE are properly substituted at As sites (C_{As}) for T_g=550 °C and work as acceptors. One notes that T_g=550 °C is the standard growth temperature for MBE of GaAs but is a quite lower one than the necessary annealing temperature that is required for electrical (for obtaining saturated value at least higher than ~750°C) and optical activation (at least higher than ~850°C) of dopants introduced by HE-I^2 method [14]. The above results describe that in-growth activation of C atoms is realized by CIBMBE and no post-growth heat treatment is needed.

For $|N_A-N_D|$ >2.15x10^{17} cm^{-3}, a broad emission denoted by [g-g]$_\beta$ is observed at 1.49 eV which exhibits a red shift with growing $|N_A-N_D|$. A similar emission has been reported for $|N_A-N_D|$ >~3x10^{17} cm^{-3} in acceptor-doped GaAs prepared by conventional methods [17, 23-27]. The results suggest that the introduced C atoms are optically activated as acceptors even at this high $|N_A-N_D|$. In HE-I^2 GaAs layers, this energy shift was not observed due to uneliminated radiation damage which hinders significantly the radiative optical transitions [14, 23-26]. These results again apparently state that the radiation damage owing to 100 eV C$^+$-ion irradiation is considerably small.

CIBMBE layers grown at T_g= 500 °C presents PL spectra and emission intensities well comparable to those grown at T_g= 550 °C, indicating that introduced C atoms are efficiently substituted at As sites even at T_g= 500 °C. This T_g is an extremely low temperature not only to activate electrically and optically the dopant impurities incorporated by HE-I^2 method but also to eliminate lattice damage induced by the identical method [6,9,14]. These properties affirmatively mention that CIBMBE is superior to other doping methods in view of mass-purity, high activation rate and highly-suppressed radiation damage.

The emission energies obtained for as-irradiated GaAs layers are plotted in Fig. 7 as a function of $|N_A-N_D|$. 'g' and (e, C_{As}) emissions present no obvious energy dependence, while [g-g] and [g-g]$_\beta$

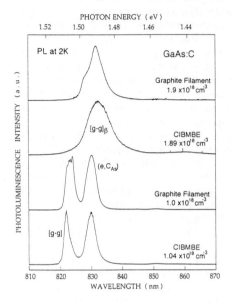

Fig.8 Low temperature (2 K) photoluminescence (PL) spectra from C⁺-irradiated GaAs layers made by CIBMBE method and those from C-doped GaAs prepared by graphite filament method.

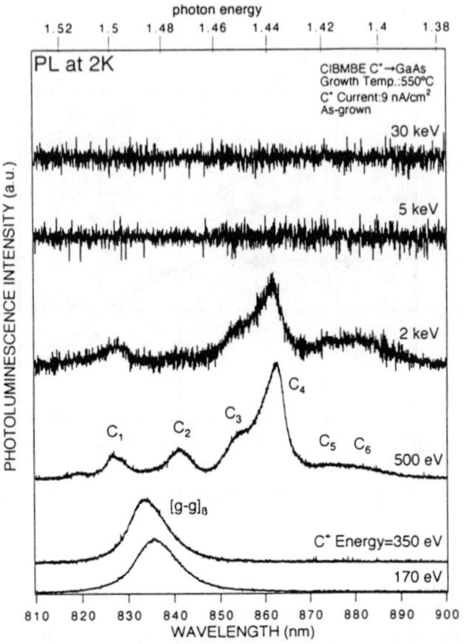

Fig.9 Low temperature (2 K) photoluminescence (PL) spectra from C⁺-irradiated GaAs as a function of C⁺ energy (E_C^+). All samples were grown using CIBMBE method with a constant C⁺ ion beam current density of 9 nA/cm² at 550°C. Samples in this figure are as-irradiated ones.

reveal red-shift with increasing $|N_A-N_D|$.

The 2 K PL spectra for C-doped GaAs prepared by graphite-filament (GF) method are slightly different from those grown by CIBMBE method. These features are shown in Fig. 8 in which the PL spectra for GF and CIBMBE films having nearly the same value of $|N_A-N_D|$ are demonstrated [28]. One remarks that GF-GaAs layers have fine structures in [g-g] and [g-g]$_\beta$ emissions. The carbon beam produced by GF method is known to contain C-clusters. The formation of fine structures can be partly attributed to the incorporation of those clusters which are supposed to induce the inhomogeneous pairing of acceptor impurities.

3.3.2. Dependence of PL features on E_C^+

The 2 K PL spectra of C⁺-irradiated GaAs layers as a function of E_C^+ are shown in Figs.9 and 10. Figure 9 exhibits the spectra for as-irradiated layers, while Fig.10 those for layers furnace-annealed at 850°C for 20min.

In the 2K PL spectra of as-irradiated samples for E_C^+ =30 eV ~170 eV (not shown in Fig.9), a broad emission denoted by [g-g]$_\beta$ is predominant and shows energy shift with increasing E_C^+ [7,11]. The observation of [g-g]$_\beta$ proclaims the efficient activation of incorporated acceptor impurities and the formation of acceptor-acceptor pairs with $|N_A-N_D| > \sim 1\times10^{17}$cm⁻³ [17,23-26]. For all the shallow

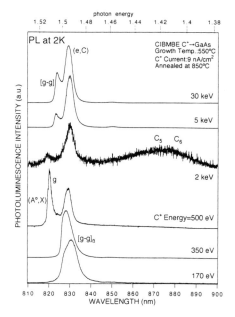

Fig.10 Same as in Fig.9 but samples are furnace-annealed at 850°C for 20min in purified H_2 ambient.

acceptor species, $[g\text{-}g]_\beta$ was found to reveal a steep red energy-shift and a moderate increase of its emission energy half-width (FWHM) with increasing $|N_A\text{-}N_D|$. For $E_C^+ < 170eV$, $[g\text{-}g]_\beta$ shows a red-shift with increasing E_C^+, but it presents blue-shift for $E_C^+ > 170$ eV with growing E_C^+. These features imply that the amount of optically activated C atoms increases for $E_C^+ < 170$ eV and decreases for $E_C^+ > 170$ eV with increasing E_C^+. From PL measurements the most effective E_C^+ to get maximum concentration of C-acceptors is identified to be located at ~170 eV. PL data accordingly describe that radiation damage owing to C^+ irradiation is small for $E_C^+ < 170$ eV, but for $E_C^+ > 170$ eV it is gradually enhanced with increasing E_C^+ and it eventually quenches very effectively most radiative transitions. The abrupt decrease of $|N_A\text{-}N_D|$ for $E_C^+ > 170$ eV is supposed to be mainly attributed to the enhanced sputtering effect. The formation of interstitial C atoms and antisite atoms, and further the increasing desorption of adsorbed ions with increasing E_C^+ are also considered to be responsible for the above observation [14-16].

The 2 K PL spectra of as-irradiated layers for $E_C^+ = 170$ eV~30 keV are presented in Fig.9. For $E_C^+ > 350$ eV, the intensity of $[g\text{-}g]_\beta$ decreases steeply with increasing E_C^+. For $E_C^+ = 500eV$ ~ 2keV in as-irradiated samples, multiple emissions denoted by C_i (i=1 ~ 6) are reproducibly observed. C_i emissions have been not identified in any shallow acceptor-doped GaAs prepared by other methods [14,17,23-26]. No appreciable PL emission is seen for $E_C^+ > 5$ keV in which the dominant ion-beam solid-surface interaction is ion implantation process. This observation evidently states that the damage produced by these energies is not eliminated at $T_g = 550°C$ by the in-growth annealing process endowed to CIBMBE method. The strongly changing PL features observed in Fig.9 indicate that the interaction of energetic ion-beam with solid surface is crucially dependent upon E_C^+.

The 2 K PL spectra for C^+-irradiated GaAs annealed at 850°C for 20 min are displayed in Fig.10 as a function of E_C^+. PL spectra for $E_C^+ < 170$ eV (not shown in Fig.10) reveal that no conspicuous change of spectra is presented after high-temperature annealing. PL spectra for $E_C^+ > 170$ eV indicate dramatically changed features as are demonstrated in Fig.10. All C_i emissions for $E_C^+ = 500$ eV are found to be completely quenched by heating at ~ 600°C. In the sample with $E_C^+ = 2$ keV, C_5 and C_6 seem to be not suppressed by heat treatment. To the best of our knowledge, only DIBE emissions are directly related with point defects that can become radiative recombination centers in the vicinity of band-edge [29]. The origins of extremely sharp emissions (except 'g') in DIBE series, are ascribed to the complex between C (carbon) acceptor atom and various types of point defects such as As and Ga vacancies and their interstitial atoms. The individual emission in DIBE series except 'g' is completely extinguished by heat treatment above ~550°C. In analogy of DIBE series, the origins of C_i emissions, at least that of C_i (i=1 ~ 4) can be rationally interpreted to be related with a variety of point defects [29]. The spectral features of C_i series are suggested to be affected by As_4/Ga flux ratio. For the thorough elucidation of these novel radiative emissions associated with point defects, systematic fabrication of CIBMBE-samples should be performed as a function of crucial parameters such as As_4/Ga flux ratio, substrate temperature (T_g) for ion-beam irradiation, surface orientation, ion beam current density (I_c) and mass of ion species.

For $E_c'>5$ keV, a steep enhancement of PL signals were obtained for annealed layers. In this figure, reflecting the well-recovered electronic data demonstrated in Fig.5, the formation of [g-g] is pronouncedly identified at the emission energy slightly below 'g'. The red-shift of [g-g] emission is also appreciably recognized with increasing $|N_A-N_D|$.

4. Electrical properties of C^+-irradiated $In_{0.53}Ga_{0.47}As/InP$ prepared by CIBMBE

It is well known that C in GaAs is exclusively a shallow acceptor, whereas C in InAs is a shallow donor [30]. This difference comes from the wide discrepancy of bonding energies (E_b) between dopant C atom and one of the constituent host atoms such as Ga, In and As. E_b between C and Ga atoms is greater than that between C and As atoms, which leads to the preferential substitution of C atoms at As sites (C_{As}) in GaAs that work as acceptors. On the contrary, E_b between C and In atoms is weaker than that between C and As atoms, which results in the exclusive substitution of C atoms at In sites in InAs that work as donors [31]. It is accordingly of great importance to study the incorporation mechanism of C atoms for InAs-GaAs alloy system using CIBMBE. We here present preliminary results of C^+-irradiation onto the (100) surface of InGaAs as a function of E_c^+. CIBMBE growth of lattice-matched $In_{0.53}Ga_{0.47}As$ on InP substrate was made at $T_g=350°C$ with a constant ion beam current density of 9 nA/cm².

$|N_A-N_D|$ and hole mobilities obtained from Hall effects measurements at RT are shown as a function of E_c^+ in Figs.11 and 12, respectively. A striking feature in Figs.11 and 12 is the type-conversion of electrical conduction in $In_{0.53}Ga_{0.47}As$ layers. For lowest E_c^+, samples indicate n-type conduction, while for $E_c^+ >\sim100$ eV they exhibit p-type conduction. It is interesting to note that this critical value (~100 eV) is fairly greater than the presumed displacement energy ($E_{ds}=\sim35$ eV) of

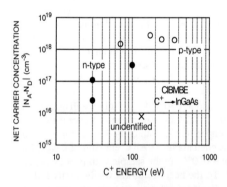
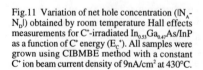
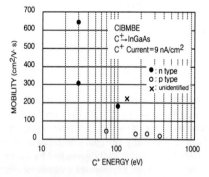

Fig.11 Variation of net hole concentration ($|N_A-N_D|$) obtained by room temperature Hall effects measurements for C^+-irradiated $In_{0.53}Ga_{0.47}As/InP$ as a function of C^+ energy (E_c^+). All samples were grown using CIBMBE method with a constant C^+ ion beam current density of 9nA/cm² at 430°C.

Fig.12 Variation of Hall mobilities obtained by room temperature Hall effects measurements for C^+-irradiated $In_{0.53}Ga_{0.47}As/InP$ as a function of C^+ energy (E_c^+). Samples were prepared as are described in Fig. 11.

InAs by C^+ ion impinging and is not significantly different from the value of the optimum $E_C^+ (=130\sim170 \text{ eV})$ to obtain the highest activation rate of irradiated C^+ ions into GaAs. In Fig.12, the mobilities for both n- and p-type samples are not so high compared with C-doped $In_{0.53}Ga_{0.47}As$ prepared by others methods. This observation is partly due to the compensation effect of existing C acceptors and donors [30]. In order to have low-compensated samples with high mobilities, optimization of CIBMBE growth-parameters such as $As_4/In+Ga$ flux ratio and T_g is strongly needed.

5. Conclusions

C^+ ion-beam doping into GaAs was performed using combined ion beam and molecular beam epitaxy method (CIBMBE) as a function of beam-irradiation substrate-temperature (T_g), C^+ ion-beam current density (I_C^+) and irradiation-energy (E_C^+). The 2 K PL spectra from as-irradiated CIBMBE GaAs layers revealed that in-growth optical activation of irradiated C^+ ions is realized at T_g as low as 500°C. The activation rate of incorporated C atoms was found to be strongly dependent upon E_C^+, and the highest activation rate of 88% was obtained for $E_C^+ = \sim170$ eV. For $E_C^+ > 170$ eV, the activation rate steeply decreases with increasing E_C^+, noting that the sputtering effect becomes the dominant process. Further for $E_C^+ > 2$ keV, owing to the significant radiation damage no in-growth activation was achieved. By furnace-annealing at 850°C for 20min, these samples presented well-recovered values of $|N_A-N_D|$ and PL intensities although the highest activation rate obtained was merely ~10 %. C^+ ion-irradiation was also made for $In_{0.53}Ga_{0.47}As/InP$ system by CIBMBE method. A conversion of electric conduction-type from n to p with increasing E_C^+ was identified in Hall effect measurements, stating that the location of amphoteric C atoms either at In (or Ga) or As sites is crucially influenced by E_C^+. It was revealed that low-energy ion beam doping by CIBMBE is an advantageous method in view of dopant purity, high controllability of dose, high activation rate and damage-free doping at low-temperature.

The authors wish to thank H.Shibata and A.C.Beye for their critical comments. One of the authors (C.W.Tu) would like to thank the support of the NSF-Japan Center for the Global Partnership Program.

References

1. E. Chason, P. Bedrossian, K. M. Horn, J. Y. Tsao, and S. T. Picraux, Appl. Phys. Lett. 57, 1793 (1990).
2. T. E. Haynes, R. A. Zuhr, S. J. Pennycook, and B. R. Appleton, Appl. Phys. Lett. 54, 1439 (1989).
3. J. -P. Noël, J. E. Greene, N. L. Rowell, S. Kechang, and D. C. Houghton, Appl. Phys. Lett. 55, 1525 (1989).
4. C. J. Tsai, H. A. Atwater, and T. Vreeland, Appl. Phys. Lett. 57, 2305 (1990).
5. W. -X. Ni, J. Knall, M. A. Hasan, G. V. Hansson, J. -E. Sundgren, S. A. Barnett, L. C. Markert, and J. E. Greene, Phys. Rev. B 40, 10449 (1989).
6. T. Iida, Y. Makita, S. Kimura, S. Winter, A. Yamada, P. Fons, and S. Uekusa, J. Appl. Phys 77, 146 (1995).
7. T. Iida, Y. Makita, S. Kimura, Y. Kawasumi, A. Yamada, S. Uekusa, and T. Tsukamoto, Proc. 8th Int. Conf. on Molecular Beam Epitaxy, Osaka, Japan, 1994, accepted for publication in J. Cryst. Growth.
8. T. Iida, Y. Makita, S. Kimura, S. Winter, A. Yamada, H. Shibata, A. Obara, S. Niki, P.

Fons, Y. Tsai and S. Uekusa, Mater. Res. Soc. Symp. Proc. 300, 357 (1993) .

9. T. Iida, Y. Makita, S. Kimura, S. Winter, A. Yamada, H. Shibata, A. Obara, S. Niki, P. Fons, Y. Tsai and S. Uekusa, Appl. Phys. Lett. 63, 1951 (1993).

10. T. Iida, K. Harada, S. Kimura, T. Shima, H. Katsumata, Y. Makita, H. Shibata, N. Kobayashi, S. Uekusa, T. Matsumori, and K. Kudo, Proc. 9th Conf. on Ion Beam Modification of Materials (IBMM'95), Canberra, Australia,

11. T. Iida, Y. Makita, S. Kimura, S. Winter, Y. Tsai, Y.Kawasumi, A. Yamada, H. Shibata, A. Obara, S. Niki, P. Fons, S. Uekusa, and T.Tsukamoto, Mat. Res. Soc. Symp. Proc. 316, 1029 (1994).

12. Y. Makita, T. Iida, S. Kimura, S. Winter, A. Yamada, H. Shibata, A. Obara, S. Niki, Y. Tsai, and S. Uekusa, Mat. Res. Soc. Symp. Proc. 316, 965 (1994).

13. D. K. Brice, J. Y. Tsao, and S. T. Picraux, Nucl. Inst. Method B44, 68 (1989).

14. S. Shigetomi, Y. Makita, A. C. Beye, A. Yamada, N. Ohnishi, and T. Matsumori, J. Appl. Phys. 69, 1613 (1991).

15. T. J. de Lyon, J. M. Woodall, M. S. Goorsky, and P. D. Kirchner, Appl. Phys. Lett. 56, 1040 (1990).

16. C.R. Abernathy, S.J. Pearton, R. Caruso, F. Ren, and J. Kovalchik, Appl. Phys. Lett. 55, 1750 (1989).

17. H. Shibata, Y. Makita, M. Mori, Y. Nakayama, T. Takahashi, A. Yamada, K. M. Mayer, N. Ohnishi, and A. C. Beye, *GaAs and Related Compounds* (Institute of Physics, Bristol, 1989), p. 245.

18. P. Fons, Y. Makita, S. Kimura, T. Iida, A. Yamada, H. Shibata, A. Obara, S. Niki, Y. Tsai and S. Uekusa, Mat. Res. Soc. Symp. Proc. 316, 1011(1994).

19. M. Nakayama, K. Kubota, T. Kanata, H. Kato, S. Chika, and N. Sano, J. Appl. Phys. 58, 4342 (1985).

20. T. Nakamura, A. Ushirokawa, and T. Katoda, Appl. Phys. Lett. 38, 13 (1981).

21. T. Motooka and O. W. Holland, Appl. Phys. Lett. 58, 2360 (1991).

22. J. D. Wiley, Semiconductors and Semimetals, edited by R. K. Willardson and A. C. Beer, (Academic Press, New York, 1975), Vol. 10, p. 91.

23. Y. Makita, H. Tanaka, M. Mori, N. Ohnishi, P. Phelan, S. Shigetomi, H. Shibata and T. Matsumori, J. Appl. Phys. 65, 248 (1989) .

24. Y. Makita, Y. Takeuchi, N. Ohnishi, T. Nomura, K. Kudo, H. Tanaka, H-C. Lee, M. Mori, Y. Mitsuhashi, Appl. Phys. Lett. 49, 1184 (1986).

25. N. Ohnishi, Y. Makita, K.Irie, K. Kudo, T. Nomura, H. Tanaka, M. Mori, Y. Mitsuhashi, J. Appl. Phys.60, 2502 (1986).

26. Y. Kawasumi, S. Kimura, T. Iida, A. Obara, H. Shibata, N. Kobayashi, T. Tsukamoto, and Y. Makita, Proc. 9th Int. Conf. on Ion Beam Modification of Materials (IBMM '95, Canberra, Australia, 1995).

27. N. Ohnishi, Y. Makita, M. Mori, K. Irie, Y. Takeuchi, and S. Shigetomi, J. Appl. Phys. 62, 1833 (1987).

28. R. J. Malik, R. N. Nottenberg, E. F. Schubert, J. F. Walker, and R. W. Ryan, Appl. Phys. Lett. 53, 2661 (1988).

29. N. Ohnishi, Y. Makita, H. Asakura, T. Iida, A. Yamada, H. Shibata, S. Uekusa, T. Matsumori, Appl. Phys. Lett., 62, 1527 (1993).

30. C.W.Tu, Proc. of the 5th International Conf. on InP and Related Materials, Paris, France, (April 18-22, 1993), p695.

31. H.Ito and T.Ishibashi, Jap. J. Appl. Phys.,30, L944 (1991).

LUMINESCENT NANOMETER-SIZED Si CRYSTALS FORMED IN AN AMORPHOUS SILICON DIOXIDE MATRIX BY ION IMPLANTATION AND ANNEALING

T.S.IWAYAMA*,Y.TERAO*,A.KAMIYA*,M.TAKEDA*,S.NAKAO** and K.SAITOH**
*Department of Materials Science, Aichi University of Education, Igaya-cho, Kariya-shi, Aichi 448, Japan
**National Industrial Research Institute of Nagoya, Kita-ku, Nagoya 462, Japan

ABSTRACT

Si ion implantation followed by thermal annealing has been used to synthesize luminescent nanometer-sized Si crystals in an amorphous SiO_2 matrix. Transmission electron microscopy indicates the formation of Si nanocrystals by annealing at 1100 °C, and the growth in average size of Si nanocrystals with increasing annealing time. The shape of the emission spectrum of the photoluminescence is found to be independent of both excitation energy and annealing time, while the excitation spectrum of photoluminescence increases as the photon energy increases and its shape depends on annealing time. The results indicate that the photons are absorbed by Si nanocrystals, for which the band-gap energy is modified by the quantum confinement effects, and the emission of photons is not due to direct electron-hole recombination inside Si nanocrystals but is related to defects probably at the interface between Si nanocrystals and SiO_2.

INTRODUCTION

In recent years, there has been considerable interest in semiconductor nanostructures, especially porous Si[1,2] and Si ultrafine particles[3-7], which exhibit strong luminescence in the visible range. Si nanocrystals have been produced by a number of methods including anodization using HF solution, CVD, co-sputtering and gas evaporation. Although a considerable amount of investigation has been devoted and many models of the light emission from these nanostructures have been proposed, the mechanism of luminescence is still controversial. The complexity of the structure and of the composition of Si nanoparticles fabricated so far may be a cause of the controversy.

Another novel approach to produce nanocrystals may be ion implantation. The ion implantation technique has the advantage that given numbers of ions can be placed in a controlled depth distribution[8]. Recently, it has been found that implantation of some ions into insulating materials gives rise to highly nonlinear optical properties[9-11] and visible luminescence behaviors[12,13], due to colloid formation. Ion beam synthesis of Si nanocrystals in a SiO_2 matrix is a potential candidate for the method of manufacturing Si nanocrystals of a pure and well-regulated condition as not only for basic research but also for application for optoelectronic devices.

The present authors have carried out studies on the luminescence behaviors of Si^+-implanted silica glasses[12,14,15] and thermal oxide films grown on crystalline Si wafer[16-18]. We have shown that SiO_2 implanted with Si ions exhibits two luminescence bands in the visible range. One peaked around 2.0 eV is observed in as-implanted specimens and annealed at about 600 °C, and the other peaked around 1.7 eV is observed after annealing specimens at 1100 °C. The latter band is similar in a spectral region to the porous Si luminescence. Moreover, the heating of the specimens at 1100 °C induces decomposition of SiO_x leading to crystalline Si precipitation in a SiO_2 matrix. Based on these studies, the 2.0 eV band can be attributed to Si excess defects in the SiO_x environment and the 1.7 eV band to the electron-hole recombination in the interface between the Si nanocrystal and the SiO_2. However, the detailed mechanism of the latter luminescence is not yet clear. The purpose of this paper is to report the correlation between the microstructure of Si^+-implanted layer and the luminescence band around 1.7 eV observed after annealing at 1100 °C, and to discuss the mechanism of the absorption of excitation photons and the emission of the luminescence.

Mat. Res. Soc. Symp. Proc. Vol. 388 © 1995 Materials Research Society

EXPERIMENT

Oxidized boron doped p-type Si wafers with a resistance of 0.01-0.02 $\Omega \cdot$cm were used. The orientation of the Si wafers was (100). Thermal oxide films on crystalline Si wafer were obtained through oxidation of the wafers by annealing at 1050 °C for 10 h in 60 % H_2 and 40 % O_2 ambient. The oxide thickness was about 1.75 μm. Si^+-implantation was carried out at an energy of 1 MeV to a fluence of 2 x 10^{17} ions/cm^2 at room temperature. Heat treatments of the ion implanted specimens were carried out in vacuum using an electric oven.

Photoluminescence spectra of specimens before and after annealing were measured at room temperature in a conventional way. An Ar-ion laser was used as an excitation source and the luminescence was detected by a cooled photomultiplier tube, employing the photon counting technique. Photoluminescence excitation spectroscopy was carried out using monochromatized light beams from a Xe-lamp as an excitation source and by detecting the luminescence through appropriate sharp-cut filters. The transmission electron micrographs were obtained by using a 300 keV electron microscope (H-9000NAR, Hitachi). For examining the microstructures of the specimens, the cross-sectioning method was employed. The specimens were first mechanically thinned down and then ion-milled to electron transparency using 5 keV Ar^+-bombardment.

RESULTS

The luminescence spectra induced before and after annealing at 1100 °C for 90 min and 300 min of the specimens implanted to a fluence of 2 x 10^{17} ions/cm^2 at room temperature are shown in fig.1, in which each curve is normalized by the peak heights. As already reported[16], the shape of the luminescence spectra is complex because of interference effects. The primary luminescence band is peaked around 2.0 eV and 1.7 eV, respectively, similar to those having been observed before and after annealing at 1100 °C in silica glass specimens. As also reported previously[14,16], the luminescence band around 2.0 eV, which is annealed after heating to about 600 °C, has been ascribed to the Si excess defects formed in SiO_2 or defects in SiO_x. Moreover, the peak energies of the 2.0 eV luminescence depend on both the ion fluences and the temperature of substrates during implantation.

In this paper we are concerned with the luminescence band around 1.7 eV, which is observed after annealing specimens at 1100 °C. It is clear from fig.1 that the shape of the luminescence spectra is not affected by the annealing time at 1100 °C. It is noted that the shape of the luminescence spectra after annealing is also found to be independent of the ion fluences and the temperature of substrates during ion implantation. However, the intensity of the luminescence is affected by the annealing time. The results of evolution and annealing of the luminescence band during annealing at 1100 °C are shown in fig.2. Evidently the luminescence intensity grows and then decreases, as the annealing time increases. The maximum luminescence intensity is obtained with annealing for 90 min.

The excitation spectra for the photoluminescence of the specimens implanted with Si ions were obtained after heat treatments at 1100 °C for 90 min and 240 min. The results are shown in fig.3. It is clear that the emission intensities increase with increasing incident photon energy in both specimens, although the increase is more significant in the specimen annealed for 90 min. It is noted that the shape of the photoluminescence spectra is found to be independent of the excitation energies, as shown in fig.4.

We obtained the cross-section high-resolution transmission electron micrographs for specimens before and after heat treatment at 1100 °C, around the depth of the projected range of implanted Si estimated by using the TRIM code[19]. The results are shown in fig.5. No trace of the formation of crystalline Si are indicated before the heat treatment, as shown in fig.5(a). However, the micrograph after heat treatment at 1100 °C for 90 min indicates Si nanocrystals ranging in size of a few nanometers in an amorphous SiO_2 matrix, as shown in fig.5(b). Moreover, we observed the growth in size and the coalescence of Si crystals with increasing in annealing time. The micrograph of the specimen annealed for 300 min is shown in fig.5(c).

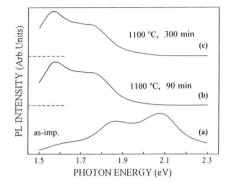

Fig. 1. Photoluminescence spectra of 1 MeV Si⁺-implanted thermal oxide film grown on crystalline Si wafer to a fluence of 2×10^{17} ions/cm² at room temperature (a) before annealing, and after subsequent annealing at 1100 °C for (b) 90 min and (c) 300 min. The zero line of curves (b) and (c) are shifted vertically to the position of the horizontal dashed line.

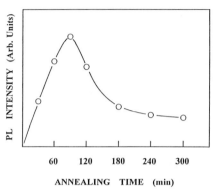

Fig. 2. Change of photoluminescence intensity as a function of annealing time at 1100 °C. 1 MeV Si ions were implanted into thermal oxide film grown on crystalline Si wafer to a fluence of 2×10^{17} ions/cm² at room temperature.

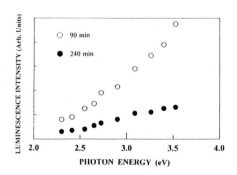

Fig. 3. Excitation spectra for the photoluminescence of 1 MeV Si⁺-implanted thermal oxide films grown on crystalline Si wafers to a fluence of 2×10^{17} ions/cm² at room temperature, after annealing at 1100 °C for 90 min and 240 min.

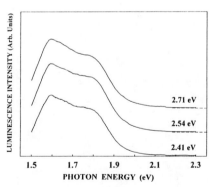

Fig. 4. Photoluminescence spectra of 1 MeV Si⁺-implanted thermal oxide films grown on crystalline Si wafers to a fluence of 2×10^{17} ions/cm² at room temperature, after annealing at 1100 °C for 90 min. The excitation energy is indicated in the figure.

(a)

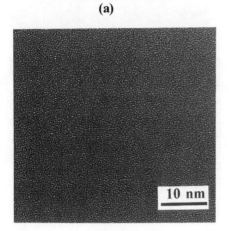

10 nm

Fig. 5. Cross-section transmission electron micrographs of the specimen implanted with 1 MeV Si ions to a fluence of 2×10^{17} ions/cm^2 at room temperature (a) without annealing, and annealed at 1100 °C for (b) 90 min and (c) 300 min.

(b)   **(c)**

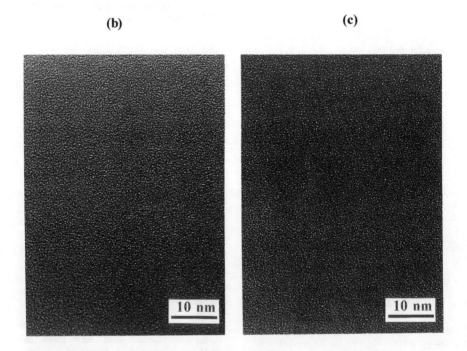

10 nm   10 nm

DISCUSSION

As shown elsewhere[14,15], after annealing at 600 °C of Si^+-implanted SiO_2, most defects generated by ion implantation as well as the defect-related luminescence band around 2.0 eV are annihilated. No luminescence is observed after annealing between 600 °C and 1000 °C. It is most likely that most of the implanted Si atoms are stabilized in the form of SiO_x annealed between 600 °C and 1000 °C. On the other hand, indirect evidences have been accumulated that annealing at 1100 °C produces Si segregates: the observation of a periodic interference pattern in the optical absorption spectra and of Raman spectroscopy indicates the growth of Si crystals in a SiO_2 matrix[14] by the following chemical reaction:

$$SiO_{2-\delta} \rightarrow (1 - \delta / 2) SiO_2 + \delta / 2 Si \qquad (0 < \delta < 2)$$

However, in the present study we confirmed directly the formation of nanometer-sized Si crystals in an amorphous SiO_2 matrix after annealing at 1100 °C by transmission electron microscopy.

In this section, we discuss the correlation between the microstructure of Si^+-implanted layer and the luminescence band around 1.7 eV observed after annealing at 1100 °C. The luminescence intensity grows at the initial stage and then decreases during annealing, without changing the shape of the luminescence spectra. Since the size of the Si nanocrystals evidently grows as the annealing time increases, it is most reasonable to consider that the existence of the lower and upper limits of the size of Si crystals which can make contribution to the luminescence around 1.7 eV. According to Takagahara et al.[20], the band-gap energy strongly depends on the size of the semiconductor particle, in the order of a few nanometers. However, the absence of the dependence of the shape of luminescence spectra on the annealing time excludes the possibility that the luminescence is due to the direct recombination between electrons and holes confined in the inside of Si nanocrystals.

The results in the present experiments are consistent with the presumption that the absorption of photons induces the generation of electron-hole pairs confined in Si nanocrystals, and the emission of photons probably originate from processes at the interface between crystalline Si nanoparticles and SiO_2. As the average size is smaller, photons with higher energies can contribute to the luminescence. However, as the annealing time increases, the average size increases and hence the absorption of photons at higher energies will be diminished, because of the smallness of the band-gap energy. The energy of photons imparted to the kinetic energies of the electrons and holes is not transferred to luminescence centers. It is considered to be immediately transferred to phonons. Thus, only the Si nanocrystals having a band-gap energy smaller than the excitation energy and larger than the luminescence energy can make contribution to the luminescence.

According to the theoretical calculation of band-gap energies for Si quantum dots by Takagahara et al.[20], the diameter of Si particles corresponding to 2.5 eV (excitation energy employed in the present experiments) and 1.7 eV (luminescence energy) is about 3 nm and 4 nm, respectively. In the model denoted above, we expect the Si nanocrystals of a average diameter between 3 nm and 4 nm embedded in a SiO_2 matrix induce emission at the highest efficiency by excitation with an Ar-ion laser (2.5 eV). This estimation is consistent with the experimental results of transmission electron microscopy, as shown in fig.5(b).

Now, we consider the luminescence centers. It is known that the luminescence emitted by recombination of the self-trapped excitons in crystalline SiO_2 has been shown to be peaked at 2.8 eV[21-23]. This luminescence is thought to be due to the recombination of electron on the Si dangling bond and hole on the O dangling bond, which are generated by breaking of Si-O bond[24,25]. The similar luminescence band in amorphous SiO_2 has been shown to be peaked around 2.3 eV[26]. Moreover, the luminescence band observed around 2.0 eV in Si^+-implanted SiO_2 before annealing, in which the peak energy of the luminescence depends on both the implantation fluence and the substrate temperature during implantation, can be attributed to the Si excess defects formed in SiO_2 or defects in SiO_x[14,16]. It is suggested that the peak energy of the luminescence from Si-O systems is sensitive to a variety of local disorders[15]. The local disorders include both structural and bond disorder. The arguments above suggest that the 1.7 eV emission observed in the present experiments, is due to the recombination of electrons and holes at the interface between Si nanocrystal and SiO_2, with relatively lower oxygen concentration. The exact picture of the luminescence center is under investigation.

CONCLUSIONS

The present experimental results give the most conclusive evidence that Si nanocrystals in an amorphous SiO_2 matrix can be the source of the absorption of photons of the energies above the band-gap energy and are capable of emitting photons, if the size is sufficiently small. The incident photons are absorbed by Si nanocrystals with modified band-gap governed by the quantum confinement effects. The emission, however, is not originated from the quantum confinement effects but probably due to defects at the interface between nanometer-sized Si crystals and SiO_2. The Si nanocrystals of a average diameter between 3 nm and 4 nm embedded in an amorphous SiO_2 matrix induce emission at the highest efficiency by excitation with an Ar-ion laser (2.5 eV). The method developed in this study, the formation of Si nanocrystals formed in an amorphous SiO_2 matrix by ion implantation and annealing is the cleanest among the nanostructures developed so far and has possibilities for optoelectronic application.

ACKNOWLEDGEMENTS

The authors would like to express their gratitude to Prof. Noriaki Itoh for valuable discussions. This work has been partly supported by a Grant-in-Aid for Science Research from the Ministry of Education, Science and Culture of Japan, by the Tatematsu Foundation, and by the Iketani Science and Technology Foundation.

REFERENCES

1. L.T.Canham, Appl. Phys. Lett. **57**, 1046 (1990).
2. V.Lehmann and U.Gosele, Appl. Phys. Lett. **58**, 856 (1991).
3. D.J.DiMaria, J.R.Kirtley, E.J.Pakulis, D.W.Dong, T.S.Kuan, F.L.Pesavento, T.N.Theis, and J.A.Cutro, J. Appl. Phys. **56**, 401 (1984).
4. S.Furukawa and T.Miyasato, Jpn. J. Appl. Phys. **27**, L2207, (1988).
5. H.Takagi, H.Ogawa, Y.Yamazaki, A.Ishizaki, and T.Nakagiri, Appl. Phys. Lett. **56**, 2379 (1990).
6. H.Morisaki, F.W.Ping, H.Ono, and K.Yazawa, J. Appl. Phys. **70**, 1869 (1991).
7. S.Hayashi, T.Nagareda, Y.Kanzawa, and K.Yamamoto, Jpn. J. Appl. Phys. **32**, 3840 (1993).
8. J.F.Ziegler, *Ion Implantation Technology*, J.F.Ziegler (ed.) p.1 (North-Holland, Amsterdam, 1992).
9. K.Becker, L.Yang, R.F.Haglund Jr, R.H.Magruder, R.A.Weeks, and R.A.Zuhr, Nucl. Instrum. Methods **B59/60**, 1304 (1991).
10. H.Hosono, Y.Abe, Y.L.Lee, T.Tokizaki, and A.Nakamura, Appl. Phys. Lett. **61**, 2747 (1992).
11. Y.Takeda, T.Hioki, T.Motoshiro, and S.Noda, Appl. Phys. Lett. **63**, 3420 (1993).
12. T.Shimizu-Iwayama, M.Ohshima, T.Niimi, S.Nakao, K.Saitoh, T.Fujita, and N.Itoh, J. Phys.: Condens. Matter **5**, L375 (1993).
13. H.A.Atwater, K.V.Shcheglov, S.S.Wong, K.J.Vahala, R.C.Flagan, M.L.Brongersma, and A.Polman, Mat. Res. Soc. Symp. Proc. **316**, 409 (1994).
14. T.Shimizu-Iwayama, K.Fujita, S.Nakao, K.Saitoh, T.Fujita, and N.Itoh, J. Appl. Phys. **75**, 7779 (1994).
15. N.Itoh, T.Shimizu-Iwayama, and T.Fujita, J. Non-cryst. Solids **179**, 194 (1994).
16. T.Shimizu-Iwayama, S.Nakao, and K.Saitoh, Appl. Phys. Lett. **65**, 1814 (1994).
17. T.Shimizu-Iwayama, S.Nakao, K.Saitoh, and N.Itoh, J. Phys.: Condens. Matter **6**, L601 (1994).
18. T.Shimizu-Iwayama, S.Nakao, and K.Saitoh, Jpn. J. Appl. Phys. **Suppl. 34-1**, 86 (1995).
19. J.F.Ziegler, J.P.Biersack, and U.Littmark, *The Stopping and Range of Ions in Solids vol.1* (Pergamon, New York, 1985).
20. T.Takagahara and K.Takeda, Phys. Rev. **B46**, 15578 (1992).
21. D.L.Griscom, Proc. 3rd Int. Frequency Control Symposium p.98 (Electronic Industries Association, Washington D.C., 1979).
22. A.N.Trukhin, Sov. Phys. Solid State **21**, 644 (1979).
23. C.Itoh, K.Tanimura, N.Itoh, and M.Itoh, Phys. Rev. **B39**, 11183 (1989).
24. A.L.Shluger, J. Phys. **C21**, L432 (1988).
25. A.Shluger and E.Stefanovich, Phys. Rev. **B42**, 9664 (1990).
26. N.Itoh, K.Tanimura, and C.Itoh, *The Physics and Technology of Amorphous SiO_2*, R.A.B.Devine (ed.) p.135 (Plenum, New York, 1988).

ANION EXCHANGE REACTIONS AND INITIAL GaN EPITAXIAL LAYER FORMATION UNDER NITROGEN PLASMA EXPOSURE OF A GaAs SURFACE

R. J. HAUENSTEIN[*], D. A. COLLINS[**], M. L. O'STEEN[*], Z. Z. BANDIC[**], AND T. C. MCGILL[**]
[*]Oklahoma State University, Dept. of Physics, PS-145, Stillwater, OK 74078
[**]California Institute of Technology, Dept. of Applied Physics, 128-95, Pasadena, CA 91125

ABSTRACT

Initial nitridation of molecular beam epitaxially (MBE) grown GaAs surfaces by means of an electron cyclotron resonance (ECR) microwave plasma source is investigated *in situ* through time-resolved reflection high energy electron diffraction (RHEED), and *ex situ* high-resolution X-ray diffraction (HRXRD) techniques. Brief (< 8-10 s) plasma exposure of GaAs (100) surfaces results in surface N-for-As anion exchange and a new, specular, commensurate (3x3) RHEED pattern which appears to correspond to up to 1 monolayer of coherently strained GaN on GaAs. Anion exchange kinetics is studied through the time-dependence of the onset and decay of the (3x3) RHEED pattern as a function of substrate temperature. For the first time, coherently strained GaN_yAs_{1-y} /GaAs heterostructures are grown and characterized. Direct evidence for thermally activated processes from both RHEED and HRXRD observations is presented, and N desorption and surface-segregation phenomena are proposed to explain the experimental results.

INTRODUCTION

GaN-based wide-gap III-V refractory semiconductors have come under intensely renewed interest as a prospective short-wavelength optoelectronic device material. Several reviews of GaN-based materials have been given recently,[1-3] and a number of successful demonstrations of GaN-based blue-green[4] and blue[5] light-emitting diode (LED) devices have been reported. Owing to the considerable technological demand for blue-to-UV semiconductor LEDs and lasers, most research on GaN-based wide-gap semiconductors has emphasized light-emitting device structures. In contrast, relatively few reports have concentrated on fundamental investigation of the epitaxial growth process itself. In particular, for the case of electron-cyclotron-resonance microwave plasma assisted molecular beam epitaxial (ECR-MBE) growth of GaN/GaAs structures, the exchange of N and As atoms ("anion exchange") plays a crucial role in the initial stages of epitaxy, but, nonetheless, has received little attention.

In this paper, we investigate anion exchange processes at an initial GaAs (100) surface upon nitrogen plasma (N*) exposure in an ECR-MBE system. Anion-exchange processes are monitored *in situ* through time-resolved reflection high energy electron diffraction (RHEED) techniques. Brief N* exposure is shown to result in the formation of a comparatively stable, coherently strained nitrided surface layer on GaAs, due to N-for-As surface exchange. This layer exhibits a previously unreported (3x3) surface reconstruction, and can be overgrown commensurately with GaAs. In this way, coherently strained GaNAs/GaAs strained-layer superlattices are fabricated for the first time, and characterized through high resolution X-ray diffraction (HRXRD). Finally, two important but previously unreported thermally activated

259

processes, nitrogen desorption and surface segregation, which can be expected to have direct significance to the initial stages of GaN epitaxial growth on GaAs, are identified and examined quantitatively.

EXPERIMENTAL

In this study, high quality GaAs (100) surfaces are first prepared through conventional III-V MBE growth techniques, and subsequently, are exposed *in situ* to a flux (N* plasma or cracked-As_2) as a function of temperature and exposure time. In one set of experiments, the anion exchange (onset and decay of the nitrided-surface reconstruction) is assessed qualitatively and quantitatively through time-resolved *in situ* RHEED observations; in a second set of experiments, nitrided surfaces are buried under GaAs and are subsequently characterized structurally through four-crystal high resolution X-ray diffraction (HRXRD). In all cases, samples are prepared as follows. First, GaAs buffer layers(~1 μm) are epitaxially deposited onto GaAs (100) epi-ready substrates at 590°C and a growth rate of ~1 monolayer (ML) per sec. Next, the As-stabilized surface is exposed to the N* plasma which is produced by the flow of N_2 through an ECR microwave plasma source (Wavemat, Model MPDR 610iA) at a forward power of 100 W. During exposure, temperature is carefully maintained at a constant value in the range, 540 to 620°C. Precise exposure-time control is effected by means of a computer which controls both the plasma power and the action of a shutter which is positioned at the mouth of the ECR source. Time- and spatially resolved RHEED data (Perkin-Elmer, 10 keV) is acquired with the use of a CCD camera and S-VHS VCR unit, and is subsequently processed digitally on playback through the use of a frame grabber and custom-developed image processing and analysis software.[6] Further details of our ECR-MBE growth system and procedures can be found elsewhere.[7]

For the HRXRD study, a set of 36-period (to improve signal-to-noise) nitrided-ML/GaAs (GaN_yAs_{1-y}/GaAs) superlattices was grown as a function of temperature and nitridation time, as shown in Table I. All other procedures and conditions were nominally held constant in order to minimize run-to-run variation of our temperature calibration. Each period involves: (i) N* exposure without As_2 flux for the time shown in the table; (ii) *immediate* overgrowth of GaAs at ~0.75 ML/s for 100 s; (iii) an "As-soak" (As_2 flux only) for 30 s.

Table I Indicated are growth temperature, temperature uncertainty, and plasma exposure time for each sample in the set. *Uncertainties listed below indicate only reproducibility from run to run within the sample set.

Sample Number	Substrate Temperature (°C)	Relative Uncertainty* (°C)	Plasma Exposure Time (sec)	Nitrided Layer Composition, y (%)
OS95.002	560	+/-5	4	16.5
OS95.003	570	+/-2	4	7
OS95.004	580	+/-5	4	4
OS95.005	550	+/-2	4	26
OS95.006	540	+/-5	4	26.5
OS95.009	550	+/-5	6	33

The HRXRD measurements are performed with the use of a Philips Materials Research Diffractometer (MRD) operated in the 4-crystal mode using (220) Ge reflections. Both $\omega/2\theta$ scans, and "area" scans (reciprocal-space mapping), about the substrate (004) and (115) Bragg peaks, respectively, are performed on each sample. Additionally, dynamical simulations of the $\omega/2\theta$ scans, to aid in compositional and strain profile analysis of our superlattices, are computed with the use of the Philips High Resolution Simulation (HRS) software package.

RESULTS AND DISCUSSION

Figure 1 shows schematically the effect of the N* flux on initial GaAs (100) RHEED pattern. Near 600°C, the characteristic As-stabilized (2x4) GaAs surface reconstruction [Fig. 1(a)] upon N* exposure immediately changes to the *specular* (3x3) pattern depicted in Fig. 1(b). This reaction is reversible: upon replacement of the N* flux with an As flux, the pattern again returns to the original (2x4) though *much more slowly* than the forward reaction. We have first reported this behavior elsewhere[7] where it has been argued that the observed RHEED pattern changes correspond to surface-layer anion exchange of N and As species, for sufficiently brief (< 8-10 s) N* exposure times. In contrast, we have shown that longer exposure times (up to 1 min) result in the formation strain-relaxed islands of zinc blende GaN.[7] Together, these observations suggest a *nitridation* of the As-terminated GaAs surface due to incident N* flux such that first rapid surface, followed by slower subsurface, anion exchange processes, take place.

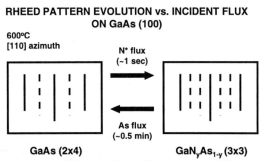

RHEED PATTERN EVOLUTION vs. INCIDENT FLUX ON GaAs (100)

600°C
[110] azimuth

N* flux (~1 sec)

As flux (~0.5 min)

GaAs (2x4)   GaN$_y$As$_{1-y}$ (3x3)

Fig. 1 RHEED Pattern evolution under N* and As fluxes. Note time scales. GaNAs pattern is *commensurate w/* substrate. Reaction is reversible, if N* exposure time is limited to a few sec.

The RHEED pattern dynamics have also been examined quantitatively through time-resolved measurements. The pattern evolution is quantitatively assessed as a function of time through intensity of the 1/3-order reconstruction line. This line, present in the (3x3), but absent from the (2x4) pattern, is taken as a measure of the extent of surface reaction. Assessment of the (3x3) onset and decay are presented in Fig. 2(a). From the inset, it is seen that the presence (absence) of As flux has negligible effect on the rate of surface nitridation (N-for-As replacement), and that the time scale for nitridation is ~1 s. In contrast, the surface, once nitrided, is quite stable, even when all fluxes are removed. Note that the time scale for decay of the (3x3) pattern is *much* longer than that of As-terminated GaAs (2x4) at similar temperature. This result suggests strongly that the (3x3) pattern is primarily associated w/ a N-terminated surface.

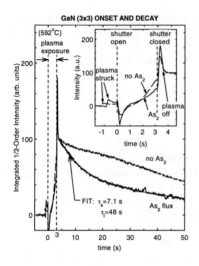
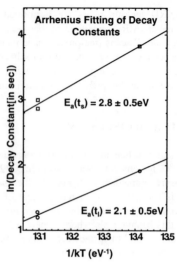

Fig. 2 (a) 1/3-order RHEED streak intensity decay, corresponding to N desorption from the (3x3) nitrided surface. Decay rates are approximately exponential, as shown. (b) Temperature dependence of decay rates. Results suggest thermally activated processes, with activation energies indicated.

Decay of the (3x3) surface in the presence of an As flux is well described by a superposition of short (τ_s) and long (τ_l) exponential decay terms, as seen in Fig. 2(a). The corresponding lifetimes, τ_s and τ_l, respectively, each exhibit temperature-dependent behavior, as shown in Fig. 2(b). Assuming a Boltzmann-factor dependence, a preliminary assessment of thermal activation energies is possible. The results of this analysis are given in the figure. Considering the large difference in Ga-N and Ga-As bond strengths (heats of formation 6.81 and 5.55 eV respectively[8]) compared to kT, the above results are consistent with thermally activated surface N or As desorption processes, with a much higher kinetic barrier present for the case of N desorption.

An important characteristic of the (3x3) pattern of Fig. 1(b) is that it is *commensurate* with the GaAs substrate. In fact, we have found that it is possible to overgrow commensurate GaAs over a nitrided surface layer as described above. This heterostructure can be viewed as a buried monolayer of GaN_yAs_{1-y} in GaAs. Since the resultant *coherently strained* GaN/GaAs heterostructure is amenable in principle to X-ray determination of y, this suggests the following experimental sequence to help ascertain the surface composition which is associated with the (3x3) surface reconstruction: (i) (3x3) surface nitridation; (ii) GaAs overgrowth; (iii) HRXRD assessment of composition and strain.

This sequence was performed for the samples listed in Table I. The resultant x-ray rocking curves, about the substrate (004) reflection, are presented in Fig. 3(a). All samples were grown under nominally identical conditions (4 s exposure), apart from growth temperature. (Scans for growth at 550 and 570°C have been omitted for clarity.) All samples are coherently strained, as confirmed through (115) area scans. Inspection of the figure clearly reveals a very strong dependence on growth temperature T of the epitaxial structure. Two-dimensional equivalent alloy compositions (y defined above) are determined from XRD peak positions and verified through dynamical simulations; final results are summarized in Table I. Figure 3(b) shows an Arrhennius plot of the resultant ML alloy compositions. The results, within experimental error, clearly suggest two distinct regimes of behavior which act to limit N incorporation: (i) *dose-limited* (low-T); and (ii) *kinetically limited* (high-T). This finding suggests that a thermally activated N surface-segregation process, in addition to the thermally activated N desorption process found above from RHEED data, is active, and can be significant at typical GaN ECR-MBE growth conditions. Extracting an activation energy from the linear portion of the data yields a value which is greater than that found for the desorption processes alone from Fig. 2(b) above, consistent with the existence of an additional thermally activated process. We have developed a kinetic model to account for the observed results, which also quantitatively explains the increased composition observed for the 6 sec exposure (sample OS95.009); this kinetic model will be described in detail elsewhere.[9]

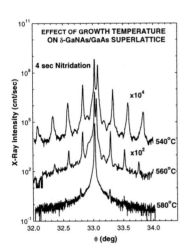 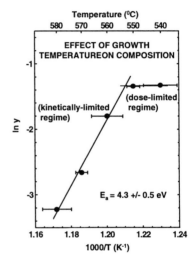

Fig. 3 (a) HRXRD about GaAs (004) for GaN_yAs_{1-y}/GaAs superlattice structures. Results indicate a strong dependence of y on growth temperature. (b) Arrhenius plot, showing the effect of y growth temperature. A thermally activated loss mechanism for the N is suggested. Low-T (dose-limited) and high-T (kinetically-limited) regimes are also seen.

CONCLUSION

In conclusion, we have studied the interaction of a nitrogen ECR microwave plasma and a GaAs surface under typical ECR-MBE growth conditions, and have observed and characterized a number of important new phenomena which are of significance in understanding the initial nucleation kinetics of GaN films on GaAs. Sufficiently brief N* exposure results in the formation of a commensurate, planar surface layer of GaN_yAs_{1-y}, associated with a characteristic (3x3) pattern observable through RHEED, which can be overgrown with GaAs to form a coherently strained GaN/GaAs heterostructure. Finally, direct evidence is presented for the existence of thermally activated N desorption and surface-segregation processes, and preliminary activation energies for these processes are given.

ACKNOWLEGEMENTS

The authors gratefully acknowlege the support of the Air Force Office of Scientific Research under Contract Nos. F49620-93-1-0211 and F49620-93-1-0389. Additionally, we wish to acknowledge the partial support of the Office of Naval Research under Contract No. N000014-92-J-1845.

REFERENCES

1. R. F. Davis, Proc.of the IEEE **79**, 702 (1991).

2. S. Strite and H. Morkoç, J. Vac. Science Tech B **10**, 1237 (1992).

3. I. Akasaki, H. Amano, J. Electrochem..Soc. **141**, 2266 (1994).

4. H. Morkoç and B. Sverdlov (unpublished).

5. S. Nakamura, T. Mukai, M. Senoh, Appl. Phys. Lett. **64**, 1687 (1994).

6. D. A. Collins, M. W. Wang, R. W. Grant, and T. C. McGill, J.Appl. Phys. **75**, 259 (1993).

7. R. J. Hauenstein, D. A Collins, X. P. Cai, M. L. O'Steen, and T. C. McGill (unpublished).

8. *CRC Handbook of Chemistry and Physics*, ed. by R. C. West, (CRC Press Inc, Boca Raton, 1987), p. E102.

9. R. J. Hauenstein, Z. Z. Bandic, M. L. O'Steen, and T. C. McGill (unpublished).

DEPOSITION OF GaN FILMS USING SEEDED SUPERSONIC JETS

H. HENRY LAMB,* KEN K. LAI,* VICTOR TORRES,** AND ROBERT F. DAVIS**
*Department of Chemical Engineering, North Carolina State University, Raleigh, NC 27695-7905
**Department of Materials Science and Engineering, North Carolina State University, Raleigh, NC 27695-7907

ABSTRACT

GaN films were deposited on sapphire (0001) from triethylgallium (TEG) and NH_3 seeded in separate He free jets. As the dissociative chemisorption of NH_3 is expected to be rate-limiting in GaN growth, the NH_3/He nozzle temperature was varied to control the incident kinetic energy of the NH_3 molecules. Using an NH_3/TEG flow ratio of 205 and an NH_3/He nozzle temperature of 400°C, stoichiometric films were deposited at substrate temperatures ≥530°C. Scanning electron microscopy revealed that films deposited at 580°C are polycrystalline α-GaN with randomly oriented 0.5-μm grains. Films with a closely similar morphology are deposited by using an NH_3/He nozzle temperature of 90°C. In contrast, films deposited at 580°C using an NH_3/He nozzle temperature of 510°C consist of whiskers (aspect ratio = *ca.* 4) which exhibit α-GaN (0001)/sapphire (0001) heteroepitaxy, as evidenced by X-ray diffraction and reflection high-energy electron diffraction.

INTRODUCTION

Among the III-V nitrides, gallium nitride (GaN) has been the most studied for optoelectronic applications. The potential for fabrication of a variety of near-UV optoelectronic devices, especially light-emitting diodes and blue semiconductor lasers, has stimulated interest in the growth, doping and physical properties of GaN thin films.[1,2] Moreover, GaN is predicted to have a high ballistic electron drift velocity and, consequently, may prove useful for transit-time-limited (high-frequency) microwave devices. The principal obstacle to the technological utilization of GaN is the growth of monocrystalline, epitaxial films with low defect densities. The lack of high-quality substrates that provide a good lattice match for heteroepitaxial growth of α-GaN is an unresolved problem. Sapphire (α-Al_2O_3) is the most widely used substrate, although there is a 16% lattice mismatch at the α-GaN (0001)/sapphire (0001) interface. To date, the best crystal quality has been obtained by employing a low-temperature AlN or GaN buffer layer between the substrate and the film.[3]

GaN films have been grown using a variety of techniques, including chemical vapor deposition (CVD), metal-organic chemical vapor deposition (MOCVD), molecular beam epitaxy (MBE), and plasma-assisted processes.[1,2] Typical growth temperatures for monocrystalline GaN are 800-1000°C for CVD using $GaCl_3$ and NH_3[4] and 750-1050°C for MOCVD using trimethylgallium (TMG) and NH_3 in H_2.[5] GaN growth temperatures of 700-850°C are reported for gas-source MBE using elemental Ga and NH_3.[6] Single-crystal growth can be achieved at lower temperatures (600-700°C) by plasma-assisted MBE, in which a nitrogen plasma is used to generate highly reactive species such as atomic N or N_2^+.[7,8]

The rate-limiting step in GaN deposition by MOCVD or gas-source MBE is believed to be the *dissociative chemisorption* of NH_3 on a Ga-rich growth surface. In this regard, Ceyer and coworkers have demonstrated that the direct dissociation probability of CH_4 on Ni(111) correlates with the normal component of the incident kinetic energy.[9] More recent results have indicated that similar effects can be expected for Group IV hydrides on semiconductor surfaces, e.g., the direct dissociation probability of Si_2H_6 on Si(100) increases nearly linearly with incident kinetic energy for mean energies >1 eV.[10] This fundamental work reveals an opportunity for using energetic neutral molecule beams for effecting chemistry on semiconductor surfaces. In particular, the use of suprathermal kinetic energies to increase the dissociation probability of hydrides, such as Si_2H_6, SiH_4, NH_3 and CH_4, is relevant to the growth of wide bandgap semiconductors.

In this paper, we report initial results on the use of a seeded supersonic jet to produce suprathermal NH_3 molecules for GaN deposition on sapphire. TEG was chosen as the Ga precursor, since it decomposes at low temperature via β-hydride elimination with negligible co-deposition of carbon. The deposited films were characterized by Auger electron spectroscopy (AES), X-ray diffraction (XRD), reflection high-energy electron diffraction (RHEED) and scanning electron microscopy (SEM).

EXPERIMENTAL METHODS

The supersonic jet deposition apparatus comprises a two-compartment vacuum chamber, a heated nozzle assembly (for NH_3/He), an unheated tubular nozzle (for TEG/He), a gas delivery system, and a substrate heater assembly. The 10-in. cylindrical chamber is divided into two compartments by a removable skimmer plate. The source chamber is pumped by a CVC PVMS-6 diffusion pump backed by a Tuthill 3206 Roots blower and a Welch 1374 mechanical pump with a liquid nitrogen cold trap. The deposition chamber is pumped by a Varian VHS-6 diffusion pump backed by a Varian 2033 mechanical pump with a liquid-nitrogen-cooled baffle. The base pressure in each chamber is less than $3x10^{-7}$ Torr. In the experiments described herein, the skimmer plate was removed, allowing the heated nozzle to be placed within 1 in. of the substrate. The heated nozzle assembly was fabricated from a 1-in. O.D. high-pressure filter body (HiP, Pittsburgh). A thin Pt disk containing a 200-μm laser-drilled pinhole (Energy Beam Sciences) is mechanically sealed to a short piece of 1/8-in. OD tubing which protrudes from one end of the filter body. The disk is centered and compressed by a threaded cap with a 1-mm exit aperture. The nozzle assembly is sheathed by a 275-W annular nozzle heater (Chromalox HBA-103027, inner diameter: 1 in., length: 3 in., and max. attainable temperature: 600°C). The unheated tubular nozzle was constructed from 1/8-in. OD high-pressure tubing with an inner diameter of approximately 0.4 mm.

The pressure in the heated nozzle is controlled by a multi-channel flow/pressure controller (MKS 647) using pressure feedback from a capacitance manometer. The NH_3 and He flow rates are regulated by MKS 1259 mass flow controllers (MFCs); in operation, the NH_3: He flow ratio is fixed, and the total flow rate is adjusted to maintain a pressure setpoint. The He flow rate through the metal-organic bubbler is controlled using an MFC. The bubbler is immersed in a constant-temperature bath, and the bubbler pressure is controlled by a downstream needle valve and monitored by capacitance manometer. As TEG is pyrophoric, the process effluent is passed through a canister scrubber (Novapure).

The substrate heater assembly is mounted on a rotary-linear feedthrough which is positioned along the centerline of the deposition chamber. The substrate is heated from the backside by a quartz halogen lamp (max. attainable temperature: 750°C). The temperature is measured using a thermocouple and controlled by a Yokogawa UP-25 PID controller that actuates a Eurotherm 832 solid-state relay.

Semiconductor-grade NH_3 (Matheson) was further purified by passage through a Pall Gaskleen purifier. Electronic-grade TEG was purchased from Air Products. The sapphire (0001) substrates (about 5 x 5 mm²) were cleaned using the following procedure: 5-min ultrasonic degreasing cycles in trichloroethylene, acetone and methanol, a 10-min DI water rinse, a 10-min dip in hot, 1:1 H_3PO_4:H_2SO_4, a 10-min DI water rinse, a 10-min dip in a 10% aqueous HF solution, and drying in flowing dry nitrogen. The substrate was supported by a larger piece of clean Si that was mounted in a 30-mm stainless steel holder. The holder is transferred into the deposition chamber via a loadlock, which is evacuated by an Alcatel Drytel 31 molecular drag/membrane pump.

Typical nozzle conditions of 1000 Torr and 450°C resulted in a 5% NH$_3$/He flow rate of 340 sccm and a background pressure of 5×10^{-3} Torr. The Mach disk location was estimated as 60 mm from the nozzle using the following equation:

$$X_M = 0.67d\sqrt{\frac{P_0}{P_b}} \qquad (1)$$

where P$_0$ is the nozzle pressure, P$_b$ is the background pressure, and d is the orifice diameter.[11] As the supersonic free-jet expansion terminates at the Mach disk, a nozzle-to-substrate distance of approximately 38 mm was employed to ensure that the flow remained supersonic. The discharge from the unheated TEG nozzle was directed toward the substrate from a similar distance. Typically, the TEG bubbler pressure was 280 Torr, and the He flow rate was 14 sccm. The TEG concentration was varied by adjusting the bubbler temperature.

After deposition, the sample was removed and transferred through air to the loadlock of a multi-technique surface analysis chamber. Surface elemental analysis was performed using a PHI Model PC 3017 AES subsystem comprising a 10-155 single-pass cylindrical mirror analyzer, coaxial electron gun and associated electronics. N(E) spectra were recorded using a primary beam energy of 3 keV and differentiated numerically. To establish the stoichiometry of the GaN films, the AE spectra were compared to that of a GaN standard prepared by MBE. Film morphology and thickness were determined using a JEOL 6400 field-emission scanning electron microscope. A Rigaku X-ray diffractometer was used to establish the crystallographic orientation of the films by thin-film X-ray diffraction.

RESULTS

Sapphire (0001) substrates were exposed to continuous jets of NH$_3$ and TEG seeded in He, using an NH$_3$/He nozzle temperature of 450°C and an NH$_3$:TEG ratio of 175. Ga was deposited on the substrate via TEG decomposition at 320°C, as evidenced by AES. In contrast, a minimum substrate temperature of 450°C was required for nitrogen incorporation. Continuous GaN films containing excess Ga were deposited at substrate temperatures greater than or equal to 530°C.

The influence of NH$_3$:TEG ratio on the composition of films deposited at 580°C was investigated by adjusting the TEG bubbler temperature while holding the remaining process variables constant. Figure 1 shows the AE spectra of substrates that were exposed to continuous jets providing a range of NH$_3$:TEG ratios. Only a trace of Ga and little if any N was deposited by using a NH$_3$:TEG ratio of 480, as evidenced in Figure 1a. By decreasing the NH$_3$:TEG ratio to 240, Ga and N were co-deposited, but the resultant film is very thin and/or discontinuous, as evidenced by the Al *KLL* AES peak arising from the substrate (Figure 1b). As the ratio was further decreased, more Ga and concomitantly more N were deposited. An NH$_3$:TEG ratio of 205 resulted in the deposition of stoichiometric GaN, as evidenced by comparison of the AE spectrum (Figure 1c) with that of a GaN standard deposited by MBE. As expected, when NH$_3$:TEG ratios less than 200 were employed, GaN films containing excess Ga were deposited, as evidenced by the AE spectra in Figure 1d and 1e. Ar$^+$ sputtering revealed that the majority of the carbon and oxygen contamination is localized near the surface of the films, suggesting that it arises from air exposure prior to analysis.

The effects of NH$_3$/He nozzle temperature (and thereby the mean kinetic energy of the NH$_3$ molecules) on the GaN deposition rate and film morphology were investigated. The substrate temperature was maintained at 580°C. Because the total flow rate (He bath gas and NH$_3$) decreased with increasing nozzle temperature, it was necessary to adjust the TEG bubbler temperature to achieve stoichiometric GaN deposition; the required NH$_3$:TEG ratio varied with nozzle temperature from 250 at 90°C to 190 at 600°C. The GaN growth rate, as estimated from the film thickness and deposition time, increased with NH$_3$/He nozzle temperature from approximately 1 μm/h at 90°C to 3.3 μm/h at 510°C.

SEM images of GaN films which were deposited using different NH_3/He nozzle temperatures revealed significant differences in film morphology. Films that were deposited using nozzle temperatures of 90 and 400°C consist of randomly oriented 0.5-µm grains, as illustrated in Figure 2a. XRD confirms that the films are polycrystalline wurtzitic α-GaN without a preferred grain orientation. In contrast, a film that was deposited using a nozzle temperature of 510°C consists of α-GaN whiskers with dimensions of about 1 x 0.25 x 0.25 mm^3 (Figure 2b). The XRD pattern (Figure 3) is indicative of α-GaN with a strong (0001) orientation. The RHEED pattern is consistent with a preferred in-plane orientation of the whiskers relative to the sapphire (0001) substrate.

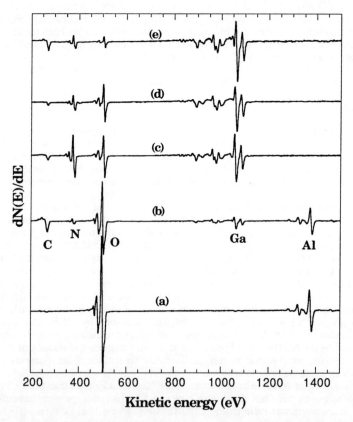

Figure 1. AE spectra of sapphire (0001) substrates after exposure at 580°C to continuous jets of NH_3 and TEG in He, providing the following NH_3:TEG ratios: (a) 480 , (b) 240, (c) 205, (d) 175, (e) 130.

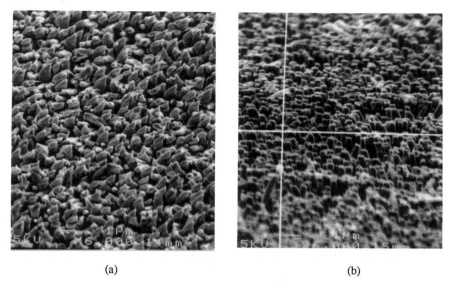

(a)   (b)

Figure 2. GaN films deposited on sapphire (0001) at 580°C using NH₃/He nozzle temperatures of (a) 400°C and (b) 510°C.

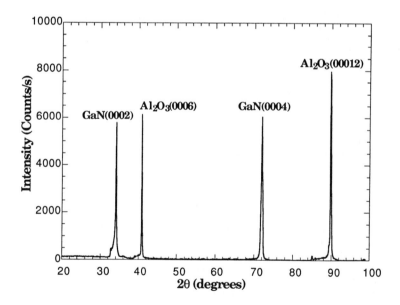

Figure 3. XRD pattern of GaN deposited on sapphire (0001) at 580°C using a NH₃/He nozzle temperature of 510°C and an NH₃:TEG ratio of 200.

DISCUSSION

The primary goal of this investigation was to define a process window for deposition of GaN films using seeded supersonic jets. In summary, we found that stoichiometric α-GaN films can be deposited on sapphire (0001) at substrate temperatures $\geq 530°C$ by using an NH_3:TEG ratio of approximately 200. In comparison, deposition can be achieved at 400-600°C by plasma-assisted MBE,[7,8] but much higher temperatures (900-1000°C) and larger NH_3:Ga ratios are typically employed for GaN deposition by MOCVD.[5]

The rate-limiting step in GaN deposition from TEG and NH_3 is believed to be dissociative chemisorption of NH_3 on a Ga-rich growth surface. The NH_2-H bond strength is large (4.66 eV), but the barrier to dissociative chemisorption is undoubtedly much smaller owing to surface chemical bonding of the NH_2 and H moieties. The results demonstrate that at 580°C dissociative chemisorption of NH_3 on sapphire does not occur in the absence of Ga. Conversely, elemental Ga has a very low melting point (29°C) and tends to form droplets on a heated substrate unless it reacts with N-bearing species.[7] The influence of NH_3 kinetic energy on GaN growth was investigated by varying the NH_3/He nozzle temperature. The mean kinetic energies of NH_3 molecules seeded in the He supersonic free jet were not measured, but they were estimated (assuming zero velocity slip) as 0.3, 0.6, and 0.7 eV for nozzle temperatures of 90, 400 and 510°C.[11] The NH_3:TEG ratio required for GaN deposition decreases from of 190 to 250 as the nozzle temperature is increased, suggesting a relatively weak effect of incident kinetic energy on NH_3 reaction probability. Nonetheless, the GaN deposition rate increases dramatically from approximately 1 μm/h to 3 μm/h on increasing the nozzle temperature from 90 to 510°C. Measurements made using an axial quadrupole mass spectrometer indicated that the ratios of the NH_3^+, NH_2^+ and NH^+ peak intensities in the NH_3 cracking pattern did not change with nozzle temperature. Consequently, the possibility of NH_3 cracking in the heated nozzle was discounted.

The morphology and crystallographic orientation of the deposited GaN films changed markedly with NH_3/He nozzle temperature. Films deposited at the lower nozzle temperatures contain randomly-oriented α-GaN grains, whereas films deposited using a nozzle temperature of 510°C, consist of α-GaN whiskers with an aspect ratio of about 4. We infer from XRD and RHEED that each whisker is a single crystal which is epitaxially aligned with the sapphire (0001) substrate. These results suggest that increasing the incident kinetic energy of the NH_3 molecules leads to preferential growth in the α-GaN <0001> direction, which in this case is oriented parallel with the substrate normal.

ACKNOWLEDGMENTS

This work was supported by an NSF Presidential Young Investigator Award (CTS-8958350) and by the Office of Naval Research under contract N00014-90-J-1427.

REFERENCES

1. J. I. Pankove, Mater. Res. Soc. Proc. **162**, 515 (1990).
2. R. F. Davis, Proc. IEEE **79**, 702 (1991).
3. S. Yoshida, S. Misawa and S. Gonda, Appl. Phys. Lett. **42**, 427 (1983).
4. H. P. Maruska and J. J. Tietjen, Appl. Phys. Lett. **15**, 327 (1969).
5. S. Nakamura, Japan. J. Appl. Phys. **30**, L1705 (1991).
6. R. C. Powell, N. -E. Lee and J. E. Greene, Appl. Phys. Lett. **60**, 2505 (1992).
7. C. R. Eddy, Jr., T. D. Moustakas, and J. Scanlon, J. Appl. Phys. **73**, 448 (1993).
8. C. Wang and R. F. Davis, Appl. Phys. Lett. **63**, 990 (1993).
9. S. T. Ceyer, Science **249**, 133 (1990).
10. J. R. Engstrom, D. A. Hansen, M. J. Furjanic, and L. Q. Xia, J. Chem. Phys. **99**, 4051 (1993).
11. D. R. Miller, in *Atomic and Molecular Beam Methods*, edited by G. Scoles (Oxford Press, New York, 1988) pp.14-53.

GROWTH AND PROPERTIES OF CARBON NITRIDE THIN FILMS

Z. JOHN ZHANG, PEIDONG YANG AND CHARLES M. LIEBER
Division of Applied Sciences and Department of Chemistry
Harvard University, Cambridge, MA 02138

ABSTRACT

Recent research on carbon nitride thin films grown using pulsed laser deposition combined with atomic beam techniques is reviewed. The composition, growth mechanism and phases of these films have been systematically investigated. The nitrogen composition was found to increase to a limiting value of 50% as the fluence was decreased for laser ablation at both 532 nm and 248 nm wavelengths. Time of flight mass spectroscopy investigations of the ablation products have shown that the fluence variations affect primarily the yield of the carbon reactant. These experiments demonstrate that the overall film growth rate determines the average nitrogen composition, and furthermore, suggest that a key step in the growth mechanism involves a surface reaction between carbon and nitrogen. Infrared spectroscopy has been used to assess the phases present in the carbon nitride thin films as a function of the overall nitrogen content. These measurements have shown that a cyanogen-like impurity occurs in films with nitrogen compositions greater than 30%. Studies of thermal annealing have shown, however, that this impurity phase can be eliminated to yield a single phase C_2N material. In addition, systematic studies of the electrical resistivity and thermal conductivity of the carbon nitride films are discussed.

I. INTRODUCTION

Carbon nitride materials represent an exciting challenge to both fundamental and applied research [1]. Basic research is needed to develop new preparative methods for combining rationally carbon and nitrogen into solid materials and to elucidate the structures and physical properties of these new materials. In addition, applications of carbon nitride solids are being actively pursued since simple bonding arguments and more rigorous theoretical calculations suggest that these materials should have attractive properties, such as extreme hardness and thermal conductivity.

Mat. Res. Soc. Symp. Proc. Vol. 388 © 1995 Materials Research Society

Early theoretical investigations of carbon nitride solids focused on a hypothetical binary carbon-nitrogen compound, β-C_3N_4, that has the β-Si_3N_4 structure [2]. The results from this theoretical work suggested that β-C_3N_4 should have a reasonably large cohesive energy and thus should be metastable. More recent theoretical studies have also investigated the stability and properties of C_3N_4 compounds that have structures distinct from that of the β-Si_3N_4 structure [3]. Significantly, these latter studies have found that carbon nitride C_3N_4 materials having either defect zinc-blende or rhombohedral graphite-like structures should have stabilities comparable to or greater than β-C_3N_4. Because there is also no reason to assume that carbon nitride must have a C_3N_4 stoichiometry, there is many metastable C-N phases that must be carefully considered in experimental studies of this new class of materials.

Experimental approaches to the synthesis of carbon nitride have utilized a variety of techniques, including plasma-assisted chemical vapor deposition [4], thermal pyrolysis [5], shock wave compression [6], reactive sputtering[7-9], and laser ablation [1,10-12]. Although evidence for nanocrystalline β-C_3N_4 has been obtained for materials prepared by laser ablation [10,12] and reactive sputtering [7], the overall nitrogen composition of these materials has been significantly less than the C_3N_4 stoichiometry [13]. Furthermore, we have noted that the reported diffraction data are not unique to the β-C_3N_4 structure but can also be explained by a tetragonal-CN phase analogous to the high-pressure form of GeP [1,11]. To clarify this uncertain experimental situation requires systematic studies of the composition and phases of carbon nitride. Furthermore, to advance rationally this material for applications requires an understanding of the relationships of nitrogen composition and key physical properties.

This brief review summarizes our recent systematic studies of the nitrogen composition range, phases, and physical properties of thin film carbon nitride materials prepared using a laser ablation/atomic beam deposition technique. We find that the nitrogen composition increases to a limiting value of 50% as the laser fluence is decreased at both 532 nm and 248 nm. Analysis of these experiments and time of flight mass spectroscopy results show that the growth rate determines the overall nitrogen composition, and furthermore, suggests that a key step in the growth mechanism involves a surface reaction between carbon and nitrogen. Infrared (IR) spectroscopy has been used to study the phase purity in these carbon nitride films, and show that a cyanogen-like impurity occurs in films with nitrogen compositions greater than 30%. The cyanogen-like impurity phase can, however, be eliminated by thermal annealing to yield a single phase C_2N material. In addition, systematic measurements of the electrical resistivity and thermal conductivity of the carbon nitride films indicate that this material may be an attractive candidate for electronic applications.

II. EXPERIMENTAL

Our approach to the synthesis of carbon nitride materials combines pulsed laser ablation with an atomic beam source. A schematic diagram of our experimental set-up is shown in Figure 1. Carbon fragments were produced by ablation of high-purity pyrolytic graphite in a vacuum chamber using either a frequency-doubled Nd:YAG laser (532 nm) or a KrF excimer laser (248 nm). A N_2-seeded He flow was passed through a radio frequency (rf) discharge source to generate a reactive nitrogen beam consisting primarily of relatively high kinetic energy atomic nitrogen [14]. The reactive nitrogen atom beam ($>10^{18}$ atoms/sr-s) intersects the carbon ablation plum at the substrate surface. Using this experimental apparatus only pure carbon and nitrogen reactants are produced, and thus elemental impurities such as hydrogen and oxygen are eliminated.

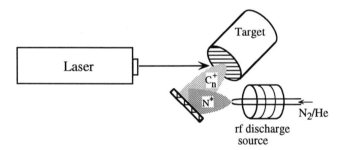

Figure 1. A schematic diagram of the experimental set-up for the carbon nitride growth. The substrate, rotating graphite target and atomic nitrogen source are all contained within a vacuum chamber that is evacuated by a liquid nitrogen trapped 5000 l/s diffusion pump.

The carbon nitride thin films were grown on quartz or HF-etched Si (100) surfaces at ~200 °C. The nitrogen beam conditions used in these studies were optimized through variations in the N_2:He ratio and rf power to yield the maximum nitrogen composition under fixed ablation conditions [10,11]. The base pressure of deposition chamber is less than 10^{-8} torr; however, with the nitrogen beam source operating the pressure usually rises to between 10^{-5} to 10^{-4} torr during film growth.

III. RESULTS AND DISCUSSION

We have systematically grown a series of carbon nitride films under different the laser fluences and wavelengths, and employed Rutherford backscattering spectroscopy (RBS) to determine the chemical composition in these films.   In general, we find that the nitrogen composition of the film increases as the laser fluence decreases.   This trend is illustrated in Figure 2.   These RBS spectra show different composition carbon nitride films grown by 532 nm wavelength laser ablation of a graphite target at different fluences.

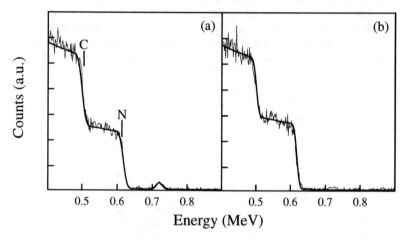

Figure 2. RBS spectra (General Ionics Model 4117, 2.0 MeV He ion beam) of carbon nitride films.   The C and N compositions were determined from fits (smooth lines) to the experimental data.   The films in (a) and (b) are 2.0 and 2.2 μm thick, respectively.

Figure 2a corresponds to data obtained from a carbon nitride film that was prepared using a laser fluence of 1.4 J/cm^2; this film contains 37% nitrogen.   When the fluence is decreased to 0.9 J/cm^2, RBS data shows the nitrogen composition increased to 45% (Fig. 2b).   No further increase in nitrogen composition was observed, however, for fluences less than 0.8 J/cm^2.   Similar results have also been obtained for carbon nitride films grown by laser ablation at 248 nm [11].

Time-of-flight (TOF) mass spectroscopy has been employed to explore the origin of the dependence of the nitrogen composition on laser fluence.   Preliminary results show that the carbon fragments in our laser ablation plum consist primarily of one to four carbon atom species, and of these species one carbon fragments are dominant.   Hence, we have used this dominant species as an indicator of the total carbon yield under different ablation conditions. We find that the carbon yield for ablation using a 1.4 J/cm² fluence is significantly greater than the yield obtained using 0.9 J/cm² laser fluence (Fig. 3). More generally, these experiments show that the carbon yield increases systematically with laser fluence, although the increase is non-linear [15]. Preliminary analysis of this data also indicates that there is very little change in the kinetic energy of the carbon fragments for the range of fluences studied.

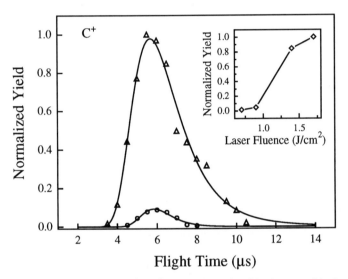

Figure 3. TOF distributions for C⁺ produced with laser ablation fluences of 0.9 J/cm² (O) and 1.4 J/cm² (Δ). The inset is a plot of total yield of C⁺ versus laser fluence.

The correlation between carbon yield and laser fluence indicates that the growth rate may determine in large part the carbon nitride composition. For example, the carbon nitride films grown with 1.4 and 0.9 J/cm² have the growth rates of 3.1 and 1.4 Å/s and compositions of 37% and 45% nitrogen, respectively. To confirm that the growth rate determines composition and not for example the energy of the carbon fragments, we have also prepared carbon nitride films using a constant laser fluence at different pulse repetition rates. RBS analyses of two carbon nitride films prepared using 248 nm ablation at a constant laser fluence of 3.4 J/cm² and different repetition rates of 9 and 27 Hz are shown in Figure 4. Significantly, the film grown using the lower repetition rate (growth rate of 1.6 Å/s) had a nitrogen composition of 45% (Fig. 3a), while the film grown using a higher 27 Hz repetition rate (i.e., and higher growth rate of 4.0 Å/s) had a lower nitrogen composition of 36% (Fig. 3b).

Our investigations of the laser fluence and repetition rate effects and TOF mass measurements indicate clearly that the overall growth rate determines the nitrogen composition in these carbon nitride films. We suggest that this correlation arises from a growth mechanism that involves as a key step the reaction between carbon and nitrogen at the growth surface. If on the other hand gas phase reactions were important, the nitrogen composition would not have changed with variations in the repetition rate since the reaction time between carbon and nitrogen in the gas phase is same for both repetition rates.

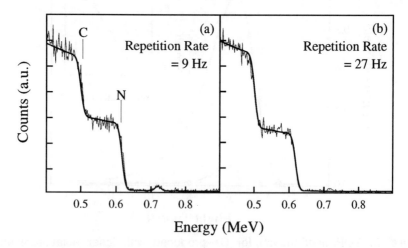

Figure 4. RBS spectra of carbon nitride films grown at 248 nm laser wavelength with repetition rates of (a) 9 and (b) 27 Hz. The films were 2.3 and 2.1 µm thick, respectively.

We have used this knowledge of the growth mechanism to explore systematically the nitrogen composition limits of carbon nitride. Significantly, we have found that the nitrogen composition in carbon nitride films saturates at a value close to 50% [11]. This saturation value implies that a possible stoichiometry of carbon nitride is CN (and perhaps not C_3N_4), although this assignment requires that the 50% nitrogen thin films contain a single phase. To evaluate the phases present in these carbon nitride materials we have used infrared (IR) spectroscopy. A typical IR spectrum recorded on a carbon nitride film containing 41% nitrogen is shown in Figure 5a. This spectrum exhibits two overlapping bands at ~1500 and 1350 cm^{-1} and a third stretching mode at ~2200 cm^{-1}. The overlapping bands at ~1500 and 1350 cm^{-1} are believed to correspond to C=N and C-N stretching modes [16], and are consistent with an extended inorganic carbon nitride solid.

The third band occurring at ~2200 cm⁻¹ corresponds to a -C≡N or cyanogen-like stretching mode [17]. This -C≡N bonding arrangement precludes an extended inorganic carbon nitride solid and represents a cyanogen-like impurity phase in these materials. The cyanogen-like impurity increases very little as the nitrogen composition increases from 30 to 35%. However, as the nitrogen composition increases above 35%, the intensity of the ~2200 cm⁻¹ band increases dramatically. These observations imply that increases in the carbon nitride nitrogen composition above about 35% occur in large part at the expense of forming a -C≡N impurity phase. Hence, we conclude that the nitrogen composition of the dominant carbon nitride phase present in these thin films is less than 50%.

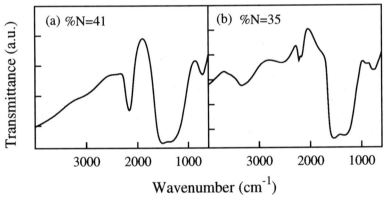

Figure 5. IR spectra of carbon nitride films recorded before (a) and after vacuum annealing at 550 °C (b). The nitrogen compositions both before and after the annealing procedure were determined by RBS.

We have also investigated the effects of thermal annealing to our carbon nitride thin films since cyanogen-like materials, which represent the impurity phase, typically have low thermal stabilities [17]. The effect of thermal annealing on the structure and composition of the carbon nitride thin films was evaluated by recording IR and RBS spectra before and after the annealing process. For example, Figure 5b shows the IR spectrum of the film used to record Figure 5a after it was annealed in vacuum at 550 °C for three hours. Notably, we find that the annealing process nearly eliminates the intense C≡N mode at 2200 cm⁻¹ which was found in the as-grown film. RBS analysis also shows that the nitrogen composition in the annealed film is reduced from 41% to 35%. Systematic RBS studies of the film composition before and after annealing demonstrates that the nitrogen composition in films initially containing more than 40% nitrogen decreases during the annealing process and reaches a common, stable composition of ~35%. Furthermore, IR spectra recorded on these same materials show that the -C≡N impurity phase is essentially eliminated in all of the films that have been annealed to yield the common, stable composition of ~35% nitrogen. These results indicate that a nearly single phase C_2N material containing 35% nitrogen and both C-N and C=N bonding can be prepared by our post-annealing process.

The thermal annealing process also results in profound changes in electrical properties of our carbon nitride films. Resistivity measurements made on thin films that were deposited on insulating quartz substrates have shown that the resistivity of the carbon nitride films increases systematically with nitrogen composition, although this increase is highly nonlinear. The resistivity at 200 K initially increases slowly until the nitrogen composition reaches 30%, and then increases much more rapidly until it ultimately saturates at $1-5 \times 10^3$ Ω-cm in films with a nitrogen composition of ~45%. We believe that the initial slow increase in resistivity corresponds to transport dominated by amorphous carbon, and that the saturation value corresponds to the resistivity of the carbon nitride (but a form that also contains a -C≡N impurity phase).

Significantly, carbon nitride films containing initially ≥40% nitrogen exhibit a much high resistivity after our vacuum annealing treatment. Temperature dependent resistivity measurements made on a carbon nitride film containing 45% nitrogen before annealing and 35% nitrogen after vacuum annealing at 550 °C are shown in Figure 6.

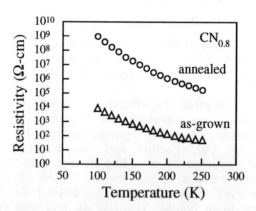

Figure 6. Temperature dependent resistivity measurements recorded on a carbon nitride thin film with 45% nitrogen before (Δ) and after vacuum annealing (O) at 550 °C.

The resistivity of the film increased 3-5 orders of magnitude after the annealing process. We believe that the enhancement in resistivity values for the post-annealed carbon nitride films is due to the elimination of the cyanogen impurity phase since this phase can increase the density of states at the Fermi level [18]. Importantly, the stability of the carbon nitride electrical properties to thermal processing make this material an attractive candidate for microelectronics applications.

We have also investigated the thermal conductivity of these films since the ability of carbon nitride to transport heat efficiently is an important consideration for potential applications such as microelectronics. The thermal conductivity of these carbon nitride thin films has been measured using an well-established ac (3ω) electrical technique [19]. In our measurements, a thin and narrow gold line was deposited onto carbon nitride thin films grown on silicon substrates, and this metal line, which served as both a heater and temperature sensing element, was used to determine the thermal conduction through the film. Significantly, the thermal conductivity calculated for the carbon nitride film from this data, 0.8 to 1.3 W/m-K, is among the highest observed for any type of amorphous material [19,20].

IV. SUMMARY

Carbon nitride thin films have been prepared by combining pulsed laser deposition and atomic beam techniques. The nitrogen composition was found to increase to a limiting value of 50%, and the composition was shown to be controlled by the overall film growth rate. Analysis of these results suggests that a key step in the growth mechanism involves a surface reaction between carbon and nitrogen. IR spectroscopy was used to elucidate the phases present in these carbon nitride films, and demonstrated that a cyanogen-like impurity occurs in films with nitrogen compositions greater than 30%. A thermal annealing process has been developed, however, to eliminate this impurity and yield a single phase C_2N material. Systematic investigations of the electrical resistivity and thermal conductivity of the carbon nitride films reveal that this material is a good electrical insulator and thermal conductor. Furthermore, thermal annealing process was found to enhances significantly the resistivity without degrading the thermal conductivity. The new understanding of carbon nitride properties obtained from these studies provides a strong basis with which to assess the use of this material in high-performance application.

ACKNOWLEDGMENTS. We acknowledge some experimental help from Drs. Shoushan Fan and Jinlin Huang, and thank Dr. Arthur Moore of Union Carbide for providing pure graphite. This work was supported primarily by the MRSEC Program of the National Science Foundation under Award Number DMR-9400396.

REFERENCES

1. C. M. Lieber and Z. J. Zhang, Adv. Mater. **6**, 497 (1994).
2. A. Y.Liu and M. L. Cohen, Science **245**, 841 (1989).
3. A. Y. Liu and R. M. Wentzcovitch, Phys. Rev. B **50**, 10362 (1994).
4. H.-X. Han and B. Feldman, Solid State Commun. **65**, 921 (1988).
5. L. Maya, D. R. Cole, and E. W. Hagaman, J. Am. Ceram. Soc. **74**, 1686 (1991).
6. M. R. Wixom, J. Am. Ceram. Soc. **73**, 1973 (1990).
7. K. M. Yu, M. L. Cohen, E. E. Haller, W. L. Hansen, A. Y. Liu, and I. C. Wu, Phys. Rev. B **49**, 5034 (1994).
8. S. Kumar and T. L. Tansley, J. Appl. Phys. **76**, 4390 (1994).
9. M. Y. Chen, D. Li, X. Lin, V. P. Dravid, Y.-W. Chung, M.-S. Wong, and W. D. Sproul, J. Vac. Sci. Technol. A **11**, 521 (1993).
10. C. Niu, Y. Z. Lu, and C. M. Lieber, Science **261**, 334 (1993).
11. Z. J. Zhang, S. Fan, and C. M. Lieber, Appl. Phys. Lett. (in press).
12. J. Narayan, J. Reddy, N. Biunno, S. M. Kanetkar, P. Tiwari, and N. Parikh, Mater. Sci. Eng. B **26**, 49 (1994).
13. Z. J. Zhang, J. Huang, S. Fan, and C. M. Lieber, Mater. Sci. Eng. (in press).
14. J. E. Pollard, Rev. Sci. Instrum. **63**, 1771 (1992).
15. P. Hess, in <u>Photoacoustic, Photothermal and Photochemical Ptocesses at Surfaces and in Thin Films</u>, edited by P. Hess (Springer-Verlag, Berlin, 1989), p. 55.
16. J. Kouvetakis, A. Bandari, M. Todd, B. Wilkens, and N. Cave, Chem. Mater. **6**, 811(1994).
17. J. J. Cuomo, P. A. Leary, D. Yu, W. Reuter, M. Frisch, J. Vac. Sci. Technol. **16**, 299 (1979).
18. Z. J. Zhang, and C. M. Lieber (unpublished results).
19. D. G. Cahill and T. H. Allen, Appl. Phys. Lett. **65**, 309 (1994).
20. C. J. Morath, H. J. Maris, J. J. Cuomo, D. L. Pappas, A. Grill, V. V. Patel, J. P. Doyle, K. L. Saenger, J. Appl. Phys. **76**, 2636 (1994).

NITROGEN PLASMA ION IMPLANTATION INTO CARBON FILMS DEPOSITED BY THE ANODIC VACUUM ARC

IMAD F. HUSEIN, FAN LI, YUANZHONG ZHOU, RYNE C. ALLEN, AND CHUNG CHAN
Plasma Science and Microelectronics Research Laboratory, Department of Electrical and Computer Engineering, Northeastern University, Boston, MA, 02115, USA

ABSTRACT

Amorphous carbon films (a-C) deposited by the anodic vacuum arc on Si substrates were implanted with nitrogen using the Plasma Immersion Ion Implantation (PIII) technique to form carbon nitride films (CN_x). Scanning Electron Microscopy (SEM) of the a-C films show a surface morphology with maximum grain size in the order of a few nanometers and the exclusion of macroparticles. Increasing the nitrogen content of the CN_x films increased the intensity of the X-ray Photoelecton Spectroscopy (XPS) C 1s peak at 286.6 eV and formed a new peak at 285.6 eV which both can be associated with the carbon-nitrogen bond formation. Nanoindentaiton measurements showed that the hardness of the a-C films increased after implanting nitrogen into them. These CN_x films exhibited a hardness of 19 GPa.

1. INTRODUCTION

The possibility of forming solids with covalent C-N bonds, such as β-C_3N_4, was proposed by Liu and Cohen in 1989 based on theoretical calculations [1]. They predicted that carbon nitrides with composition C_3N_4 in the β phase, β-C_3N_4, has a high bulk modulus comparable to diamond and its cohesive energy is moderately large. More recently, Liu et al. [2] identified a zinc-blende-like cubic and graphitelike metastable carbon nitrides with a composition of C_3N_4. Cohesive energy calculations showed that the β and graphitelike structures have very close cohesive energies which suggest that in forming the β phase we have to overcome the competition from the energetically favorable sp^2 bond found in the graphitelike phase [2]. Many experimental methods have been tried to synthesize β-C_3N_4 using different film deposition techniques: ion beam deposition [3], RF diode sputtering [4-6], pulsed laser ablation [7], ion and vapor deposition [8], dc magnetron sputtering [3,9], and plasma decomposition of CH_4 and N_2 [10].

In this study amorphous carbon films (a-C) were deposited on Si wafers by the anodic vacuum arc technique [11-13]. Carbon nitride thin films (CN_x) were prepared by implanting the a-C films with nitrogen using the Plasma Immersion Ion Implantation (PIII) technique [14]. The films were characterized by X-ray Diffraction (XRD), X-ray Photoelectron Spectroscopy (XPS), Scanning Electron Microscopy (SEM), and nanoindentation measurements.

2. EXPERIMENTAL PROCEDURES

The anodic vacuum arc technique was used to deposit carbon films on silicon wafers. An arc is ignited in vacuum (pressure of 1x 10^{-5} Torr) by an electric trigger between two graphite electrodes of cylindrical geometry and sustained by a consumable anode. The anodic vacuum arc produces a partially ionized carbon vapor plasma (less than 20% ionized) [11]. Films with thickness around 0.8-1.7 µm (measured by a Dektak3 surface profilometer) were deposited with a deposition rate of 0.5 µm/min. using currents between 50-60 A. Higher deposition rates can be achieved depending on factors such as, the geometry of the electrodes, arc current, and distance between the arc and the substrate. A detailed description of the anodic arc system is found in [11,12].

The carbon films were immersed in a nitrogen plasma with a working gas pressure maintained on the order of 1.3x10^{-2} Pa (1x10^{-4} Torr). Negative voltage pulses were applied to the substrate in order to extract the nitrogen ions from the plasma. Nitrogen ions also diffuse into the films between the high voltage pulses via thermal diffusion [14]. Voltages of 1-2 kV were used with a pulse length of 4-8 µs and a repetition frequency of 6-14 kHz.

The composition and the carbon-nitrogen bonding nature of the films were investigated by XPS. This analysis was carried out on a Leybold MAX 200 X-ray photoelectron spectrometer using non-monchromatised Mg Kα exciting radiation for general survey spectra. A power of 15kV and 20mA was used with a survey spectra of 192 eV pass-energy. The sample area was 4x7 mm^3. The survey spectra was collected in the 1000 - 0 eV range. The spectral manipulation involved subtraction of the x-ray satellites. The spectra was normalized to account for spectrometer transmission function. The peak areas were measured and sensitivity factors were used to obtain relative atomic percentage from survey spectra. For high resolution spectra monochromatized Al x-ray radiation was used. A power of 15 kV and 30 mA was used with a Pass-energy of 36 eV. The calibration of the scale involved Cu sample, Ar$^+$ sputtered to remove any C or oxide. Cu 2p and 3p regions were used (Cu 2p$_{3/2}$ - 932.7 eV; Cu 3p-75.1 eV). To correct the binding scale for possible variable sample potential arising from charging, the high resolution peaks were shifted to place C maxima at 284.6 eV.

A D-500 X-ray Diffraction (XRD) Siemens equipment using CuKα line was used to investigate the crystal structure of the films. The films hardness was measured by a Nanoindenter1 equipment with a 3 sided diamond-Berkovick indentor. The rate of loading and unloading was 400µN/sec.

3. RESULTS AND DISCUSSION

The scanning electron micrographs showing the surface morphology of the a-C and CN$_x$ films on silicon substrates are shown in figure 1a and 1b respectively. Figure 1a show a uniform and dense coating with no macroparticle contamination. Fine grain sizes of the order of few nm was observed. While figure 1b shows a rough surface after

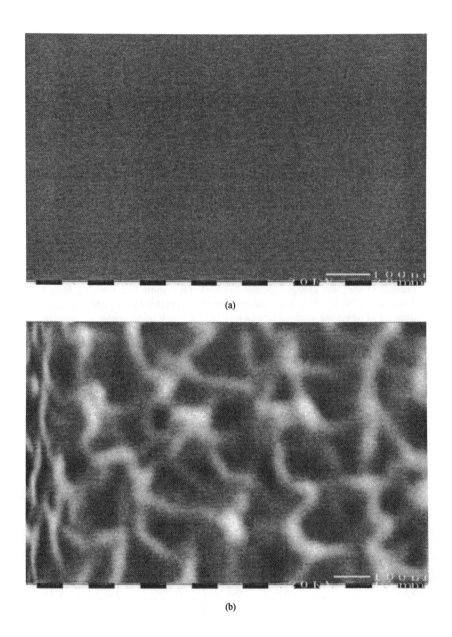

(a)

(b)

Figure 1: SEM surface micrographs of (a) *a*-C thin film and (b) CN$_x$ thin film on silicon substrates.

nitrogen implantation. Both a-C and CN_x films were found to be amorphous by x-ray diffraction analysis.

The elemental composition of the films was obtained from the survey spectra. The nitrogen to carbon ratios (N/C) are considerably lower than the optimal ratio predicted for β-C_3N_4. Most CN_x films had N/C ratios between 0.2 - 0.25. A small amount of oxygen contamination was detected in the films. A typical high resolution C 1s spectra of an a-C and CN_x films are shown in figures 2a and 2b respectively. The peak deconvolution results of an a-C film (of figure 2a) and two films with N/C ratios of 0.25 (of figure 2b) and 0.2 are summarized in Table I. The intensities of the peaks and their ratio to the main peak are also included for analysis. Carbon bound to itself, no matter how hybridized, has a characteristic peak at 285.0 eV which is usually used as the binding energy reference [15]. Thus, the two peaks at 284.3 eV and 284.9 eV in both the a-C and CN_x films indicates the formation of -C-C-C-C- groups. The peaks at 288.3 eV and 289.5 eV can be attributed to some C(O) type groups on the surface. The peak at 291.2 eV identified in the a-C film is most probably due to plasmon.

Two peaks of interest are identified at the binding energies 285.6 eV and 286.6 eV. Analysis of the ratio changes for the 286.6 eV peak show that the CN_x films have higher ratios than the a-C film. Thus, this peak can be attributed to C-N and C-C bonding. The presence of new peak after implantation at 285.6 eV of binding energy suggests that it can be associated with the formation of carbon-nitrogen bonds.

Nanoindentation measurements showed that the a-C films have a hardness of 17 ± 0.5 GPa. A hardness of 19 ± 0.6 GPa was obtained for the CN_x films.

Table I. XPS C 1s peaks deconvolution of a-C and CN_x films with different N/C ratios.

Binding	Peaks Deconvolution					
Energy	a-C Film		CN_x (N/C = 0.2)		CN_x (N/C=0.25)	
(eV)	at.%	Ratio	at.%	Ratio	at.%	Ratio
284.3 eV	50.4	1.5	4.7	0.12	5.4	0.15
284.9 eV	33.9	1.0	38.6	1.0	36.4	1.0
285.6 eV	----	----	23.3	0.6	23.7	0.65
286.6 eV	7.3	0.2	19.7	0.5	18.3	0.5
288.3 eV	4.5	0.13	7.4	0.2	11.1	0.3
289.5 eV	2.2	0.06	6.3	0.2	5.2	0.14
291.2 eV	1.7	0.05	----	----	----	----

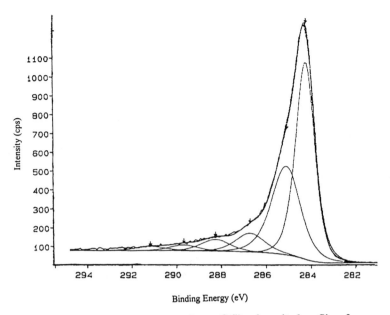

Figure 2a. XPS C 1s spectra of an a-C film deposited on Si wafer by the anodic vacuum arc.

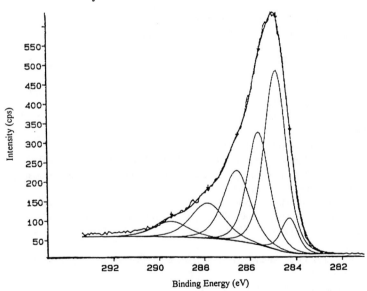

Figure 2b. A typical C 1s spectra of a CN_x film prepared by nitrogen plasma ion implantation into a-C films. This film has N/C ratio of 0.25.

4. CONCLUSION

Amorphous carbon nitride (CN_x) films with N/C ratio of 0.2 to 0.25 were prepared by nitrogen plasma ion implantation into amorphous carbon films (a-C). The a-C films were deposited on Si utilizing the anodic vacuum arc technique. Comparison of the XPS analysis of the a-C and CN_x suggest the formation of carbon-nitrogen bonds. Nanoindentation measurements show an increase in the hardness of the a-C films after nitrogen implantation.

ACKNOWLEDGMENTS

The authors would like to thank Dr. Jacob I. Kleiman president of Integrity Testing Laboratory, Ontario, Canada for the XPS analysis. And special thanks to Mr. Joe A. Genevich for helping construct much of the apparatus used in this experiment.

REFERENCES

[1].   A.Y. Liu and M.L. Cohen, Science **245**, 841 (1989); Phys. Rev. B **32**, 7988 (1985).
[2].   A.Y. Liu and R.M. Wentzcovitch, Phys. Rev. B **50**, 10362 (1994).
[3].   D. Marton, K.J. Boyd, A.H. Al-Bayati, S.S. Todorov, and J.W. Rabalais, Phys. Rev. Lett. **73**, 118 (1994).
[4].   Tse-An Yeh, C.L. Lin, J.M. Sivertsen, and J.H. Judy, IEEE Trans. Magn. **27**, 5163 (1991).
[5].   K.M. Yu, M.L. Cohen, E.E. Haller, W.L. Hansen, A.Y. Liu, and I.C. Wu, Phys. Rev. B **49**, 5034 (1994).
[6].   C.J. Torng, J.M. Sivertsen, J.H. Judy, and C. Chang, J. Mater. Res. **5**(11), 2490 (1990).
[7].   C. Niu, Y.z. Lu, and C.M. Lieber, Science **261**, 334 (1993).
[8].   K. Ogata, J. Chubaci, and F. Fujimoto, J. Appl. Phys. **76**(6), 3791 (1994).
[9].   D. Li, E. Cutiongco, Y. Chung, M. Wong, and W. Sproul, Surface and Coating Technology, **68/69**, 611 (1994).
[10].   H. Han and B.J. Feldman, Solid State Communications **65**(9), 921 (1988).
[11].   S. Meassick, C. Chan, and R. Allen, Surface and Coatings Tech. **54**, 343 (1992).
[12].   S. Meassick, J. Kumpf, R. Allen, C. Chan, and T. Sroda, Materials Letters **14**, 63 (1992).
[13].   V. Buck, J.B. Orde, and M. Mausbach, Materials Science and Engineering A**140**, 770 (1991).
[14].   Z. Xia and C. Chan, J. Appl. Phys. **73**, 3651 (1993).
[15].   C. D. Wagner, *Handbook of X-Ray Photoelectron Spectroscopy*, (Perkin-Elmer, Physical Electronics Division, Eden Prairie, MN, 1978.

CARBON NITRIDE FILMS USING A FILTERED CATHODIC ARC

SCOTT.M. CAMPHAUSEN*, D.A. TUCKER*, J. BRULEY**, AND J.J. CUOMO*
*North Carolina State University, Department of Materials Science and Engineering, Raleigh, NC 27695
**Lehigh University, Department of Materials Science and Engineering, Bethlehem, PA 18015

ABSTRACT

Carbon nitride films were deposited on sapphire, silicon, molybdenum and quartz substrates using a filtered cathodic arc with a radio frequency (rf) inductively coupled plasma. Previous researchers have used the cathodic arc but obtained low concentrations of nitrogen in their films. The addition of the rf coils was used to enhance dissociation of molecular nitrogen in the vicinity of the substrate. The resultant films were analyzed for chemical bonding and composition by several methods including Electron Energy Loss Spectroscopy (EELS), Auger Electron Spectroscopy (AES), and infrared spectroscopy (IR). EELS and IR provided bonding information and showed that the films were largely composed of sp^2 bonding and contained carbon-nitrogen double bonds, respectively. Also, chemical composition of the films was determined using AES, IR and EELS.

INTRODUCTION

In 1989, Liu and Cohen [1] predicted through an empirical model that β-C_3N_4 should have a bulk modulus greater than diamond. This structure is based on the known structure of β-Si_3N_4 with the replacement of silicon atoms by carbon. This presented a major challenge for the materials science community to synthesize carbon and nitrogen in the β-Si_3N_4 structure. Since this theoretical prediction, several processing techniques have been used in an attempt to produce β-C_3N_4 and have included DC magnetron sputtering [2, 3], rf magnetron sputtering [4, 5], ion and vapor deposition [6], dual ion beam sputtering [7], rf diode sputtering [8], and cathodic arc deposition [9]. Figure 1 presents a graph of the maximum atomic percentage of nitrogen achieved for these different investigators. It shows that many researchers have attained films with low concentrations of nitrogen relative to the β-C_3N_4 structure although Yu et al. [8] employed a rf diode sputtering technique and reported the formation of β-C_3N_4 particles.

In this paper, carbon nitride films were deposited using a filtered cathodic arc system. Davis et al. [9] used a filtered cathodic arc with a nitrogen background gas to produce carbon nitride films. In their experiments, the flow rate of the nitrogen was varied to achieve different amounts of nitrogen in the CN_x films and concentrations up to 30 at.% nitrogen were obtained.

Mat. Res. Soc. Symp. Proc. Vol. 388 © 1995 Materials Research Society

In an attempt to synthesize β-C$_3$N$_4$, a metastable phase structure, we utilized the concept of energetic condensation. Previously, a filtered cathodic arc system had been used by Lossy et al. [10] to deposit amorphous diamond. They found that quenching of the energetic ions was necessary to synthesize this metastable phase of carbon.

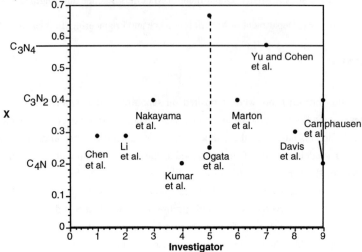

Figure 1. Plot of the C$_{1-x}$N$_x$ composition versus various investigators.

EXPERIMENTAL

A series of experiments was carried out using the filtered cathodic arc with a nitrogen background gas. The run pressures were varied to obtain different nitrogen contents in the films. The highest concentration of nitrogen was obtained with a run pressure of 1 x 10^{-3} Torr. However, these films contained only 5 at.% nitrogen. Next, a Kaufman ion source was used in conjunction with the filtered cathodic arc to increase the activated nitrogen species at the substrate. These films had a nitrogen content of 17 at.%. Since both of these processing techniques produced low concentrations of nitrogen, it was inferred that a higher concentration of activated nitrogen species was required in the vicinity of the substrate. This was accomplished by using a radio frequency inductively coupled nitrogen plasma.

The filtered cathodic arc was slightly modified in order to obtain a high nitrogen plasma density in front of the substrates. RF copper coils were placed in front of the substrate (see Figure 2) and were powered by a RF 1 kW power generator at a frequency of 13.56 MHz. A series of films were grown under different substrate bias conditions and Table I summarizes the process parameters. The target for the cathodic arc was made of 99.9995% pure graphite and nitrogen (99.9995% purity) was introduced as the processing gas. The substrate was water cooled to quench the energetic ions. During the process, the water temperature ranged from 10 to 30° C. A thermal compound, part number 120-8 manufactured by Wakefield Engineering, was applied to the back of the substrates in order to increase thermal contact with the water cooled substrate holder. The experiments were run for 10 minutes with a duty cycle of 1 minute on and one minute off. Films were grown on single crystal sapphire, silicon (100), molybdenum and quartz. The silicon and molybdenum substrates were cleaned in an unbuffered HF bath and the sapphire and quartz substrates were cleaned in an alcohol bath. The composition of these films was obtained by Auger Electron Spectroscopy (AES) and Electron Energy Loss Spectroscopy (EELS). Bonding information was obtained by EELS and IR spectroscopy.

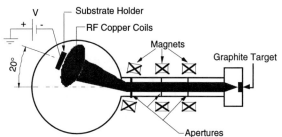

Figure 2. Schematic diagram of the filtered cathodic arc system modified by the addition of rf coils.

Table I. Process parameters used for deposition of CN_x films.

Sample ID	A	B	C	D
Base Pressure (Torr)	1×10^{-6}	1×10^{-6}	1×10^{-6}	1×10^{-6}
Run Pressure (Torr)	1×10^{-3}	1×10^{-3}	1×10^{-3}	1×10^{-3}
Forward Power (Watts)	500 - 700	500 - 700	500 - 700	500 - 700
Reflected Power (Watts)	75	75	75	75
DC Bias (Volts)	0	-25	-200	-300

RESULTS AND DISCUSSION

Surface compositions were obtained by AES. Samples were fractured into $1cm^2$ pieces and loaded into a UHV chamber. The films all showed a minimal amount of oxygen contamination at the surface. It was found that increasing the DC bias to -200 Volts that the composition of nitrogen in the films decreased from 40 at.% for the grounded case to 20 at.%. The composition results are tabulated in Table II and suggest that the carbon nitride stoichiometries are C_3N_2 in the case of zero and -25 Volts biasing and C_4N in the case of -200 and -300 Volts biasing. However, the results show an increase in nitrogen concentration with the addition of the rf inductively coupled nitrogen plasma as compared to just having nitrogen as a background gas.

To perform EELS analysis, the preparation of the specimen foils went as follows. Small ($1mm^2$) pieces were fractured from the CN_x coated silicon substrate. The silicon was next etched in an HF/HNO_3 mixture. The thin film was then rinsed in deionized water and floated onto a 3mm Cu grid. The films were not continuous but small pieces could be collected.

The thin films were examined in a VG microscopes' HB501 dedicated FEG scanning transmission electron microscope fitted with a Gatan parallel electron energy-loss spectrometer (PEELS). The microscope was operated at 100 keV with a beam current of about 0.2 nA. The microscope chamber vacuum was better than 9 x 10^{-10} Torr. Spectra were recorded with the beam scanning over an area of about 9 nm^2, the beam convergence was about 8 mrad and the spectrometer acceptance angle was about 16 mrad. Table II, along with the AES compositional analysis, presents the results of the EELS analysis.

Table II. Composition of CN_x films as obtained by AES and EELS.

Sample ID	A	B	C	D
DC Bias (Volts)	0	-25	-200	-300
N composition (at.%)	40	40	20	20
C (π/σ)	0.9	-	0.8	0.8
N (π/σ)	0.8	-	0.9	1.2

Figures 3 and 4 show the carbon and nitrogen edges, respectively, for samples A, C and D. The possible sources of error are: inhomogeneities in the film composition, surface contamination by carbon, and background removal. Both films C and D contained a smaller (indistinguishable) concentration of nitrogen than film A as evident in Figure 4. The pi/sigma ratio quoted in Table II reflects the ratio of the pi to sigma intensities in 4 eV and 10 eV windows, respectively. The ratios have been normalized to the value of 1.4 extracted from the spectrum of graphitized carbon which was recorded under similar experimental conditions. The high numbers (close to 1) suggest the films are almost 100 % sp^2 bonded. Figure 5 shows the carbon and nitrogen K edges of sample A overlapped for comparison. Even though there is only a small concentration of nitrogen, its presence disrupts any amorphous diamond continuous sp^3 network resulting in the π-bonded carbon clusters. Similar results were found by Davis et al.[9]. It was observed that the difference between the nitrogen and carbon edges in film A is very similar to the nitrogen edges of films C and D. This suggests that a high percentage of nitrogen occupies similar atomic environments as the carbon atoms, i.e. nitrogen is substitutional for carbon.

Transmission Fourier transform infrared spectroscopy measurements were performed on an Analect (Model FX6360) instrument. The films were crushed and mixed into a KBr powder and scanned over the region between 3200 and 800 cm^{-1}. The results gave evidence for the existence of C=N and some C-H bonds. The presence of hydrogen is most probably due to the high base pressure.

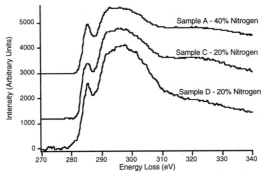

Figure 3. Carbon K edges for Samples A, C and D.

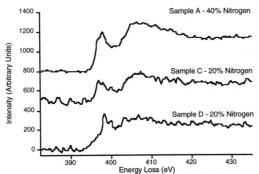

Figure 4. Nitrogen K edges for Samples A, C and D.

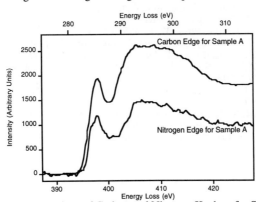

Figure 5. Comparison of Carbon and Nitrogen K edges for Sample A.

CONCLUSIONS

A radio frequency inductively coupled plasma in conjunction with a filtered cathodic arc has been utilized to obtain carbon nitride stoichiometries of C_3N_2 at zero and -25 Volts bias. Thus, the addition of the rf plasma to the cathodic arc system has resulted in an increase in the nitrogen concentration over that produced in carbon nitride films without the use of the inductively coupled plasma. At -200 and -300 Volts, the stoichiometry was found to be approximately C_4N. This shows that increasing the bias has decreased the nitrogen concentration in the films. EELS analysis showed that all the films were predominantly sp^2 bonded and IR provided evidence for carbon-nitrogen double bonds in the films.

ACKNOWLEDGMENTS

We are grateful to Robert Johnson who helped maintain the filtered cathodic arc. Many thanks to Dr. C.R. Guarnieri of IBM Watson Research for technical advice. Infrared spectroscopy was obtained with the help of Robert Ratway. We thank IBM Watson Research Center, Yorktown Heights for their donation of the filtered cathodic arc system. Also, we appreciate useful discussions and assistance by members of the Center for Advanced Manufacturing Processes and Materials.

REFERENCES

1. A.Y. Liu and M.L. Cohen, Science **11**, 841 (1989).
2. M.Y. Chen, D. Li, X. Lin, V.P. Dravid, W.C. Yip, S.W. Ming, and W.D. Sproul, J. Vac. Sci. Technol. A **11**, 521 (1993).
3. D. Li, Y.W. Chung, M.S. Wong, and W.D. Sproul, J. Appl. Phys. **74**, 219 (1993).
4. N. Nakayama, Y. Tsuchiya, S. Tamada, K. Kosuge, S. Nagata, K. Takahiro, and S. Yamaguchi, Jpn. J. Appl. Phys. **32**, L1465 (1993).
5. S. Kumar and T.L. Tansley, Solid State Commun. **88**, 803 (1993).
6. K. Ogata, D.C.J. Fernando, and F. Fujimoto, J. Appl. Phys. **76**, 3791 (1994).
7. D. Marton, B.A.H. Al, S.S. Todorov, K.J. Boyd, and J.W. Rabalais, Nucl. Instr. and Meth. **B90**, 277 (1994).
8. K.M. Yu, M.L. Cohen, E.E. Haller, W.L. Hansen, A.Y. Liu, and I.C. Wu, Phys. Rev. B **49**, 5034 (1994).
9. C.A. Davis, D.R. McKenzie, Y. Yin, E. Kravtchinskaia, G.A.J. Amaratunga, and V.S. Veerasamy, Philos. Mag. B **69**, 1133 (1993).
10. R. Lossy, D.L. Pappas, R.A. Roy, J.J. Cuomo, and V.M. Sura, Appl. Phys. Lett. **61**, 171 (1992).

HIGH-PRESSURE PULSED PLASMA SYNTHESIS OF CARBON-NITRIDE THIN FILMS

V. N. Gurarie, A. V. Orlov, L. A Bursill, Peng JuLin, K. W. Nugent and S. Prawer
School of Physics, University of Melbourne, Parkville, VIC, 3052, Australia

ABSTRACT

In this study a high N_2 pressure has been used to enhance the N incorporation into CN films produced by a shock plasma deposition method. Auger spectroscopy indicates that increasing the nitrogen pressure from 0.1 atm to 10 atm results in an increase of nitrogen incorporation into CN films to a maximum of 43 at.%. Nitrogen distribution varies across the surface of the deposit, showing an increase of nitrogen content with depth in the center of the deposition and a decrease with depth at points away from the center. SEM and optical microscopy indicate that under increased nitrogen pressure the grain structure becomes finer. Raman spectra contain sharp peaks characteristic of a distinct crystalline CN phase. TEM diffraction patterns for the films produced under N_2 pressure in the range of 0.05-0.1 atm unambiguously show the presence of micron-sized crystals displaying a cubic symmetry, and not the predicted β-Si_3N_4 type structure. PEELS data suggest that in the crystalline phase a significant fraction of the nitrogen atoms have sp^2 trigonal bonds and there is a significant degree of sp^3 character for the carbon atoms.

INTRODUCTION

Recent theoretical work has predicted that a new ultra-hard compound should be able to be synthesised with a composition C_3N_4 and a hardness to rival that of diamond [1]. These predictions sparked considerable interest in attempts to experimentally realise this compound [2-11]. The majority of the experimental work reports the formation of amorphous nitrogen rich carbon films. The investigations show that an enhanced nitrogen content improves wear resistance, hardness, tribological and other properties of these films. In [10,11] nitrogen rich carbon depositions were reported to contain a crystalline component related to the β-C_3N_4 phase.

Recently we reported the production of C-N films containing up to 32% nitrogen using a shock-plasma method in a nitrogen ambient [12]. The method employed a capacitive discharge of 4 μs duration between carbon electrodes in a nitrogen gas ambient of variable pressure in a range of 10 Torr to 100 Torr. Both nanocrystalline and amorphous morphologies were produced. In the present work a new deposition chamber has been constructed which allows the plasma to be produced in a high pressure N_2 ambient (up to 20 atm). The aim of the present experiments is to increase the nitrogen content in the CN plasma and to investigate how this affects the nitrogen incorporation into the CN films, as well as the structure and morphology of the films produced. In the present study we used the Auger spectroscopy, Raman spectroscopy, SEM and optical microscopy to study films deposited in this high nitrogen pressure ambient. We also report TEM diffraction and PEELS data on a crystalline phase containing C and N produced by shock plasma deposition, which, however, has a structure different to that predicted for the ultrahard C_3N_4 compound.

EXPERIMENTAL

The experimental arrangement used for the deposition of nitrogen rich carbon films is shown in Fig. 1. Carbon electrodes of 4.5 mm in diameter are placed in a discharge chamber filled with nitrogen gas. A capacitor battery is discharged through a gap between the carbon electrodes. The chamber is designed to use relatively high nitrogen pressure (up to 20 atm). The discharge is controlled by introducing a third electrode near the discharge gap which is connected to a Rhumkorff coil. This third electrode produces a background ionisation near the electrode space which is necessary to trigger the discharge of the capacitor. The discharge voltage of the capacitor is controlled by the relative position of the third electrode and the power supplied by

Mat. Res. Soc. Symp. Proc. Vol. 388 © 1995 Materials Research Society

the Rhumkorff coil. A diode is inserted in the Rhumkorff coil circuit to avoid the capacitor discharge through the coil. The use of high nitrogen pressure and the triggering electrode are the new features of this installation. In other respects the method is similar to that described earlier [12].

The CN depositions were obtained using nitrogen gas pressure of 8 atm and the interelectrode separation of 2 mm. This allowed the amount of nitrogen in the interelectrode space to be increased by two orders of magnitude compared to the deposition conditions used earlier [12]. The discharge time was ~ 4 μsec. The capacitance used was 12.5 μF and the voltage 3 kV, giving a released power of approximately 14 MW. Estimates similar to those conducted in [12]

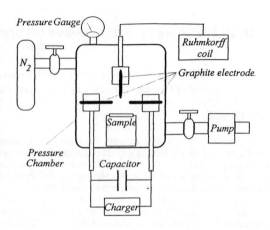

Fig. 1. Experimental arrangement.

indicate that the nitrogen in the plasma under these conditions is almost completely dissociated and one fold ionised.

The pressure in the carbon nitrogen plasma can be roughly estimated as $P = 4 P_o T / T_o$, where P_o and T_o are the initial breakdown pressure and temperature and the factor of 4 takes into consideration the dissociation of nitrogen molecules and the release of carbon in an amount equal to that of nitrogen. An estimate of peak plasma temperature gives $T \approx 3 \times 10^4$ K [12]. Thus, the plasma shock pressure is of the order of 400 P_o, which at $P_o = 1.0$ MPa gives the shock pressure $P \approx 400$ MPa. Because of the high shock pressure, substrates such as silicon and glass are easily fractured especially when placed close to the electrodes. Moving them away from the electrodes drastically reduces the deposition rate, but then quartz glass, sapphire crystals and glassy carbon substrates all proved to be mechanically sufficiently stable for the deposition.

However, the high temperature and high pressure plasma flow causes ablation of the substrate surface. The analysis of sapphire crystal and quartz glass used as substrates placed at a distance of 2.3 mm from the electrode axis has demonstrated that the substrate material is removed from the central part of the plasma affected area and partly re-deposited at the edges of the area. Raman spectroscopy of the deposition at the edges clearly demonstrates that the substrate material dissociates forming pure silicon in the case of the quartz glass substrate and pure aluminium in the case of the sapphire substrate.

To avoid the substrate ablation the distance h between the substrate and the electrodes was optimised. At h = 4 mm ablation was not observed in the central hot spot of the deposition where the CN films of ~ 10 μm thick were obtained after ~2000 pulses. A further increase of h results in a significant drop in the deposition rate since the plasma flux is reduced as ~ 1 / h³.

It is necessary to note that the high pressure N_2 ambient also enhances the ablation rate of the carbon electrodes during the discharge. The loss of weight by the electrodes is increased by ~40 to 50 times when the nitrogen pressure is increased from 10^2 Torr to 10^4 Torr. Thus, because of the increased rate of ablation of the carbon electrodes, the increase of nitrogen pressure in the chamber by two orders of magnitude increases relative nitrogen concentration in the CN plasma by only a factor of about 2. The enhanced ablation of the electrodes obviously results from the high energy concentration in the plasma formed by a high pressure discharge.

RESULTS AND DISCUSSION

Fig. 2. demonstrates the variation of the film composition in the surface layer as a function of the radial distance, r, from the centre of the deposit as measured by Auger spectroscopy. The nitrogen concentration increases from about 18 at% in the centre to about 36 at% towards the periphery of the deposition. A similar effect was previously observed for depositions using lower nitrogen pressures [12]. The effect is likely to be related to the nonuniform distribution of nitrogen in plasma. The ablation products from the carbon electrodes tend to oust the nitrogen to the front of the compressive shock wave, where nitrogen concentration is expected to be higher.

The distribution of nitrogen also varies with depth from the surface. At the periphery of the deposition the nitrogen content decreases with depth as shown in Fig. 3 (a). As noted in [12] this effect can be related to a destruction of C-N bonds by the sputter depth profiling which allows the nitrogen to escape. However, the Auger

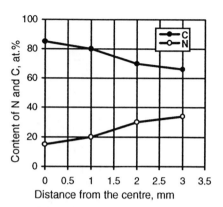

Fig. 2. Surface nitrogen content versus distance r from deposition centre.

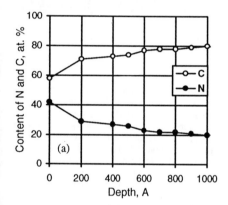

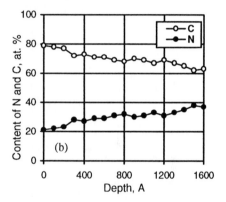

Fig. 3. Film composition versus depth as determined by Auger spectroscopy; (a) r = 3 mm, (b) centre, r = 0 mm.

depth profiling in the central area of the deposition shows an increase of nitrogen concentration with depth, as illustrated in Fig. 3 (b). This result suggests that if the depth profiling sputtering does bring about the nitrogen escape, then the actual nitrogen content should be even higher than that displayed in Fig. 3 (b).

The Auger data suggests that under a N_2 pressure of ~8 atm the nitrogen concentration at the periphery of the deposition before sputtering reaches 43 at.%, which is considerably higher than the maximum 32 at.% obtained for the depositions under the N_2 pressure of ~0.1 atm. As indicated above a high nitrogen pressure ambient considerably enhances electrode ablation during the discharge. As a result the actual increase in nitrogen concentration in plasma is much less than it is expected from the increase in nitrogen gas pressure in the chamber. However, in these preliminary experiments the voltage, capacitance and electrode separation were not optimised, so the method apparently has a potential for a further increase of nitrogen concentration in the plasma by optimising the discharge parameters and reducing the carbon ablation. In previous work at lower N pressure, the depositions were found to exhibit both a columnar structure and cauliflower morphology. The morphologies are not as pronounced for the high pressure depositions. The grain size of the deposit is ~ 0.2-0.3 μm which is significantly smaller than that for the low pressure depositions.

Microhardness for these films was impossible to measure since they proved to be very brittle and cracked before any impression of the indentor appeared. Fig. 4 shows cracking in a section of the CN deposition and the film profile.

The Raman spectrum from various areas of the deposition is typical of those seen for graphitic carbon and also observed for the depositions obtained at a relatively low nitrogen pressure [12]. The spectrum in Fig.

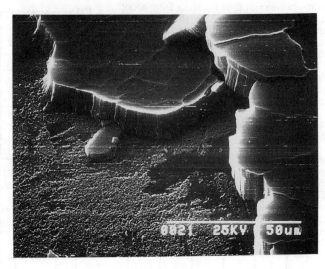

Fig. 4. SEM image of CN deposition. $P_o = 8$ atm, discharge power 5 MW, 2000 pulses.

5 was recorded using the exciting wave length of 514.5 nm in a region of the deposit adjacent to the central ablated area of the quartz substrate. This region was morphologically distinct from the rest of the deposit. It appeared lighter under the optical microscope and smoother under the SEM, compared to the rest of the film. The spectrum shows lines from the silicon along with a rising background from the luminescence of the disordered material. In this sample silicon originates from the ablation and decomposition of the quartz (SiO_2) substrate in the central most hot plasma affected area. The peaks seen on this background above 1200 cm^{-1} are probably due to the presence of amorphous carbon - nitrogen compounds. The spectrum also contains a number of sharp peaks at 1600, 1030, 1000 and 620 cm^{-1} with the strongest one at 1000 cm^{-1}. The sharpness of these peaks indicates that these lines arise from a well defined crystalline material. The occurrence of the peak at 1600 cm^{-1} strongly suggests that the peaks originate from the phase consisting of light elements such as CN, rather than from that which could possibly be formed by silicon with carbon, nitrogen or oxygen such as SiC, Si_3N_4, or polycrystalline SiO_2. The phase which gave rise to the sharp peaks was found to be unstable under the laser beam, decomposing completely after around 10 minutes exposure to this laser power. This may be due to heating from the absorption of the laser by the dark coloured material surrounding this region. The difficulty in observing such a phase over much of the sample may be due to its being mixed with amorphous carbon. This will firstly result in heating and rapid decomposition of the sample, and also in a reduction of the penetration of the laser beam into the sample reducing the volume of material sampled and making the spectrum more difficult to see over the greater background from the amorphous carbon. High resolution TEM and electron diffraction data have not yet been done on these samples; however, previous low pressure depositions were found to contain a small fraction of crystalline material.

The electron diffraction (ED) pattern of this material is shown in Fig. 6, which was obtained from the thin edge of a micron-sized crystal. The latter was more difficult to find and presumably constituted the minor fraction of the specimen with respect to the amorphous component and graphite. The axial ratio of $\sqrt{2}$ and the spot spacings and interplanar angles are characteristic of those expected for the [011] zone axis pattern of a face-centred cubic crystal having cell parameter approximately a = 0.630 nm. ED patterns showing the corresponding [001] and [111] zone axis patterns were also obtained confirming the presence of a f.c.c. structure

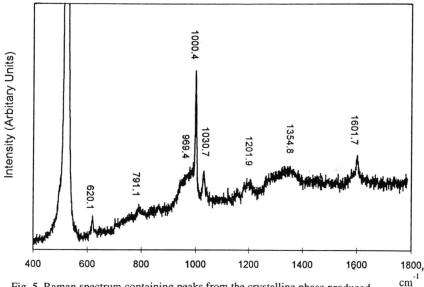

Fig. 5. Raman spectrum containing peaks from the crystalline phase produced by high-pressure pulsed plasma CN deposition; $P_0 \approx 8$ atm, 2000 pulses.

(a = 0.630 nm). This structure exhibited observable reflections and overall intensity typical of the NaCl-type space group Fm3m. We were unable to find either single crystal patterns or diffuse powder patterns consistent with the β-C_3N_4 phase predicted by Liu and Cohen. These results indicate that further structural complexity for CN compounds may exist.

Crystals giving such patterns tended to degrade in the electron beam. As a result the ED patterns formed circular arcs centred on the originally sharp spots. This effect is due to the original single crystal being partially degraded into polycrystalites, which were clearly observed in the high resolution TEM images of the crystalline phase. Prolonged observation

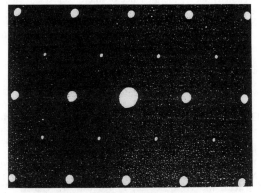

Fig. 6 Electron diffraction pattern from a 1 micron crystallite; note axial ratio √2 characteristic of a f.c.c. crystal viewed along [110]. $P_0 \approx 0.05$ atm, 3000 pulses.

resulted finally in degradation into an amorphous structure. We suppose that the crystalline phase which gives rise to the sharp peaks in the Raman spectrum in Fig. 5 is the same as that which has the ED pattern shown in Fig. 6. However, a further analysis is required to confirm it.

Fig. 7 shows typical parallel-electron-energy-loss spectrum (PEELS) for this type of crystalline (a = 0.630 nm) material. A Gatan Parallel Detection Electron Spectrometer (Model 666) was used at 400 keV; the electron probe diameter was typically in the range 10-40 nm for the PEELS analysis. The fine structure of the carbon and nitrogen edges [13], starting at approximately 285 and 400 eV respectively, indicates that there is some pi* structure contribution to the PEELS results for both the carbon and nitrogen edges of this crystalline material. For the carbon edge this is less pronounced, implying that there is a significant degree of sp^3 character for the carbon. For the nitrogen we can conclude that the sp^2 character is dominant. PEELS analysis also shows small amounts of calcium and oxygen present in the crystalline material suggesting that this structure may be stabilised by calcium and oxygen at about 1-2 at.% levels.

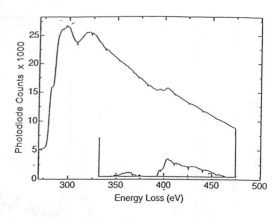

Fig. 7. Raw PEELS spectrum from the crystalline phase; note the presence of leading edge fine structure characteristic of sp^2 bonding, for both carbon and nitrogen.

CONCLUSIONS

Carbon-nitride films were prepared using a high intensity plasma discharge between carbon electrodes in a high-pressure nitrogen ambient. The distribution of nitrogen in the deposition was analysed by Auger spectroscopy which showed up to 43 at.% N in the films. Raman spectra indicated the occurrence of a definite crystalline phase in the deposition. The TEM diffraction unambiguously shows the presence of micron-sized crystals, which display a cubic symmetry, and not the predicted β-Si_3N_4 type structure. These results suggest that further structural complexity for CN compounds may exist. There is PEELS evidence that in the crystalline phase the carbon atoms appear to have both trigonal and tetrahedral bonding while a significant fraction of the nitrogen atoms have sp^2 trigonal bonds.

REFERENCES

1. Amy Y. Liu and Marvin L. Cohen, Physical Review B, **41** (15), 10727-34 (1990).
2. M. Iwaki and K. Takahashi, J. Mater. Res., **5** (11), 2562-66 (1990).
3. C.G. Torng, J.M. Siversten, J.H. Judy and C. Chang, J. Mater. Res., **5** (11), 2490-96 (1990).
4. He-Xiang Han and B. J. Feldman, Solid State Communications, **65** (9), 921-23 (1988).
5. J.F.D. Chubaci, T. Sakai, T. Yamamoto, K. Ogata, A. Ebe and F. Fujimoto, Nuclear Instr. and Met. in Phys. Res., B80/81, 463-66 (1993).
6. D. Li, Y. Chung, M. Wong and W.D. Sproul, J.Appl. Phys., **74** (1), 219-23 (1993).
7. L. Maya, D.R. Cole and E.W. Hagaman, J. Amer. Cer. Soc., **74** (7), 1686-88 (1991).
8. M. Ricci, M. Trinquecoste, F. Auguste, R. Canet, P. Delhaes, C. Guimon, G. Pfister-Guillouze, B. Nysten and J.P. Issi, J. Mater. Res. **8** (3), 480-88 (1993).
9. F. Fujimoto and K. Ogata, Jpn. J. Appl. Phys. **32** (3B), 420-23 (1993).
10. C. Niu, Y.Z. Lu and C.M. Lieber, Science, **261**, 334-37 (1993).
11. O. Matsumoto, T. Kotaki, H. Shikano, K. Takemura and S. Tanaka, J. Electrochem. Soc., **141** (2), 16-18 (1994).
12. V. N. Gurarie, A. V. Orlov, K. W. Nugent, P. Weiser and S. Prawer in Novel Forms of Carbon II, edited by C.L. Renschler, D.M. Cox, J.J. Pouch and Y. Achiba (Mat. Res. Soc. Proc. **349**, Pittsburgh, PA, 1994) pp. 37-42.
13. S.D. Berger, D.R. McKenzie and P.J. Martin, Phil. Mag. Letts., **57**, 285-290 (1988).

FABRICATION OF A FREE-STANDING, SYNTHETIC, SINGLE CRYSTAL DIAMOND PLATE USING ION IMPLANTATION AND PLASMA-ENHANCED CHEMICAL VAPOR DEPOSITION

J.B. POSTHILL, D.P. MALTA, T.P. HUMPHREYS, G.C. HUDSON, R.E. THOMAS, R.A. RUDDER, and R.J. MARKUNAS
Research Triangle Institute, Research Triangle Park, North Carolina 27709-2194

ABSTRACT

Using a specific combination of energetic and chemical processes we have grown homoepitaxial diamond on and lifted it off of a type Ia natural C(100) crystal. Before growth, the C(100) crystal is exposed to a self implant of 190keV energy and dose of $1E16$ cm^{-2}. Low temperature (~600°C) homoepitaxial diamond growth conditions were used that are based on water-alcohol source chemistries. To achieve layer separation (lift-off), samples were annealed to a temperature sufficient to graphitize the buried implant-damaged region. Contactless electrochemical etching was found to remove the graphite, and a transparent synthetic (100) single crystal diamond plate of 17.5μm thickness was lifted off. This free-standing diamond single crystal plate was characterized and found to be comparable to homoepitaxial films grown on unimplanted single crystal diamond.

INTRODUCTION

Research efforts are underway world-wide in order to develop a low cost, single crystal diamond wafer that could be used for active electronic device fabrication or other technological purposes. One approach has been to develop a heteroepitaxial diamond technology using recent advances in thin film diamond deposition processes -- in particular various chemical vapor deposition (CVD) techniques. Different non-native substrates have been examined, with the greatest success being found with: cubic BN single crystals [1, 2], BeO single crystals [3], Ni single crystals [4], SiC single crystals [5] and Si single crystals [e.g., 6]. Efforts are underway to improve these heteroepitaxial diamond processes using these respective substrates in order to increase the quality and value of the diamond layers.

A second approach involves employing techniques to make a diamond single crystal template by careful positioning/bonding of multiple single crystal diamonds on a non-native substrate combined with homoepitaxial diamond growth in order to epitaxially join the diamonds. If successful, this could result in a diamond boule technology. Underlying this approach to making a single crystal diamond template, is prior research in this laboratory which demonstrated that epitaxial diamond could be grown laterally and vertically at comparable rates [7]. Specially patterned Si(100) wafers were made in order to hold small (< 100μm), similarly-faceted seed diamonds in an array of {111}-faceted pyramidal pits in the Si, and after growing ~240μm of CVD diamond on such a unique structure, a large-area mosaic diamond film was created [8]. Another version of this method involves starting with considerably larger diamond single crystals (2mm×2mm) that have had their edges and faces oriented and polished to (100) [9]. Two diamond crystals were soldered in close proximity to each other onto a Mo substrate and placed in a hot filament diamond CVD reactor for epitaxial joining. Reported most recently were results from this laboratory which utilized edge-oriented 3mm×3mm (100) diamonds that were bonded to a Si substrate and then epitaxially joined [10]. In addition to these methods of making a large-area single crystal diamond template, a method must be developed to cut wafers that minimizes kerf losses.

We briefly describe our own demonstration of a method to "cut" a diamond single crystal combined with diamond homoepitaxial deposition based on plasma-enhanced CVD using water-alcohol mixtures in order to fabricate a free-standing, synthetic, single crystal diamond plate.

EXPERIMENTAL PROCEDURE AND RESULTS

Diamond Substrate Preparation

Type Ia (100) natural diamond single crystals were used for these proof-of-concept experiments. They were procured from the vendor with adequate polishing to enable good quality homoepitaxial diamond growth [11]. Different diamonds were implanted with C^{12} at 190keV commercially. Doses varying from 5E15 cm^{-2} to 5E16 cm^{-2} were examined, with the 1E16 cm^{-2} dose thus far representing the best combination of providing a surface to grow good quality homoepitaxial diamond on and enabling lift-off to be performed later. This will be the only sample from which results will be presented.

Diamond Growth

An rf-driven plasma enhanced CVD process that utilizes water and alcohol (ethanol) to grow the homoepitaxial films was used in this study. The use of alcohol/water mixtures for diamond growth was pioneered in this laboratory [12], and the development and details of this homoepitaxial diamond deposition process are reported elsewhere [13]. Briefly, the conditions of homoepitaxial diamond growth used were: Reactant flow was 18 sccm ethanol and 12 sccm water; Pressure = 1 Torr; Power (rf) $\cong$ 1.5kW; Growth temperature $\cong$ 600°C; Growth rate $\cong$ 0.5 μm-hr^{-1}. Diamond homoepitaxial growth was done on the implanted side of each diamond, with periodic interruption to inspect the progress of diamond homoepitaxy on the implanted substrates. Fig. 1 shows surface topographies as observed by scanning electron microscopy (SEM) after 1.6μm growth and 17.5μm growth on the same substrate (previously implanted with 1E16 cm^{-2} dose). The surfaces appear reasonably smooth, and the unidirectional "ridges" that are believed to be artifacts of the polishing process are no longer seen after the thickness of 17.5μm is reached. Epitaxy was verified using low energy electron diffraction (LEED) after 1.6μm growth. These results are very similar to those achieved upon diamond homoepitaxy on as-received diamond substrates using this growth process.

Separation Methods

The diamond was annealed in flowing N_2 at ~1000°C to graphitize the buried, implanted region, and the sides of the diamond, which had considerable growth on them, were ground to expose the edges of the buried graphitic region. Fig. 2 shows a side (cross-section) of the ground diamond with buried graphite layer and homoepitaxial film. Different methods were attempted to lift off the plate, which were (in the order they were employed): (1) heating in a CrO_3/H_2SO_4 solution (failed), (2) heated in flowing O_2 at up to 600°C (failed) [14], and contactless electrochemical etching in DI water (successful). This successful method was developed earlier by Marchywka, et al. [15]. One unique aspect of this etching process is that it occurs in a front which begins from a side of the wafer. When the buried graphitic region is progressively etched, that area no longer appears darkened to the unaided eye. Although not permitted to go to completion, it was clear when much of the original 2mm×2mm diamond had the buried graphitic

300

layer removed by etching. A portion of the layer was then pried off mechanically. The resulting 2mm×0.5mm×17.5μm single crystal diamond plate is shown in Fig. 3 with the RTI logo as a back drop. The transparent nature of the free standing diamond plate is evident.

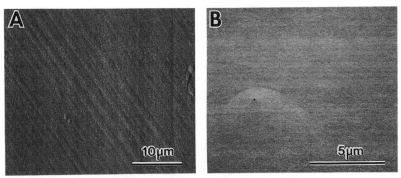

Fig. 1   SEM of homoepitaxial diamond grown on implanted diamond (100) crystal; (A) after 1.6μm growth and (B) after 17.5μm growth.

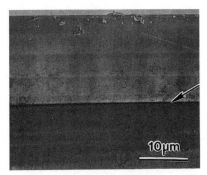

Fig. 2   SEM of ground edge after 17.5μm homoepitaxial diamond growth and 1000°C anneal. The location of the buried graphitic layer is delineated with an arrow.

Fig. 3   Free-standing, synthetic, single crystal diamond plate which appears transparent over the RTI logo after contactless electrochemical etching of the buried graphitic region.

Characterization of Free-Standing Diamond Plate

The top surface of the lifted-off diamond plate was roughened; this had been earlier seen to be caused by the exposure of the diamond to flowing O_2 when heated. Additionally, the underside of the plate had some small pits. MicroRaman spectroscopy was performed on the region of the original diamond substrate on which implantation/growth/lift-off had occurred and on the lifted-off diamond plate. The spectra are shown in Fig. 4. Both had identical diamond LO phonon peak positions at 1332 cm^{-1} with full-width at half-maximum (FWHM) of 3.0 cm^{-1}. The background of the spectrum from the substrate is featureless, but the spectrum from the lifted-off plate shows a broadened peak at 2076 cm^{-1}. There is also broadband luminescence observed at increasing wave number. The 2076 cm^{-1} feature appears to be a luminescence peak (λ = 576nm; E = 2.153eV) that has been observed previously in nitrogen-containing CVD polycrystalline diamond using optical and electron excitation [16, 17]. It should be mentioned that no attempt was made to specifically remove or reduce the concentration of nitrogen in the liquid sources used for diamond growth in this study. Also, the first anneal in flowing N_2 may have contributed to the presence of this Raman feature. This point requires further investigation.

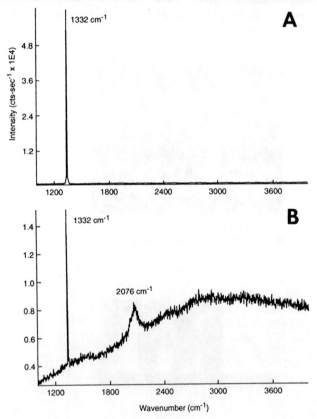

Fig. 4 MicroRaman spectra from; (A) natural type Ia substrate in region from which diamond plate was removed and (B) the lifted-off diamond plate as seen in Fig. 3.

302

The lifted-off diamond plate was then subjected to an oxidizing flame to assess the difference in defect densities by etch pit formation. This characterization technique has been used previously to delineate defects in natural single crystal diamonds, CVD polycrystalline diamond films [18], and homoepitaxial diamond. Fig. 5 shows an etch pit density (EPD) which is approximately 3×10^6 cm^{-2}. This is quite comparable to the EPD observed in other homoepitaxial diamond films grown in this laboratory [13].

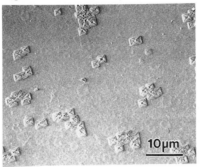

Fig. 5   SEM showing etch pit density on the top surface of the lifted-off diamond plate after 2 sec. exposure to an oxidizing flame. Density is ~ 3×10^6 cm^{-2}.

DISCUSSION AND SUMMARY

The separation (or lift off) of a single crystal homoepitaxial diamond film has been demonstrated at dimensions of ~ 2mm×0.5mm×17.5μm. This diamond plate could be handled and is transparent to the unaided eye. Both the surface topography and the defect density appear not to be compromised by the ion implantation required for lift-off. The ion implantation of carbon at 190keV is straight forward and can be done in a commercially-available, easily-accessible ion implanter. This multi-step process, as presented above, uses a relatively low temperature homoepitaxial diamond deposition process, but this may not be required. The water/alcohol growth chemistry is clearly compatible with good homoepitaxial diamond growth on ion implanted surfaces. The apparent presence of nitrogen in the lifted-off layer remains a concern, but the cause of this can be isolated and eliminated if it proves problematic. The future challenge of this technology is to ensure that the diamond lift-off process can be adequately scaled to larger area diamonds. If so, and when combined with a diamond boule technology [10], a synthetic diamond single crystal wafer fabrication process can be envisaged.

ACKNOWLEDGMENTS

We wish to thank M. Marchywka (Naval Research Laboratory) and R. Kessler (Univ. of North Carolina - Chapel Hill) for their scientific expertise and helpful discussions. The authors gratefully acknowledge support of this work by BMDO/IST through the Office of Naval Research (Contract No. N00014-92-C-0081).

REFERENCES

1. S. Koizumi, T. Murakami, T. Inuzuka, and K. Suzuki, Appl. Phys. Lett. **57**, 56 (1990).
2. L. Wang, P. Pirouz, A. Argoitia, J.S. Ma, and J.C. Angus, Appl. Phys. Lett. **63**, 1336 (1993).
3. A. Argoitia, J.C. Angus, L. Wang, X.I. Ning, and P. Pirouz, J. Appl. Phys. **73**, 4305 (1993).
4. W. Zhu, P.C. Yang, and J.T. Glass, Appl. Phys. Lett. **63**, 1640 (1993.
5. B.R. Stoner and J.T. Glass, Appl. Phys. Lett. **60**, 698 (1992).
6. X. Jiang, C.-P. Klages, R. Zachai, M. Hartweg, and H.-J. Fusser, Appl. Phys. Lett. **62**, 3438 (1993).
7. R.A. Rudder, J.B. Posthill, G.C. Hudson, D.P. Malta, R.E. Thomas, R.J. Markunas, T.P. Humphreys, and R.J. Nemanich, Proc. 2nd Intl. Conf. New Diamond Science and Technology, 1991 MRS Int. Conf. Proc., 425 (1991).
8. M.W. Geis, H.I. Smith, A. Argoita, J. Angus, G.-H.M. Ma, J.T. Glass, J. Butler, C.J. Robinson, and R. Pryor, Appl. Phys. Lett. **58**, 2485 (1991).
9. G. Janssen and L.J. Giling, Diamond and Related Materials, Diamond Films '94, (Il Ciocco, Italy, 1994) in press.
10. J.B. Posthill, D.P. Malta, G.C. Hudson, R.E. Thomas, T.P. Humphreys, R.C. Hendry, R.A. Rudder, and R.J. Markunas, submitted for publication - in review (1995).
11. Vendor for polished and oriented diamond single crystals: Harris Diamond Corp., Mount Arlington, New Jersey, USA; distributor for: Drukker International, Cuijik, The Netherlands.
12. R.A. Rudder, G.C. Hudson, J.B. Posthill, R.E. Thomas, R.C. Hendry, D.P. Malta, R.J. Markunas, T.P. Humphreys, and R.J. Nemanich, Appl. Phys. Lett. **60**, 329 (1992).
13. J.B. Posthill, T. George, D.P. Malta, T.P. Humphreys, R.A. Rudder, G.C. Hudson, R.E. Thomas, and R.J. Markunas, Proc. 51st Ann. Meeting Microsc. Soc. of America, edited by G.W. Bailey and C.L. Rieder (San Francisco Press, 1993) 1196.
14. N.R. Parikh, J.D. Hunn, E. McGucken, M.L. Swanson, C.W. White, R.A. Rudder, D.P. Malta, J.B. Posthill, and R.J. Markunas, Appl. Phys. Lett. **61**, 3124 (1992).
15. M. Marchywka, P.E. Pehrsson, D.J. Vestyck, Jr., and D. Moses, Appl. Phys. Lett. **63**, 3521 (1993).
16. L. Bergman, M.T. McClure, J.T. Glass, and R.J. Nemanich, J. Appl. Phys. **76**, 3020 (1994).
17. R.J. Graham and K.V. Ravi, Appl. Phys. Lett. **60**, 1310 (1992).
18. D.P. Malta, J.B. Posthill, R.A. Rudder, G.C. Hudson, and R.J. Markunas, J. Mater. Res. **8**, 1217 (1993).

PREPARATION AND CHARACTERIZATION OF CHROMIUM CONTAINING AMORPHOUS HYDROGENATED CARBON FILMS (a-C:H/Cr)

R. GAMPP, P. GANTENBEIN, AND P. OELHAFEN
University of Basel, Institute of Physics, Klingelbergstrasse 82, CH-4056 Basel, Switzerland

ABSTRACT

Chromium containing amorphous hydrogenated carbon films (a-C:H/Cr) were prepared in a process that combines rf plasma activated chemical vapor deposition of methane and magnetron sputtering of a chromium target. During the deposition the silicon substrates were kept at 200°C and dc biased at -200 V in order to obtain films with high chemical stability which is required for the application as solar selective surfaces. The films with different Cr concentrations (5 to 49 at.%) were characterized by in situ x-ray photoelectron spectroscopy (XPS). Up to 40 at.%, chromium proves to be built into the cermet-like films in the form of chromium carbide clusters. Above 40 at.%, chromium is partly metallic. A modification of the a-C:H matrix in the vicinity of the chromium carbide clusters has been observed.

INTRODUCTION

Metal-containing amorphous hydrogenated carbon films (a-C:H/Me) can be prepared in a process combining plasma activated chemical vapour deposition of a hydrocarbon monomer (e.g. methane or acetylene) and sputtering of a metal target[1]. It has been found that such films consist of an amorphous hydrogenated carbon matrix and metallic or carbidic inclusions[1-10]. This material has a high potential for the application in solar selective surfaces because of the possibility to vary the complex refractive index of a coating from pure metallic to dielectric values by changing the composition. Such absorber coatings did reveal excellent optical performance in evacuated tubular collectors[11-13]. However, the utilization of these coatings (typically prepared by deposition on grounded and unheated substrates) in flat plate collectors is not possible because of the poor thermal stability in air[9,14]. Recently, we succeeded in preparing more stable a-C:H/Cr films by heating and dc biasing the substrates. In this study an in situ x-ray photoelectron spectroscopy (XPS) investigation of these a-C:H/Cr films is presented.

EXPERIMENTAL

The experimental set up for the deposition of the a-C:H/Cr films is schematically shown in Fig. 1. A rf plasma was located between two parallel electrodes with a distance of 75 mm apart from each other. One electrode was a water-cooled magnetron on which the chromium target (90 mm diameter) was mounted. The other was the substrate holder. A dc power supply, shielded from rf signals, was used to set a dc bias voltage of -200 V to the holder. Si (100) wafers with a specific room temperature resistivity of between 12 and 20 $\Omega\cdot$cm have been chosen as substrates. They are sufficiently conductive to transmit the dc voltage to the growing film and to avoid charging during the electron spectroscopy measurements. On the other hand they are transparent in the near infrared region which allows transmission measurements. The films for the optical analysis were deposited on both sides polished wafers (40x40 mm^2) and those for the photoelectron spectroscopy measurements on one side polished wafers (10x15 mm^2). During the deposition the temperature of the substrate was kept at 200°C with resistive heaters and thermocouple readings. The flow of the process gases Ar (40 sccm) and CH$_4$ (0.8$\cdots$4.0 sccm) was maintained by controllers establishing a working pressure of about 2.8 × 10^{-2} hPa. The rf power (13.56 MHz, 200 W) was capacitively coupled to the magnetron yielding a typical dc self bias voltage of -435 V.

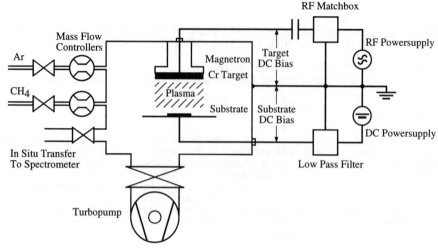

Fig. 1: Experimental set up for the dc biased rf sputtering deposition.

A pure a-C:H film has been deposited for comparison by replacing the Cr target by a copper electrode. This electrode was completely coated with hydrocarbons in a pure methane plasma at a pressure of about 8.0×10^{-3} hPa and a rf power of 250 W. Therefore, no copper was sputtered off the target and a pure a-C:H film was deposited on the substrate. The temperature and bias voltage of the substrate were also kept at 200°C and 200 V, respectively.

In all cases above, the chamber was pumped down to a pressure of about 1×10^{-6} hPa before the deposition.

The samples for the photoelectron spectroscopic measurements were *in situ* transferred to the spectrometer (Leybold EA10/100) after the deposition. X-ray photoelectron spectra were recorded with Mg Kα radiation (hv = 1253.6 eV) and an energy resolution of about 0.9 eV. The binding energy scale was calibrated by setting the binding energy of the Au 4f7/2 core line of a gold reference sample at $E_B = 84.0$ eV. The thickness of the samples was in the range of 5 to 20 nm. No surface charging effects were observed.

XPS spectra of a polycrystalline chromium reference sample were also recorded. The sample was sputter-cleaned with Ar ion bombardment before the measurement. The residual contaminations of oxygen and carbon were below 3 at.% and 1 at.%, respectively.

The base pressure of the measuring chamber during the data acquisition was about 1×10^{-9} hPa.

RESULTS AND DISCUSSION

Determination of Cr Concentration and Deposition Rate

The Cr, C, O and Ar concentration of the films were determined by XPS. The oxygen contamination was found to be below 0.3 at.% and the Ar content between about 0.2 and 0.5 at.% in all films. The Cr concentration as a function of the methane flow is shown in Fig. 2. As expected, more and more hydrocarbons are incorporated into the films by increasing the flow of methane and therefore, the relative Cr concentration decreases. In Fig. 3 the deposition rate is drawn as a function of the methane flow. The rate increases with increasing methane flow again due to the higher hydrocarbon incorporation. The deposition rate was determined from the optical analysis of a-C:H/Cr films which will be discussed in detail elsewhere[15].

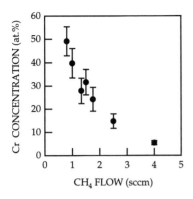
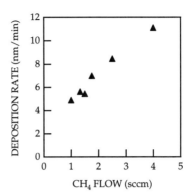

Fig. 2: Chromium concentration as a
function of methane flow.

Fig. 3: Deposition rate versus methane flow.

Compared to a-C:H/Cr films deposited under similar conditions but unheated substrates[9] the concentrations of carbon and argon are significantly lower. Obviously, the higher desorption and higher mobility of loosely bound atoms and molecules at higher substrate temperature lead to a reduction of hydrocarbon and argon.

Chemical Analysis

The evaluation of the core level spectra provides information about the chemical interaction of carbon and chromium in the a-C:H/Cr films. Fig. 4 displays the C 1s and Cr 2p3/2 core level spectra of a-C:H/Cr films with different Cr contents. For comparison the spectra recorded from the a-C:H film and the polycrystalline Cr sample are also plotted in this figure.

The C1s spectrum of the a-C:H film exhibits a single line at a binding energy of 284.9 eV, characteristic of hydrogenated carbon films[16]. With rising Cr content an additional carbon state at a binding energy of 282.8 eV becomes apparent. This binding energy is consistent with carbon in a carbidic bonding state with chromium[17-19]. The formation of carbidic bonds is further evidenced in the Cr 2p3/2 core level spectra in Fig. 4b). Compared to metallic chromium the Cr 2p3/2 core level spectra of the a-C:H/Cr films are significantly broader and shifted to higher binding energy. The inset of this figure shows the change of the Cr 2p3/2 binding energy as a function of the Cr content. Within the range of 20 to 40 at.% the Cr 2p3/2 binding energy is almost constant at 574.7 eV, which we assign to carbidic chromium. Below 20 at.% the peaks are located at higher binding energies. This phenomenon is explained by size effects. Metal containing a-C:H films consist of metallic or carbidic clusters which are embedded in an amorphous hydrogenated carbon matrix[1-10]. With decreasing metal content the cluster size reduces down to the nanometer range[4, 7, 8], resulting in an increase of the binding energy of the metal core levels[2, 7, 9, 10]. Above 40 at.% the Cr 2p3/2 binding energy approaches towards the position of pure Cr, indicating partially metallic chromium. Similar results have been obtained in the case of tantalum and titanium containing a-C:H[7].

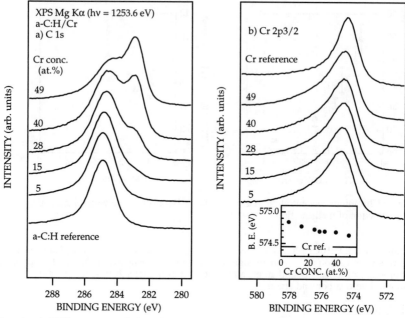

Fig. 4: a) XPS spectra of the C 1s core level of a-C:H/Cr films with different Cr concentrations and a pure a-C:H film. b) Cr 2p3/2 line of a Cr sample and the a-C:H/Cr films as in a). The inset of the figure shows the Cr 2p3/2 binding energy as a function of the chromium content. The horizontal line indicates the binding energy of the Cr reference sample.

In order to obtain a more detailed analysis of the chemical bonding of carbon in the a-C:H/Cr films a deconvolution of the C 1s spectra has been done. The numerical fit of the C 1s spectrum of the a-C:H/Cr film with 40 at.% Cr is displayed in Fig. 5. This spectrum is the same as that in Fig. 4a) but drawn within a wider binding energy range. Asymmetric Doniach-Sunjic curves have been used to describe the line shapes of the core levels[20]. Inelastically scattered photoelectrons were taken into account by a Shirley background[21] and two lorentzian-like plasmons. The plasmon located at a binding energy of about 294 eV can be assigned to chromium carbide. A plasmon energy loss of about 11 eV relative to the energy of the carbidic C 1s state has been confirmed by electron energy loss spectroscopy. The other plasmon at a binding energy of about 310 eV corresponds to the $\sigma + \pi$ plasmon loss of a-C:H. In addition, the Ar 2s line at a binding energy of about 320 eV was taken into account.

The binding energy and the full width at half maximum (FWHM) of the two carbon states determined from the numerical fits are drawn as a function of the Cr concentration in Fig. 6. At low Cr content the contribution from carbidic carbon states is too small to be fitted. The binding energy and the FWHM of the carbidic C 1s state remain constant at about 282.8 eV and 1.2 eV, respectively. In contrast, the binding energy of the carbon state in the a-C:H matrix decreases and the peak width increases with increasing Cr concentration. This behavior can be explained by a modification of the chemical environment of the carbon atoms in the a-C:H matrix. With increasing Cr content, more and more C atoms in the matrix come near the chromium carbide clusters and therefore have a smaller binding energy. Raman spectroscopy measurements are currently being performed in order to answer the question which structural modifications of the a-C:H matrix result from the change in the chemical environment.

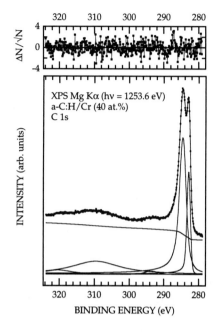

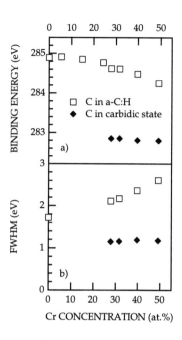

Fig. 5: Deconvolution of the electron distribution curve of the C 1s spectrum of the a-C:H/Cr film with 40 at.% Cr. Three peaks are used to describe carbon in carbidic state, carbon in the a-C:H matrix and the Ar 2s core line. The inelastically scattered electrons are modeled with two plasmon losses and a Shirley background.

Fig. 6: a) Binding energy and b) FWHM of the two carbon 1s lines as a function of the chromium concentration.

CONCLUSION

The importance of the substrate bias for the film properties has been well recognized for pure metal[22] and pure a-C:H films[23, 24]. However, few work has been done to characterize the influence of this parameter onto the properties of metal-containing a-C:H films. In this study we presented the experimental set-up to deposit a-C:H/Cr films on negatively biased substrates and the results of an *in situ* photoelectron spectroscopy characterization. Up to 40 at.%, chromium proves to be built into the cermet-like films in the form of chromium carbide clusters. Above 40 at.%, chromium is partly metallic. A chemical modification of the a-C:H matrix in the vicinity of the chromium carbide clusters has been observed. A realistic model of the structure of a-C:H/Cr films should include these different phases. A simple two phase model, which is often used to describe the structure of cermets, is not sufficient.

ACKNOWLEDGMENTS

Financial Support from the Swiss Bundesamt für Energiewirtschaft under grant nr. EF-REN(92)075 is gratefully acknowledged. We like to thank Dr. H.-G. Boyen for his software engineering and Dr. Xianyue Xie for helpful discussions and her critical reading of the manuscript.

BIBLIOGRAPHY

1. C.P. Klages and R. Memming, in Properties and characterization of amorphous carbon films, edited by J.J. Pouch, S.A. Alterovitz, (Trans Tech Publications, Zürich, 1989) p. 609.
2. H. Koeberle, M. Grischke, F. Thieme, and C. Benndorf, Surf. Coat. Tech. **39/40**, 275 (1989).
3. C. Benndorf et al., Synth. Met. **41-43**, 4055 (1991).
4. C. Benndorf, M. Fryda, C.-P. Klages, K. Taube, and H.-G. Haubold, Mater. Sci. Eng. A **140**, 795 (1991).
5. M.C. Sainte-Catherine and G. Farges, Surf. Coat. Technol. **54/55**, 266 (1992).
6. M. Wang, K. Schmidt, K. Reichelt, H. Dimigen, and H. Hübsch, J. Mater. Res. **7**, 667 (1992).
7. M. Grischke, Fortschr.-Ber. VDI Reihe 5 Vol. **179** (VDI-Verlag GmbH, Düsseldorf, 1989).
8. M. Fryda, Fortschr.-Ber. VDI Reihe 5 Vol. **303** (VDI-Verlag GmbH, Düsseldorf, 1993).
9. R. Gampp et al. in Optical Materials Technology for Energy Efficiency and Solar Energy Conversion XIII, edited by V. Wittwer, C.G. Granqvist, C.M. Lampert, (SPIE **2255**, Freiburg, FRG, 1994) p. 92.
10. P. Oelhafen, P. Gantenbein, and R. Gampp in Optical Materials Technology for Energy Efficiency and Solar Energy Conversion XIII, edited by V. Wittwer, C.G. Granqvist, C.M. Lampert, (SPIE **2255**, Freiburg, FRG, 1994) p. 64.
11. G.L. Harding and B. Window, J. Vac. Sci. Technol. **16**, 2101 (1979).
12. G.L. Harding, B. Window, D.R. McKenzie, A.R. Collins, and C.M. Horwitz, J. Vac. Sci. Technol. **16**, 2105 (1979).
13. D.R. McKenzie, B. Window, G.L. Harding, A.R. Collins, and D.W.J. Mackey, J. Vac. Sci. Technol. **19**, 93 (1981).
14. G.L. Harding, J. Vac. Sci. Technol. **13**, 1070 (1976).
15. R. Joerger and R. Gampp, to be published.
16. P. Oelhafen, D. Ugolini, S. Schelz, and J. Eitle, in Diamond and Diamond-like films and coatings, edited by R.E. Clausing et al., (Plenum Press, New York, 1991) p. 377.
17. F. Schuster, F. Maury, and J.F. Nowak, Surface and Coatings Technol. **43/44**, 185 (1990).
18. F. Maury and F. Ossala, Thin Solid Films **207**, 82 (1992).
19. F. Maury and F. Ossala, Thin Solid Films **219**, 24 (1992).
20. S. Doniach and M. Sunjic, J. Phys. C **3**, 385 (1970).
21. D.A. Shirley, Phys. Rev. B **5**, 4709 (1972).
22. O. Christensen, Solid State Technol. **13**, 39 (1970).
23. P. Koidl, C. Wild, B. Dischler, J. Wagner, and M. Ramsteiner, in Properties and characterization of amorphous carbon films, edited by J.J. Pouch, S.A. Alterovitz, (Trans Tech Publications, Zürich, 1989) p. 41.
24. D. Ugolini, J. Eitle, and P. Oelhafen, Vacuum **41**, 1374 (1991).

DEPOSITION OF ALUMINUM OXYNITRIDE FILMS BY MAGNETRON SPUTTERING: EFFECT OF BOMBARDMENT AND SUBSTRATE HEATING ON STRUCTURAL AND MECHANICAL PROPERTIES

RUSSELL V. SMILGYS*, ERIC TAKAMURA**, IRWIN L. SINGER+, STEVEN W. ROBEY++, AND DOUGLAS A. KIRKPATRICK*
*Science Applications International Corp., 1710 Goodridge Dr., McLean, VA 22102
**GeoCenters, 10903 Indian Head Highway, Fort Washington, MD 20744
+Naval Research Laboratory, Washington, D.C. 20375
++National Institute of Standards and Technology, Gaithersburg, MD 20899

ABSTRACT

Aluminum oxynitride films, 1μm thick, are deposited onto glass substrates by planar magnetron sputtering from an alumina target in a mixture of nitrogen and argon. One set of films is deposited onto glass substrates that are heat sunk to a holder, whose temperature is held below 100°C. A second set of films is deposited onto glass substrates that are mechanically clamped to a holder, whose temperature is allowed to rise up to 250°C. Characterization by continuous indentation testing, secondary electron microscopy, and x-ray diffraction reveals significant differences in mechanical properties and surface structure between the two sets of films. Films deposited with holder cooling have a smooth surface and no evidence of crystallinity; films deposited without holder cooling have etch pits on their surface that vary with position across the substrate. The later films show crystallinity and have twice the hardness and a 60% greater elastic modulus.

INTRODUCTION

Alumina is a hard, inert, transparent ceramic. These properties are often desirable in a protective coating. Alumina films with bulk-like properties can be deposited by chemical vapor deposition (CVD) but high (>800°C) deposition temperatures are required [1]. Alumina films can also be produced at lower temperatures by physical vapor deposition (PVD) techniques such as sputtering. In general the properties of alumina films deposited by PVD techniques are inferior [1]. Many properties of films, such as hardness, adhesion to substrate and crystallinity depend critically on deposition conditions. One of the most difficult conditions to control is substrate temperature, especially when the substrate is a relatively poor conductor, such as glass.

In this study we examine the mechanical and structural properties of sputter-deposited aluminum oxynitride films subjected to two commonly used methods of thermal management: simple clamping of substrates to an uncooled holder and pasting down substrates to an intentionally cooled holder. Indentation testing is used to determine hardness, elastic modulus (defined as $E/(1-\upsilon^2)$, where E is Young's modulus and υ is Poisson's ratio) and indentation fracture toughness. Secondary electron microscopy (SEM) and x-ray diffraction (XRD) are used to characterize film morphology and crystallinity. Changes in the properties of the films are discussed in terms of bombardment and thermal effects.

FILM PREPARATION AND CHARACTERIZATION

Films are prepared by the technique of planar magnetron sputtering. The sputtering system is built around a stainless steel chamber approximately 15 liters in volume pumped by a 50 l/s turbomolecular pump backed by a mechanical pump. It achieves a base pressure of 4×10^{-5} Pa (3×10^{-7} torr). Nitrogen and argon gas are introduced into the chamber through leak valves. The gas pressure is monitored by either a pirani gauge or a capacitance manometer.

Mat. Res. Soc. Symp. Proc. Vol. 388 © 1995 Materials Research Society

A 2" diameter planar magnetron source is mounted in the chamber with the target horizontal. The magnetron is driven by a 500 W RF power supply (13.56 MHz) with a matching network. The target is alumina of purity 99.99%. Substrates are held horizontally facing the magnetron target at a distance about 4 cm. When the goal is to maintain a low substrate temperature, the substrate is mounted with silver paste to an aluminum block. The block is clamped with silver paste to a commercial cold finger (15 cm long, 2.5 cm in diameter). When compressed air is forced into the cold finger during deposition, the block temperature does not rise above 100°C. If neither silver paste nor compressed air are used, the block temperature rises to nearly 250°C over the course of an hour long deposition. These two conditions will be referred to as cool and hot deposition. The holder is electrically grounded.

A typical deposition begins by mounting a glass substrate (Erie Scientific microscope slide) 2-6 cm^2 in size on the holder. The glass is cleaned by rinsing in high purity alcohol. After loading, the chamber is pumped down below 1×10^{-2} Pa (8×10^{-5} torr) before beginning deposition. Substrates are not sputter cleaned prior to deposition. First nitrogen is leaked into the chamber to reach a pressure from 3.3×10^{-2} to 0.13 Pa (0.25-1 millitorr). Then argon is leaked to bring the total pressure up to 0.67 Pa (5 millitorr). In this way the percentage of nitrogen in the gas mixture is set from 5 to 20%. The RF generator is set to maintain a constant power, typically 400 W. Sputtering for one hour deposits a film about 1 μm thick. The film composition is not determined, but we refer to the films as being aluminum oxynitrides.

A SEM is used to view the films. Films are analyzed by XRD in the Seeman-Bohlin geometry.

Hardness and elastic modulus are measured by continuous microindentation [3] with a Fischerscope H100 using a Vickers indentor [4]. The shape function of the indentor (the area as a function of depth) was established by performing test indents on Si(111). In order to minimize the effect of the substrate on film tests, plastic indents are kept below 20% of the film thickness. In this study we report two hardness numbers. The first is the universal hardness (Hu), defined as the load divided by area of contact; Hu incorporates both elastic and plastic resistance to deformation. The second is the more traditional plastic hardness (Hp), which here is calculated as the load over the area of contact subtracting the elastic contributions to the area of contact. Elastic modulus is derived from the unloading portion of the load-displacement curves.

Another mechanical test based on indentation measures the ability of a film to resist fracture. Vickers indentation is performed with incrementally increasing maximum loads until radial cracks appear at the corners of the indent. This load is termed the critical load (Lc).

RESULTS

Surface morphology, structure, and mechanical properties are found to depend on substrate temperature. Films deposited cool appear smooth and uniform, whereas those films deposited hot have a roughly textured appearance. By eye the morphology manifests itself as a diffuse white ring approximately 2 mm in width and up to 2 cm in diameter. The center coincides with the axis of the magnetron. The remainder of the film, inside and outside of the ring, is clear and transparent.

SEM images show that films with a ring have a pitted surface morphology on a microscopic scale. Figure 1a-c shows that inside the ring the surface is completely peened, but with increasing distance from the center of the ring the density of pits drops. The ring seems to be an optical effect caused by light scattering from pitting of a certain density (see Figure 1b).

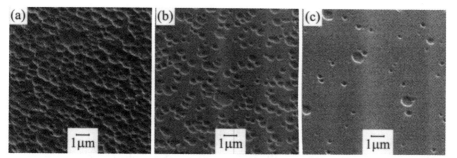

Figure 1. SEM images of surface of aluminum oxynitride film deposited hot. The density of pits varies with position across the film (a) position inside ring visible by eye, (b) position on ring, (c) position outside ring.

Figure 2 compares XRD spectra of representative films deposited cool and hot. Only the film deposited hot has features that indicate crystallinity. Line broadening prevents identification of the phase(s) present. Several aluminum oxide and aluminum nitride polymorphs are consistent with the features. Based on the width of the features we can conclude that either the crystallites are as small as 10 nm and/or are highly stressed.

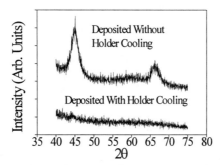

Figure 2. X-ray diffraction spectra shows crystallinity for film deposited without holder cooling.

Films deposited hot have mechanical properties that vary with position across the ring. Figure 3a shows that the hardness of the film increases as one moves closer to the center of the ring. The location and width of the ring is represented by gray bars at the bottom of the plot. Each data point represents the average of four continuous indentations, made in a row along a ring diameter; therefore the ring is crossed twice. The data points at the two ends of each plot correspond to regions of bare glass where clamps used for mounting the substrate mask the film. The same trends in film mechanical properties are observed in a row away from the clamps.

Figure 3b shows that the elastic modulus has less variation along the same row. We find that hardness and elastic modulus values across the ring have greater dispersion than outside the ring. This may be caused by the rough surface texture inside the ring. Films deposited cool have constant hardness and modulus values at all positions. They are also much softer and less stiff .

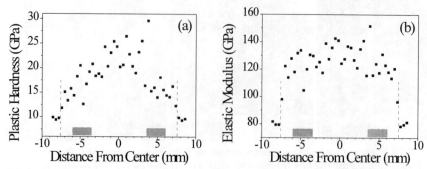

Figure 3. Microindentation measurements show that plastic hardness and elastic modulus are position dependent for films deposited without holder cooling. Measurements are made along a diameter of the ring. The position and width of the ring is identified by the gray bars at the bottom of each plot. The measurements farthest from the ring center correspond to masked regions of the glass substrate.

Films deposited hot also have a position dependent critical load. The critical load within the ring is above 1400g. Outside the ring the critical load drops to below 600g. Films deposited cool have critical loads below 200g. Table 1 compares Hp, Hu, E, and Lc for films, deposited cool and hot, with Si(111), glass and sapphire.

Table I. Mechanical properties of aluminum oxynitride films deposited cool and hot are compared with reference materials. 25 mN load used for hardness measurements.

	Hp (GPa)	Hu (GPa)	E (GPa)	Lc (gf)	% N*
Si(111)	19	7.8	190	<100	-
Sapphire	39	15	400	<100	-
Glass	9.2	3.7	83	200	-
Hot deposition#					
AR11-01	38	8.4	180	2500	5
AR11-05	30	7.4	160	1700	15
AR20-11	21	6.2	130	1400	10
AR20-13	24	6.2	140	na	10
AR20-08	19	5.5	110	1500	10
Cool deposition					
AR02-19	9	3.9	85	200	20
AR02-17	8.6	3.8	86	<100	20
AR02-21	8.9	4	89	<100	20
AR02-22	9.7	4.3	95	<100	20
AR02-20	8.5	3.8	87	<100	10
AR02-18	10	4	88	100	10

*Percent nitrogen in gas mixture during sputter deposition
#Measurements performed within ring

DISCUSSION

In this study we find that substrate temperature strongly influences the properties of aluminum oxynitride films deposited by sputtering. Films deposited cool are smooth with no evidence of crystallinity, whereas those deposited hot are pitted and have some crystallinity.

The development of surface morphology on films deposited by sputtering is well established in the literature. Depending on gas pressure, target to substrate distance, power, and total deposition time, surface features such as ripples, rods, cones, pyramids, and pits have been documented [5]. Energetic bombardment by ions and neutral particles activates the surface and causes resputtering. Etch pits, similar to those seen in Figure 1, are often initiated by preferential sputtering at dislocations or other heterogeneous features on an otherwise homogeneous material surface. Other researchers have also noted the development of a ring on films deposited by planar magnetron sputtering. Tominaga *et al.* observed a ring of discoloration concentric with the magnetron axis on their AlN films [6]. Based on time-of-flight measurements the authors identified bombardment by NO⁻ and O⁻ with development of the ring.

Films deposited either cool or hot are presumably subjected to the same bombardment conditions, yet a pitted surface morpology only forms when films are deposited hot. This suggests that pit formation is influenced by the substrate temperature. The connection may be crystallization that occurs at a higher temperature. Figure 2 shows that only the films deposited hot are crystalline. Bombardment can introduce defects into a crystalline film, and continued bombardment can lead to etch pits at the defects by preferential sputtering [7]. Pitting, similar to that shown in Figure 1, can be seen in ion bombarded natural alumina [8]. In contrast, the entire surface of a growing amorphous film is defective; therefore no selective etching can occur under the same bombardment conditions.

Films deposited at above 1/3 of the melting temperature are expected to be crystalline. Franze *et al.* reported that alumina films deposited by ion beam assisted deposition on glass show some crystallinity at 400°C [9]. This temperature is within the range possible for the unpasted substrate on an uncooled holder.

The mechanical properties can also be explained in terms of crystallinity. Studies of alumina films by continuous indentation testing indicate that the plastic hardness of amorphous films is about half that of crystalline films (about 7 to 10 GPa and 20 GPa, respectively [1]); elastic modulus values of amorphous films are from half to a third of crystalline films (140-250 GPa and 400 GPa, respectively [1,10]). While the absolute values for hardness of films studied in this experiment are somewhat higher, the ratios are consistent with literature values. The Vickers indentor and the machine used in this study were not designed to measure mechanical properties accurately at depths of 0.1μm; moreover, no special precautions were taken to minimize vibrations and thermal fluctuations during data acquisition. The increase in critical load can be attributed to a combination of increased compressive stress applied to the surface and improved adhesion of the film at higher temperatures. The thermal coefficient of alumina is greater than that of glass. Therefore, on cooling the film applies a compressive stress to the glass surface. The stress is proportional to the temperature drop and higher deposition temperatures apply a higher stress. Stress may also be introduced by energetic bombardment during deposition.

In summary, we find that the structural and mechanical properties of aluminum oxynitride films depend on the thermal behavior of glass substrates. We record the highest hardness and elastic modulus where resputtering, and by extension bombardment and substrate heating, is most intense.

ACKNOWLEDGMENTS

The authors are grateful to Dr. L. Seitzman for discussions and x-ray diffraction spectra, and to Mr. R. Oprison and Mr. T. Long for technical support. This work was supported by the Advanced Research Projects Agency under Contract MDA972-93-C-0022.

REFERENCES

1. J. -E. Sundgren and H. T. G. Hentzell, J. Vac. Sci. Technol. A **4**, 2259 (1986).
2. B. Lawn, Fracture of Brittle Solids (Cambridge University Press, Cambridge 1993) Chap 8.
3. W. C. Oliver and G. M. Pharr, J. Mat. Res. **7**, 1564 (1992).
4. W. W. Weiler, Amer. Soc. for Testing and Eval. **18**, 229 (1990).
5. D. J. Kester and R. Messier, J. Mater. Res. **8**, 1938 (1993).
6. K. Tominaga, S. Iwamura, Y. Shintani, and O. Tada, Jpn. J. Appl. Phys. 22, 418 (1983).
7. O. Auciello, and R. Kelly, eds. Ion Beam Bombardment Modification of Surfaces (Elevier, New York, 1984).
8. B. A. Banks, NASA Tech. Memo. #81721 (NASA, 1981).
9. W. Franze, T. Tetreault, W. Kosik, W. Croft, and J. K. Hirvonen, Mat. Res. Soc. Proc. **279**, 825 (1993).
10. D. Dongfeng and K. Kato, Thin Solid Films **245**, 104 (1994).

HIGH-RATE DEPOSITION OF THIN FILMS BY HIGH-INTENSITY PULSED ION BEAM EVAPORATION

A.N.ZAKOUTAYEV, G.E.REMNEV, YU.F.IVANOV, M.S.ARTEYEV, V.M.MATVIENKO, AND A.V.POTYOMKIN
Nuclear Physics Institute, Tomsk Polytechnic University, Tomsk 634050, Russia

ABSTRACT

A high-intensity ion beam (500 keV, current density 60 - 200 A/cm^2, power density $(0.25 - 1) \cdot 10^8$ W/cm^2, pulse duration 60 ns, pulse repetition rate 4-6 min^{-1}) was used to deposit thin metal and carbon films by evaporation of respective targets. The instantaneous deposition rate was 0.6 - 5 mm/s. The films were examined using transmission electron microscopy and transmission electron diffraction. The metal films had a polycrystalline structure with the grains measuring from 20 to 100 nm, the lower the melting point the greater the grain size. The carbon films contained 25 - 125 nm diamonds. The ablation plasma was studied employing methods of pulsed spectroscopy.

INTRODUCTION

In recent years, many research groups have been carrying out investigations using concentrated energy flux having power density of 10^7 W/cm^2 (mainly pulsed laser radiation and high-intensity pulsed ion beams of nanosecond duration) for the deposition on a substrate of thin metallic and multicomponent films [1-3].

The given method is superior to other techniques in that it affords high deposition rates (0.1 - 1.0 cm/s). This is also characterized by high temperatures of ablation plasma (several thousands of degrees) and by deposition time of a few microseconds. An added advantage of the method is the preservation of the stoichiometric composition of the film on a substrate and the formation of films having a meta-stable structure, in particular a diamond-like one [3]. The films are distinguished by high density and excellent quality. This line of investigation has many potential fields of application owing to low specific expenditure of energy required for the production of films, advantageous use of beam energy and of plasma sputtering products because of their narrow angular directivity, use of a high-intensity pulsed ion beam (HIPIB) in a single vacuum cycle for other operations (e.g., cleaning the substrate surface prior to film deposition, pulsed high-temperature annealing of defects at lower current density of HIPIB), the possibility of applying several layers of different materials on one and the same substrate. The process is amenable to automation. The results of investigations employing HIPIB for the deposition of thin films of high-melting metals, i.e. W, Ta, Mo, Nb, as well as of Al and C are presented. Target sputtering is carried out by HIPIB of nanosecond duration which is generated on a Temp-2 accelerator (ion energy 500 keV; current density on the target varying from 60 to 200 A/cm^2; current pulse duration on the target 60 ns; beam composition: carbon ions and protons).

DEPOSITION

The experiments were carried out on a Temp-2 supercurrent ion accelerator [4]. A schematic diagram of the deposition process by HIPIB evaporation is shown in Fig. 1. A magnetically insulated diode served as a HIPIB source. Magnetic insulation was used for electron flux cut-off; this was applied by conducting a current pulse having t/4 of about 20 microseconds through a coil cathode. The anode surface on the side facing the

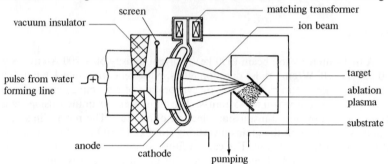

Fig. 1. Schematic diagram of the HIPIB deposition system.

cathode slots had 2-mm thick polyethylene with equally spaced holes of small diameter. When a high-voltage positive pulse was fed to the anode, on the same surface there formed anode plasma from which ion acceleration was carried out. The diode has a high energy efficiency and owing to optimization of its geometry, it also has a long lifetime (it is good for several thousands of pulses); instability of HIPIB current density on the target has been reduced to several percent. The central problem in our case was the sputtered material of the target and of the target positioner getting on to the anode and the vacuum insulator surface. The insulator surface was to a large extent protected by installing a special-purpose screen. Fig. 2 shows typical waveforms of the diode voltage and ion current density. The ion beam current density on the target was varied in the

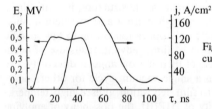

Fig. 2. Typical waveforms of diode voltage and ion current density.

range from 60 to 200 A/cm^2 by shifting the target relative to the focal plane. The local current density of HIPIB current was measured by means of a collimated Faraday cylinder having a transverse magnetic field. This version of ion diode afforded the accelerated ions having energy exceeding 0.5 MeV. To match a low-induction coil cathode and a capacitor bank, a special matching transformer was used. The targets were positioned near the beam focus with its surface aligned 35 to 45° to the beam axis. The substrate was positioned at an angle of 0 to 30° with the target surface and 30 to 60 mm away from the target. The base pressure in the vacuum chamber was about 10^{-4} Torr.

RESULTS AND DISCUSSION

1. Transmission Electron Microscopy and Transmission Electron Diffraction Observations

Transmission electron microscopy (TEM) and transmission electron diffraction (TED) were used to characterize the films. TEM/TED results were obtained on an EM-125 microscope using an electron beam of 125 kV. All the TED data were collected under a selected-area mode with a sampling diameter of about 1 micrometer. The average sizes of the structural components of the films were determined from microphotographs using methods of planimetry. MgO and NaCl substrates were used for the observation since they are convenient for sample preparation and stable to heating and ion bombardment during deposition. TEM/TED samples of the carbon films were prepared by etching a MgO substrate in H_3PO_4 solution.

2. Metal films

Thin W, Ta, Mo, Nb and Al metal films were deposited on freshly chipped (001)NaCl surface at room temperature. All the films obtained were non-porous and had a mirror-smooth surface. Film deposition rate was 0.6 - 4.0 mm/s. The table shows parameters of film thickness deposited with a single pulse. Fig. 3 presents transmission electron images of film structure.

TABLE. Parameters of the film deposition.

Material	Al	Nb	Mo	Ta	W
Deposition rate, nm/pulse	12,9	4,5	7,7	5,1	1,8

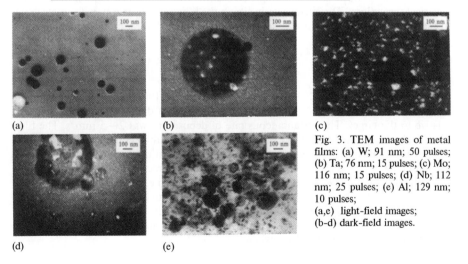

(a)　　　　　(b)　　　　　(c)

(d)　　　　　(e)

Fig. 3. TEM images of metal films: (a) W; 91 nm; 50 pulses; (b) Ta; 76 nm; 15 pulses; (c) Mo; 116 nm; 15 pulses; (d) Nb; 112 nm; 25 pulses; (e) Al; 129 nm; 10 pulses;
(a,e) light-field images;
(b-d) dark-field images.

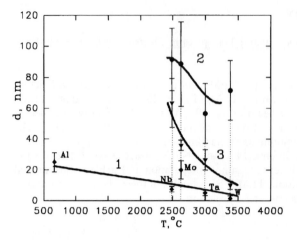

Fig. 4. Melting point dependence of
(1) grain size in films;
(2) size of spherical formations;
(3) size of crystallities in
spherical formations.
Graphs 2 and 3 give values
for W (dense spherical formations)
and for Mo, Nb, Ta (hollow spherical formations).

An inverse dependence is observed between grain size and melting point of metal, i.e., grain size increases from 1.0 - 1.5 nm to 25 nm on going from W to Al (see Fig. 4, Graph I). Film structure varied from quazi-amorphous for W to polycrystalline for Al.

The surface of the W, Ta, Mo and Nb films shows spheroidal formations which constitute up to 20% by volume; their size decreases from 91 to 71 nm on going from Nb to W as the melting point rises (Fig. 4, Graph 2). The same formations occurring in the W films have a dense monocrystalline structure, while among those occurring in the Ta, Mo and Nb films alongside dense ones there are hollow formations very much like blisters having a polycrystalline shell. With decreasing melting point the size of the crystallites in the shells increases from 26 nm for Ta to 64 nm for Nb. Thus, the grain size of the shells exceeds considerably that of the film itself (Fig. 4, Graph 3). This suggests that crystallization in the shells occurs at a different rate as compared to that of the films. The Al films have a homogeneous structure; these show neither blisters nor spherical formations. The average grain size in the film is about 25 nm, with single grains in the surface layers measuring up to 110 nm. The grains have a monocrystalline struc-ture.

3. Carbon Films

The deposition of C films was performed on a (100) MgO substrate at room temperature at a rate of 10 to 15 nm/pulse depending on the position of the substrate relative to the target. The instantaneous rate of deposition was 2 - 5 mm/s. The films were analysed by the TEM/TED method and were found to consist mainly of amor-phous carbon. Besides, crystalline graphite modification in the form of acicular crystals was observed. TED analysis revealed polycrystalline diffraction reflexes that could be assigned as diamond. Dark-field imaging carried out in the diffraction reflexes of dia-mond showed that the diamond crystals varied in size from 25 to 125 nm with a popu-lation of up to 10% in the film. The crystals were spherically-shaped. The average size of the diamond particles was assessed at about 63 nm. More complete discussion on carbon film preparation may be found elsewhere [5].

4. Time-Resolved Spectroscopic Measurements of Ablation Plasma

The ablation plasma used for film deposition was investigated using a self-insulated diode under a two-pulse mode [4]. The beam generated had the following parameters: $E = 220 - 240$ keV; $j = 90 - 120$ A/cm²; $\tau = 50$ ns (Fig. 5), i.e., the peak value of power density on to the target was up to ~$2.8 \cdot 10^7$ W/cm².

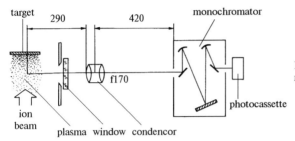

Fig. 5. Scheme of spectroscopic measurements.

The spectroscopic measurements were carried out using the ablation plasma that was generated by ion beam bombardment of a copper target. The scheme of measurements is shown in Fig. 5. The central area of the ablation plasma was examined at various distances from the target, the spatial resolution along the axis of ablation plasma propagation being equal to about 2 mm. The integrated plasma glow in the visible and the UV region of the spectrum vs time was registered using a PEC-22 photocathode. The rate of propagation of the ablation plasma front was determined using a standard time-of-flight procedure; near the target it was equal to $1.8 \cdot 10^6$ cm/s, while at ~4 cm from the target it was equal to $7.2 \cdot 10^5$ cm/s. The total spectra of ablation plasma radiation were obtained on an ISP-30 spectrograph. Analysis of the spectra revealed the occurrence of several tens of Cu I and a few Cu II lines at 0.2 cm from the target. However, at >1 cm from the target a dramatic depletion of the spectrum was observed owing to the disappearance of the Cu II and of the highly excited Cu I transitions lines, while at >2 cm the spectrum showed the Cu I 4s²S - 4p²P resonance transitions lines alone (3273.96 Å and 3247.54 Å). A more sophisticated spectral analysis was performed using a MDR-23 monochromator equipped with a photographic attachment whose performance was comparable to that of a spectrograph having a sufficiently high resolution. The temperature of the plasma was determined for the whole observation region (at up to 5 cm from the target) using measurements of the Doppler broadening of the Cu I 4s²S - 4p²P resonance lines with readsorption taken into account. The concentration of normal atoms was determined from the same lines employing the two-mirror method [6]; the effective reflectance coefficient of the dielectrical mirrors used was determined experimentally using a low-pressure hollow-cathode discharge He lamp. The temperature of electrons was determined using the ratio of the total intensities of the CuI 4p²P - 5d²D (4022.63 Å) and the Cu I 4p²P - 4d²D (5153.0 Å) lines and their concentration from the ratio of the total intensities of the forbidden and the allowed lines, 4p²P - 5F (3686.0 Å) and 4p²P - 6d²D (3687.44 Å), respectively [7]. The latter measurements were made in the region where the respective lines were recorded (at <2 cm from the target). The error of measurement of the unknown quantities was about 40%, the spectroscopic constants required were borrowed from [8,9].

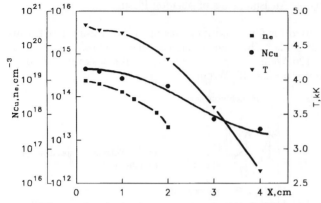

Fig. 6. N_{Cu} atom and (n_e) electron concentration and ablation plasma temperature (T) vs distance from the target

The results obtained suggest that the ablation plasma is virtually at equilibrium, at least it is so near the target (here the Boltzmann and the Saha law are satisfied). The following results were obtained at 0.2 cm from the target: $T_{pl} = T_e = 4800$ K; $N_{Cu} = 1.8 \cdot 10^{19}$ cm^{-3}; $n_e = 2.4 \cdot 10^{14}$ cm^{-3}. The spatial evolution of these parameters are presented in Fig.6.

CONCLUSION

The method of deposition using high-intensity pulsed ion beam evaporation was employed to prepare thin metal (W, Ta, Mo, Nb and Al) and carbon films. The rate of film deposition was 0.6-5 mm/s. TEM and TED analyses were used to investigate the films. The structure of the metallic films ranged from quazi-amorphous to polycrystalline on going from W to Al depending on the melting point of the metal. The deposited carbon films were a mixture of amorphous carbon, crystalline graphite and diamond, the diamond particulates ranging from 25 to 125 nm in size with a population of up to 10% in the film.

Acknowledgment
This work was funded under Los Alamos National Lab Contract No. 1631Q0014-35.

References

[1] Y.Shimotori, M.Yokahama, H.Isobe et al., J. Appl. Phys. **63**, 968 (1988).
[2] G.E.Remnev and V.A.Shulov, Laser and Particle Beams **11**, 707 (1993)
[3] G.P.Johnston, P.Tiwari, D.J.Rej et al., J. Appl. Phys. **76**, 5949 (1994).
[4] I.F.Isakov et al., Vacuum **42**, 159-162 (1991).
[5] D.J.Rej, G.E.Remnev et al., submitted to the 3rd International Conference on the Applications of Diamond Films and Related Materials, 1995.
[6] S.E.Frish in Spectroscopy of Gas-Discharge Plasma (Nauka, Moscow, 1970) p.7.
[7] L.I.Grechikhin et al. in Low-Temperature Plasma (Nauka, Moscow, 1967) p.287.
[8] G.A.Kasabov and V.V.Eliseyev, Spectroscopic Tables for Low-Temperature Plasma (Nauka, Moscow, 1973).
[9] A.A.Radshchen and S.M.Smirnov, Parameters of Atoms and Ions (Nauka, Moscow, 1986).

ELECTRON BEAM ASSISTED CHEMICAL VAPOR DEPOSITION OF GOLD IN AN ENVIRONMENTAL TEM

JOHN KOUVETAKIS, RENU SHARMA AND B. L. RAMAKRISNA
Department of Chemistry and Biochemistry, Arizona State University
Tempe, Arizona 85287-1604

JEFF DRUCKER
Department of Physics, University of Texas at El Paso
El Paso, Texas 79968

PAUL SEIDLER
IBM Research Division, Thomas J. Watson Research Center
Yorktown Heights, New York 10958

ABSTRACT

We demonstrate a novel technique for in situ observation of the chemical vapor deposition of high purity gold using ethyl(trimethylphosphine)gold(I). An environmental transmission electron microscope with 3.8 eV resolution was used to observe and compare the growth of the material with or without electron beam irradiation (120 keV) with Si (100) substrate temperatures ranging from 125-200 °C. Typical precursor pressures of 10^{-4} Torr and E-beam irradiation resulted in rapid growth of virtually continuous gold films. Thermal deposition without the beam resulted in low nucleation densities, low deposition rates, and island-like growth. Images and diffraction patterns acquired during the deposition process indicated polycrystalline gold and elemental analysis at the nanometer scale showed that the films had excellent chemical purity. Atomic force microscopy was also used to investigate the three dimensional morphology of the materials. The most notable result of the deposition process is the dramatic enhancement of the growth rate due to the beam irradiation.

INTRODUCTION

In situ, real time studies of thin film growth by chemical vapor deposition (CVD) are generally hindered by the corrosive chemical environment and the hostile conditions such as high temperatures and pressures employed during this process. In MOCVD (metal-organic CVD) of metals such as Cu, Al, and Au, understanding the initial stages of microstructural evolution on substrate surfaces is important because these elements are widely used in the microelectronics industry for bonding and wiring.[1] High purity films of these metals can be selectively grown by CVD techniques through insulators for multilevel interconnects in a very conformal manner.[2] The role of metal film microstructure on interconnect reliability will become more of an issue as feature sizes shrink. CVD processes are potentially quite attractive for wafer scale metallizations, thus an in situ technique for investigating the early stages of CVD metallization can be highly advantageous in terms of assessing the precursor selection and design as well as determining conditions for optimum film growth.

In this report, we focus on the use of a new technique we recently developed to observe the early growth stages of Au films on silicon substrates in an environmental transmission electron microscope (TEM). A Phillips 400T TEM operating at 120 keV and equipped with a differentially pumped environmental cell was used in this study. The original function of the cell was to study interactions between gases and solid specimens at elevated temperatures and pressures while imaging the reaction process. We have successfully applied this apparatus to

Mat. Res. Soc. Symp. Proc. Vol. 388 © 1995 Materials Research Society

perform chemical vapor deposition experiments with Al utilizing simple chemical precursors as feedstock gases.[3] In a typical deposition experiment, the gaseous molecular species were allowed to react on a resistively heated substrate in the cell chamber while continuously observing the growth of the resulting thin film in real time. The reaction conditions and the cell arrangement approximate a cold wall CVD reactor. We deposited gold from the thermal decomposition (125-200 °C) of ethyl(trimethyphospine)gold(I), a stable low-volatility (vapor pressure 10^{-4} Torr at 22 °C) compound. The thermal decomposition of this precursor over a temperature range of 125-200 °C has been previously reported to deposit high quality gold selectively on atomically clean surfaces.[4]

Energetic beams such as laser beams have been used to deposit gold from organometallic precursors in order to repair open defects on multichip packaging modules. In this process a laser is used as a localized heat source to pyrolize the adsorbed precursor molecule to pure metal, thus achieving selective metallization in small dimensions.[5] We envisioned using the electron beam in the transmission electron microscope to study deposition of material of comparable dimensions and quality. The thermal CVD of gold using this precursor has been extensively characterized by other studies under conditions easily obtained by our apparatus. Therefore, a comparison of these studies with ours is important to the continued development of this new technique.

EXPERIMENTAL

The E cell is schematically depicted in Figure 1 and it is described in detail elsewhere.[3] The base pressure of the E-cell is typically the microscope vacuum of 10^{-7} Torr. The $CH_3CH_2AuP(CH_3)_3$ precursor, a liquid at room temperature, was loaded into a stainless steel ampoule under argon attached to the gas manifold of the cell apparatus and degassed under high vacuum with several freeze-pump-thaw cycles. The deposition experiments were performed at 10^{-4} Torr which is also the room temperature vapor pressure of the source material.

The substrates were 3mm in diameter (100)Si discs prepared for plan view type microscopic studies using standard TEM sample preparation techniques. These discs were first treated with dilute nitric acid to dissolve any Cu contamination from the sample preparation, degreased in organic solvents and then dipped in 5% hydrofluoric acid until the surface became hydrophobic. They were introduced into the microscope via a heating holder with a single tilt axis and evacuated to 10^{-7} Torr. Substrate temperatures during CVD ranged from 125-200 °C.

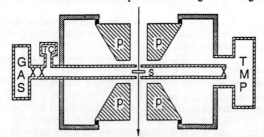

Figure 1. Schematic of the E cell/CVD reactor used in this work. The precursor is introduced from the gas ampoule through a shut-off valve and a leak valve onto the heated substrate (S). (S) sits in the pole piece (P) of the microscope and is isolated from the microscope vacuum by a 200 mm in diameter aperture. When the valve between the substrate space and the turbomolecular pump (TMP) is open the TMP evacuates the cell chamber. During growth this valve is closed and the TMP differentially pumps the gas escaping through the apertures.

During deposition, the valve to the turbo pump (TMP) was closed, as shown in Figure 1, and the reaction chamber was pumped by the same turbo through the differentially pumped apertures. The major disadvantage of our system was that the flow of the gas phase precursor was unthrottled. However, we have been able to achieve remarkably stable deposition pressures by controlling the precursor flow from the gas ampoule by a shut off valve and a high precision leak valve. The pressure is measured by a thermocouple gauge situated near the sample chamber of the environmental cell and is reliable to a few mTorr. Images and diffraction patterns can be acquired either using standard photographic techniques or using a TV camera at video rates of 30 frames per second.

Atomic force microscopy (AFM) images of the particles were obtained in ambient air in constant force mode with a commercial Nanoscope III Digital Instruments microscope using silicon nitride cantilever probes having a force constant of approximately 0.1 Newtons/meter.

RESULTS AND DISCUSSION

We have found that metallic gold deposits can be grown in the E cell of the TEM using $CH_3CH_2AuP(CH_3)_3$ at moderately low temperatures on Si substrates. In situ selected area diffraction patterns revealed polycrystalline FCC gold and ex situ microprobe analyses indicated that no carbon or phosphorous is trapped in the gold films. These results are consistent with the results obtained by Holl, et al. from depositions involving the same precursor on heated Cu and Cr surfaces. Our experiments demonstrate that our technique works and we can actually perform chemical vapor depositions on miniature wafers in the TEM. Therefore, this technique offers the potential for studying a variety of processes involving MOCVD.

Beam irradiation of the deposition surface during imaging resulted in much higher growth rates than those observed in the absence of the beam. We believe that this enhancement of growth is not because the beam is heating the substrate but a consequence of a beam assisted decomposition reaction. Moreover, in previous deposition studies of Al films from trimethylamine alane we did not observe any significant beam induced enhancement of the growth under similar conditions. The effect of the beam on the gold compound was by comparison significant.

In order to quantify the effect of the electron beam on the film growth, deposition sequences for both beam-induced and beam-free growth were acquired at 125, 150, and 200 °C. The ratio of the beam-induced growth rate to the beam-free growth rate was determined by comparing images digitized from the real-time video at several stages during growth. This growth rate ratio is simply the ratio of the deposition times for digitized images deemed comparable from the beam-induced vs. beam-free experiments. To ensure that no beam-induced growth occurred during the beam-free growths, the following procedure was used. First, the Si sample was raised to the deposition temperature. Next, the sample was exposed to the precursor for the desired time. After the exposure, the precursor was pumped from the cell and video was acquired. Sufficient video was acquired to ensure that the beam did not cause any growth due to decomposition of remaining precursor adsorbed during exposure. For example, at 125 °C, the beam-induced experiment showed significant growth during the first minute of deposition. However, during the beam-free experiment, no deposition was observed after acquiring more than one minute of video while imaging at each sample time. In Figure 2, we show electron micrographs obtained from digitized frames of video recorded for the beam-assisted decomposition of the precursor at 200 °C. The micrographs indicate that after one minute of deposition the substrate is nearly covered with gold particles and virtually continuous films are obtained after four minutes of deposition. In contrast, as visible in Figure 3, some scattered particles begin to appear after 15 minutes of deposition and substantial deposition is only observed after 75 minutes.

Growth rates for beam-induced deposition at 125, 150, and 200 °C were greater by 150, 40, and 10 times, respectively, than the corresponding growth rates of thermally grown material in

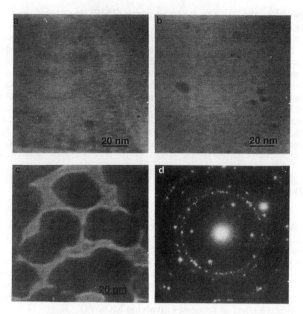

Figure 2. Electron micrographs show (a) the clean Si substrate, (b) scattered Au particles after 15 minutes of deposition, and (c) large gold islands after 75 minutes of deposition, respectively. The corresponding electron diffraction pattern is illustrated in (d).

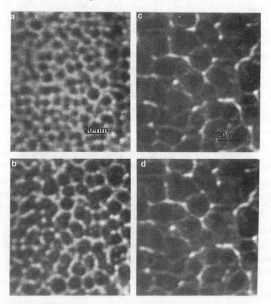

Figure 3. Electron micrographs (a), (b), (c), and (d) illustrate beam assisted deposition after 1, 2.5, 4, and 9 minutes respectively.

the absence of the beam. The trend of beam induced growth rate enhancement vs. substrate temperature is clear. The most dramatic growth rate enhancement occurred at the lowest substrate temperature and then fell monotonically as the substrate temperature was increased. We explain this result as follows. Since the pressure of the precursor was identical for all experiments, the flux of molecules to the surface was always the same. However, in the absence of the beam, the decomposition of adsorbed molecules is entirely a thermally activated process. During the time that a precursor molecule is in contact with the surface it may decompose into a deposited atom and devolve volatile byproducts. At lower temperatures, this decomposition rate is slower so the impinging electron flux has a higher probability of assisting the decomposition of adsorbed molecules. This contrasts with higher temperatures where the rate of electron-induced decomposition becomes proportionally less significant as the rate of thermal decomposition rises.

The three dimensional morphology of the films was determined by using atomic force microscopy (AFM). This technique is sensitive to the surface topography and hence provides information about both the lateral and the vertical size of the gold islands. With regard to height information, AFM is a very valuable and complementary tool to electron microscopy and has the advantage that the investigations can be carried out in the ambient without extensive sample preparation. We have compared the results for a sample grown in the absence of the beam with a sample prepared by beam-assisted growth. Without the beam, the particles appear to be nearly uniform for any given growth time. The ratio of the lateral to the vertical dimension was found to be approximately 10:1 (Figure 4). For beam-assisted growth we observed a nonuniform size distribution of particles. However, for each individual particle the ratio of the lateral to the vertical dimension was about 10:1. This indicated that the relative growth rates in the horizontal and vertical directions were nearly identical in both beam-assisted and beam-free growth conditions.

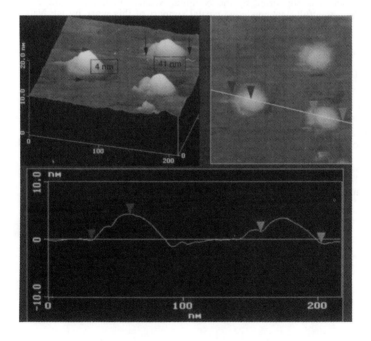

Figure 4. AFM repulsive force mode image of gold particles at 200 °C. The ratio of the lateral to the vertical dimension is found to be 10:1.

CONCLUSIONS

We have grown pure Au films on resistively heated Si substrates at 125-200 °C from $CH_3CH_2AuP(CH_3)_3$ in an environmental TEM with an arrangement that approximates a cold wall CVD reactor. In this novel reactor system, we observed enhanced growth of the material when the microscope electron beam was used in conjunction with the resistive heating. We determined that this enhancement was 160x at 125 °C, but it decreased monotonically to 10x as the substrate temperature rose to 200 °C. As the temperature increases, the rate of growth by thermal means also increases thus rendering the beam enhancement less dramatic. In situ images and diffraction patterns revealed that the material is polycrystalline gold and ex situ microanalysis indicated that the films did not contain any carbon or phosphorus impurities for samples grown with or without the aid of the electron beam. AFM inspections indicated that the ratio of the lateral to vertical dimensions was always 10:1. We envision that this technique for observing Au CVD in situ will be invaluable in studying the selective growth of this metal on various metallic and semiconductor substrates.

ACKNOWLEDGEMENTS

This work is based on research conducted at the Center for High Resolution Electron Microscopy, which is supported by the National Science Foundation under Grant (DMR-9115680). We greatfully acknowledge partial support of this work by the National Science Foundation through the NYI award to J.K. (DMR 9458047).

REFERENCES

1. David B. Beach, Samuel E Blum, and Francoise K. LeGoues, J. Vac. Sci. Technol. A 7 (5), 3117 (1989)
2. D. B. Beach, IBM J. Res. Develop. 34(6) 795 (1990)
3. Jeff Drucker, Renu Sharma, Karl Weiss, and John Kouvetakis, J. Appl. Phys. 76(12), 8198 (1994)
4. M. M. Banasznak Holl, P. F. Seidler, S. P. Kowalczyk, and F. P. Mc Feely, Appl. Phys. Lett. 62(13), 1475 (1993)
5. T. H. Baum, P. B. Comita, and T. T. Codas, Proc. SPIE, 123, (1991)

LOW TEMPERATURE DEPOSITION OF DIAMOND THIN FILMS USING ELECTRON CYCLOTRON RESONANCE PLASMAS

DONALD R. GILBERT AND RAJIV K. SINGH
University of Florida, Department of Materials Science and Engineering, Gainesville, FL 32611

ABSTRACT

We have used an electron cyclotron resonance enhanced plasma system to deposit films from methyl alcohol at a pressure of 1.00 Torr. Depositions occurred on silicon substrates at temperatures from 550 - 750 °C. Film morphology was examined using SEM. Structural characterization was conducted using Raman spectroscopy and x-ray diffraction. Based on these analyses, certain trends in film quality as a function of deposition temperature were determined.

INTRODUCTION

Research on chemical vapor deposition (CVD) of diamond in recent years has been driven by the desire to exploit diamond's many exemplary attributes. Along with diamond's extreme hardness, it is also a low friction material with excellent chemical inertness, making it a prime candidate for wear-resistant coatings. Diamond is also an attractive optical coating material because of its excellent infrared transmission characteristics and relatively low dielectric constant.

Deposition of diamond thin films has been well established in many different gas activation systems (e.g., hot filament, R.F. and microwave plasma, DC plasma jet, and oxy-acetylene flame). Typical gas phase compositions for diamond growth have been widely surveyed in the carbon-hydrogen-oxygen (C/H/O) system and have been used to determine the well-known Bachmann-triangle.[1] Although there is still debate concerning the specific roles of different monomer species in the growth process, certain aspects of diamond deposition are widely accepted. Hydrogen is critical in preferentially removing non-diamond-bonded carbon which is codeposited along with diamond. It also serves to stabilize the forming diamond surface by occupying dangling bond sites. Oxygen plays a similar role to hydrogen in preferentially etching graphitic carbon, either alone in its atomic state or as part of an OH molecule. At the same time, oxygen acts to effectively reduce the amount of gas phase carbon available for deposition by forming CO, which is very thermally stable. It has been reported that oxygen addition to a diamond CVD system resulted in higher growth rates and better structural quality of deposited films, as well as facilitating deposition at lower substrate temperatures.[2-6]

Despite general success in the development of diamond CVD, there are still many obstacles to the general application of diamond films. The barrier to nucleation of diamond on non-diamond substrates is a common problem. The most successful results have been obtained using carbide-forming materials with relatively high melting points such as silicon and molybdenum. Even on these materials, results are significantly improved by using pre-deposition treatments such as scratching of the growth surface with diamond grit or "seeding" with diamond particles.[7] Another problem in diamond deposition is thermal expansion mismatch between the deposited diamond film and the substrate material. The resulting stress due to this mismatch can result in cracking and delamination of the film. Disregarding thermal mismatch, the typical substrate temperature used during diamond deposition (> 800 °C) can be detrimental to many materials. Therefore, it is desirable to be able to deposit diamond at as low a temperature as possible while retaining good crystalline quality and reasonable growth rates.

The electron cyclotron resonance enhanced plasma system is well suited to the task of low temperature deposition. By operating at sufficiently low pressures, the free electron behavior is separated from the behavior of the ions and neutral species in the plasma. Application of a magnetic field of the appropriate strength (875 G for 2.45 GHz microwaves) in the plasma generation region induces cyclotron (helical) motion of a specific frequency. Matching of the cyclotron frequency to the microwave frequency produces resonant coupling, which greatly increases the energy of the electrons. This effectively maintains the free electrons at higher temperatures than the surrounding ions and neutral species. As a result, the electrons interact with other species via impact ionization and dissociation, which increases the plasma density. This process is very effective at pressures below 100 mTorr, where the electron mean-free-path is sufficient to allow several cycles before impacting other species. However, the magnetic enhancement is beneficial to pressures greater than 1.0 Torr.

In accordance with the Bachmann triangle of viable gas phase composition, methanol (CH_3OH) has been shown to be a suitable precursor for diamond film growth. Methanol is easily handled as a liquid and can be obtained in high purity. In this paper, we present the results of experimental deposition of diamond films from methanol gas precursors in an ECR plasma system at low substrate temperatures.

EXPERIMENT

Deposition was performed in a Plasma-Therm BECR-6 electron cyclotron resonance system, which is shown schematically in Figure 1. The necessary magnetic field is produce by an arrangement of rare-earth permanent magnets. Single crystal silicon substrates were seeded with sub-micron diamond particles, as shown in Figure 2. Substrates were placed on a variable position mount with independent heating capabilities. Films were deposited for five hours from gas mixtures of 1% methanol (CH_3OH) by volume in hydrogen. The total gas pressure used was 1.00 Torr. A 2.45 GHz microwave source was operated at 1000 watts input power.

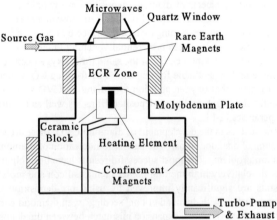

Figure 1 Schematic of ECR experimental system.

Figure 2 SEM showing typical diamond seeding on silicon substrate.

RESULTS

Depositions were performed at substrate temperatures ranging from 550 - 750 °C. The resulting film morphologies were examined using scanning electron microscopy. The resulting micrographs are presented in Figure 3. All films show the formation of facetted crystal structures commonly associated with diamond growth. The film formed at 550 °C shows a few irregular micron-sized crystals. These irregularities are likely due in part to secondary nucleation and growth at surface defects. At 600 °C, the deposited film shows greater uniformity of crystal structure. This trend continues for the film deposited at 700 °C, which shows less evidence of secondary nucleation than the previous films. The 700 °C film also appears to have smaller average-sized crystals than the 600 °C. Deposition at 750 °C resulted in similar features, although there is more evidence of secondary nucleation and twin structures.

Structural characterization of these films was conducted using Raman spectroscopy. The resulting spectra are presented in Figure 4. All spectra were taken using the 514.5 nm emission line of an argon ion laser. The diamond peak near 1332 cm^{-1} is seen in all spectra. These spectra showed little evidence of the peak shifting commonly associated with residual film stress. The film grown at 600 °C showed the smallest full-width-at-half-maximum (FWHM) with a value of approximately 8 cm^{-1}. Films grown at 700 and 750 °C had slightly larger FWHM values of up to approximately 10 cm^{-1}. Deposition at 550 °C resulted in a FWHM of nearly 16 cm^{-1}, which indicates a substantial reduction in the crystalline quality compared with the other films. All films also showed broad peaks near 1355, 1440, and 1540 cm^{-1}, with some small variations in the peak positions for each film. These peaks indicate the presence of graphite and diamond-like-carbon in the deposited films. The relative height of the diamond peak versus the background non-diamond spectral features is greatest for the 600 and 700 °C films, suggesting a better over-all quality film with less non-diamond content.

Figure 3 SEM micrographs of films deposited at (a) 550, (b) 600, (c) 700, and (d) 750 °C.

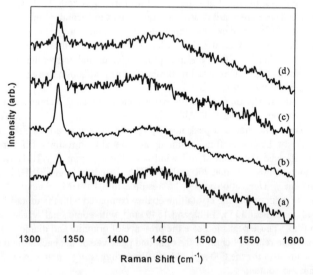

Figure 4 Raman spectra of films deposited at (a) 550, (b) 600, (c) 700, and (d) 750 °C.

To further corroborate the presence of diamond in these films, x-ray diffraction spectroscopy was performed on each sample. Each film showed the diamond (111) peak at a 2θ value of 43.9°. A representative spectrum is shown in Figure 5, taken from the film deposited at 700 °C. The silicon (400) peak, resulting from the substrate, is seen at 69.2°. Because the diamond films all have thicknesses of approximately one micron or less, the sampling volume for diffraction is quite small. This, combined with carbon's low atomic weight, results in a weak diamond peak compared to the peak of the underlying silicon substrate.

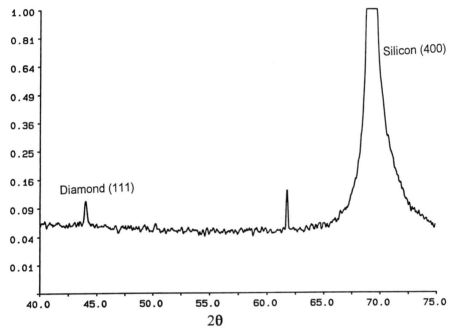

Figure 5 XRD spectrum of film deposited at 700 °C.

CONCLUSIONS

Diamond films were successfully deposited from methanol/hydrogen gas mixtures at temperatures as low as 550 °C. Raman analysis indicated the best quality films were grown at 600 and 700 °C, based on FWHM values and relative peak heights. The diamond quality appeared to drop off below 600 °C, based on the much broader Raman diamond peak observed in the film deposited at 550 °C. This correlated with film morphology as determined by SEM, which showed the most regular and uniform morphologies at 600 and 700 °C. As the deposition temperature was increased above 700 °C, the film quality appeared to decrease again, although less drastically than at 550 °C.

REFERENCES

1.   P. Bachmann, D. Leers, and H. Lydtin, Diamond and Related Materials **1**, 1 (1991).

2.   M. Nunotani, M. Komori, M. Yamasawa, Y. Fujiwara, K. Sakuta, T. Kobayashi, S. Nakashima, S. Minomo, M. Taniguchi, and M.Sugiyo, Jpn. J. Appl. Phys. **30**, L1199 (1991).

3.   S. Harris and A. Weiner, *Diamond, Silicon Carbide and Related Wide Bandgap Semiconductors*, J. Glass, R. Messier, and N. Fujimori (eds.), MRS Symposia Proceedings **162**, Materials Research Society, Pittsburgh, PA, 103 (1990).

4.   J. Chang, T. Mantei, R. Vuppuladhadium, and H. Jackson, J. Appl. Phys. **71**, 2918 (1992).

5.   Y. Liou, A. Inspektor, R. Weimer, D. Knight, and R. Messier, J. Mater. Res. **50**, 2305 (1990).

6.   O. Sanchez, C. Gomez-Aleixandre, F. Agullo, and J. Albella, Diamond and Related Materials **3**, 1183 (1994).

7.   P. Pehrsson, F. Celii, and J. Butler, in *Diamond Films and Coatings*, R. Davis (ed.), Noyes Publications, Park Ridge, NJ, 92 (1993).

Part IV

Beam-Induced Defects, Chemical Effects, and Characterization

SURFACE DAMAGE DURING keV ION IRRADIATION: RESULTS OF COMPUTER SIMULATIONS

R.S. AVERBACK, MAI GHALY and HUILONG ZHU
Department of Materials Science and Engineering, University of Illinois at Urbana-Champaign, Urbana, IL.

ABSTRACT

MD simulations have been employed to investigate damage processes near surfaces during keV bombardment of metal targets. For self-ion implantation of Au, Cu, and Pt in the range of 5-20 keV, we have found that the proximity of the surface leads to significantly more damage and atomic mixing in comparison to recoil events occurring in the crystal interior. In some cases, large craters are formed in a micro-explosive event, while in others a convective flow of atoms to the surface creates adatoms and leaves dislocations behind. Both the amount damage created in the surface and its morphology depend sensitively on the details of the energy deposition along individual ion trajectories. The results of these simulations will be summarized and compared to recent scanning tunneling microscopy studies of individual ion impacts in Pt and Ge.

INTRODUCTION

Most past descriptions of radiation effects in solids have been built on the assumption of two-body atomic collisions, and for the most part, they neglect the detailed physical and chemical properties of the solid. Part of the reason for this development stems from the pioneering work on defect production by Vineyard and co-workers at Brookhaven National Laboratory using molecular dynamics (MD) computer simulations [1]. They showed that the threshold energy for defect production in metals was greater than ≈ 25 eV so that the thermodynamics of the solid could presumably be safely ignored. Models based on the assumptions of binary collisions (BC) and linear cascades, such as MARLOWE [2], TRIM [3], and Boltzmann transport theory [4] are justified on this premise. The early Brookhaven work, however, was restricted to very low energy recoil events where thermal spike effects are negligible. Many experiments and computer simulations have since established that the BC model breaks down in predicting defect production in high energy cascades, and that the collective atomic motion in the thermal spike, and in many cases local melting, strongly influence the primary state of damage, including defect production [5], atomic mixing [5], and sputtering [4]. Fig. 1 illustrates this point; it shows the evolution of a cascade in the ordered intermetallic compound β-NiAl. Clearly seen is a region of high structural and chemical disorder that has been found to have many similarities with a liquid [6]. The boundary between the structurally ordered and disordered zones, in fact, is sharp, suggesting the co-existence of two distinct phases. The observation of local melting is now a common finding in MD simulations of energetic cascade events. The kinetic energy distribution of those atoms in the disordered zone provides a convenient means to characterize different time regimes in the cascade dynamics, as shown in Fig. 2, again for β-NiAl. At times earlier than ≈ 0.5 ps, the number of atoms in the melt is small, the effective temperature, T, is high, and the distribution of kinetic energies, KE, deviates markedly from Maxwellian. By ≈ 1 ps, however, the distribution of kinetic energies becomes nearly Maxwellian, and the number of

Mat. Res. Soc. Symp. Proc. Vol. 388 © 1995 Materials Research Society

atoms in the melt, denoted by N in the figure, becomes a maximum. Simulations like that shown in Fig. 1, have now been conducted in several metals [7], and semiconductors [8], and a rather complete understanding of cascade processes is beginning to emerge. Surprisingly, nearly all MD simulations of cascades have examined energetic recoils initiated in the crystal interior, and only recently have ion implantations through a surface become under investigation by this technique.

How thermal spikes and local melting influence surface damage has long been a matter of conjecture. It was assumed for many years that the high (non-linear) sputtering yields during self-ion bombardment of materials like Au was somehow due to thermal spikes [9], and models based on evaporation from hot surfaces were constructed. Uncertainties about heat conduction away from the cascade and surface binding energies in cascades, however, have

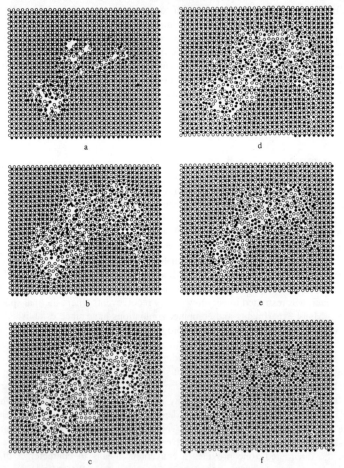

Fig. 1 Evolution of the cascade dynamics of a 10 keV cascade in "β-NiAl." The symbols represent the locations of Ni(o) and Al(o) atoms within a cross sectional slab, two atomic planes thick, projected onto the (100) plane. (a) 0.1 ps.; (b) 0.3 ps; (c) 0.5 ps; (d) 1.0 ps; (e) 2.0 ps; and (f) 6.0 ps. Ref. [6].

prevented reliable calculations of this type. Furthermore, it had not been appreciated until very recently [10], that the surface can strongly influence the dynamics of cascade processes and thereby influence the damage structure produced at surfaces. We will show here that the surface can have a profound influence on defect production mechanisms, and these surface mechanisms can lead to unusually large changes in the surface morphology.

Model for Molecular Dynamics Simulations:
Most of the simulations to be described here were performed with the code MOLDYCASK, which is described in the literature [11], but adapted at Illinois [12] to treat surfaces. Embedded Atom Method (EAM) potentials were employed for all of the simulations [13]. The potentials of Foiles et al. [14] were employed, but modified at short range to merge smoothly with the Universal potential described by Ziegler et al. [15]. It is difficult to estimate how well these potentials represent their real counterparts for the phenomena described here, since atomic collisions occur far from equilibrium while the EAM potentials are fitted and tested close to equilibrium. Preliminary studies on Cu comparing

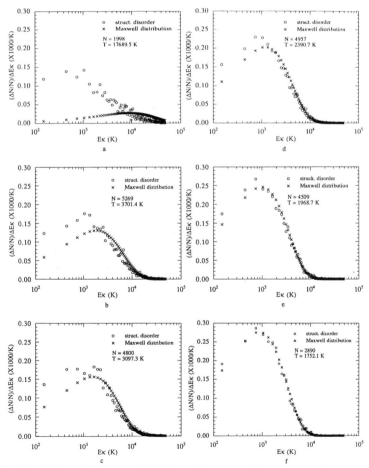

Fig. 2 Kinetic energies of the atoms within the melt zone at various instants of time. Ref. [6].

calculations of energies at close separations by pseudopotentials methods and modified EAM potentials, however, do show good agreement [12]. For the present work on surface interactions, it should be noted that: the surface energies predicted by EAM appear too low for most metals; the melting point is nearly correct for Cu, but 300 K too low for Au and 450 K too low for Pt; the point defect properties and activation enthalpies for self diffusion are about correct for these same metals [14]. Since the surface interactions investigated here occur at high energies, we do not expect that minor deficiencies in these potentials, which might influence more sensitive phenomena such as surface reconstruction, will much affect the qualitative results of our simulations. Coupling between electron and phonon systems is neglected in these simulations since its importance is still uncertain; it is not, in any case, yet possible to reliably include such coupling in MD codes. The large size of the computational cell, 1.5 - 3.0 x10^5 movable atoms, and damping at the boundary layers, prevent serious problems with energy dissipation.

RESULTS:

Cascades in the crystal interior: 10 keV cascades in Au:

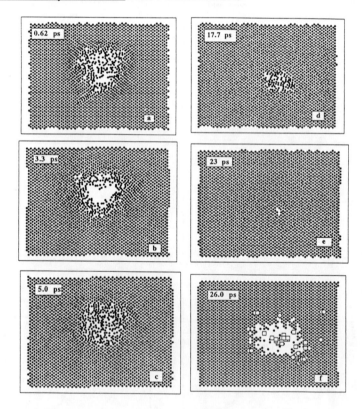

Fig. 3. Snapshots of atom locations with a cross sectional slab of thickness a$_0$/2 viewed in the <100> direction, for a 10 keV cascade in Au. Ref. [12].

Before discussing the interactions of energetic ions with surfaces, it is convenient to first highlight the behavior of collision cascades in the bulk. Fig. 3 shows snapshots of the atom locations during the evolution of a 10 keV cascades in Au [12]. The general features of this cascade event are similar to those seen for β-NiAl in Fig. 1. In this case, however, the energy is confined to a smaller volume, since the cross sections for atomic collisions are larger, i.e., higher atomic number. Unlike the cascade in β-NiAl, the very high energy density leads to cavitation in the central core of the cascade. This means that there is both a high temperature and high pressure in the cascade. In Fig. 4 the average temperature and pressure in the Au cascade are plotted as functions of distance from the center of the cascade at various instants of time. Temperatures are seen to reach ≈ 6,000 K, which is well above the melting temperature, T_m, of Au and virtually all other materials. As time evolves, the cascade cools by heat dissipating to the surrounding lattice. Notice that the radius at which the temperature drops below T_m corresponds reasonably well with the disordered zone in Fig. 3. The hydrostatic pressure in the cascade core is also revealing; after 1.37 ps, it is a few GPa. Since the relevant shear modulus in Au is ≈ 28 GPa, the pressure is too low for general yielding, for example by "loop punching." Near surfaces, where the elastic constraint of a surrounding crystal is largely removed, the high pressure can induce mass flow. In Fig. 3d, the resultant defect structure of the cascade, or primary state of damage, is illustrated. Here, the point defects are projected onto the (100) plane. The unfilled region corresponds to a projection of the once melted region at its largest extent. From this figure, it can be seen that the vacancies are clustered in the center of the cascade and the interstitials are located at the periphery.

Twenty Frenkel pairs were produced by this event. The amount of atomic in the event can be quantified by the mixing parameter,

$$Q = <R^2>/6n_0E \tag{1}$$

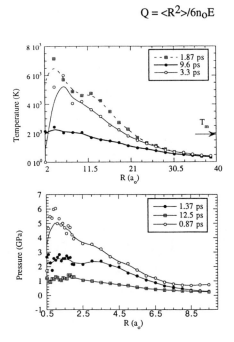

Fig. 4a Temperature profile during the evolution of a 10 keV bulk cascade in Au. Ref. [12].

Fig. 4b Pressure profile during the evolution of a 10 keV bulk cascade in Au. Ref. [12].

where $<R^2>$ is the sum of the mean square displacements of all atoms in the cell, n_0 is the atomic density and E is the initial recoil energy. For this bulk cascade in Au, $Q = 59$ $\text{Å}^5/\text{eV}$. This is ≈ 50 times larger than predictions based on the binary collision approximation, thus illustrating the importance of the thermal spike behavior.

Effects of the surface:

10 keV self-bombardment of Au:

The thermal spike conditions that are established in cascades have profound influence on surface damage. Earlier MD investigations of surface effects were mostly interested in the so-called "prompt phase" of the collision dynamics, i.e., times less than a few tenths of ps. [16], when nearly all sputtering occurs; these simulations were not run to sufficiently long times to see the full extent of the cascade dynamics on surface damage. The thermal spike phase of the cascade, in fact, only influences sputtering in very dense cascades, but it plays a more general role in determining the morphology of the surface damage. This is illustrated in Fig. 5 for a MD event representing 10 keV self-atom bombardment of Au.

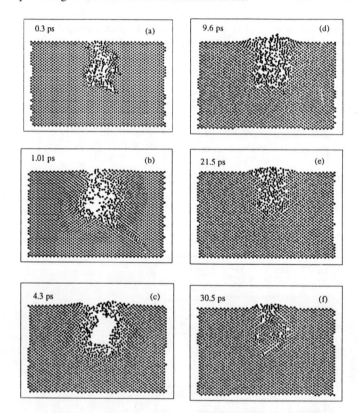

Fig. 5 A cross sectional view of atom locations for 10 keV self-atom bombardment of Au. Ref. [9].

The development of this 10 keV surface event begins similarly to the corresponding one in the bulk. Local melting and cavitation are seen to take place early in the event. Some differences in the two events begin to become apparent after ≈ 5 ps. For example, the cavity grows larger and persists for a longer time in the surface event. As the two events evolve further, it becomes more and more obvious that surface and bulk cascades are quite different.

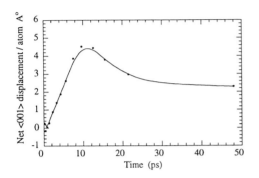

Most striking is the convective flow of liquid Au onto the surface in the surface event. This is driven by the high pressure developed in the cascade. The motion of atoms to the surface can be seen in Fig. 6 where the net displacement per atom in the melt zone is shown as a function of time. Atoms are seen to move continuously to the surface over a period of ≈ 15 ps. The net downward motion after 15 ps is due to resolidification and volume contraction. In all, ≈ 4,000 atoms moved a distance of ≈ 2 Å, upward.

Fig. 6. Net displacement of atoms in <100>. Ref. [12]. In the crystal interior, such flow would require the punching out of dislocation loops. In amorphous materials, density fluctuations can be established by this mechanism in the bulk, and possibly lead to dimensional changes [17]. For mass flow at the surface, the pressure need only to exceed the surface tension..

A comparison of the primary states of damage produced in the surface and bulk 10 keV cascades in Au also reveals significant differences. As noted above, both interstitial and vacancy clusters are formed in the bulk cascade, with a net total of 20 Frenkel pairs in all. The vacancies are located in the core of the cascade and the interstitials on the periphery (see Fig. 3). For the surface cascade, approximately 554 vacancies are created within a complex dislocation arrangement [9]. It is noteworthy that only four interstitial atoms are produced in this surface event, since MD simulations of bulk cascade events in several metals all find that the number of interstitials is equal to ≈ 0.25 the number deduced from the Kinchin-Pease formula, ≈ 25 in this case. This observation further illustrates that defect production at surfaces is much different than in the bulk.

The mixing parameter in this event is ≈ 80 $Å^5$/eV, which is somewhat larger than in the bulk. The reason is not yet certain, but it appears that it is a consequence of the lower density in this cascade event. Note that the cavity is larger and lasts longer. The loss of atoms to the surface and elastic deformation of the surface layer are seemingly responsible for the lower densities. It is clear that atomic motion is easier in regions of lower density. This result has practical consequences for techniques that employ sputter-depth profiling since the depth resolution of these methods is usually limited by atomic mixing from the sputtering beam.

Self-ion bombardment of Au has long been known to give rise to non-linear cascades, but not of most others metals, raising the question whether the surface effects discussed above are relevant for other materials. This was addressed by a simple model of damage based on viscous flow [18]. The model assumes that an ion deposits its energy along a linear track normal to the surface. Energy spreads radially outward, creating a cylindrical tube of liquid.

Pressure is developed in the liquid by the volume change on melting and the elastic confinement of the surrounding solid medium, and this pressure forces liquid to flow onto the surface according to Pouiselle's eqn. The flow of liquid to the surface in this model [18],

$$Q = \frac{1}{8\pi n_o}\left(\frac{\frac{\Delta n_o}{n_o}\frac{1}{\kappa}\frac{\Delta V}{V}}{\mu D(1+\frac{2(1+v)}{E\kappa})}\right)^{1/2}\left(\frac{F_D}{\varepsilon_m}\right)^2$$

(2)

where n_o is the atomic density, V is the molar volume, κ is the compressibility of the liquid, μ is the liquid viscosity, D is the thermal diffusivity, E is the elastic modulus, v is Poisson's ratio, F_D is the energy deposition per unit length normal to the surface, and ε_m is the energy per atom in the melt. The last term in this expression is simply proportional to the ratio of the nuclear stopping power to the cohesive energy of the material, and thus it has the same general form as the equation for sputtering yields, except Q is proportional to the square of this ratio and sputtering is linearly proportional to it. We thus expect that the surface damage created by viscous flow should scale with sputtering yields. For a particular material, the flow should simply scale with the nuclear stopping power and melting temperature. Consequently, we examined self-atom bombardment of Pt, since it has nearly the same nuclear stopping as Au, but a much higher melting temperature, and self-atom bombardment of Cu, since it has nearly the same melting temperature as Au but a much smaller nuclear stopping power.

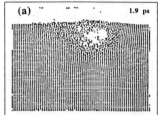

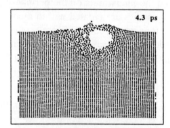

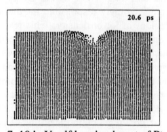

Self-bombardment of Pt and Cu:

Various 10 keV events were simulated for Pt. The features are quite similar to those reported for the 10 keV event in Au: local melting, cavitation and convective mass flow to the surface [12]. It was found that the deposited energy distribution relative to the surface plays a critical role in determining the final defect structure. When the energy spike is located near the surface, for example, a crater is formed, as shown in Fig. 7. Table 1 further illustrates that when the energy is deposited close to the surface, defect production, ion beam mixing, and sputtering all increase. Notice in Table 1 that for the event where the ion penetrated rather deeply into the Pt, the mixing and defect production approach the bulk values. Finally, a comparison between cascade events in Au and Pt (i) in the bulk and (ii) where the energy deposition is very similar, illustrates that the thermal spike behavior is more intense in Au, i.e., there is more mixing and more viscous flow.

Fig. 7. 10 keV self bombardment of Pt. [12].

Evidence for the structure shown in Fig. 7 has been obtained by STM studies of 5 keV Xe bombardment of Pt [19]. Results are shown in Fig. 8. Here surface craters with surrounding adatoms are clearly visible. Note that the dimensions are quite similar to those in Fig. 7.

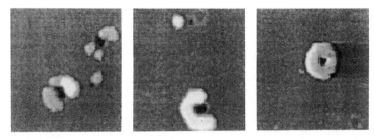

Fig. 8. STM images of Pt (111) after irradiation with 5 keV Xe. Each frame is 30x30 nm^2 Ref. [19].

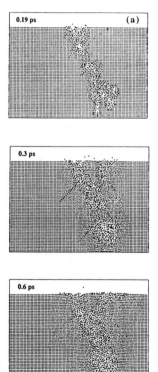

The cascade development of a 15 keV event in Cu is shown in Fig. 9. This event is interesting because it illustrates the formation of subcascades and how these subcascades melt together to form a continuous liquid, tube-like structure. This makes it possible for atoms to flow from far below the surface and create damage at these depths. Since other MD simulations have shown that the number of defects produced in bulk Cu cascades is about 0.2 the Kinchin-Pease number, it can be deduced from Table 1 that vacancy production in the surface events can be much larger than in bulk events. For 15 keV cascades in Cu, 40 Frenkel pairs are expected in the bulk. For the Cu event listed in Table 1 that had deeper penetration, the defect production was only 42. Another 15 keV in Cu with less penetration, however, showed considerably more defect production. Mixing is also larger in the surface cascades than in the bulk, as noted in the table. The reason is presumably the same as mentioned above for Au.

Fig. 9. 15 keV self bombardment of Cu. Ref. [12].

High energy bombardment:.

Although very high energies have not yet been realized by MD, a simulation of a 20 keV

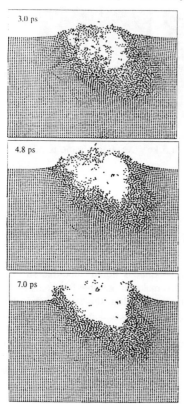

Fig. 10. 20 keV bombardment of Au. Ref. [12].

cascade in Au shows a rather spectacular phenomenon, as seen in Fig. 10. Here, the cascade cavity intersects the surface and creates a large crater. Sputtering by direct collision processes is only partially responsible for the crater. The more important mechanism occurs several picoseconds later. The large pressure in the cavity cannot be mechanically constrained by the thin membrane of material covering it, and so it undergoes general yielding in a "microexplosive" event. It was found that the membrane tears in three places and that the three segments peel backward over the surface. These kinds of structures have been observed by TEM during Bi and Bi_2 irradiation of Au with energies greater than ≈ 20 keV [20]. The yield of such craters in these experiments was low, on the order 10^{-2} craters/ion, but it was a strongly increasing function of energy. Recently, STM has been used to examine 20 keV Ga bombardment of Ge with a similar observation, i.e., the formation of large craters, but with low probability [21]. This differs from the case of Xe bombardment of Pt, discussed above, where the craters were small but their probability formation near unity [18].

SUMMARY AND CONCLUSIONS:

These results of MD simulation illustrate the important role of the surface in defect production. It was shown that both the magnitude of the damage, its morphology, and the mechanisms of producing it are far different at the surface than in the crystal interior. The mechanisms responsible for this difference were also clearly elucidated. In some cases, the damage was due primarily to local melting and viscous flow of atoms onto the surface. In other more special cases, when the energy density is high and the location of the cascade is just below the surface, a micro-explosion is possible in which large surface craters are produced and many atoms and clusters are sputtered. The flow of material of atoms onto the surface has implications for determining the stress state in the sample, as the loss of atoms to the surface places the surface layer in a state of tensile stress. The results also have important implications for sputtering mechanisms and sputtering depth profiling. The MD simulations showed that atomic mixing is enhanced in the surface region, and this would lead to reduced resolution in depth profiles.

Table 1.

Target (energy in keV)	Depth of incident ion (Å)	# of FP's vacs SIA's	# atoms above surface	Sputtering yield	Mixing parameter $eV/Å^5$
Au (10)	40.5	554 4	554	4	79.9
Au (bulk-10)		20 20			51.6
Au (20)	36.8	crater 8	2058	350	278.6
Cu (15)	70.0	42 28	14	4	9.5
Cu (15)	30.1	109 25	84	29	11.7
Pt (bulk-10)		32 32			15.4
Pt (10)	58.8	50 28	22	3	18.8
Pt (10)	24.0	crater 6	384	22	83.7
Pt (10)	26.4	crater 5	537	26	78.4
Pt (10)	43.2	153 12	149	5	43.8

ACKNOWLEDGMENTS

The authors are grateful to Dr. Th. Michely for stimulating discussions and for Fig. 8. The research was supported by the DoE, Basic Energy Sciences (Grant DEFG02-91ER45439). Grants of computer time from the National Energy Research Computer Center at Livermore, CA., the National Center for Supercomputing Applications at UIUC, and the Materials Research Laboratory at UIUC are gratefully acknowledged.

REFERENCES

1. J.B. Gibson, A.N. Goland, M. Milgram, and G.H. Vineyard, Phys. Rev. 120, 1229 (1960).
2. M.T. Robinson and I.M. Torrens, Phys. Rev. B 9, 5008 (1974).
3. J.P. Biersack and L.G. Haggmark, Nucl. Instr. and Meth. B 174 257 (1980).
4. See e.g., P. Sigmund, in *Sputtering by Particle Bombardment*, ed. R. Behrisch, (Springer-Verlag, Berlin, 1981) Chapter 1.
5. See e.g., R.S. Averback, T. Diaz de la Rubia and R. Benedek, Nucl. Instr. Meth. B 33, 693 (1988).
6. H. Zhu, R.S. Averback, and M. Nastasi, Philos. Mag. A 71, 735 (1995).
7. D. Bacon and T. Diaz de la Rubia, J. Nucl. Mater. 216, 275 (1994).
8. T. Diaz de la Rubia and G.H. Gilmer, Phys. Rev. Lett. 74, 2507 (1995).

9. See e.g., P. Sigmund and C. Claussen, J. Appl. Phys. 52, 355 (1981).
10. M. Ghaly and R.S. Averback, Phys. Rev. Lett. 72, 364 (1994).
11. MOLDYCASK is a derivative of MOLDY, see M.W. Finnis, Atomic Energy Authority Report No. AERE-R-13182, (1988).
12. M. Ghaly, Ph.D. thesis, University of Illinois Urbana-Champaign, (1995).
13. M.S. Daw and M.I. Baskes, Phys. Rev. Lett. 50, 1285 (1983).
14. S.M. Foiles, M.I. Baskes, and M.S. Daw, Phys. Rev. B 33, 7983 (1986).
15. J. Ziegler, J.P. Biersack, and U. Littmark, in *Stopping and Range of Ions in Solids*, (Pergamon Press, New York, 1985) Vol. 1.
16. See e.g., D. Harrison, Radiat. Effs. 70, 1 (1983).
17. H. Trinkaus and A.I. Ryazanov, Phys. Rev. Lett. (in press).
18. R.S. Averback and Mai Ghaly, J. Appl. Phys.**76**, 3908 (1994).
19. C. Teichert, M. Hohage, Th. Michely, and G. Comsa, Phys. Rev. Lett. 72, 1682 (1994).
20. K.L. Merkle and W. Jäger, Philos Mag. A. 44, 741 (1981)
21. P. Bellon, S.J. Chey, E. Van Nostrand, Mai Ghaly, D.G. Cahill, and R.S. Averback, Surf. Sci. (in press).

SMOOTHING DURING ION-ASSISTED GROWTH BY
TRANSIENT ION BEAM-INDUCED DEFECTS

B.K. Kellerman, E. Chason, J.A. Floro, and S.T. Picraux
Sandia National Laboratories, Albuquerque, New Mexico 87185

J.M. White
Science and Technology Center at the University of Texas at Austin, Austin, Texas 78712

ABSTRACT

Several studies have shown that the surface morphology can be smoother during simultaneous ion bombardment and growth than during growth alone, however, the atomistic mechanism responsible for the smoothing effect has been difficult to determine. We have developed Monte Carlo simulations of growth and defect diffusion to model the interaction between growth atoms and ion-induced defects and to present a simple atomistic mechanism that describes the effects of low-energy ion bombardment during ion-assisted growth of germanium. Measurements of ion-induced point defect production indicate that a large number of defects exist only temporarily on the surface at typical growth temperatures, because the defects have sufficient mobility to recombine and annihilate. We propose that this ion-induced transient defect population plays a significant role in modifying the dynamic surface morphology. The simulations support a surface smoothing mechanism that involves the destabilization of adatom islands by the transient ion-induced defects. The optimum simulated steady-state surface morphology can be achieved with ion-induced defect production rates less than or equal to 10 defects/ion. We find that low-energy ion bombardment during growth effectively lowers the temperature at which step-flow growth can be achieved.

INTRODUCTION

Low-energy ion beams are well suited to the modification of surfaces because they do not introduce substantial bulk damage. Low-energy ion bombardment during molecular beam epitaxy has been shown to control and/or improve several aspects of film growth[1]. For example, studies have reported larger average island sizes in the presence of ion bombardment due to the depletion of small clusters by sputtering and ion-induced dissociation[2,3]. Also, we have shown that during ion-assisted homoepitaxy of Ge(001) simultaneous low-energy ion bombardment and growth produces a smoother surface morphology than either growth or ion bombardment alone[4]. We interpreted these results in terms of several mechanisms including the destabilization of small clusters by direct ion impingement and the annihilation of adatoms with ion beam-induced surface vacancies. However, the atomistic mechanism responsible for the ion-induced effects was difficult to confirm. In general, a fundamental understanding of the interactions of low-energy ions with surfaces and the role of ion-induced defects is still lacking. In fact, several investigations have used computer simulations to explain the effects of low-energy ion bombardment in ion beam assisted deposition and ion beam deposition, because the effects of low-energy ion bombardment with semiconductor surfaces are often complex[5,6].

In order to provide an atomistic mechanism describing the effects of low-energy ions with germanium surfaces and the role of ion-induced defects we have developed kinetic Monte Carlo simulations. We present a simple atomistic mechanism describing the effects of low-energy ion bombardment of dynamic germanium surfaces in terms of the production and interaction of growth-induced and ion-induced defects. We show that a simple ion-induced defect mechanism can explain experimental observations of surface smoothing during simultaneous low-energy ion bombardment and growth. Low-energy ion bombardment produces a smoother surface morphology through the production of adatoms and vacancies that help destabilize adatom islands thereby enabling adatoms to reach step edges. In this delicate balance, the optimum simulated steady-state surface morphology occurs when the net production of ion-induced defects is the same as the production of growth-induced defects. Also, we find that the optimum steady-state surface morphology can be achieved with ion-induced defect production rates as high as 10 defects/ion.

PREVIOUS EXPERIMENTAL OBSERVATIONS

This work was motivated by two previous experimental studies of low-energy ion bombardment. The first investigation monitored the evolution of the Ge(001) surface morphology during low-energy ion bombardment (200 eV Xe) and homoepitaxy using Reflection High Energy Electron Diffraction (RHEED)[4]. We demonstrated that a smoother surface can be obtained using a combination of ion bombardment and growth than can be obtained by ion bombardment or growth alone. We also showed that the smoothest steady-state surface morphology occurred when the ratio of net ion-induced defect production rate to deposition rate was approximately unity. As described earlier, we proposed several mechanisms for describing the enhancement of surface smoothness, however, the atomistic mechanism responsible for producing the surface smoothing enhancement was difficult to verify. Our attempt to better understand the mechanism of how low-energy ion bombardment affects the *dynamic* surface morphology began with an investigation of low-energy ion bombardment on *static* Ge(001)-2x1 surfaces[7]. RHEED was used to quantify the number of surface vacancies and adatoms produced per incident ion at different substrate temperatures. For 200 eV Ar/Xe ion bombardment, the observed surface defect yield (number of surface vacancies and adatoms produced per incident ion) decreased abruptly by an order of magnitude as the substrate temperature was increased. We developed Monte Carlo simulations of defect production and diffusion and showed that surface recombination of point defects explained the observed temperature dependence[8]. All that remains after complete recombination of vacancies/adatoms at typical growth temperatures are the uncompensated sputter vacancies or sputter yield, one defect/ion. Although the observed defect yield may be on the order of one defect/ion at typical growth temperatures, an order of magnitude more defects transiently reside on the surface prior to recombining or annihilating. The following experimental observations are central to our current simulation study: simultaneous low-energy ion bombardment and growth produces a smoother surface morphology than growth alone and a large population of ion-induced defects exist temporarily on the surface at typical growth temperatures.

MONTE CARLO SIMULATIONS

To understand the enhancement in surface morphology in terms of the interaction of the transient ion-induced defects with growth atoms, we have developed simple kinetic Monte Carlo simulations of growth and surface defect diffusion to model the interaction between growth atoms and ion-induced defects. The Monte Carlo simulations used in this study have been described in detail previously[9]. We employed the following procedure. Each deposition event was represented by placing a single adatom at a random location on the surface, while each ion impingement event was represented as the production of multiple adatoms and monovacancies (typically ten defects per ion), with a sputter yield of one defect per ion, unless otherwise noted[5]. Our modeling of the interaction and surface diffusion of adatoms and monovacancies involved the random walk of point defects on a square surface mesh (64 atom by 256 atom surface consisting of 4 equivalent terraces). The surface defects are very mobile at typical growth temperatures and migrated isotropically on the surface with an activation energy for diffusion of 0.8 eV[10] until they clustered, recombined or annihilated at a step edge. Recombination events occurred when an adatom and vacancy occupied the same lattice site. When an adatom landed next to an existing step edge, it was incorporated into the island. However, adatoms were allowed to detach from islands with an activation energy that depended upon its coordination. The activation energies for jumping across steps, jumping to and from an island, edge diffusion, and detachment of one-, two- and three-coordinated atoms could be varied independently but were typically held at the same values of 0.0 eV, 0.8 eV, 1.0 eV, 1.0 eV, 1.2 eV, 1.4 eV, respectively.

In our earlier experiments, RHEED was used to measure changes in the surface morphology *in situ* and in real time during the ion beam-assisted deposition of germanium[4]. To provide a means of direct comparison between experimental results and Monte Carlo simulations, we calculated RHEED intensities as the surface evolved during each simulation run. We computed the intensity of the out-of-phase specular RHEED spot using either the kinematic approximation[11] or the step density[12]; similar results were obtained using either method. All simulated RHEED intensities presented here were calculated using the kinematic approximation. We are not attempting to quantitatively interpret absolute values of RHEED intensities but compare relative RHEED intensities as a measure of surface roughness. Thus, smoother surfaces are represented by higher RHEED intensities while rougher surfaces have lower intensities.

RESULTS AND DISCUSSION

The simulations are able to account for the enhancement in the surface smoothness during simultaneous ion bombardment and growth. The enhancement is demonstrated in Figure 1. In Fig. 1 the relative RHEED intensity is plotted as a function of time. Fig. 1 contains both simulated data and experimental data from ref. 4 for a surface initially exposed to growth followed by simultaneous growth and ions. The x-axis in the experimental data has been re-scaled so the data sets may be overlaid. During the simulation, an initially smooth surface is exposed to either growth atoms (nucleation/coalescence regime) at a flux corresponding to 0.1 monolayers/sec until approximately 0.50 monolayers (ML) has been deposited. Following the deposition of 0.50 ML, a combination of growth and ion bombardment are introduced in order to allow a direct comparison of surface morphologies obtained during growth and during a combination of growth and ion bombardment. A similar procedure was followed during the experimental investigation[4]. The surface immediately starts to roughen as indicated by a decrease in the RHEED intensity. At the time indicated by the arrow, ion bombardment is added to the flux. As shown in Fig. 1, the simulations revealed that a combination of growth and ion bombardment produce a smoother surface than growth alone, consistent with experimental observations. It is somewhat remarkable that a simple mechanism can explain this effect. In addition, we conducted a simulation in which the surface was initially exposed to ion bombardment followed by simultaneous growth and ions, and we found that a combination of growth and ion bombardment produce a smoother surface morphology than ions alone.

A comparison between ion-assisted and thermal smoothing following the deposition of 0.50 ML is shown in Figure 2. Figure 2 contains plan view images of three simulated surfaces. Fig. 2(a) depicts a surface following the deposition of 0.50 ML. Fig. 2(b) and 2(c) are two surfaces that were treated differently following the deposition of 0.50 ML. The surface in Fig. 2(b) was allowed to thermally smooth with no further growth, while the surface in Fig. 2(c) was exposed to simultaneous growth and ion bombardment for an equivalent amount of time. [The evolution of the RHEED intensity corresponding to the surface in Fig. 2(c) is found in Fig. 1.] The surface in Fig. 2(b) still shows the presence of large adatom islands on the terraces. However, on the surface exposed to low-energy ions, all islands have disappeared and the adatoms have reached step edges.

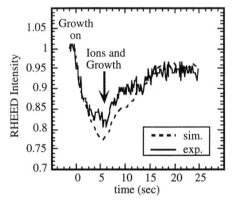

FIG 1. RHEED intensity (simulated and experimental[4]) of a smooth surface exposed initially to growth (0.10 ML/sec) followed by simultaneous ion bombardment and growth. The arrow indicates the initiation of concurrent ion bombardment and growth. The substrate temperature was 250 °C.

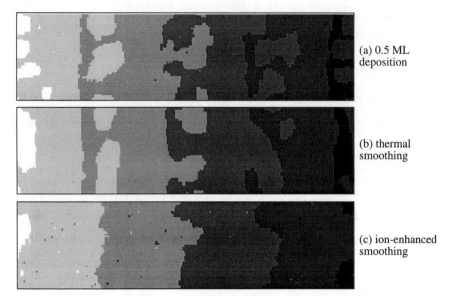

(a) 0.5 ML
deposition

(b) thermal
smoothing

(c) ion-enhanced
smoothing

FIG 2. Gray-scale plan view images of simulated surfaces (darker shades represent lower features). The surface in (a) was exposed to 0.50 ML growth. The other surfaces were initially exposed to 0.50 ML growth followed either by: no growth (b) or simultaneous ion bombardment and growth (c) for the same length of time. The substrate temperature was 250 °C in each case.

The terrace edges are clearly evident on the surface exposed to low-energy ions but more difficult to identify on the other surface. Signs of point defects (adatoms and vacancies) exist on the ion-enhanced surface, but the concentration of these defects is much less than the concentration of adatoms that constitute the islands on the surface that evolved thermally. Unlike the large adatom islands, these point defects contribute very little to the decay in the RHEED intensity. Thus, the rate of thermal smoothing is much lower than the rate of low-energy ion bombardment enhanced smoothing.

Our kinetic Monte Carlo simulations demonstrate that the enhancement of surface smoothness during simultaneous growth and ion bombardment occurs primarily via the destabilization of adatom islands by ion-induced defects. The annihilation of ion-induced vacancy defects with growth-induced adatom defects represents another, less significant, mechanism of smoothing. The population of ion-induced defects primarily enhances the smoothing process by reducing large islands into small adatom clusters allowing adatoms to reach step edges. The success of this mechanism lies with the relative stability of small adatom clusters compared to large islands. Small adatom clusters are less stable than large adatom islands, especially at elevated substrate temperatures. Thus, low-energy ion bombardment is most effective is when a large difference exists between the stability of small adatom clusters and large adatom islands.

Figure 3 illustrates how important the instability of small clusters is to the "effectiveness" of low-energy ion bombardment during growth. Each curve in Fig. 3 represents 0.50 ML of growth followed by simultaneous growth and ion bombardment at different substrate temperatures. As the substrate temperature is decreased, the RHEED intensity drops to a lower value following 0.50 ML of deposition, representative of the stability of island formation on terraces. In addition, the value of the steady-state RHEED intensity as well as the rate of recovery of the dynamic surface morphology increases with increasing substrate temperature. Our proposed mechanism produces the most significant enhancement of the surface morphology at 250°C (Fig. 3) when the growth is near step flow. In this regime, the relative stability of small adatom clusters compared to large

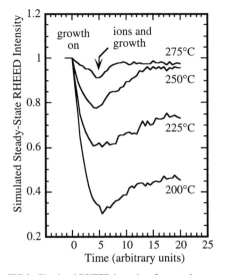

FIG 3. Simulated RHEED intensity of a smooth surface initially exposed to growth (0.50 ML) followed by simultaneous growth and ion bombardment at substrate temperatures of 200°C, 225°C, 250°C, and 275°C. The arrow indicates the initiation of concurrent ion bombardment and growth.

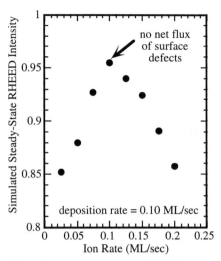

FIG 4. Comparison of steady-state RHEED intensity obtained during simultaneous ion bombardment and growth for different incident ion-to-vapor flux ratios. The deposition rate was held fixed at 0.10 ML/sec while the ion bombardment rate was varied from 0.00 to 0.20 ML/sec. The arrow indicates when the net creation of adatoms and vacancies is balanced.

islands can be exploited. Low-energy ion bombardment reduces large islands into small clusters. The small clusters are sufficiently unstable so that they can spontaneously decompose permitting the adatoms to reach step edges. Thus, low-energy ion bombardment during growth can effectively lower the temperature at which step-flow growth can be achieved. The enhancement is not as pronounced at lower substrate temperatures, because surface defects become less mobile and small clusters become more stable.

As described earlier, the ion-induced defects reside transiently on the surface at growth temperatures prior to recombination with other defects or annihilation at step edges. If the ion-induced defects do not recombine with opposite-type defects generated within the same cascade, they can migrate away from the original point of ion impact and interact with other defects. Within the framework of this mechanism, the effects of low-energy ion bombardment are long range and direct impingement of an existing island is not necessary for its breakup.

In the experimental study, a maximum in the steady-state RHEED intensity was found by varying the incident ion-to-vapor flux ratio. Several simulations (similar to experimental procedure) were performed in which the deposition rate was held fixed at 0.1 ML/sec while the ion bombardment rate was varied from 0.0 to 0.2 ML/sec (following the deposition of 0.5 ML at 0.1 ML/sec). Each simulation run was carried out until a steady-state RHEED intensity was reached. Figure 3 shows the steady-state RHEED intensity achieved as a function of the ion bombardment rate (with a constant deposition rate). Using the kinetic Monte Carlo simulations, we found that a maximum in the simulated steady-state RHEED intensity was obtained when the incident ion-to-vapor flux ratio was approximately one, consistent with experimental observations. This result can be interpreted in the following manner. Each growth atom produces one adatom defect on the surface, while each ion impingement event contributes one vacancy defect. Recall that the sputter yield for 200 eV Ar/Xe is approximately one defect/ion. After complete recombination of ion-induced vacancies and adatoms, all that can remain on the surface is the uncompensated sputter vacancy, one defect/ion. As a result, the optimum surface smoothness is obtained when the net production of surface defects is balanced.

The simulations have also been used to address the question of whether the minimization of defect production is necessary within the framework of this mechanism. The effects of producing multiple surface defects during each ion impingement event can be addressed by comparing the steady-state surface morphology obtained as a function of the number of defects produced per ion. The number of defects was varied from 1 to 20 keeping the difference between the number of vacancies and adatoms at one, thereby maintaining a constant sputter yield. The optimum steady-state surface morphology (corresponding to the peak in Fig. 4) can be achieved using defect production rates as high as 10 defects/ion. We have previously shown during 200 eV Ar/Xe ion bombardment of germanium surfaces that all the ion-induced damage (approximately 10 defects/ion) is confined to the near surface region and that all ion-induced defects reach the surface[7,8].

We have presented a simple atomistic mechanism describing the effects of low-energy ion bombardment of semiconductor surfaces that is consistent with experimental observations of the evolution of surface morphology during low-energy ion bombardment of dynamic and static Ge(001) surfaces. The mechanism includes the production and interaction of growth-induced and ion-induced defects. Our representation of low-energy ion bombardment involves the production of multiple adatom and vacancy surface defects. These defects exist only transiently on the surface but are able to produce a smoother surface morphology during ion beam-assisted deposition primarily through the destabilization of adatom clusters enabling adatoms to reach step edges. The significant ion-enhancement occurs with the destabilization of the more stable large adatom islands. In addition, we have shown that the optimum steady-state dynamic surface morphology occurs when a balance exists between the growth-induced and net ion-induced defects. We found that the optimum steady-state surface morphology can occur with ion-induced defect production rates less than or equal to 10 defects/ion.

This work was performed at Sandia National Laboratories and supported by the United States Department of Energy under Contract No. DE-AC04-94AL85000. JMW acknowledges the support of the National Science Foundation under Grant No. CHE 8920120.

REFERENCES

email:{bkkelle,ehchaso,jafloro,picraux}@sandia.gov
1. J.K. Hirvonen, Mater. Res. Soc. Symp. Proc. **316**, 545 (1994).
2. M. Marinov, Thin Solid Films **46**, 267 (1977)
3. J.E. Greene, S.A. Barnett, J.-E. Sundgren, and A. Rockett, in *Ion Beam Assisted Film Growth*, edited by T. Itoh (Elsevier, Amsterdam, 1989), p. 101.
4. E. Chason, P. Bedrossian, K.M. Horn, J.Y. Tsao, and S.T. Picraux, Appl. Phys. Lett. **57**, 1793 (1990). and E. Chason, K.M. Horn, J.Y. Tsao, and S.T. Picraux, Mater. Res. Soc. Symp. Proc. **128**, 35 (1989).
5. E. Chason, P. Bedrossian, J.E. Houston, J.Y. Tsao, B.W. Dodson, and S.T. Picraux, Appl. Phys. Lett. **59**, 3533 (1991).
6. M. Kitabatake and J.E. Greene, J. Appl. Phys. **73**, 3183 (1993). and O. Vancauwenberghe, N. Herbots, and O.C. Hellman, J. Vac. Sci. Technol. B **9**, 2027 (1991).
7. J.A. Floro, B.K. Kellerman, E. Chason, S.T. Picraux, D.K. Brice, and K.M. Horn, J. Appl. Phys. **77**, 2351 (1995).
8. B.K. Kellerman, J.A. Floro, E. Chason, D.K. Brice, S.T. Picraux, and J.M. White, J. Vac. Sci. Technol. A (in press).
9. E. Chason and B.W. Dodson, J. Vac. Sci. Technol. A **9**, 1545 (1991).
10. E. Chason, J.Y. Tsao, K.M. Horn, S.T. Picraux, and H.A. Atwater, J. Vac. Sci. Technol. A **8**, 2507 (1990).
11. J.M. van Hove, P.R. Pukite, P.I. Cohen, and C.S. Lent, J. Vac. Sci. Technol. A **4**, 1251 (1986).
12. S. Clarke and D.D. Vvedensky, Phys. Rev. B **36**, 9312 (1987).

THE ROLE OF IONS FOR THE DEPOSITION OF HYDROCARBON FILMS, INVESTIGATED BY IN-SITU ELLIPSOMETRY

A. VON KEUDELL
Max-Planck-Institut für Plasmaphysik, EURATOM Association
Boltzmannstr.2, 85748 Garching, Germany

ABSTRACT

The growth mechanisms for the deposition of hydrocarbon films (C:H-films) from a methane electron cyclotron resonance (ECR) plasma are investigated by means of in-situ ellipsometry. Ion bombardment during plasma-enhanced chemical vapor deposition of hydrocarbon films mainly governs the properties of the films and the total growth rate. The role of ions for the growth rate and the film properties is discussed in this paper. Films were deposited with varying RF-bias, resulting in a DC self-bias ranging from floating potential up to 100 V. The ion-induced modification of the film properties was investigated by a new technique using a double layer consisting of a polymer-like film with low optical absorption and a hard carbon film with high absorption on top. The interface between these layers was analysed after deposition by a layer-by-layer etching in an oxygen plasma at floating potential. From these data it is possible to determine with high accuracy the range of the ion-induced modification of the optical properties in the underlying polymer-like film. The thickness of this modified layer ranges from 6 Å at 30 V self-bias to 40 Å at 100 V self-bias, which is consistent with the range of hydrogen ions in polymer-like films as calculated by the computer code TRIM.SP.
Based on the presented results, the growth of C:H-films and the resulting film properties can be modelled by the growth at activated sites at the film surface. These activated sites are represented by dangling bonds, induced by the ion bombardment. They also show up in the ellipsometric results during the deposition of C:H-films by a change of the optical response of the film surface.

INTRODUCTION

Hydrocarbon films exhibit a wide range of applications due to their large variety of properties [1,2,3]. They are used as infrared transparent films or as hard coatings for tools and optical devices. C:H-layers are usually prepared by plasma enhanced chemical vapor deposition from hydrocarbon source gases. In these plasmas, the hydrocarbon molecules are dissociated and ionized and the radicals as well as the ions impinging on the substrates lead to the growth of the films. But despite the important role of a-C:H-films in plasma technology, the growth mechanisms are not well known. Only main features can be postulated: The deposition rate is decreasing with increasing substrate temperature, independent of the source gas and the deposition device. Further, the total growth rate cannot be explained by the simple incorporation of ions during deposition, because the incoming flux of ions, as can be measured by mass spectrometry, is in general lower than the observed deposition rate [4,5]. This lack in the total deposition rate cannot be satisfied by neutral species from the plasma, because the cross section for the formation of a new chemical bond for an impinging radical on a hydrogen terminated C:H-film surface is small [6]. This was shown by Lange et al. [7].
Thus, it is necessary to introduce a synergetic effect between ions and neutrals for a growth mechanism to explain the total growth rate of C:H-films. This synergetic effect has been recently modelled by an 'adsorbed layer' model [8,9,10]: neutral species impinge on a growing film surface and get physisorbed in a first step. This results in a dynamic coverage of the C:H-film, which is balanced by the thermal desorption of the adsorbed species. The physisorbed neutral species may chemisorb in a second step by the interaction with impinging ions. The impinging ions transfer

their kinetic energy to the adsorbed species to overcome the activation barrier for a new chemical bond. The temperature dependence of this coverage leads to a temperature dependent growth by incorporated neutral species. In a combined modelling for the surface and the plasma good agreement is achieved with the measurement of deposition rates and the measurement of CH-densities by laser induced fluorescence (LIF) [11].

However this model has been contradicted by recent experiments: The temperature dependence of the net deposition rate is due to a temperature dependent erosion by atomic hydrogen [12,13]. This temperature dependent erosion step has been neglected for simplicity in the modelling by an 'adsorbed layer'. Based on these new results it is not necessary to postulate an 'adsorbed layer' of radicals on the growing C:H-film to describe the temperature dependence of the measured total growth rate. If we assume that this adsorbed layer does not exist, the synergetic effect between ions and neutrals on the growing film may be modelled alternatively by the creation of 'activated sites' on the film surface: The impinging ions create in a first step dangling bonds on the growing film by the displacement of bonded hydrogen. In a second step neutral species immediately chemisorb at these 'activated surface sites'.

The film properties are strongly dependent on the ion energy during deposition. Hard carbon films with a low hydrogen content are deposited using high ion energies (>100 eV) and soft polymer-like films with a large hydrogen content are deposited using low ion energies (<30 eV) [1,2,14]. Only main features can be postulated to describe the influence of the ion bombardment on the film properties. As can be seen from a compilation of Jacob and Möller [15], all C:H-films exhibit a specific range of film compositions, independent of the deposition device and the hydrocarbon source gas. This indicates that C:H-films have an equilibrium configuration, which can be well described by the random covalent network model [16]. Another class of carbon films with an extreme low hydrogen and a high sp^3/sp^2-ratio can be deposited (ta-C), using an ion dominated deposition process [17]. The formation of these films was successfully explained by the subplantation model, which is only based on ion kinetics [18,19,20]. This demonstrates that on the one hand the equilibrium state and on the other hand ion kinetic has to be used to predict the film composition. But it is still unclear, how the equilibrium composition of C:H-films is induced by the ion bombardment.

To answer these questions, the growth mechanisms during the deposition of C:H-films from an ECR-plasma of methane are investigated in-situ by means of ellipsometry, by measuring the optical response of a growing film. Ellipsometry is an especially sensitive method to investigate small changes in the optical response of film surfaces [21]. This paper deals with the ion induced modification of C:H-films and the ion induced formation of dangling bonds at the film surface. These dangling bonds represent the 'activated sites' for C:H-film growth.

EXPERIMENT

An electron-cyclotron resonance plasma was used for the deposition of C:H-films from methane and for the erosion of the deposited films with hydrogen or oxygen-plasmas. The gas flow was measured by flow controllers and ranges from 15 sccm to 20 sccm. A magnetic field was applied to obtain the resonant field of 87.5 mT. The absorbed microwave power in the plasma was controlled by directional couplers, yielding 0.01 Wcm^{-3}. Films were deposited on silicon substrates mounted on a RF-driven electrode. The use of a DC-Bias to vary the energy of the ions impinging on the growing film surface is not practical, because a charging of the insulating films reduces the sheath potential between plasma and film surface. The total pressure was about 0.2 Pa. At these low pressures the sheath is considered to be collisional free, so the measured self bias corresponds directly to the ion energy. This DC-self-bias was ranging from floating potential to 100 V. The substrate temperature can be changed by an external heating of the RF-electrode and is controlled by a thermocouple. To obtain a good thermal contact between sample, electrode and thermocouple colloidal carbon was used as bonding substance. During the heating of a Si-substrate excellent agreement is achieved between the temperature as measured by the thermocouple and the temperature as derived from ellipsometry based on the temperature dependent change in the optical constants of silicon [22,23]. To obtain a decoupling of the plasma as the source for the ions and radicals and the growth mechanisms on the film surface, a metallic

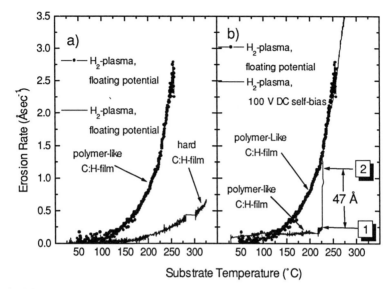

Fig.1a): Polymer-like and hard carbon films were eroded with a hydrogen plasma at floating potential. The erosion rate increases with substrate temperature and is larger for polymer-like films. **1b):** Polymer-like films are eroded with a hydrogen plasma at floating potential and at a self bias of 100 V. Ion bombardment during erosion reduces the erosion yield. At point 1 the additional RF-bias during erosion is switched off.

cage was used for the confinement of the plasma. Through an aperture in the cage (diameter 35 mm) a plasma beam is extracted for the deposition of the C:H-layers. In this view, any changes in the RF-Bias or the substrate temperature do not influence the production of ions and radicals in the bulk plasma.

We use a rotating analyser ellipsometer (RAE) to measure the ellipsometric angles Ψ and Δ [24,25,26]. As light source, an Xe-short arc lamp is used and is monochromatised by an acousto-optic-filter. The whole setup is calibrated by 'residuum calibration' [27] and the residual errors are 0.0038° in Ψ and 0.009° in Δ for Ψ=12.2° and Δ=112.7°. The details of the ellipsometric setup are presented elsewhere [28]. All the experiments presented here are performed at a wavelength of 600 nm.

THE INTERACTION OF HYDROGEN IONS WITH POLYMER-LIKE FILMS

The temperature dependence of the growth of C:H-films from a methane plasma is mainly caused by the temperature dependent erosion by atomic hydrogen [12,13], which is generated by the dissociation of methane in the plasma. The erosion of hard carbon films and of polymer-like films by a hydrogen plasma at floating potential is shown in Fig.1a. The etch rate increases with increasing substrate temperature, and the temperature dependence of the erosion is larger for the polymer-like film than for the hard carbon film. In addition, the etch rate by thermal atomic hydrogen, which was generated by a hot filament in the reactor, was investigated. The temperature dependence of the erosion for a polymer-like film by a hydrogen plasma or a hot filament as the source for thermal hydrogen is identical. This indicates that the erosion is caused by a reaction with thermal hydrogen atoms. The difference between the erosion of the hard carbon film and the polymer-like film shows that the etch rate depends strongly on the film composition.

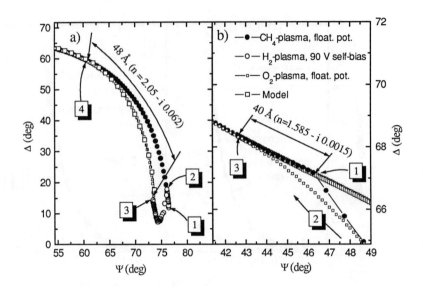

Fig.2a): Ellipsometric results for the deposition and etching of a C:H film. A polymer-like C:H film was etched in a hydrogen plasma at floating potential (point 1). At point 2 an additional RF bias was switched on. At point 3 the resulting film system was etched in an oxygen plasma with no bias. **2b):** Ellipsometric results for the deposition of a double-layer system comprising a hard carbon film on top of a polymer-like film. During the deposition process an additional RF bias leading to 90 V DC self bias was switched on at point 1. By etching the interface between the two films in an oxygen plasma (curve 2), the range of the ion induced modification of the film properties was determined.

Up to now the erosion has been investigated using low ion energies in the range of the floating potential and atomic hydrogen has been identified as the main reactant. In Fig.1b the temperature dependent erosion of a polymer-like film by a hydrogen plasma at floating potential is compared with the erosion of the same film by a hydrogen plasma at a self-bias of 100 V. It can be seen that the use of an additional biasing does not increase the erosion yield. The temperature dependent erosion of the polymer-like film by a hydrogen plasma with a self bias of 100 V is similar to the erosion of a hard carbon film with thermal atomic hydrogen, as presented in Fig1a. This leads to the assumption that the ion bombardment with hydrogen ions during the erosion transforms the film surface of the polymer-like film into a hard carbon top layer: as the etch rate by thermal hydrogen atoms depends on the film composition it is reasonable that the formation of a hard carbon toplayer reduces the erosion yield. This assumption is supported by the observation that the erosion rate increases by a factor of 4 after switching off the additional RF-bias at a substrate temperature of 220°C (Point 1 in Fig. 1b). At this point the erosion rate accelerates and coincides with further increasing substrate temperatures with the erosion rate as measured for the erosion of the polymer-like film with a hydrogen plasma at floating potential (Point 2 in Fig.1b). This can be explained as follows: by switching off the additional RF-bias the film surface is no longer transformed into a hard carbon top layer. After the erosion of this toplayer the underlying polymer-like film is reached. From the integration of the erosion rate, the thickness of this toplayer can be derived, yielding 47 Å.

The transformation of the C:H-film surface due to the ion bombardment during erosion can also be observed by a change of the optical response as shown in Fig.2a: a polymer-like-film has been deposited until Δ reaches values of about 15 deg. At this point the sensitivity of Ψ and Δ on

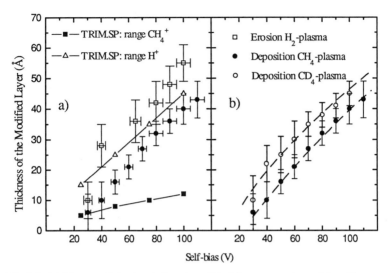

Fig.3a): Thickness of the ion-induced modified layer as measured by ellipsometry. Also shown is the range of carbon and hydrogen ions in a polymer-like C:H film as calculated by TRIM.SP. **3b):** The thickness of the ion-induced modified layer is compared for the deposition of C:H-films from pure methane and deuterated methane.

any changes in thickness or optical constants is maximal. After switching off the methane plasma (Point 1 in Fig. 2a), the film is etched by a hydrogen plasma at floating potential. The identical values in Ψ,Δ for deposition and erosion imply that the erosion process does not change the morphology or the optical constants of the surface of a polymer-like film. At point 2 in Fig. 2a an additional RF-bias is switched on, leading to an increased ion bombardment with kinetic energies corresponding to the self-bias of 90 V. The Ψ,Δ-values are shifted towards lower values of Ψ, indicating the formation of a top layer with a larger extinction coefficient on top of the polymer-like film. After switching off the hydrogen plasma (point 3 in Fig.2a), the film is etched by an oxygen plasma at floating potential. This etching leads to a layer-by-layer removal of the deposited film system. The uniformity of the erosion by an oxygen plasma was verified independently by the deposition of polymer-like films and their erosion by an oxygen plasma at floating potential. At point 4 in Fig.2a the curve in Ψ,Δ for the erosion by the oxygen plasma coincides with the curve for the initial deposition of the polymer-like film. This erosion curve can be modelled by an optical two-layer model. The optical constants are n=1.585-i 0.0015 for the polymer like film and n=2.050-i 0.062 for the top layer with a thickness of 48 Å. The optical constants for the toplayer were determined independently by the measurement of the optical properties of C:H-films, which were deposited at a self-bias of 90 V.

The ion film interaction can not only be observed during the erosion of polymer-like films by a hydrogen plasma but also during the deposition of these films from methane as shown in Fig. 2b: a polymer-like film has been prepared up to a thickness of 70 nm (point 1 in Fig. 2b). At this point an additional RF-bias is switched on resulting in a DC self-bias of 90 V. The impinging ions lead to the growth of a hard carbon film on top of the polymer-like film. The interface between polymer-like film and hard carbon film can be investigated in situ by etching the deposited film system with an oxygen plasma. This leads to a layer-by-layer removal of the deposited film system (curve 2 in Fig. 2b). As shown in Fig 2b., the Ψ,Δ-curve for the erosion by the oxygen plasma reaches the Ψ,Δ-curve for the deposition of the polymer-like film at a different point (Point 3 in Fig.2b) as compared to the starting point (Point 1 in Fig.2b) of the ion bombardment. This leads to the conclusion that the impinging ions lead not only to the growth of a hard carbon film on top of the polymer-like film at the onset of the ion bombardment, but also to a modification of the underlying polymer-like film due to the penetration of the ions. The range of this ion-induced modification of the film properties can be determined by an optical modelling of the polymer-like film with n=1.600-i 0.002. From the difference in Ψ,Δ between the onset of the ion bombardment (point 1 in Fig. 2b) and the coincidence of the deposition and the erosion curve (point 4 in Fig.2b) a modified layer of 40 Å for a self bias of 90 V is determined.

In Fig. 3a., the thickness of the ion-induced modified layer is shown to depend on the self-bias for deposition from methane and erosion by a hydrogen plasma. In addition, computer simulations for ion-solid interactions with the code TRIM.SP [29,30] are performed for the ranges of hydrogen ions and CH_4^+-ions in polymer-like films. The comparison of the measured thickness of the ion-induced modified layer and the calculated range for the ions show that only the penetration of hydrogen ions can explain the observed range of the ion-induced modification of polymer-like films. Especially for the interaction of a methane plasma with a polymer-like film, the thickness of the modified layer cannot be explained with the range of CH_4^+-ions or any other carbon carrying ion, according to the TRIM.SP calculations.

To check whether the range of the ion-induced modification of polymer-like films has to be attributed to the range of hydrogen ions or to chemical effects, equivalent experiments as shown in Fig. 2b were performed using deuterated methane. As shown in Fig.3b, the range of the ion induced modification of a polymer-like film using deuterated methane is significantly higher than the range using normal methane as hydrocarbon source gas. This is consistent with the predictions of the TRIM.SP calculations. The range of deuterium ions is larger in C:D-films in comparison to the range of hydrogen ions in C:H-films: for energies below 150 eV the determining factor for the range of hydrogen ions is not the energy transfer in a collision processes between hydrogen and carbon, but the scattering angle. As deuterium has the double mass of hydrogen, it is scattered to lower angles in a collision with carbon atoms yielding a larger penetration depth compared to hydrogen ions with the same kinetic energy. This shows that the range of the ion-induced modification of polymer-like films can be explained by ion kinetics. We therefore conclude that the range of hydrogen ions is responsible for the ion-induced modification of polymer-like films.

The following two-step process is proposed to describe the interaction of hydrogen ions with C:H-films. In a first step, ions create free bonds in the carbon network by physical displacement [31] or chemical abstraction of bonded hydrogen atoms. In a second step these free bonds may vanish by several reactions. On the one hand they can recombine with atomic hydrogen from the plasma. On the other hand, they can either relax in forming sp^2-coordinated carbon [6], which leads to a larger absorption coefficient of the modified layer, or these free bonds can be saturated by other dangling bonds in the carbon network yielding a larger density of the modified layer, which leads to a larger index of refraction. This formation of new carbon-carbon bonds implies a reconfiguration of the carbon network, which might be sterically hindered. However, this activation barrier for a re-arrangement of the carbon network may be overcome by a transfer of kinetic energy from the impinging ions to the carbon network The assumption of this two step process implies a dynamic coverage of dangling bonds at the film surface during deposition of C:H-films. The possible formation and experimental identification of these dangling bonds will be discussed as follows.

THE FORMATION OF DANGLING BONDS AT THE C:H-FILM SURFACE

To investigate the formation of dangling bonds at the film surface during growth, C:H-films were deposited with an additional self-bias, as shown in Fig. 4a: a C:H-film is deposited with a self bias of 80 V with a thickness of 80 nm. After switching off the plasma (point 1 in Fig. 4a), the ellipsometric angle Ψ increases by 0.3 deg ($\delta\Psi_1$) on a time scale of 1000 sec (point 2 in Fig.4a). Then the film is etched by an oxygen plasma at floating potential. The Ψ,Δ–curve for the erosion by an oxygen plasma is shifted to larger values of Ψ with respect to the Ψ,Δ–curve for the deposition of the C:H film (deposition-erosion-cycle 1).

The difference in the optical response of the C:H-film during deposition and erosion can also be observed after switching off the methane plasma for the growth of the C:H-film and starting the oxygen plasma for the erosion immediately after the deposition (erosion cycle 2 in Fig. 4a). The shift $\delta\Psi_2$ between deposition and erosion after etching a toplayer (~50Å between points 1

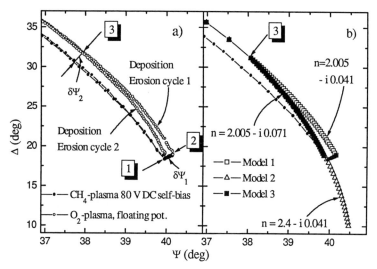

Fig.4a):After switching off the plasma for the deposition of a C:H-film at 80 V self bias (point 1), the ellipsometric angle Ψ shifts to larger values ($\delta\Psi \sim 0.3$ deg) on a time scale of 1000 sec. Afterwards the film is etched with a oxygen plasma at floating potential (point 2, deposition erosion cycle 1). In deposition-erosion cycle 2 the film is etched with an oxygen plasma at floating potential immediately after switching off the deposition plasma. The shift $\delta\Psi_2$ at point 3 is equivalent to the shift $\delta\Psi_1$ at point 2 **4b):** Model curves for the interpretation of the ellipsometric results from Fig. 4a). The models represent a homogeneous film with n= 2.005 - i 0.041 (model 1) and a toplayer of 50 Å with n = 2.005-i 0.071 (model 3) and n = 2.4 - i 0.041 (model 2), respectively.

and 3 in Fig. 4a) by the oxygen plasma is equivalent to the shift $\delta\Psi_1$ after a delay of 1000 sec. This leads to the conclusion that the difference in Ψ,Δ between the deposition with additional RF-bias and the erosion by an oxygen plasma at floating potential, can be equivalently monitored by etching the first monolayers of the deposited films or by observing the relaxation of the deposited film after switching off the plasma. This observation can be easily explained by the assumption that the source for this shift is removed by the erosion of the surface. In this view, $\delta\Psi$ represents a change in the optical response of only the film surface.

In Fig. 4b. three optical models for the change in the optical response of the film surface are presented. These models are fitted to the measured Ψ,Δ-values, as shown in Fig.4a). The scaling of Fig. 4a and 4b is identical. Model 1 represents the optical response of a homogeneous film with n=2.005-i0.041, model-curve 2 and 3 represent a homogeneous film with a toplayer of 50 Å with n=2.4-i0.041 and a toplayer of 50 Å with n=2.005-i 0.071, respectively. It can be clearly seen that only a change in the extinction coefficient of the toplayer during the relaxation of the deposited film is consistent with the observed shift $\delta\Psi$. With a decreasing extinction coefficient of the toplayer model curve 3 changes into model curve 1. This variation of the extinction coefficient results in a shift of the end point of model curve 3 which coincides with the measured shift in Ψ. Any change in the index of refraction would also lead to a change in the optical thickness of the films and thus to a change in Δ, as shown by model curve 2. However, during the relaxation of the films no shift in Δ is observed, which implies that any changes in the film thickness after switching off the methane plasma can be excluded. In addition, any changes in the surface morphology like surface roughness can be excluded, because this would also show up in a change in Δ. The observed shift $\delta\Psi$ might be explained by an increase in the substrate temperature of 50°C, but the substrate temperature as measured by thermocouples decreases after deposition by 5-10°C. We

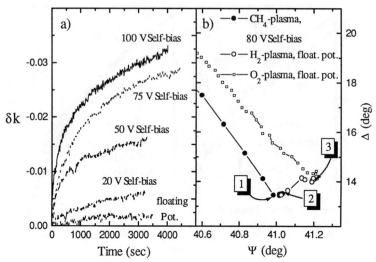

Fig.5a): The shift in Ψ after switching off a methane plasma at a varying self bias is monitored. The shift δΨ is transformed into a shift δk of a toplayer of 50 Å, to obtain comparable results for different points in the Ψ,Δ-plane. The shift dk is increasing with increasing self-bias during deposition. **5b):** After switching off a methane plasma at 80 V DC self-bias (point 1), a hydrogen plasma at floating potential (point 2) is switched on. Afterwards the film is etched with a oxygen plasma at floating potential (point 3).

conclude that the change of the optical response of the surface is due to a change of the extinction coefficient of a top layer.

In Fig. 5b. the reaction of the deposited films in the presence of atomic hydrogen is presented. A C:H-film was deposited with a self-bias of 80 V. After switching off the methane plasma (point 1 in Fig.5b), a hydrogen plasma at floating potential is switched on for 10 sec (point 2 in Fig.5b). As can be seen from Fig.5b the shift δΨ accelerates at this point, which indicates that the extinction coefficient of the film surface is decreasing very fast due to the interaction with atomic hydrogen. After switching off the hydrogen plasma, the deposited film is etched by an oxygen plasma at floating potential (point 3 in Fig.5b). This result shows, that the change of the optical response of the film surface reacts very strongly to the presence of atomic hydrogen.

The shift δΨ after switching off the methane plasma was investigated for the deposition with varying bias potential. As the index of refraction of C:H-films depends strongly on the ion energy the shift δΨ has to be monitored at different points in the Ψ,Δ-plane. To allow a comparison of the results for different ion energies, the shift δΨ was transformed in a shift δk of the extinction coefficient k of a toplayer with a thickness of 50 Å. As shown in Fig.5a, the shift δk increases with increasing self-bias during the deposition.

To interpret the observed change of the optical response of the film surface and its dependence on the ion energy as well as on the presence of atomic hydrogen, we propose the following: The impinging ions create dangling bonds at the film surface. The number of created dangling bonds is presumed to be equal to the displacement yield for bonded hydrogen atoms in a hydrocarbon network [31]. The formation of dangling bonds at the film surface leads to an increased extinction coefficient, because these free bonds correspond to states in the mobility gap of amorphous films [32]. The absorption of C:H-films is very sensitive to these states [33]. The relaxation of these free bonds or their possible saturation by radicals from the gas phase leads to a decreasing density of states in the mobility gap and thus to a decreasing extinction coefficient of the film surface.

The cross section for the recombination of atomic hydrogen with a free bond in the carbon network is very large. In this view, the fast decrease of the extinction coefficient of the film surface is consistent with the fast recombination of atomic hydrogen with the dangling bonds.

The amount of these dangling bonds increases with increasing ion energy, because higher ion energies lead to a larger displacement yield of bonded hydrogen atoms and thus to a larger extinction coefficient of the film surface. Based on this, we consider $\delta\Psi$ as a quantity, which is correlated to the total amount of free bonds at the film surface. The influence of the coverage of dangling bonds on the growth process is discussed as follows.

GROWTH MECHANISMS FOR THE DEPOSITION OF C:H-FILMS

The presence of dangling bonds at the film surface during C:H-film growth has consequences on the one hand for the synergetic effect between ions and neutrals during C:H-growth and on the other hand for the formation of film properties due to the ion bombardment.

According to the deposition model by 'activated surface sites', neutral radicals from the plasma can chemisorb on the C:H-film surface by their recombination with a dangling bond. To obtain a quantitative contribution to the total growth rate by this incorporation of impinging radicals from the plasma, a dynamic coverage of these activated sites on the film surface is required. The time constant of the change in the optical response of a C:H-film surface, as interpreted by the change of the total amount of dangling bonds on the growing film is long, so the dynamic coverage is mainly balanced between the creation of active sites due to the displacement yield by the ion bombardment and the recombination of these active sites with impinging radicals from the plasma. In a balance equation for the coverage of a growing film by dangling bonds the decay time for these bonds may be neglected.

The modification of a polymer-like film due to interaction with impinging hydrogen ions, was explained by a two step process: In a first step, free bonds are created in a toplayer of the polymer-like film according to the range of the impinging ions. In a second step, these free bonds can recombine in forming new carbon-carbon bonds by a rearrangement of the carbon network. The reconfiguration of the carbon network may be sterically hindered, but the activation barrier for this rearrangement may be overcome by the transfer of kinetic energy to the network due to the ion bombardment.

The same two step process can also be used to describe the growth process of C:H-films: we consider a fully saturated C:H-film surface. In a first collision cascade ions create free bonds in a toplayer according to the range of the ions and the displacement yield for bonded hydrogen and carbon atoms. In such a collision process new carbon-carbon bonds are created instantaneously, which implies the abstraction of two hydrogen atoms to obtain two free sites and a reconfiguration of the network to recombine these bonds. As this process is sterically hindered, this reconfiguration is not always successful and a certain amount of free bonds remains unsaturated. These free bonds can be saturated by thermal hydrogen atoms from the plasma. This results in a balance between the creation of dangling bonds by the ion bombardment and their recombination with hydrogen atoms from the plasma. The activation barrier for a reconfiguration of the network and thus the recombination of the remaining free bonds can also be overcome by the transfer of kinetic energy to the carbon network in a second collision cascade. Summarizing, the ability of the ion bombardment to create new C-C-bonds in the growing C:H-film is not only a function of the ion energy but also a function of the density of free bonds in the carbon film. As shown by the experiments on the interaction of a C:H-film surface with ions and hydrogen atoms, the dynamic coverage of free bonds on the film surface is a function of the ion energy and the incoming flux of hydrogen atoms.

From this follows that we can influence the amount of new carbon-carbon bonds, created by the ion bombardment by modifying the coverage of dangling bonds at the C:H-film surface. This can be achieved by the addition of hydrogen to the methane source gas, because this results in a higher flux of hydrogen atoms towards the surface and thus to a decreasing coverage of dangling bonds due to their recombination with hydrogen atoms.

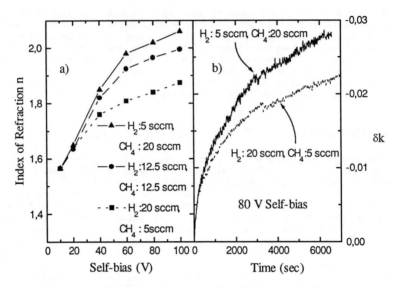

Fig.6a): The index of refraction is investigated for different hydrogen to methane ratios in the source gas and a varying self-bias during deposition. **6b):** The shift δk after switching off the plasma for the deposition at 80 V self-bias is larger for a low hydrogen content in the source gas compared to a large hydrogen content in the source gas.

As the index of refraction of C:H-films is directly correlated to the density of the films and thus to the amount of carbon-carbon bonds in the film, we investigated the deposition of C:H-films in dependence on the ion energy and different hydrogen to methane ratios in the source gas. In Fig.6b the relaxation after the deposition of C:H-films at a self-bias of 80 V with a low hydrogen to methane ratio in the source gas and a high hydrogen to methane ratio in the source gas is shown. It can be seen that δk is larger for a low hydrogen content in the source gas compared to the δk for a large hydrogen content in the source gas. This is consistent with the assumption that the increased flux of atomic hydrogen leads to a lower dynamic coverage of dangling bonds on the film surface and thus to smaller δk.

In Fig.6a the index of refraction of C:H-films is shown in dependence on the hydrogen to methane ratio in the source gas and in dependence on the RF-bias used during the deposition. It can be seen, that the index of refraction is not influenced by the hydrogen to methane ratio in the source gas for a low ion energy used during deposition. At high ion energies, however, films exhibit a low index of refraction if they are deposited from a source gas with a large hydrogen to methane ratio. If the hydrogen to methane ratio in the source gas is low the index of refraction of the deposited films is increased. This dependence of the index of refraction on the gas mixtures and the ion energy can easily be explained in terms of the above proposed two-step growth mechanism. At low ion energies, the dynamic coverage of free bonds is low, due to the low displacement yield for bonded hydrogen atoms. This implies that the probability of an impinging atomic hydrogen atom to recombine with a surface site, which is unsaturated, is low. In this view, the variation of the dynamic coverage of dangling bonds only little varies with the hydrogen flux towards the substrate. Therefore, in a growth regime using low ion energies the index of refraction is governed by the instantaneous formation of new carbon-carbon bonds due to the ion bombardment and is independent of the content of atomic hydrogen in the plasma. At larger ion energies the dynamic coverage of dangling bonds on the film surface is increased due to the increasing displacement yield. It follows that the probability of an impinging hydrogen atom to recombine with an unsaturated surface site is high. In this growth regime, the content of hydrogen atoms in the plasma influences the total amount of dangling bonds at the film surface. According to the two step process, the ability of the formation of new carbon-carbon bonds due to the ion bombardment depend on the presence of dangling bonds at the film surface. If the dynamic coverage of dangling bonds at the surface is lowered by the recombination with atomic hydrogen, films with a low index of refraction are obtained.

In this view, the proposed two step process can successfully describe the dependence of the film properties not only on the ion energy, but also on the composition of the incoming flux of radicals and ions from the plasma. The observed effects can easily be explained by a dynamic coverage of unsaturated surface sites, which can be quantified by the investigation of the change of the optical response of the C:H-film surface.

Summarizing, the growth of C:H-films has to be described: with the ion kinetic to describe the displacement yield of bonded atoms in the carbon network due to the ion bombardment and also with the equilibrium state of C:H-films to describe the recombination of unsaturated bonds with hydrogen atoms and the reconfiguration of a sterically hindered carbon network.

CONCLUSION

It has been shown, that the growth mechanisms for the deposition of C:H-films by plasma enhanced chemical vapor deposition can be explained by the formation of dangling bonds at the surface of the growing film. These dangling bonds can be monitored by means of in-situ ellipsometry. The total amount of these dangling bonds is determined by the balance between their creation due to the displacement of bonded hydrogen and carbon atoms by the ion bombardment and by their recombination with atomic hydrogen from the plasma. The density of the deposited films depend strongly on the amount of these dangling bonds, because the formation of new carbon-carbon bonds is more efficient in the presence of unsaturated bonds in a toplayer. Based on these results, it is possible to estimate the film structure of the deposited films by the knowledge of the flux and the energy distribution of incoming species or by the direct monitoring of the film surface with optical methods. The proposed growth mechanims can serve as a basis for a new attempt to model the growth of C:H-films either by rate equations for the film composition or by molecular dynamics. But nevertheless more specific beam experiments are required to identify the proposed mechanisms unambiguously.

The author would like to thank W. Jacob, W. Möller, W. Fukarek and J. Biener for helpful discussions. Appreciation is due to M. Mayer for the TRIM.SP calculations.

1. J.C. Angus, P. Koidl and S. Domitz in: 'Plasma deposited thin films', edited by J. Mort and F. Jansen (CRC Press, Boca Raton, 1986), p. 89
2. Y. Catherine, Materials Science Forum **52&53**, 175 (1989)
3. G. A. Bootsma, F. Meyer, Surf. Sci. **14**, 52 (1969)
4. H.Sugai, H. Kojima, A. Ishida, H. Toyoda, Appl. Phys. Lett. **56**, 2616 (1990)
5. Y. Catherine, P. Couderc, Thin Solid Films **144**, 265 (1986)
6. J. Biener. A. Schenk, B. Winter, J. Küppers, J. Electr. Spectr. Rel. Phen., **64/65**, 331, (1993)
7. K. Lange, W. Jacob, W. Möller, Proc. 11th Int. Symp. on Plasma Chemistry, ed. by J.E.Harry (IUPACV 1993), p. 1266
8. A. v.Keudell, W. Möller, R. Hytry, Appl. Phys. Lett. **62**, 937 (1993)
9. A. v.Keudell, W. Möller, J. Appl. Phys. **75**, 7718 (1994)
10. W. Möller, Appl. Phys. A **56**, 527 (1993)
11. M. Engelhard, W. Jacob, W. Möller, A. Koch, Proc. 11th Int. Symp. on Plasma Chemistry, ed. by J.E.Harry (IUPACV 1993), p. 1382
12. A. Horn, A. Schenk, J. Biener, B. Winter, C. Lutterloh, M. Wittmann, J. Küppers, Chem. Phys. Lett. **231**, 193 (1994)
13. W, Möller, W. Fukarek, K. Lange, A. v. Keudell, W. Jacob, Jap. J. Appl. Phys. (in press)
14. P. Koidl, Ch. Wild, B. Dischler, J. Wagner, M. Ramsteiner, Materials Science Forum **52&53**, 41 (1989)
15. W.Jacob, W. Möller, Appl. Phys. Lett. **63**, 1771 (1993)

16. J.C. Angus, Thin Solid Films **142**, 145 (1986)
17. M. Weiler, S. Sattel, K. Jung, H. Ehrhardt, V.S. Veerasamy, J. Robertson, Appl. Phys. Lett. **64**, 2797 (1994)
18. Y. Lifshitz, S.R.Kasi, J.W. Rabalais, W. Eckstein, Phys. Rev. B **41**, 10468 (1990)
19. J. Robertson, Diam. Rel. Mater. **2**, 984 (1993)
20. J. Robertson, Phil. Trans. R. Soc. Lond. A **342**, 277 (1993)
21. R.W. Collins, Materials Science Forum **52&53**, 342 (1989)
22. G.E. Jellison, H.H. Burke, J. Appl. Phys. **60**, 841 (1986)
23. G.M.W. Kroesen, G.S. Oehrlein, T.D. Bestwick, J. Appl. Phys. **69**, 3390 (1991)
24. D.E. Aspnes, A.A. Studna, Appl. Opt. **14**, 220 (1975)
25. P.S. Hauge, Surface Science **96**, 108 (1980)
26. R.W. Collins, Rev. Sci. Instrum. **61**, 2029 (1990)
27. J.M.M. de Nijs, A.H.M. Holtslag, A. Hoeksta and A. van Silfhout, J. Opt. Soc. Am. **A5**, 1466 (1988)
28. W. Fukarek, A. von Keudell, Rev. Sci. Instrum. (in press)
29. W. Möller in 'Materials modification by high fluence ion beams', edited by R. Kelly and M.F. da Silva, NATO ASI Series E, **155**, (Kluwer Academic Publishers, Dordrecht (1989), p. 151
30. W. Eckstein, 'Computer Simulation of Ion Solid Interactions', Springer Series in Materials Science, **10** (Springer-Verlag, Berlin, Heidelberg 1991)
31. W. Möller and B.M.U. Scherzer, Appl. Phys. Lett. **63**, 1870 (1987)
32. N.F. Mott, E.A. Davis, 'Electronic Processes in Non-Crystalline Materials', Oxford Univ. Press 2nd ed. (1979)
33. J. Robertson, Materials Science Forum **52&53**, 125 (1989)

CHEMICAL EFFECTS OF SUBSTRATE TEMPERATURE AND FEED GAS COMPOSITION ON ION BEAM DEPOSITED AlN and AlN:H

L. HUANG, X. D. WANG, K. W. HIPPS, U. MAZUR, J. T. DICKINSON, AND R. HEFFRON
Materials Science Program, Washington State University, Pullman, WA 99164-4620, USA

ABSTRACT

Direct ion beam sputter deposition of AlN is studied using both pure nitrogen and a 75% N_2/25% H_2 mixture as the feed gas for the ion gun. The chemical characteristics of these films are probed using infrared spectroscopy and chemical etching. The presence of Al-N_2 species is associated with reactive and highly defective films. The presence of NH_x species increases by several orders of magnitude the rate at which AlN film are etched by base. However, stoichiometric and low defect AlN films prepared by depositing AlN:H onto substrates heated to 200 °C have reactivities similar to the best AlN films produced in the absence of hydrogen.

INTRODUCTION

Aluminum nitride has a large band gap,[1] a high density, large thermal conductivity, a small thermal expansion coefficient, and a very large volume resistivity.[2] Aluminum nitride is a hard material.[3] Thin films of freshly prepared AlN and AlN:H are even harder, with the hydrogenated film being about as hard as alumina.[4] AlN is chemically stable to attack by atmospheric gasses at temperatures less than 700 C.[5-8] Most of the work on AlN film growth to date has resulted in films that are crystalline and preferentially oriented relative to the substrate surface, although changes in deposition conditions can change which axis is preferred.[9-12] While oriented crystalline films are needed for piezoelectric devices, they are not always the most desirable form for thin film applications. Aita has suggested that nanocrystalline AlN films might serve as excellent corrosion inhibitors.[13] Hasegawa has suggested that an amorphous form of AlN might have superior properties for electronic applications, since a mismatch between the substrate and AlN lattice constants would not result in high levels of strain.[14] Mazur and coworkers found that single ion-beam sputter deposited (IBSD) AlN films were oriented microcrystalline, extremely inert chemically, but adhered poorly to substrates such as glass or mica - where extended stress relief patterns could be observed.[9,15,16,17] The addition of as little as 5% hydrogen to the ion beam feed gas eliminated these stress failures and atomically smooth coatings resulted on virtually any substrate at room temperature. It has been shown that these AlN:H and AlN:H derived films are nanocrystalline.[18,19]

Impurities can play a significant role in degrading the long term performance of these amorphous films. For example, the AlN:H films produced by ion beam deposition onto room temperature substrates are extremely susceptible to attack by water vapor in air. After 10 days, they blister due to conversion of AlN to aluminum oxyhydroxide, which is easily monitored by infrared spectroscopy.[9] In this paper we correlate the presence of specific defects in AlN films with the chemical reactivity of these films. We also show that nanocrystalline and stoichiometric AlN film formed at a substrate temperature of 200°C are almost as inert to attack by base as are good quality microcrystalline films.

EXPERIMENTAL

Films were made in a cryopumped UHV chamber capable of attaining <10^{-9} Torr without baking. The ion gun was a 2.5 cm source purchased from Ion Tech. The gas mixtures consisted of 100% N_2 or of 25% H_2 mixed with 75% N_2. All gas was admitted through the ion gun. IBSD was performed at 5×10^{-5} Torr total pressure with a net gas flow rate of 8 sccm. A nominal beam current of 28±2 mA was used with various beam voltage settings. The target employed was a 3" 99.999+ purity aluminum disk, situated 6" from the gun and 3" from the substrate at an angle of 30° to the substrate normal. The films reported here were deposited onto substrates heated to various temperatures by direct contact with a copper block heated with an electrical cartridge heater.   Cu coated microscope slides (glass) were used for IR spectroscopy, etching studies, and electron microprobe analysis. AlN and AlN:H films were also deposited onto fused quartz substrates for UV-Visible spectroscopy. The freshly prepared films were either used directly for spectroscopic studies or subjected to chemical etching and studied microscopically with an AFM.

In order to measure chemical etch rates, several small dots of 5-minute epoxy were placed on the sample surface and the epoxy rapidly cured at 60 °C. These samples were then cooled to room temperature and immersed in a stirred solution of either NaOH at 22±1 °C , or H_3PO_4 at 75±2 °C, for a controlled period of time. The sample was removed from the etchant and rinsed first with water and then with ethanol. The 5-minute epoxy was then removed with chloroform and the resulting contours were measured with a Nanoscope III AFM in contact mode.

Reflectance absorption infrared (RAIR) spectra were obtained with an IR/98 FT-IR vacuum bench spectrometer using 84° grazing incidence radiation. All FT-IR data were acquired with 8 cm^{-1} resolution and are the result of 1,024 scans. UV-Visible spectra were acquired using on a Perkin-Elmer 330 interfaced with an IBM PC. A matching quartz plate served as the reference. Stoichiometry of the AlN and AlN:H films was determined by electron microprobe analysis utilizing a wavelength dispersive spectrometer (Camebax, made by Cameca). The results were compared with those from a bulk AlN standard. Because of the difficulty in obtaining a reliable high purity standard, the precision of the microprobe results is better than its accuracy.

RESULTS AND DISCUSSION

Even when pure N_2 is used as the feed gas, varying beam parameters produce films of varying composition. Figure 1 presents the microprobe-determined composition of AlN films formed by IBSD with pure nitrogen under conditions of varying beam voltage. The films are clearly non-stoichiometric when prepared with a beam voltage less than 1100 volts and greater than 1400 volts. The region over which near stoichiometry occurs is a function of several variables including beam current and gas flow rate. For the present study, the focus is on the nature of the species present in the non-stoichiometric films rather than the exact conditions under which they arise. Consider the RAIR spectra shown in Figure 2, corresponding to the film stoichiometries shown in Figure 1. The 700 V film has a number of defects, as indicated by the N-N stretching band of Al-N$_2$ (near 2130 cm^{-1}), and by combination bands near 1500 cm^{-1}.[15-17] Also, the principal Al-N phonon (near 910 cm^{-1}) is significantly broadened indicating a strongly disordered film.[15-17]

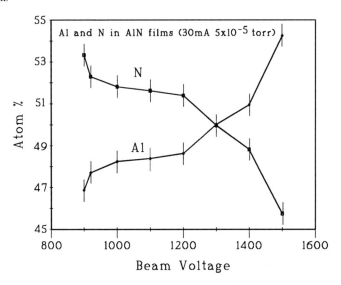

FIG. 1. Atomic percentage of Al and N in an AlN film produced in a pure nitrogen ion beam.

Raising the beam voltage to 1000 V produces a more ordered film, but the Al-N band is still broad and combination a bands is still observed near 1500 cm^{-1}. The 1300 V RAIR spectrum shows a sharp Al-N band indicative of well crystallized material. The 1500 V RAIR spectrum again shows a broadened Al-N band and also has a sloping background often associated with the presence of IR opaque particulates. Thus, the stoichiometric data and the IR spectra parallel each other nicely. When the film is nitrogen rich there is a significant amount of Al-N$_2$ present and the material is poorly ordered. When the film is Al rich the AlN is also defective, but less so than when nitrogen rich. The sloping IR background suggests that the excess Al is present in the form of particulates. This chemical and compositional change is also reflected in the rate at which the AlN film is attacked by base. Figure 3 displays the results of literally hundreds of measurements of etch step profile as a function of etch time. Note that the most chemically resistant film is the one showing the highest crystallinity and the fewest chemical defects.

When hydrogen is added to the beam feed gas, the films formed on room temperature substrates have a more complex chemical composition. As can be seen in Figure 4, the RAIR spectrum of AlN:H contains Al-N$_2$ defects, has a very broad Al-N stretching band, and also displays combination band structure just as the low voltage AlN films do. What is distinctly different is the presence of an N-H stretching band 3300 cm^{-1} and of N-H bending bands between 1000 and 1600 cm^{-1}.[15-18]

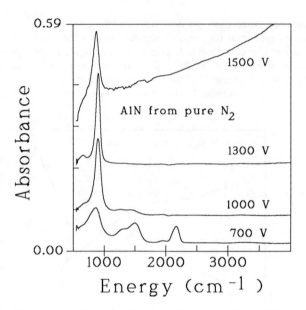

FIG. 2. RAIR spectra as a function of beam voltage. Samples are 105-nm thick AlN corresponding to those with the indicated voltage and stoichiometry in Figure 1.

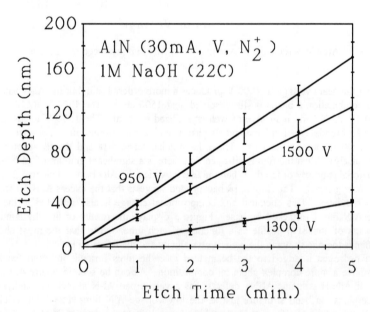

FIG. 3. Etch depth in nm versus etching time for selected AlN films.

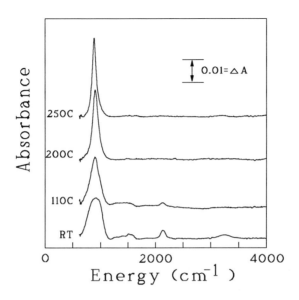

FIG. 4. RAIR spectra of 30 nm thick AlN:H films deposited on to Cu heated to the indicated temperatures.

As may also be inferred from Figure 4, heating the substrate to 200 °C or more eliminates both the Al-N$_2$ and the N-H species from the film. Electron microprobe analysis of the 200 °C heated films indicates that they are stoichiometric.[19] Based on the etching results obtained from AlN films, it is reasonable to assume that the room temperature deposited AlN:H film should be much more susceptible to etching than the heated film. What is surprising is the size of the etch rate for the unheated film. Table I collects etch rates for a number of different AlN films. The values in the first five rows were determined by us for IBSD films, while the data in the last 4 rows is taken from the CVD literature. Note that the etch rate appears to be roughly linear in the NaOH concentration in the range between 0.5 and 2 molar. A good AlN film is expected to have an etch rate of about 30 nm/min. in 1 M NaOH, and the best CVD films formed at 1000 °C by CVD have rates approaching 1 nm/min. in 1 M NaOH. As predicted from the IR spectra, the AlN:H film formed on a 200 °C substrate has an NaOH etch rate similar to the AlN films made without hydrogen. The AlN:H film made at room temperature, however, etches so fast that the base concentration must be reduced to pH 10 (0.0001 M) if non-zero etching times are to be observed. At pH 10, the room temperature grown AlN:H film etches at roughly the same rate as the 200 °C grown film does in 1 M NaOH. Thus, the hydrogen containing species affect the etch rate much more than the other species present in the defective films. This is despite the fact that all of the impurity species appear to be highly unstable thermally and can be annealed from the film at the relatively low temperature of 200 °C. The affects are qualitative as well as quantitative. The etch profile for the room temperature deposited AlN:H film etched with pH 10 solution is extremely sharp, while all the low hydrogen film etch profiles (using 0.5 M to 2M NaOH) show slow transitions over 10 microns or more. Moreover, little or no grain structure is revealed by the etching of true AlN:H, while prominent grain structure is observed in the etched AlN surface. How the NH$_x$ species catalyze the dissolution is still a mystery.

TABLE I. Etch rate in nm/min. for various AlN preparations.

Film	0.5 M NaOH (22 °C)	1.0 M NaOH (22 °C)	2.0 M NaOH (22 °C)	85% H_3PO_4 (75 °C)
AlN (700 V)	358±10			
AlN (950 V)	18±2	35±2		50±15
AlN (1300 V)		8±1	14±2	14±2
AlN (1500 V)		25±2		41±2
AlN (200 °C AlN:H)	21±2			
AlN (800 °C CVD)[a]			600	
AlN (900 °C CVD)[a]			53	174
AlN (1000 °C CVD)[a]			2	
AlN (≈700 °C CVD)[b]				125

[a] T.L. Chu and R.W. Kelm, J. Electrochem. Soc., **122**, 995 (1975).
Their data was obtained in 10% NaOH and was scaled by (2/3).
[b] Taken from L.V. Interrante, W. Lee, M. McConnell, W. Lewis, and E. Hall, J. Electrochem. Soc., **136**, 472 (1989).

AlN:H deposited on a 25 °C substrate has an etch rate of 45 nm/min in 0.0001 M NaOH.

ACKNOWLEDGMENT: We thank NSF for their support under grant DMR-9201767.

REFERENCES

1) W.M. Yim, E.J. Stofko, P.J. Zanzucchi, J.I. Pankove, M. Ettenberg, S.L. Gilbert, J. Appl. Phys. **44**, 292 (1973).
2) L.M. Sheppard, Ceramic Bulletin **69**, 1801 (1990).
3) A.A. Ivanko in "Handbook of Hardness Data"; edited by G.V. Samsonov (Keter Press, Jerusalem, 1971).
4) U. Mazur, work in progress.
5) N. Azema, R. Durand, C. Dupuy, and L. Cot, J. Euro. Ceram. Soc. **8**, 291 (1991).
6) K. Hatwar and T.R. Pian, in Ion Beam Processing of Materials., edited by G.K. Hubler and C.W. White (Mater. Res. Soc. Symp. Proc. **121**, Pittsburgh, PA, 1988), p. 557.
7) A.D. Katnani and K.I. Papathomas, J. Vac. Sci. Technol. A **5**, 1335 (1987).
8) A. Bellosi, E. Landi, and A. Tampieri, J. Mater. Res. **8**, 565 (1993).
9) X.D. Wang, K.W. Hipps, and U. Mazur, Langmuir **8**, 1347 (1992).
10) J.M. Harper, J.J. Cuomo, and H.T. Hentzell, J. Appl. Phys. **58**, 550 (1985).
11) H.T. Hentzell, J.M. Harper, and J.J. Cuomo, J. Appl. Phys. **58**, 556 (1985).
12) H. Windischmann, Thin Solid Films **154**, 159 (1987).
13) C.R. Aita and W.S. Tait, Nanostructured Materials **1**, 269 (1992).
14) F. Hasegawa, T. Takahashi, K. Kubo, and Y. Nannichi, Jap. J. Appl. Phys. **9**, 1555 (1987).
15) U. Mazur Langmuir **6**, 1331 (1990).
16) U. Mazur and A. Cleary J. Phys. Chem. **94**, 189 (1990).
17) X.D. Wang, K.W. Hipps, and Ursula Mazur, J. Phys. Chem. **96**, 8485 (1992).
18) X.D. Wang, K.W. Hipps, J.T. Dickinson, and U. Mazur, J. Mater. Res., **9**, 1449 (1994).
19) X.D. Wang, W. Jiang, M.G. Norton, and K.W. Hipps, Thin Solid Films, **251**, 121 (1994).

EFFECTS OF LASER IRRADIATION ON GROWTH AND DOPING CHARACTERISTICS OF GaAs IN CHEMICAL BEAM EPITAXY

H. K. DONG, N. Y. LI, and C. W. TU
Department of Electrical and Computer Engineering, University of California at San Diego,
La Jolla, California 92093-0407

ABSTRACT

In this paper, we report laser-assisted chemical beam epitaxy (CBE) of GaAs using triethylgallium (TEGa), tris-dimethylaminoarsenic (TDMAAs), and an Ar ion laser operating at visible or ultraviolet (UV) wavelength. The laser-assisted growth with TDMAAs, compared to As_4 or AsH_3, shows a wider range of growth enhancement at low substrate temperatures. Unlike CBE of GaAs without laser irradiation, laser-enhanced GaAs growth rate was found to be constant as the V/III incorporation ratio changes. By using diiodomethane (CI_2H_2) as a dopant gas, the GaAs films with laser irradiation show a much higher hole concentration than those grown simultaneously without laser irradiation at substrate temperatures from 460-530°C. Laser irradiation was also found to enhance silicon incorporation at low temperatures. Photothermal effects are responsible for laser-enhanced growth and silicon doping, but the wider temperature window in laser-enhanced growth and the laser-enhanced carbon incorporation are caused by additional photocatalytic or photochemical effects.

INTRODUCTION

Chemical beam epitaxy (CBE), or metallorganic molecular beam epitaxy (MOMBE), has been under active development to exploit the advantages of conventional solid-source molecular beam epitaxy (MBE) and metal-organic vapor phase epitaxy (MOVPE). One of the important issues in CBE is the selective-area growth by direct writing with an external energy source, e.g., an argon ion laser. Studies have shown that laser-assisted CBE has the potential for selective-area growth, etching, doping and compositional change, which could enable us to integrate different materials and structures on the same chip and develop novel devices, e.g., multiple-wavelength lasers [1]. Many studies have reported the selective-area growth of GaAs by laser-assisted CBE or MOMBE using triethylgallium (TEGa) and AsH_3 or As_4 [2-7]. However, the temperature window for the selective-area growth is very narrow using AsH_3 or As_4, and hence limits the application. A novel source which could provide a wider temperature window would be desirable. Another issue of great interest is replacing the highly toxic and highly-pressurized AsH_3 with novel organometallic arsenic sources. Tris-dimethylaminoarsenic (TDMAAs, $As[N(CH_3)_2]_3$), with As directly bonded to N, is a promising substitute to the highly toxic and highly pressurized AsH_3 because of its low but reasonable vapor pressure and low cracking temperature. Since there are no As-H and As-C bonds, one can expect TDMAAs to have lower toxicity and lower carbon incorporation. TDMAAs has also been successfully used in CBE/MOMBE of GaAs [8-11], and AlGaAs [8]. In this paper, we report argon ion laser-assisted CBE of GaAs using TDMAAs.

Laser processing also allows selective-area doping without prepatterning. So far, only laser-enhanced Si doping into GaAs using SiH_4 [12,13] and trimethylsilane [14] has been investigated, but no laser-modified p-type doping has been studied. In this paper we also discuss the laser-enhanced carbon doping in GaAs using a novel carbon-doping precursor, diiodomethane (CI_2H_2), and the laser-enhanced silicon doping using disilane (Si_2H_6).

EXPERIMENTAL PROCEDURE

Experiments were performed in a modified Perkin Elmer 425B MBE system, evacuated by a turbomolecular pump and equipped with gas lines for TEGa, TDMAAs, CI_2H_2, and Si_2H_6. TEGa was introduced into the growth chamber without any carrier gas through a vapor-source mass flow controller. TDMAAs, with a 20°C vapor pressure of ~1.35 Torr, was carried by hydrogen and injected into the chamber through a leak valve without precracking. The TDMAAs flow was determined by the hydrogen flow rate, which was controlled by a mass flow controller. CI_2H_2, with a 25°C vapor pressure of about 1.5 Torr, was delivered into the chamber through an ultrahigh vacuum leak valve, and its flux was adjusted using a Baratron pressure sensor. 15% Si_2H_6 (packed in hydrogen) was introduced into the chamber through an ultrahigh vacuum leak valve, and its flux was controlled by a mass flow controller.

Semi-insulating (100) GaAs substrates were chemically etched and thermally cleaned at about 600°C. The substrate temperature was calibrated using an infrared-laser interferometric technique, which gave an accuracy of about ±4°C for the substrate temperature [15]. A cw Ar^+ laser was directed at the substrate at normal incidence through a center port. Both the 488 nm and 351 nm lines of a multiline argon ion laser were used. To obtain the 351 nm line a UV optics was installed. For laser-assisted CBE of GaAs, reflection high-energy electron diffraction (RHEED) was used to determine the film growth rates without laser irradiation, while a Dektak stylus profiler was used to measure the laser-enhanced growth rates. For laser-enhanced doping experiments, the laser beam was scanned across an area of about 5 mm × 5 mm by X-Y galvanometer-controlled mirrors. Standard Van der Pauw Hall-effect measurement was used to characterize the electrical properties of the doped GaAs films.

RESULTS AND DISCUSSION

The substrate temperature dependence of the growth rates of laser-irradiated and unirradiated areas using TDMAAs and solid arsenic from our work, together with those using arsine [7], are shown in Fig. 1. The growth rate data are normalized to that at the high-temperature region (between 480 and 510°C). The TEGa and H_2 (for TDMAAs) flow rates were 0.4 and 2 sccm, respectively. The laser power density used for all cases is about 400 W/cm², except that for the 351 nm UV line it is only 20 W/cm². The substantial differences between the laser-assisted growth using TDMAAs and arsine or solid arsenic are that the threshold temperature for enhanced growth is about 100°C lower and that the temperature range is about 100°C wider (more noticeable for the 488 nm line). Significant growth enhancement was observed using the UV line while none was obtained using the 488 nm line (20 W/cm²), possibly due to the higher photolytic decomposition efficiency of TEGa in the shorter wavelength.

According to the infrared laser interferometric measurement [15], the GaAs substrate temperature rise is only about 20°C by argon ion laser irradiation with a power density of 400 W/cm². Therefore, the much wider temperature window for selective-area growth using TDMAAs is not due to the photothermal effect. Photocatalytic or photochemical reactions must have taken place, but its nature cannot be known without additional spectroscopic investigation.

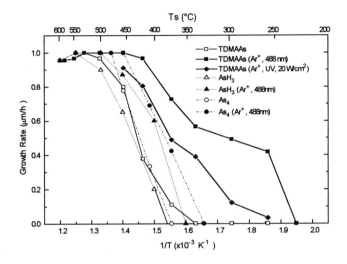

Fig. 1 The substrate temperature dependence of growth rates of laser-irradiated and non-irradiated areas. The solid lines denote the growth data obtained using TDMAAs, while the dashed and dotted lines denote the data from AsH_3 and As_4, respectively.

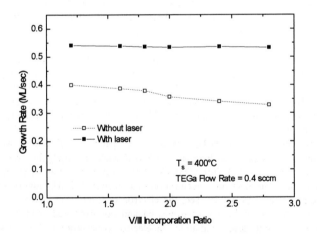

Fig. 2 GaAs growth rate as a function of V/III ratio with and without laser irradiation.

Fig. 2 shows the GaAs growth rate as a function of V/III incorporation ratio with and without laser irradiation at a substrate temperature of 400°C. As the V/III ratio increases, the GaAs growth rate decreases without laser irradiation. This is probably due to the site-blocking effect, caused by excess arsenic atoms adsorbed on the GaAs surface, or by the enhanced desorption of TEGa molecules from the GaAs surface at high arsenic surface concentration. With laser irradiation, the growth rate is enhanced throughout the V/III ratio range studied. Since the desorption of arsenic species is enhanced by laser irradiation [5], the site-blocking effect is suppressed. Therefore, the GaAs growth rate with laser irradiation is almost constant when the V/III ratio changes.

For the laser-enhanced carbon doping study, we varied the substrate temperature from 460 to 530°C while keeping the TEGa and H_2 (for uncracked TDMAAs) flow rates at 0.4 and 2 sccm, respectively. The vapor pressure of CI_2H_2 upstream to the leak valve was kept at 1 Torr. The laser power density was set at 600 W/cm², and the mirror scan rate was about 50 Hz. All of the carbon-doped GaAs samples exhibit specular surface morphology. Figure 3 shows the hole concentration, determined from Hall measurements, as a function of substrate temperature. As the substrate temperature increases from 460 to 530°C, the hole concentration in the laser-irradiated area increases monotonically up to 9×10^{19} cm^{-3} while that in the non-irradiated area decreases monotonically from 5×10^{18} to 6×10^{17} cm^{-3}. We see that the carbon incorporation into GaAs with laser irradiation is much higher than without laser irradiation.

Our previous carbon doping study [16] has shown that carbon incorporation in CBE using uncracked TDMAAs is about two orders of magnitude lower than that using cracked TDMAAs with the same CI_2H_2 settings. The low carbon level in GaAs samples grown with uncracked TDMAAs is a result of atomic hydrogen released from the efficient β-hydride elimination process on the GaAs surface [17]. The atomic hydrogen released from TDMAAs could react with cracked carbon species from CI_2H_2 and form volatile gases, e.g., methane. Therefore, the carbon incorporation is limited. As the substrate temperature increases, the surface reaction is more active, and hence the carbon concentration becomes lower. However, when TDMAAs is cracked, carbon incorporation efficiency is increased significantly because the β-hydride elimination reaction is suppressed in this case. With laser irradiation, the surface β-hydride elimination process could also be suppressed because the TDMAAs molecules are already cracked by the laser irradiation, and hence the carbon incorporation is much higher. As the substrate temperature increases, more CI_2H_2 will be decomposed, resulting in higher carbon concentration. The laser-enhanced carbon doping is definitely not due to the photothermal effect; otherwise, the carbon incorporation should be even lower because of the temperature rise caused by laser irradiation.

Silicon doping in GaAs using Si_2H_6 was also investigated in CBE grown with TDMAAs. The growth temperature was varied from 460 to 550°C while the Si_2H_6 flow rate was changed from 0.6 to 3 sccm. TEGa and H_2 (for TDMAAs) flow rates were 0.4 and 2 sccm, respectively. Figures 4a and 4b show the electron concentration as a function of Si_2H_6 flow rate and substrate temperature, respectively. As we can see, the electron concentration increases linearly as the Si_2H_6 flow rate increases for both laser-irradiated and non-irradiated samples at a substrate temperature of 500°C. With laser irradiation, the electron concentration shows a slight increase at higher Si_2H_6 flow rates. At higher substrate temperatures, the electron concentration is almost the same with and without laser irradiation. However, at lower substrate temperature, laser irradiation produces a higher electron concentration because of its thermal effect.

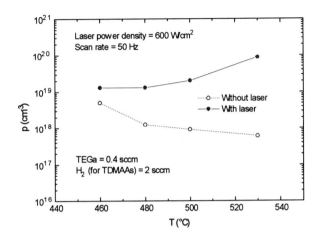

Fig. 3   Hole concentration as a function of the substrate temperature for carbon-doped GaAs samples with and without laser irradiation.

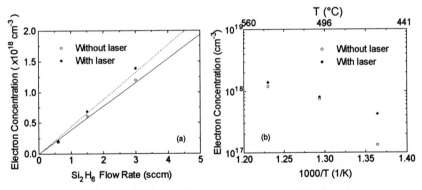

Fig. 4   Electron concentration as a function of (a) Si_2H_6 flow rate, and (b) substrate temperature.

SUMMARY

Ar ion laser-assisted CBE of GaAs using TEGa and TDMAAs, laser-enhanced carbon doping using Cl_2H_2, and laser-enhanced silicon doping using Si_2H_6 have been studied. The laser-assisted growth with TDMAAs, compared to As_4 or AsH_3, shows a wider range of growth enhancement at low substrate temperatures. The laser-enhanced growth rate is independent of the V/III ratio. The carbon doping is strongly enhanced by laser irradiation, while silicon doping is enhanced only at low temperatures. Photothermal effects are responsible for laser-enhanced growth and silicon doping, but the wider temperature window in laser-enhanced growth and the laser-enhanced carbon incorporation are caused by additional photocatalytic or photochemical effects. These laser-enhanced growth and doping with novel gas sources may provide potential applications for optoelectronic devices.

ACKNOWLEDGMENTS

This work is supported by the Air Force Wright Laboratory. The authors wish to thank W. S. Wong for his assistance. They are also grateful to the generous donation of the MBE system by Rockwell International Science Center, Thousand Oaks, California.

REFERENCE

1. R. Iga, T. Yamada, and H. Sugiura, Appl. Phys. Lett. **64**, 983 (1994).
2. T. Yamada, R. Iga, and H. Sugiura, Appl. Phys. Lett. **61**, 2449 (1992).
3. Y. Aoyagi, M. Kanazawa, A. Doi, S. Iwai and S. Namba, J. Appl. Phys. **60**, 3131 (1986).
4. V. M. Donnelly, C. W. Tu, J. C. Beggy, V. R. McCrary, M. G. Lamont, T. D. Harris, F. A. Baiochi and R. C. Farrow, Appl. Phys. Lett. **52**, 1065 (1988).
5. H. K. Dong, B. W. Liang, M. C. Ho, S. Hung and C. W. Tu, J. Cryst. Growth **124**, 181 (1992).
6. H. Sugiura, T. Yamada and R. Iga, Jpn. J. Appl. Phys. **29**, L1 (1990).
7. H. Sugiura, R. Iga, and T. Yamada, J. Cryst. Growth **120**, 389 (1992).
8. C. R. Abernathy, P. W. Wisk, D. A. Bohling, and G. T. Muhr, Appl. Phys. Lett. **60**, 2421 (1992).
9. H. K. Dong, N. Y. Li, C. W. Tu, M. Geva, and W. C. Mitchel, J. Electron. Mater. **24**, 69 (1995).
10. G. Zimmermann, H. Protzmann, T. Marschner, O. Zsebök, W. Stolz, E. O. Göbel, P. Gimmnich, J. Lorberth, T. Filz, P. Kurpas, and W. Richter, J. Cryst. Growth **129**, 37 (1993).
11. K. Fujii, I. Suemune, and M. Yamanishi, Appl. Phys. Lett. **61**, 2577 (1992).
12. K. Sugioka, and K. Toyoda, Appl. Phys. Lett. **61**, 2817 (1992).
13. J. C. Roberts, K. S. Boutros, and S. M. Bedair, and D. C. Look, Appl. Phys. Lett. **64**, 2397 (1994).
14. Y. Ban, M. Ishizaki, T. Asaka, Y. Koyoma, and H. Kukimoto, Jpn. J. Appl. Phys. **28**, L1995 (1989).
15. S. C. H. Hung, H. K. Dong, and C. W. Tu, Mater. Res. Soc. Symp. Proc. **340**, 35 (1994).
16. N. Y. Li, H. K. Dong, C. W. Tu and M. Geva, to be published in J. Cryst. Growth (1995).
17. D. A. Bohling, K. F. Jensen, and C. R. Abernathy, J. Cryst. Growth **136**, 118 (1994).

STUDY OF ANNEALED INDIUM TIN OXIDE FILMS PREPARED BY RF REACTIVE MAGNETRON SPUTTERING

Li-jian Meng[*] , A. Maçarico and R. Martins
CEMOP/UNINOVA, Quinta de Torre, 2825 Monte da Caparica, Portugal

ABSTRACT

Tin doped indium oxide (ITO) films were deposited on glass substrates by rf reactive magnetron sputtering using a metallic alloy target (In-Sn, 90-10). The post-deposition annealing has been done for ITO films in air and the effect of annealing temperature on the electrical, optical and structural properties of ITO films was studied. It has been found that the increase of the annealing temperature will improve the film electrical properties. The resistivity of as-deposited film is about 1.3×10^{-1} Ω*cm and decreases down to 6.9×10^{-3} Ω*cm as the annealing temperature is increased up to 500 °C. In addition, the annealing will also increase the film surface roughness which can improve the efficiency of amorphous silicon solar cells by increasing the amount of light trapping.

INTRODUCTION

Indium tin oxide (ITO) film has been widely used as a transparent conductor due to its high transparency to visible light and its low electrical resistivity. Applications which use ITO as a transparent conductor include liquid crystal display, solar cells and various light sensitive solid state devices (1-3). Reactive sputtering deposition is widely used for ITO film deposition. Desirable features of reactive sputtering employing alloy targets for practical use are: a high deposition rate; accurate control of the film thickness and easy fabrication of the alloy target. In general, a post-deposition annealing in air or in an oxygen free environment for reactively sputtered ITO coatings from metal alloy target is needed in order to improve the film electrical properties (4-9). In this paper, ITO films have been deposited on glass substrates by rf reactive magnetron sputtering method and the effect of the post-deposition annealing temperature (200 °C -- 500 °C) on the film structural, optical and electrical properties has been studied.

EXPERIMENTAL DETAILS

ITO films are deposited on glass substrates at ambient temperature by rf reactive magnetron sputtering technique. The target is In-Sn (90:10) alloy of 99.99% purity (125x500 mm^2 ALCATEL CIT). The distance between target and substrate is about 40 mm. During the sputtering process, the oxygen partial pressure and total pressure are $9x10^{-4}$; the rf forward and reflected powers are 1200 W and 12.3 W respectively. The deposition time is about 25 min and the film thickness is about 610 nm After the deposition, the films are annealed in a tube furnace for about 30 min in air. The annealing temperatures are 200 °C, 300 °C, 400 °C and 500 °C respectively. After annealing, the films are characterized by the X-ray diffraction (XRD), scanning electron microscopy (SEM) and the X-ray photoelectron spectroscopy (XPS) techniques

---

[*] Thankful to the **Fundação Luso-Americana para o Desenvolvimento** for providing a travel grant for this meeting.

Mat. Res. Soc. Symp. Proc. Vol. 388 © 1995 Materials Research Society

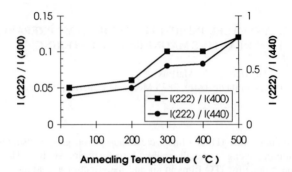

Fig. 1. The variations of peak intensity ratios of I (222)/I (400) and I (222)/I (440) as a function of the annealing temperature.

RESULTS AND DISCUSSIONS

<u>Structural Properties</u>

The XRD patterns of the ITO films contain only In_2O_3 peaks (10). All the films show the (400) plane texturing. The (002) peak intensity increases as the annealing temperature is increased although all the films have (400) preferred orientation. Figure 1 shows the variations of the I (222)/I (400) and I (222)/I (440) with annealing temperature. It is clear that the intensity ratio increases as the annealing temperature is increased.

The (400) diffraction peak width was used to estimated the crystallite dimension along the a axis (11). The results are shown in Fig.2 (a). It can be seen that the grain size along (400) direction increases as the annealing temperature is increased up to 300 °C and no clear variation after the annealing temperature is over than 300 °C. The (400) crystal plane distance was

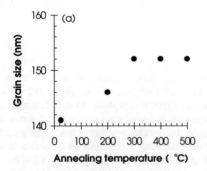

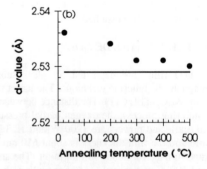

Fig. 2. (a) Grain size of ITO films vs. annealing temperature.

Fig.2.(b) d-values of ITO films vs. annealing temperature. The line gives the d-value of In_2O_3 powder.

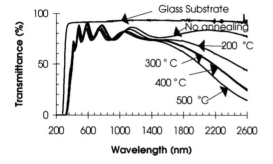

Fig. 3. Specular transmittance spectra of ITO films as-deposited and annealed at different temperatures.

calculated from the diffraction angle which has been obtained by fitting the diffraction peak. The results are shown in Fig. 2 (b). The line in the figure is the standard value d_0 (2.529 Å). It shows that the d values of all the films are larger than d_0. However, as the annealing temperature is increased, the difference between the standard value and the calculated value becomes small. That means that there are compressive stress in all the films and the stress value decreases as the annealing temperature is increased.

The SEM photographs of ITO films show that all the films have columnar structures. The average crystallite size along the sample surface is about 100 nm and no clear variation for the films annealed at different temperatures. This crystallite size is smaller than that along the direction normal to the sample surface estimated by XRD method. From SEM it can also be seen that the voids between the crystallite grain become small as the annealed temperature is increased. That means the annealing improves the film density. In addition, as the annealing temperature is increased, the film surface becomes rough which can improve the efficiency of amorphous silicon solar cells by increasing the amount of light trapping.

Optical and Electrical Properties

Figure 3 shows the transmittance of the films as-deposited and annealed at different temperatures from UV to near IR region. As shown in Fig.3., the transmittance decreases as the annealing temperature is increased at wavelengths above 1000 nm. This is because of the free-carrier absorption which increases as the carrier concentration increases. The majority carriers in ITO are electrons. That means the electron concentration in the film increases as the annealing temperature is increased.

It is well known for the ITO films that the electrons are liberated from the substitutional entered Sn atoms in the cation sublattice and from the doubly charged oxygen vacancies. Tin can exist as either SnO (valence 2) or SnO_2 (valence 4). Since indium has valence 3 in In_2O_3, the presence of SnO_2 would result in n doping of the lattice because the dopant would add electrons to the conduction band. In contrast, the presence of SnO would lower the electron density in the conduction band. Some authors believe that, at low substrate temperature, tin is present in the ITO films as SnO resulting in low carrier densities and annealing the films will transfer SnO into SnO_2 and results in the formation of an n-type semiconductor with high carrier density and low resistance (12,13). In order to know if there is also this kind of transition in our films, we have performed the X-ray photoelectron spectroscopy (XPS) measurement for all the films. XPS gives the possibility to measure the binding energy with a high energy resolution which gives the possibility to detect the different chemical states of an element. From literature it can be found that the binding energies of tin 3d5/2 for SnO and SnO_2 are 486.8 and 486.6 ev respectively (14). In order to get accurate binding energy of tin in our films, we have performed linear background subtraction and gaussian peak deconvolutions for Sn 3d5/2 photoelectron peaks using VGS 5250 software. The results show that the Sn 3d5/2 peaks for all samples can be deconvoluted by one peak which is located at 486.6eV and no clear variation of the peak shape is observed. This suggests that the tin is present in our films as SnO_2. Therefore, the increase of the carrier concentration, as the annealing temperature is increased, is not from the transition of SnO to SnO_2. We believe that the increase of the carrier concentration is due to the decrease of the donor sites trapped at the dislocations or point defect aggregates (15). As the annealing temperature is increased, the film crystallinity is improved. The improvement of the crystallinity results in a decrease in concentration of donors trapped at crystalline defects and hence a increase of carrier concentration. From Fig.1 it can be seen that the (222) diffraction peak intensity increases as the annealing temperature is increased. This variation may also contribute to the increase of the carrier concentration. However, it is also known that annealing ITO film at temperatures over 300 °C in air will cause the thermal oxidation of the oxygen vacancies and thus the extinction of associated carriers (16). This effect results in the similar free-carrier absoptions for the films annealed at 300 °C and 400 °C temperature.

The atomic concentration of the various elements in the films have been computed from measured peak areas with the relative sensitivity factors which have been given in the introduction. The results show that the ratios of O/In and Sn/In for different films are about 1.2 and 0.09, respectively. It is clear that the ratio of Sn/In in the films is slightly lower than that in the target. However, it should be noted that the ratios in the sample surface may have some differences with that in deeper layers because of the strong carbon contamination in the surface.

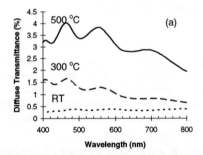 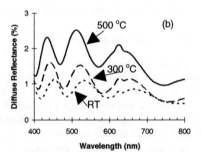

Fig. 4. (a) Diffuse transmittance spectra of ITO films as-deposited and annealed at different temperatures.

Fig. 4. (b) Diffuse reflectance spectra of ITO films as-deposited and annealed at different temperatures.

From Fig.3 it can be seen that the fundamental absorption edge of ITO film shifts to shorter wavelengths as the annealing temperature is increased. This shift resulted from the increase of the carrier concentration is known as the Burstein-Moss shift, where in the heavily doped n-type semiconductor, the Fermi level is inside the conduction band and the states near the bottom of the conduction band are filled. Therefore, the absorption edge shifts to higher energy.

Absorption coefficients of the films at different wavelengths have been calculated from transmission and reflection data. The absorption coefficient for the direct allowed transition can be written as $\alpha = (h\upsilon - Eg)^{1/2}$, where $h\upsilon$ is the photon energy and Eg is the transition energy gap (17). Extrapolations of the straight regions of the plots to $\alpha = 0$ give Eg. The results that the Eg increases as the annealing temperature is increased up to 300 °C and then has no clear variation even the annealing temperature is increased further. This variation can be related with the variation of the carrier concentration as shown in Fig.3. As the carrier concentration is increased, the optical band gap shifts to the high energy direction.

Figures 4 (a) and (b) show the diffuse transmittance and reflectance of the films as-deposited and annealed at different temperatures. As they have been shown in these figures that both the diffuse transmittance and reflectance increase as the annealing temperature is increased. From SEM it can be seen that the film surface becomes more rough as the annealing temperature is increased. The rough surface results in the increases of the diffuse transmittance and reflectance. From Fig.3 it can be seen that the film transmittance in the visible region decreases as the annealing temperature is increased. Considering the light scattering by rough surface (diffuse transmittance and reflectance), it can be concluded that the decrease of the transmittance results from the light scattering. Comparing Fig.4 (a) with (b) it can be found that, for the same film, the diffuse transmittance is higher than diffuse reflectance. Therefore, the light scattering processes in the films are mainly Rayleigh type scattering (18).

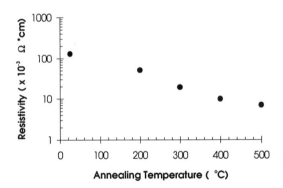

Fig. 5. Variation in resistivity of ITO films with annealing temperature.

Figure 5 shows the variation of the film electrical resitivity with the annealing temperature. As it has been shown in Fig.5 that the film resistivity decreases gradually as the annealing temperature is increased. Both the increase of the carrier density and the carrier mobility will improve the film electrical resistivity. From Fig.3 it has been seen that the carrier concentration increases as the annealing temperature is increased and will improve the film electrical resistivity. In addition, XRD and SEM's results have shown that the film structural property has been improved by increasing annealing temperature. It may improve the carrier mobility and result in the decrease of the film resistivity.

CONCLUSIONS

ITO films have been deposited on the glass substrates by rf reactive magnetron sputtering technique using metallic In-Sn (90-10) target. After the deposition, the films have been annealed in air for about 30 min. The annealing temperature is varied from 200 °C to 500 °C with an interval of 100 °C. All the films show a preferred orientation along (400) crystal plane. But the (222) diffraction peak intensity increases as the annealing temperature is increased. The grain size along (400) direction increases as the annealing temperature is increased. All the films have the compressive stress and the stress value decreases as the annealing temperature is increased. The film surface becomes more rough as the annealing temperature is increased. Tin is present in all films as SnO_2, no transition from SnO to SnO_2 has been observed in our films. As the annealing temperature is increased, the electron, which is liberated from the donor sites trapped at the dislocations or point defect aggregates due to annealing, concentration increases and results in the decrease of the near IR transmittance and increase of the optical band gap. Both the film diffuse transmittance and reflectance increase because of the increase of the surface roughness as the annealing temperature is increased. As the annealing temperature is increased, the film electrical resistivity decreases. The resistivity of as-deposited film is about 1.3×10^{-1} Ω*cm and decreases down to 6.9×10^{-3} Ω*cm as the annealing temperature is increased up to 500 °C.

REFERENCES:
1. K.L. Chopra, S. Major and D.K. Pandya, Thin Solid Films **102**, 1 (1983).
2. Z. Ovadyahu and H. Wiesmann, J. Appl. Phys. **52**, 5865 (1981).
3. M.J. Tsai, A.L. Fahrenbruch and R.H. Bube, J. Appl. Phys. **51**, 2696 (1980).
4. T. Karasawa and Y. Miyata, Thin Soild Films **223**, 135 (1993).
5. L. Davis, Thin Solid Films **236**, 1 (1993).
6. A. Kawada, Thin Solid Films **191**, 297 (1990).
7. L. Gupta, A. Mansingh and P.K. Srivastava, Thin Solid Films **176**, 33 (1989).
8. S. Chaudhuri, J. Bhattacharyya and A.K. Pal, Thin Solid Films **148**, 279 (1987).
9. Y. Shigesato and D.C. Paine, Thin Solid Films **238**, 44 (1994).
10. Powder Diffraction File, Joint Committee on Powder Diffraction Standards, (ASTM, Philadelphia, PA, 1967) Card 6-0416.
11. B. D. Cullity, Elements of X-ray Diffraction. 2nd ed. (Addison-wesley, Reading, MA, 1978).
12. G. Frank and H. Kostlin, Appl. Phys. A, **27**, 197 (1982).
13. C.H.L. Weijtens, J. Electrochem. Soc. **138**, 3432 (1991).
14. Handbook of X-ray Photoelectron Spectroscopy, edited by C.D. Wagner, W.M. Riggs, L.E. Davis, J.F. Moulder and G.E. Muilenberg (Perkin-Elmer, Eden Prairie, MN, 1979).
15. Y. Shigesato, S. Takaki and T. Haranoh, Appl. Surf. Sci **48/49**, 269 (1991).
16. M. Mizuhashi, Thin Solid Films **70**, 91 (1980).
17. D. Curie, in Luminescence of Ionrganic Solids, Edited by B. Di Bartolo (Plenum Press, New York, 1978).
18. I. Fanderlik, Optical Properties of Glass, (Elsevier, Amsterdam, 1983).

COMPETITIVE OXIDATION DURING BURIED OXIDE FORMATION USING SEPARATION BY PLASMA IMPLANTATION OF OXYGEN (SPIMOX)

Jingbao Liu*, S. Sundar Kumar Iyer, Jing Min**, Paul Chu**, Ron Gronsky*, Chenming Hu and Nathan W. Cheung

Department of Electrical Engineering and Computer Sciences, and *Department of Material Science and Mineral Engineering, Univ. of California at Berkeley, Berkeley, CA, **Department of Physics and Material Science, City Polytechnic of Hong Kong, Hong Kong

Abstract

We have recently demonstrated a new implantation technique called SPIMOX (separation by plasma implantation of oxygen) to synthesize silicon-on-insulator structures using plasma immersion ion implantation (PIII) process. The implantation is performed by applying a large negative bias to a Si wafer immersed in an oxygen plasma created by an ECR source. Since the technique has no mass analysis, coexistence of O^+ and O_2^+ ions in oxygen plasma can cause a non-Gaussian profile of the as-implanted oxygen distribution. We observed that during post-implantation annealing, the ripening process of the oxide precipitates depends on depth and concentration of the oxygen peaks. In addition, implanted oxygen can migrate towards the Si surface during annealing, preventing a continuous buried oxide layer formation. In this paper, we report our observation on the effect of the implantation profile on the competitions between internal oxidation at different depths and between internal and surface oxidation processes. With an additional He implantation, we demonstrate that the nucleation of oxide precipitation can be enhanced.

Introduction

Silicon-on-insulator (SOI) structure has many attractive advantages for high packing density and high performance IC devices[1]. In recent years, thin SIMOX[2] which is characterized by the top silicon layer being thin enough for fully depleted devices has attracted attention. Different from conventional thick SIMOX, the continuous buried oxide (BOX) layer in thin SIMOX process is not achieved during the implantation stage of the process. It forms via a ripening process of the oxide precipitates at the post-implantation annealing step during which larger precipitates grow on expense of dissolving smaller ones at high temperature. The growing precipitates will eventually coalesce to form a continuous buried oxide layer. It is obvious that the location of the nucleation sites and the post-implantation annealing sequence are extremely important for the quality of the buried oxide and top silicon layers formed.

Separation by plasma implantation of oxygen (SPIMOX)[3] is a method that can be used to form buried oxide layers in silicon wafers for SOI (silicon-on-insulator) applications. The method is based upon a combination of the separation by implantation of oxygen (SIMOX)[1] process and the plasma immersion ion implantation (PIII)[4] process, and is motivated by the desire to reduce the manufacturing cost of SIMOX wafers for SOI devices by reducing the implantation time. According to projection, a two order-of-magnitude reduction in implantation time is possible. The SPIMOX process is achieved by applying a large negative bias voltage to a Si wafer immersed in an oxygen plasma so that the ions in the plasma are

385

accelerated across the plasma sheath layer and implanted into the entire wafer surface at the same time. A large implantation current is therefore achievable with a tolerable current density. The most attractive aspect of the technique is that the implantation time is independent of the wafer area.

A broadened implanted oxygen profile due to coexistence of O^+ and O_2^+ in the plasma is generally observed in SPIMOX process[3]. This broadened oxygen profile provides a larger depth range within which nucleation of silicon dioxide precipitates is possible. This in turn can create a wavy silicon/oxide interface due to random ripening process at different depths. The oxide precipitates can sometimes nucleate and ripen very close to the top silicon surface, become a thick surface oxide layer, and compete with growth of precipitates in the buried oxide (BOX) layer. In order to obtain a high quality SPIMOX structure, which is in the thin SIMOX regime with an applied negative bias of lower than 100kV, control of both nucleation site and subsequent ripening process becomes critical.

Experimental

The experimental setup of the PIII reactor had been reported earlier[5]. Oxygen plasma was created using an ECR plasma source with 2.45GHz microwave excitation. A negative bias of 70 kV is applied to a 4-inch wafer, which was kept above 600°C during implantation. An extremely low gas pressure, below 0.1 mTorr, was used to prevent high voltage breakdown. A nominal oxygen dose of $2 \times 10^{17}/cm^2$ was achieved within three minutes of implantation time. To study oxide nucleation effects, helium implantation to the depth corresponding to projection range of O_2^+ ions at 70kV was carried out in some samples prior to oxygen implantation to create an artificial nucleation layer using damage engineering. The He plasma implantation was performed at a bias voltage of 8 kV to a dose of $10^{17}/cm^2$, using a gas pressure < 0.1mTorr. The wafer temperature was estimated to be to 400°C during He implantation. Post implantation annealing were conducted in N_2 ambient in three stages (800°C/1hr, 1000°C/1hr and 1200°C to 1250°C/2hrs). A 300nm thick silicon nitride cap layer was sputtered to the surface before the high temperature annealing to prevent surface oxidation. Cross-sectional transmission electron microscopy (XTEM) was used to characterize the microstructures of both as-implanted and annealed samples.

Results and Discussions

In Figure 1, cross-sectional transmission electron microscopy (XTEM) shows results for the oxygen plasma implanted sample. The as-implanted sample (Fig.1.a) shows a well defined oxygen-rich band, corresponding to the projection range of O_2^+ ions. Below the peak, a long profile tail due to O^+ ions is indicated by the large damage range. After annealing at 1000°C, Fig.1.b shows the removal of the implantation damage and formation of the oxide precipitates scattering in a range of depths due to the broad implanted oxygen distribution. Fig.1.c shows that most of the smaller precipitates are dissolved during high temperature annealing at 1200°C, but some of are enlarged at the position corresponding to the depth of O^+ ions. It is expected that these remaining precipitates will be dissolved if higher annealing temperatures are used and a planar BOX layer can be formed.

In addition to the competitive ripening process between precipitates at the two peaks corresponding to O^+ and O_2^+ ions, a competition process is also observed between surface and internal oxidation. The XTEM micrographs in Figure 2 demonstrate this competition. In the

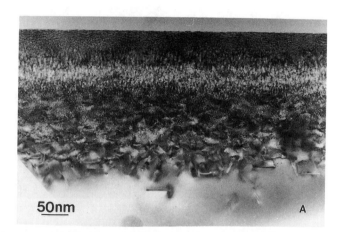

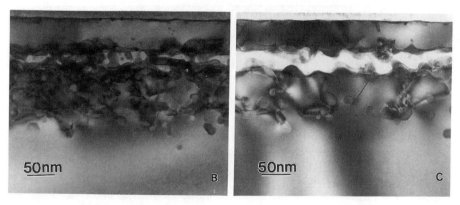

Figure 1. XTEM images of SPIMOX microstructures of (a)as-implanted at 70 kV to a dose of 10^{17} O/cm^2, (b)after annealing at 800°C/1hr and 1000°C/1hr and (c) with additional annealing at 1200°C/2hrs(c) shown the ripening process with a broad as-implanted oxygen distribution.

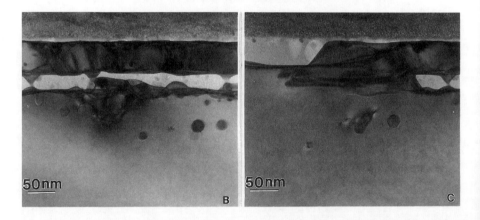

Figure 2. XTEM images of SPIMOX microstructures shown competition between internal and surface oxidation. The as-implanted sample (a) can form buried oxide (b) or surface oxide (c) after annealing due to random nucleation of the oxide precipitates.

as-implanted sample (Fig.2.a), depth fluctuation of the position where the densest precipitates are located is observed. In this sample, an amorphized region with relatively high oxygen concentration is formed during implantation; the region appears as a bright band at the surface. The interface between the brighter and darker regions is the transition between a band of crystalline silicon islands embedded in high oxygen-content amorphous Si (brighter region) and a band of oxide precipitates embedded in damaged silicon matrix (darker region). During post-implantation annealing, for most of lateral area, internal oxidation dominates and results in a continuous buried oxide layer (Fig.2.b). However, at localized pockets where a wider band of oxygen-rich Si is formed by random fluctuation, surface oxidation becomes dominant. In this situation, an oxide island becomes stable at the top surface during annealing and finally consumes all the oxygen atoms at the adjacent area to form a thick surface layer (Fig.2.c). This competing oxidation process can also happen in a random fashion if the initial annealing temperature is high enough such that all the as-implanted oxide precipitates become unstable. Even in the case that a continuous buried oxide layer is formed, diffusion of unbounded oxygen atoms towards the surface has been observed and creates the coexistence of a thickened surface oxide layer and a buried oxide layer.

In the helium implanted sample, enhanced nucleation of buried oxide precipitation is observed and shown in Figure 3. For the as-implanted sample (Fig.3.a), bubbles of 20nm to 60nm in diameter about the projected range of helium are observed. Since no bubbles are observed for samples with the He implantation alone, we have verified that these bubbles are formed during the oxygen implantation step at which the substrate temperature is kept above 600°C. The internal surfaces of this porous structure are considered to serve as nucleation sites during post-implantation annealing. After annealing at 1200°C (Fig.3.b), a buried oxide layer is formed at depth of about 50nm below a polycrystalline silicon layer although the surface of the as-implanted sample was completely amorphized. The enhancement effect of the He induced bubbles on nucleation of oxide precipitates can be seen by comparing with the sample implanted with oxygen using identical conditions but without helium implantation (Fig.4). When the surface of as-implanted sample is amorphized by oxygen implantation alone (Fig.4.a), only solid phase epitaxial growth towards the surface occurs during annealing resulting no buried oxide layer formation (Fig.4.b).

Conclusions

Since the SPIMOX technique depends on ion composition of plasma to define the oxygen distribution, nucleation sites of oxide precipitation are important to the final SOI microstructures. We have identified that surface oxidation can compete with internal oxidation process for continuous BOX formation. We have also demonstrated that artificial nucleation sites of oxide precipitates can also be created by co-implantation of He (e.g. porous structure and/or damages). This controllability suggests new opportunities for ion beam subsurface materials synthesis.

Acknowledgment

This work is supported by the Joint Services Electronics Program, contract number F49620-94-C-0038 and National Science Foundation, Grant number ECS-9202993. Microscope facilities at Berkeley is provided by funding from the US Department of Energy under contract number DE AC03 76SF00098.

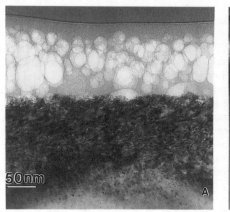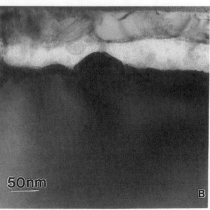

Figure 3. XTEM images of He/O sequential-implanted SPIMOX microstructure shown formation of buried oxide layer after annealing (b) from amorphized surface layer of as-implanted sample (a).

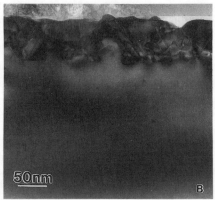

Figure 4. XTEM images of the sample implanted by oxygen alone shown no buried oxide layer formation after annealing (b) from amorphized surface layer of as-implanted sample (a).

References:

[1] J. P. Colinge, Silicon-on-Insulator Technology: Materials to VLSI, Kluwer Academic Publishers, Boston 1991

[2] A. K. Robinson, C. D. Marsh, U. Bussmann, J. A. Kilner, Y. Li, J. Vanhellemont, K. J. Reeson, P. L. F. Hemment and G. R. Booker, Nuclear Instruments and Methods, **B55**, 555 (1991)

[3] J. B. Liu, S. S. K. Iyer, J. Min, P. Chu, R. Gronsky, C. Hu, and N. W. Cheung, Proc. of Sypm.A MRS Fall meeting 1994

[4] N. W. Cheung, Nuclear Instruments and Methods, **B55**, 811 (1991)

[5] X. Y. Qian, D. Carl, J. Benasso, N. W. Cheung, M. A. Lieberman, I. G. Brown, J. E. Galvin, R. A. MacGill, and M. I. Current, Nuclear Instruments and Methods, **B55**, 884 (1991)

Characterization of Carbon Nitride Films Produced by Pulsed Laser Deposition

T. A. Friedmann, D. R. Tallant, J. C. Barbour, J. P. Sullivan, M. P. Siegal,
R. L. Simpson, and J. Mikkalson
Sandia National Laboratories, Albuquerque, NM 87185

K. F. McCarty
Sandia National Laboratories, Livermore, CA 94550

ABSTRACT

Carbon Nitride (CN_x) films have been grown by ion-assisted pulsed-laser deposition (IAPLD). Graphite targets were laser ablated while bombarding the substrate with ions from a broad-beam Kaufman-type ion source. The ion voltage, current density, substrate temperature, and feed gas composition (N_2 in Ar) have been varied. The resultant films were characterized by Raman, Fourier transform infrared (FTIR), and Rutherford back scattering (RBS) spectroscopy.

Samples with ~30% N/C ratio have been fabricated. The corresponding Raman and FTIR spectra indicate that nitrogen is incorporated into the samples by insertion into sp^2-bonded structures. A low level of $C\equiv N$ triple bonds is also found. As the ion current and voltage are increased with a pure Ar ion beam, Raman peaks associated with nanocrystalline graphite appear in the spectra. Adding low levels of nitrogen to the ion beam first reduces the Raman intensity in the vicinity of the graphite disorder peak without adding detectable amounts of nitrogen to the films (as measured by RBS). At higher nitrogen levels in the ion beam, significant amounts of nitrogen are incorporated into the samples, and the magnitude of the "disorder" peak increases. By increasing the temperature of the substrate during deposition, the broad peak due mainly to sp^2-bonded C-N in the FTIR spectra is shifted to lower wavenumber. This could be interpreted as evidence of single-bonded C-N; however, it is more likely that the character of the sp^2 bonding is changing.

INTRODUCTION

Recently, much experimental activity has been directed towards attempts to synthesize the carbon nitride compound (C_3N_4) of β-Si_3N_4 structure theoretically predicted by Liu and Cohen [1]. This compound is predicted to have hardness properties near that of diamond. Manufacture of this material either in bulk or thin film form has yet to be reported and confirmed by other investigators. Several groups have tried various techniques to synthesize this structure. Some of the attempted methods include ion-beam assisted deposition[2], reactive RF-magnetron sputtering[3, 4], electron cyclotron resonance plasma deposition[5], ion implantation into carbon films[6], e-beam evaporation of carbon with concurrent nitrogen ion bombardment[7], pulsed laser deposition (PLD) into a background gas of nitrogen[8, 9], PLD with an atomic nitrogen beam source[10], PLD with a radio frequency (RF) plasma assist[8], and ion-assisted PLD[11-13] (IAPLD).

In general, the synthesis of hard tetrahedrally-bonded compounds that are usually only stably formed in the bulk at high temperature and pressure requires the use of energetic deposition techniques. For example, we have used PLD to synthesize amorphous-tetrahedral carbon (a-tC) films with a high sp^3-fraction[9, 13, 14], and IAPLD to successfully form cubic-BN films[15]. Due to the success of PLD and IAPLD in growing similar materials, it is logical to expect this deposition technique to be a powerful tool for examining the formation and growth of CN_x materials. The motivation of this study is to controllably introduce nitrogen into a-tC films and examine the effect of ion bombardment on film structure and properties. We have deposited CN_x films under the concurrent bombardment of nitrogen and argon ions from a broad beam Kaufman-type ion source. The resultant films have been characterized by Fourier-transform infrared (FTIR), Rutherford back scattering (RBS), and Raman spectroscopies.

EXPERIMENTAL

The PLD vacuum chamber is described elsewhere[13]. It is capable of attaining a base pressure of 1 x 10^{-6} Pa and is equipped with a broad-beam Kaufman-type ion source. The pyrolytic graphite targets (Union Carbide) were ablated with a KrF (248 nm) laser capable of generating 450-mJ pulses of 17-ns duration. The deposition time was 20 minutes at a laser repetition rate of 20 Hz. Samples grown at elevated temperatures were deposited at a 40-Hz repetition rate. The laser light was focused into the vacuum chamber using a spherical lens with a 35-cm focal length. The laser beam illuminates the rotating target at a 45° angle from the target normal. The beam forms a rectangular spot on the target with an area of 0.01 cm^2, giving an energy density of 45 J/cm^2. To improve the thickness uniformity, the sample substrate was rotated, the center of the ablated plume struck the substrate off the center line, and the sample substrate was 18 cm from the target. Using these conditions, we obtain an ablation deposition with ±10% thickness uniformity over 10-cm diameter wafers at an ablation deposition rate of 0.2 Å/pulse (no ion beam) for a total sample thickness of 1250 ± 250Å. Before deposition, the uncoated silicon (100) substrates (n-doped with P to 0.02 Ωcm) were cleaned to remove the surface oxide layer by a wet dip procedure in a HF/NH$_4$F solution[16]. The "room temperature" rotating Si substrates were not actively heated during deposition, although some residual heating (< 75 °C) occurred due to the power dissipated by the normal operation of the ion gun and condensation of energetic species from the ablation plume. Sample depositions at elevated temperature were performed by resistively heating the Si substrate. The temperature was measured by a bare thermocouple pressed directly to the back of the wafer.

The 3-cm Kaufman-type ion source (mounted 10 cm from the substrate) was aimed (coincident with the laser plume) at a 25° angle from the substrate surface normal and slightly off axis from the center of the rotating Si substrate. The current density at the substrate was measured with a biased (-30 V) retractable ball of 1 cm^2 diameter. For this study, the experimentally varied parameters were the ion feed gas composition (0 - 100% flow of N$_2$ in argon at a constant total flow of 4 sccm), the ion energy (0 - 1000 eV), and the ion current density (0 - 260 µA/cm^2). During deposition the background gas pressure in the chamber was ~2.7 x 10^{-2} Pa.

To measure the effect of the ion beam on the atomic bonding, the films were characterized by Raman spectroscopy (514 nm light at 50 mW of power) and FTIR spectroscopy. Graphite has two Raman active modes at 1350 cm^{-1} and 1581 cm^{-1}. The Raman band at 1580 - 1590 cm^{-1} (the "graphite" or "G" peak) is a fundamental Brilloun-zone-center mode of graphite. The Raman band at 1350 - 1360 cm^{-1} (the "disorder" or "D" peak) is believed to be a Brilloun-zone-edge phonon mode that intensifies for graphite crystal domains less than 100 nm in size. "Glassy" carbon is a nanocrystalline graphitic phase with broad Raman peaks near 1360 cm^{-1} ("D") and 1590 cm^{-1} ("G"). Raman bands due to sp^3-bonded carbon are obscured by strongly resonance enhanced signals from sp^2 bonds. The signal from sp^2-bonded carbon atoms tends to obscure any signal due to sp^3-bonded carbon atoms[17]. In FTIR spectroscopy, these symmetric modes are normally not present; however, the presence of nitrogen incorporated into "graphite-like" rings make these modes IR active[18]. In addition, C≡N triple bonds appear near ~2190 cm^{-1}[19].

A parametric study was undertaken in order to understand the effects of varying ion beam deposition parameters. All the samples in this study were deposited using the same laser energy density (45 J/cm^2). The present paper focuses on the changes in the Raman and FTIR spectra as a function of varying ion beam deposition parameters.

DISCUSSION

Fig. 1a depicts FTIR spectra for samples grown with varying percentages of N$_2$ gas in Ar (V_b = 1000 V, I_b/A = 260 µA/cm^2). Also depicted in Fig. 1a is the nitrogen content of the samples as determined from RBS. The sample grown with a pure N$_2$ gas feed has a 28% N/C ratio. The nitrogen was distributed uniformly throughout the thickness of the film. As the nitrogen content of the samples increases, a broad peak in the FTIR appears centered near 1245 cm^{-1} that is partly due to nitrogen incorporation into sp^2-bonded structures[18]. It is possible that some of the intensity under this broad peak could be due to C-N single bonds of the type found in a β–C$_3$N$_4$ structure; however, it is not possible to make a more definite statement without other structural information (e.g., electron diffraction) or knowledge of exactly where the peak should lie in the FTIR. A more likely possibility is that there exists a broad distribution of structures that are responsible for this peak (e.g., different size sp^2-bonded clusters of varying numbers of carbon and nitrogen atoms[20]). Variation in the nitrogen content of the samples results in different local bonding conditions producing changes in the FTIR peak shape and intensity. In addition, for the sample grown with 100% N_2 in the feed gas, there is an inflection point near ~2150 cm^{-1} that is evidence of C≡N triple bonding[19]. The FTIR spectra are featureless from 2300 - 4000 cm^{-1}.

Fig. 1b depicts the Raman spectra (normalized to constant height) of the same samples of Fig 1a. For reference, the sample grown with no ion beam shows a relatively symmetric peak centered at ~1570 cm^{-1} typical of a-tC films with a high sp^3 fraction[9, 14]. In addition, a second order Si substrate peak appears near 965 cm^{-1} whose intensity is an indication of the opacity of the carbon film samples. The sample grown with a pure Ar ion beam shows a broader, more asymmetric peak that has shoulders near 1590 and 1360 cm^{-1} suggesting that this sample has more graphitic carbon. In addition the film is more opaque as indicated by the disappearance of the Si Raman peak. Clearly, the ion beam is altering the a-tC structure causing significant graphitization. This is hardly surprising, given the relatively high ion energy used (1000 eV). Typically, most a-tC films grown by other techniques involve carbon ion energies from ~30-200 eV. At lower energies, sub-surface penetration of the ions does not occur resulting in graphitic samples; and at higher energies the films also become increasingly graphitic, possibly due to increased vacancy mobility[21]. For the samples grown with nitrogen in the ion beam (Fig. 1b), the Raman spectra are not significantly different, even for the sample containing 28% nitrogen. In combination with the FTIR spectra, this suggests that the a-tC structure is being broken down by high-energy nitrogen ion bombardment and that nitrogen is incorporated into sp^2 bonds with carbon.

In an attempt to reduce the graphitic content of the samples grown at high ion energies, a series of samples was grown at lower ion energy (V_b = 250 V) and current density (I_b = 80 μA/cm^2) and varying N_2 levels in the gas feed. The FTIR spectra and the nitrogen contents as determined from RBS from these samples are shown in Fig. 2a. In the FTIR spectra, there is little evidence of nitrogen incorporation till the feed gas is 100% N_2, where the measured nitrogen

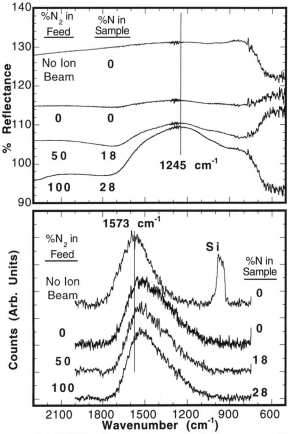

Fig. 1a) %FTIR reflectance for varying feed gas composition (N_2 in Ar) for I_b/A = 260 μA/cm^2 and V_b = 1000V.
Fig. 1b) Raman intensity for the same samples in Fig 1a).

content is 18% and a broad peak appears in the FTIR centered near 1280 cm⁻¹. For ion feed gas compositions lower than 25% N₂, the nitrogen content was not detectable by RBS (the sensitivity in this measurement was ± 1%). The FTIR spectra in Fig 1a and Fig. 2a with similar nitrogen content (~18%) are similar in appearance suggesting the reduced ion voltage and current density does not seem to have affected the film structure.

Figure 2b depicts Raman spectra (normalized to constant height) taken from the same films shown in Fig. 2a. The second order silicon peak is nearly absent for the samples grown in 100% N₂ and 100% Ar, but reappears for samples with a mixture of N₂ and argon. As discussed above, the intensity of this peak is a measure of the opacity of the samples. As has been suggested elsewhere, small amounts of nitrogen may help stabilize the sp³ structure[4] of the material thus decreasing the film opacity. However, the Raman spectrum from the sample with a high nitrogen content (18%) has shoulders near 1590 and 1360 cm⁻¹, indicating this sample is more graphitic.

The Raman spectra were fit to the sum of two

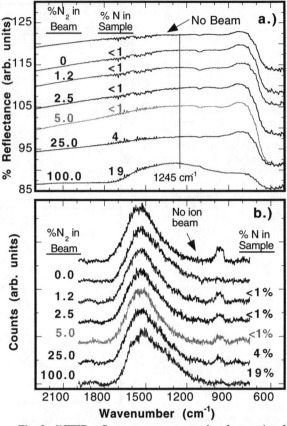

Fig. 2a %FTIR reflectance vs. wavenumber for varying feed gas composition (N_2 in Ar) for $I_b/A = 80$ $\mu A/cm^2$ and $V_b = 250V$. (The dip at ~1105 cm⁻¹ is due to SiO_2. The spectra are featureless from 2300 - 4000 cm⁻¹) Fig. 2b.) Raman intensity vs. for the same samples in Fig 2a).

gaussians centered near the positions of the G and D peaks of graphite to extract the ratio of their integrated intensities (I_D/I_G) as described elsewhere[13]. The results are plotted in Fig. 3 along with the RBS-determined nitrogen content of the films versus the percentage of N_2 in the feed gas. For low nitrogen content films, the ratio I_D/I_G shows a minimum near ~5% N_2 in the feed gas. This trend also correlates with the intensity of the second order Si peak and suggests that low levels of nitrogen may be beneficial in "healing" the a-tC network.

Amaratunga et al. have grown nitrogen-incorporated a-tC films by the filtered-cathodic arc method. It is interesting that they find low levels of nitrogen (≤1%) do not significantly alter the a-tC network and act as n-type dopants. At higher nitrogen contents (>1%) they find the a-tC network is significantly altered[22]. Fig. 3 suggests that the presence of low levels of nitrogen ions reduces changes to the a-tC network caused by the ion bombardment. As the nitrogen content increases the a-tC network breaks down leading to the creation of a more graphite-like structure and increasing the opacity of the films.

(Naively, the higher the I_D/I_G ratio, the more graphitic the microstructure of the film will be. However, the Raman spectra of a-tC and CN_x films are sufficiently different from that of nanocrystalline graphite that this interpretation is probably too simplistic. Doyle and Dennison[20] have proposed a model for the Raman spectra of nanocrystalline graphite based on distributions of sp²-bonded carbon rings with different sizes embedded in a two-dimensional continuous random network. Since the Raman signal is mainly from sp²-bonded material even for a-tC films with a high percentage of sp³ bonds, we speculate that a similar approach (modified to account for a more 3D structure) to fitting the a-tC Raman spectra may be applicable here. We

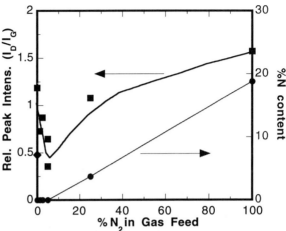

Fig. 3 I_D/I_G ratio and %N content (determined by RBS) versus %N_2 in the ion gun gas feed for the samples in Fig 2. The lines are drawn as guides to the eye. The diamond symbol is for a sample grown with no ion beam.

speculate that it may be more fruitful to think of changes in the Raman and FTIR spectra as changes in the makeup and distribution of sp²-bonded rings or clusters. Other methods of characterization must be used to probe the sp³-bonded part of the amorphous network.)

The FTIR spectra of samples deposited at elevated substrate temperatures are shown in Fig. 4. As the deposition temperature rises, the small peak due to C≡N triple bonding disappears and the broad peak due to nitrogen incorporation shifts to lower wavenumber. Since, a-tC films cannot be grown at deposition temperatures above 200 °C[21], the changes in the FTIR spectra could be related to underlying changes in the a-tC structure with temperature. We note that this shift is in the direction expected for lower C-N bond order. Interestingly, the irradiated region of these films is qualitatively much harder. A diamond scribe will easily scratch the unirradiated portion of the films, and significantly more pressure must be applied to scratch the irradiated region. At this time, more structural information must be gathered to make a definitive statement as to the reason for this increase in hardness.

CONCLUSIONS

CN$_x$ samples with $0 \leq$ N/C ≤ 0.30 have been grown using IAPLD. Effects of the addition of nitrogen to the samples can be seen in the FTIR and Raman spectra. The FTIR spectra show low levels of C≡N triple bonds and a broad peak centered near 1245 cm⁻¹ that is due to nitrogen incorporated into sp² CN bonds. These FTIR peaks

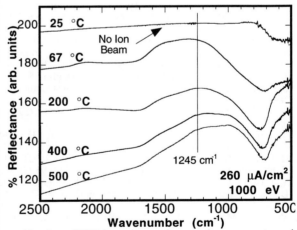

Fig. 4   FTIR intensity vs. for samples grown at varying temperatures with $I_b/A = 300$ μA/cm² and $V_b = 1000$V.

appear only in samples that have significant levels of incorporated nitrogen (as determined by RBS). In general, the Raman spectra indicate that films grown under ion bombardment near room temperature become more opaque and graphite-like. However, low levels of nitrogen incorporation (<1%) reduce the I_D/I_G ratio and decrease the opacity for the samples grown at 250 eV ion energies, indicating a possible stabilization of the a-tC structure by nitrogen. Samples deposited at elevated temperatures show no evidence of C≡N triple bonds and an FTIR peak that shifts to lower wavenumber as the substrate temperature rises.

ACKNOWLEDGMENT

This work was supported by the U.S. DOE under contract DE-AC04-94AL8500 through the Laboratory Directed Research and Development Program, Sandia National Laboratories.

REFERENCES

1. A. Y. Liu and M. L. Cohen, Phys. Rev., B **41**, 10727 (1990).
2. F. Rossi, B. André, A. v. Veen, P. E. Mijnarends, H. Schut, F. Labohm, H. Dunlop, M. P. Delplancke, and K. Hubbard, J. Mater. Res. **9**, 2440 (1994).
3. N. Nakayama, Y. Tsuchiya, S. Tamada, K. Kosuge, S. Nagata, K. Takahiro, and S. Yamaguchi, Jpn. J. Appl. Phys. **32**, L1465 (1993).
4. C. J. Torng, J. M. Silvertsen, J. H. Judy, and C. Chang, J. Mater. Res. **5**, 2490 (1990).
5. M. Diani, A. Mansour, L. Kubler, J. L. Bischoff, and D. Bolmont, Diamond and Rel. Mater. **3**, 264 (1994).
6. F. L. Freire, Jr., C. A. Achete, D. F. Franceschini, C. Gatts, and G. Mariotto, Nucl. Instrum. Meth. Phys. Res. **B80/81**, 1464 (1993).
7. K. Ogata, J. F. D. Chubaci, and F. Fujimoto, J. Appl. Phys. **76**, 3791 (1994).
8. J. Seth, R. Padiyath, and S. V. Babu, Diamond Rel. Mater. **3**, 210 (1994).
9. T. A. Friedmann, M. P. Siegal, D. R. Tallant, R. L. Simpson, and F. Dominguez, in *Novel Forms of Carbon II*, edited by C. L. Renschler, D. Cox, J. Pouch, and Y. Achiba (Materials Research Society, Pittsburgh, 1994), Vol. 349,.
10. C. Niu, Y. Z. Lu, and C. M. Lieber, Science **261**, 334 (1993).
11. J. Narayan, J. Reddy, N. Biunno, S. M. Kanetkar, P. Tiwari, and N. Parikh, Mater. Science Eng. **B26**, 49 (1994).
12. Z. M. Ren, Y. C. Du, Z. F. Ying, Y. X. Qiu, X. X. Xiong, J. D. Wu, and F. M. Li, Appl. Phys. Lett. **65**, 1361 (1994).
13. T. A. Friedmann, J. P. Sullivan, M. P. Siegal, D. R. Tallant, and R. L. Simpson, in *Beam-Solid Interactions for Materials Synthesis and Characterization*, edited by D. E. Luzzi, T. F. Heinz, M. Iwaki, and D. C. Jacobson (Materials Research Society, Pittsburgh, 1995), Vol. 354.
14. M. J. Siegal, T. A. Friedmann, S. R. Kurtz, D. R. Tallant, R. L. Simpson, F. Dominguez, and K. F. McCarty, in *Novel Forms of Carbon II*, edited by C. L. Renschler, D. Cox, J. Pouch, and Y. Achiba (Materials Research Society, Pittsburgh, 1994), Vol. 349.
15. T. A. Friedmann, P. B. Mirkarimi, D. L. Medlin, K. F. McCarty, E. J. Klaus, D. R. Boehme, H. A. Johnsen, M. J. Mills, D. K. Ottesen, and J. C. Barbour, J. Appl. Phys. **76**, 3088 (1994).
16. G. S. Higashi, Y. J. Chabal, G. W. Trucks, and K. Raghavachari, Appl. Phys. Lett. **56**, 656 (1990).
17. M. Ramsteiner and J. Wagner, Appl. Phys. Lett. **51**, 1355 (1987).
18. J. H. Kaufman, S. Metin, and D. D. Saperstein, Phys. Rev. B **39**, 13053 (1989).
19. H. Han and B. J. Feldman, Solid State Commun. **65**, 921 (1988).
20. T. E. Doyle and J. R. Dennison, Phys. Rev. B **51**, 196 (1995).
21. Y. Lifshitz, G. D. Lempert, E. Growssman, I. Avigal, C. Uzan-Sagay, R. Kalish, J. Kulik, D. Marton, and J. W. Rabalais, To appear in Diamond Relat. Mater. (1995).
22. G. A. J. Amaratunga, V. S. Veerasamy, C. A. Davis, W. I. Milne, D. R. McKenzie, J. Yuan, and M. Weiler, J. Non-Cryst. Solids **164-166**, 1119 (1993).

AFM STUDY OF SURFACE MORPHOLOGY OF ALUMINUM NITRIDE THIN FILMS

Yoshihisa Watanabe, Yoshikazu Nakamura, Shigekazu Hirayama and
Yuusaku Naota
Department of Materials Science and Engineering, National Defense Academy,
1-10-20 Hashirimizu, Yokosuka, Kanagawa 239, Japan

ABSTRACT

Aluminum nitride (AlN) thin films have been synthesized by ion-beam
assisted deposition method. Film deposition has been performed on the
substrates of silicon single crystal, soda-lime glass and alumina. The influence
of the substrate roughness on the film roughness is studied. The substrate
temperature has been kept at room temperature and 473K and the kinetic
energy of the incident nitrogen ion beam and the deposition rate have been
fixed to 0.5 keV and 0.07 nm/s, respectively. The microstructure of the
synthesized films has been examined by X-ray diffraction (XRD) and the
surface morphology has been observed by atomic force microscopy(AFM). In
the XRD patterns of films synthesized at both room temperature and 473K, the
diffraction line indicating the AlN (10•0) can be discerned and the broad peak
composed of two lines indicating the AlN (00•2) and AlN (10•1) planes is also
observed. AFM observations for 100 nm films reveal that (1) the surface of
the films synthesized on the silicon single crystal and soda-lime glass substrates
is uniform and smooth on the nanometer scale, (2) the average roughness of
the films synthesized on the alumina substrate is similar to that of the substrate,
suggesting the evaluation of the average roughness of the film itself is difficult
in the case of the rough substrate, and (3) the average roughness increases with
increasing the substrate temperature.

INTRODUCTION

Aluminum nitride (AlN) has many unique properties, and in addition, AlN thin
films are a very attractive material for making them piezoelectric substrate
because of their high ultrasonic velocity and large piezoelectric coupling factor [1].
Owing to these properties, AlN thin films are considered to be one of the most
useful for insulating layers and coating materials in microelectronics fields as well
as piezo electric devices. In applying thin films to practical usage, surface
smoothness is of great importance and atomic force microscopy (AFM) has the
potential to characterize the surface from the atomic level to the nanometer
scale.[2]

The present authors have reported AlN synthesis by evaporation of aluminum
and simultaneous irradiation with nitrogen ions, ion beam assisted deposition

(IBAD) method.[3-5] The IBAD method has the advantage of low temperature synthesis owing to the reaction of energetic ions and vaporized metal molecules.[6] In a previous paper, we synthesized AlN thin films at room temperature [3] and reported the influence of the substrate temperature and the kinetic energy of the nitrogen ion beam on the microstructure of the synthesized films.[4] In addition, the present authors have observed the surface morphology of AlN thin films by AFM and showed that the surface of the films synthesized at 473K becomes rough as compared with the films synthesized at room temperature.[5]

In the present paper, using the IBAD method, AlN thin films are synthesized on the various substrates with different surface roughness, such as silicon single crystal, soda-lime glass and alumina, at the substrate temperatures of room temperature and 473K. The influence of the substrate roughness and substrate temperature on the surface smoothness of the films is reported.

EXPERIMENTAL

Nitrogen gas (99.999% pure) and aluminum (99.99% pure) were used as ion source and target respectively. Film synthesis was carried out on the substrates of silicon single crystal (100), soda-lime glass and alumina placed into the load lock in a high vacuum chamber. Two kinds of alumina substrates with different roughness were used; one was as-fired with the average roughness of about 200 nm and the other was obtained by polishing with diamond pastes for a week and the average roughness decreases to about 7 nm. After evacuating the vacuum chamber to about 2.7×10^{-4} Pa, pure nitrogen gas was introduced to the ionization chamber and nitrogen ions were generated by an arc discharge. Then a nitrogen beam was obtained with electric field lenses for focusing and accelerating. The kinetic energy of the nitrogen ions and the deposition rate were kept at 0.5 keV and 0.07 nm/s. Aluminum was evaporated by electron bombardment from a 10 kW electron gum and the evaporation rate was monitored by a quartz sensor. The substrate temperature was kept at room temperature and 473K.

The thickness of the synthesized films was measured with a stylus device (Rank Taylor Hobson Ltd.). The crystallography of the films was determined by X-ray diffraction (XRD) using a Cu target (RINT 2500, RIGAKU Co.). The surface morphology was observed by the tapping mode with a NanoScope III (Digital Instruments) equipped with commercial silicon tips. The AFM images were acquired 256 x 256 points per frame, and were corrected by the subtraction of the background slopes.

RESULTS AND DISCUSSION

Figures 1 (a) and (b) show typical XRD patterns of films synthesized on the silicon single crystal substrate at room temperature and 473K, respectively. The thickness of these films is approximately 300 nm. In both XRD patterns, the stronger peak is assigned to the AlN (10•0) and the broad peak is composed of the AlN (00•2) and the AlN (10•1) planes. From these XRD patterns, it is confirmed that polycrystalline AlN thin films are synthesized both at room temperature and 473K, but the full width at half-maximum for the AlN (10•0) peak becomes sharper with increasing the substrate temperature.

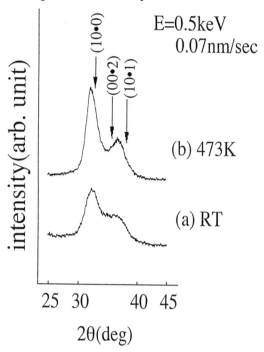

Fig. 1 X-ray diffraction patterns of films synthesized at (a) room temperature and (b) 473K. The kinetic energy of the nitrogen ion beam is kept at 0.5 keV and the deposition rate is kept at 0.07 nm/s.

Figure 2 shows a typical AFM image of a film synthesized at room temperature on the silicon single crystal. The thickness of the film is approximately 100 nm. From the image, it is found that the surface of the film is uniform and smooth on the nanometer scale. However, the surface of the film synthesized at 473K becomes slightly rough. This result is in good agreement with our first report of the AFM observations of AlN film surface.[2] The average surface roughness (R_a) is calculated to be approximately 0.4 and 0.6 nm for the films synthesized at room temperature and 473K, as a result of averaging over three different areas. The value of R_a for the silicon substrate is evaluated to be approximately 0.2 nm from AFM observation. From these results, it can be pointed out that AlN thin films with smooth surface on the nanometer scale can be synthesized by the IBAD method on silicon single crystal substrate below 473K.

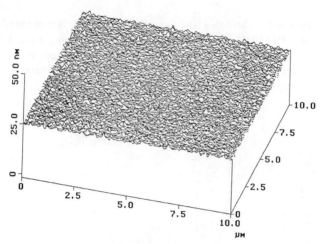

Fig. 2 Atomic force microscope image of AlN film synthesized at room temperature on the silicon single crystal substrate.

Figure 3 displays a typical AFM image of the film synthesized at room temperature on the soda-lime glass substrate. The deposition condition was the same as that for the films on the silicon single crystal substrate. Similarly, the film has a uniform and smooth surface, and the surface becomes slightly rough with increasing the substrate temperature. The value of R_a is evaluated to be approximately 1.0 nm and 1.5 nm for the films synthesized at room temperature and 473K, while the value of R_a of the substrate is approximately 0.5 nm.

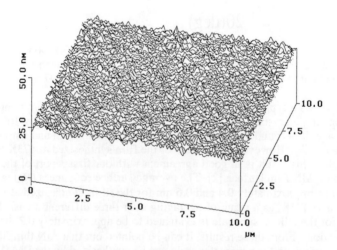

Fig. 3 Atomic force microscope image of AlN film synthesized at room temperature on the soda-lime glass substrate.

From both AFM data for the AlN films on the silicon single crystal and soda-lime glass substrates, it is found that the average roughness of the films is in proportion with the average roughness of the substrate.

Figure 4 shows a typical AFM image of the films synthesized on the as-fired alumina substrate. The substrate temperature was kept at room temperature and other deposition conditions were unchanged. It should be noted that the vertical scale of this image is eighty times as magnified as that of Figs. 2 and 3.

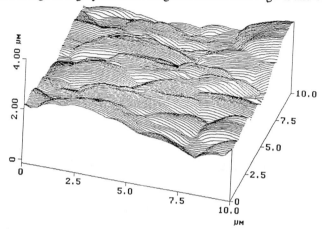

Fig. 4 Atomic force microscope image of AlN film synthesized at room temperature on the as-fired alumina substrate.

The surface roughness of the film is calculated to be approximately 170 nm and is similar to that of the substrate. The surface roughness of the film synthesized on the polished alumina substrate is found to be approximately 8 nm. These results suggest that evaluation of the surface roughness of the film itself becomes difficult in the case of the rough substrate.

The average roughness of the films synthesized at room temperature and 473K is plotted as a function of the average roughness of the substrate and shown in Fig. 5. From this figure, it can be seen that (1) the surface of the films synthesized on the smooth substrate becomes rougher than the surface of the substrate, (2) the surface roughness of the films synthesized on the rough substrate is almost same as that of the substrate, and (3) the average roughness increases with increasing the substrate temperature.

CONCLUSION

To understand the effect of the substrate roughness and substrate temperature on the surface roughness of thin films, AlN thin films were synthesized by the IBAD method on the various substrates at room temperature and 473K. The surface of the films was observed by AFM and the surface morphology was studied on the nanometer scale.

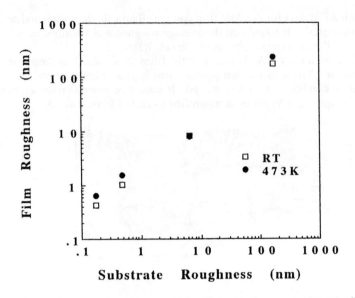

Fig. 5 Relation between the average surface roughness of AlN thin films and the average surface roughness of the substrate.

AFM observations reveal that (1) the surface of the films synthesized on the silicon single crystal and soda-lime glass substrates is uniform and smooth, and the average roughness is found to be sub-nanometer, (2) the average roughness of the films synthesized on the alumina substrates is similar to that of the substrate and it suggests that the evaluation of the average roughness of the film itself is difficult in the case of the rough substrate, and (3) the average roughness increases with increasing the substrate temperature.

References

[1] M. T. Wauk and D. K. Winslow, Appl. Phys. Lett. **13**, 286 (1968).

[2] X. D. Wang, W. Jiong, M. G. Norton and K. W. Hipps, Thin Solid Films **251**, 121 (1994).

[3] Y. Watanabe, Y. Nakamura, S. Hirayama and Y. Naota, Proc. 8th Int. Microelectronics Conf., Omiya, 1994, pp.57-60.

[4] Y. Nakamura, Y. Watanabe, S. Hirayama and Y. Naota, Surface and Coating Technologies **68/69**, 203 (1994).

[5] Y. Watanabe, Y. Nakamura, S. Hirayama and Y. Naota, Proc. 8th CIMTEC-World Ceramics Congress and Forum on New Materials, Florence, 1994, in press.

[6] Y. Andoh, Y. Suzuki, K. Matsuda, M. Satoh and F. Fujimoto, Nucl. Instruments and Methods B **6**, 111 (1985).

IN-SITU TRANSMISSION ELECTRON MICROSCOPY FOR ANALYSIS OF ION-BEAM-GROWTH PROCESSES

VLADIMIR V. PANKOV, NIKOLAI E. LEVCHUK and ANATOLY P. DOSTANKO
Belarusian State University of Informatics and Radioelectronics,
Modern Electronics Technologies Department, Minsk 220027, Belarus

ABSTRACT

Two types of systems for in situ transmission electron microscopy analysis of ion-beam etching, ion-beam sputtering and ion-beam assisted deposition are reported. Their design, operational features and some applications are presented. Radiation-stimulated diffusion in Mo-Si heterostructure, early growth of ion-beam sputtered In-Sn, In-Sn-O, ZnS:Mn films and recrystallization of In-Sn-O films during vacuum post-annealing are studied.

INTRODUCTION

Energetic ion beams are employed in a number of thin film formation processes to obtain coatings properties that cannot be achieved by using thermal deposition techniques. In situ transmission electron microscopy is one of the powerful methods to study thin film growth kinetics and analyse surface reactions. This information may be used to interpret thin-film properties and to optimize the conditions of film formation.

One can distinguish two general types of equipment for in situ transmission electron microscopy (TEM) analysis. The first-type of equipment implements dynamic in situ analysis and permits the studying of film formation processes in real time. But if physical or technical incompatibility of analytical and processing equipment occurs, the static in situ analysis technique may be used. The equipment of this type implies the interruption of the film formation process to carry out film analysis. There are two principal approaches to designing in situ TEM systems for analysis of ion-beam-growth processes. One is the combination of electron microscope with a processing equipment[1-6]. And the other is the incorporation of the miniature processing device inside the microscope[7-9]. It should be noted that at the present there are no systems for in situ TEM analysis of the ion-assisted deposition processes. In this paper the design, operational features and some applications of dynamic and static systems for in situ TEM investigation of ion-beam etching, ion-beam sputtering and ion-beam assisted deposition processes are reported.

SYSTEM FOR DYNAMIC IN SITU TEM ANALYSIS OF RADIATION-STIMULATED PHASE FORMATION AND MASS TRANSPORT

A system for dynamic in situ TEM investigation has been created on the base of a commercial transmission electron microscope - X-ray microanalyser. The schematic of the system is shown in Fig. 1. The basic element of the analysing system is an ion-beam sputtering system (IBSS) built into the microscope specimen chamber. The IBSS consists of two ion sources of closed-drift type. Axial configuration of sources provides minimal distortional action on the electron beam. The system operates at the residual gas pressure of $2\mu10^{-4}$ torr, operational gas pressure of $3-9\mu10^{-4}$ torr and provides the generation of inert and reactive gas ion beams with the energy range of 0.5-1.5 keV and current density up to 3 mA/cm^2.

The system was used for investigation of radiation-stimulated phase formation and mass transport in Mo-Si heterostructures during their bombardment by Ar ions with energy of 1 keV and current density of 0.05 mA/cm^2. As the experiments have shown there are essential differences in phase formation and mass transport during ion bombardment of the Mo-Si-structure from both the side of silicon and of molybdenum. In the case of the bombardment of Si (10 nm)/Mo (20 nm)-structure from the side of Si initially some

405

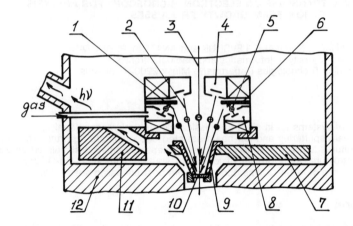

Fig. 1. Schematic of dynamic in situ TEM analysis system: 1 - condensed atoms flux; 2 - etching ion-beam; 3 - electron-beam; 4 - etching ion source; 5 - sputtering ion-beam; 6 - water-cooled target; 7 - specimen stage; 8 - sputtering ion source; 9 - specimen holder; 10 - film-substrate; 11 - adjusted stage; 12 - specimen chamber.

structural ordering of Si has been detected. Microanalysis and electron diffraction confirmed the presence of silicon after reduction of heterostructure thickness up to 8 nm. The appearance of Mo traces corresponded to the theoretical thickness calculated in accordance with Si and Mo etching rates. Polymorphic transformation from molybdenum with body-centred cubic (BCC) lattice to the face-centred cubic (FCC) modification with lattice constant of 0.401 nm has occurred beginning from heterostructure thicknesses of 13 nm. Complete transition terminated at a heterostructure thickness of 3 nm.

During ion bombardment of Mo (20 nm)/Si (20 nm)-heterostructure from the side of Mo both etching of upper layer and transition of BCC-lattice molybdenum to the FCC-modification took place. It was found that the structural ordering of the Si layer occured even at a heterostructure thickness of 33 nm whereas the mean free ion path in Mo was 5 nm. Microanalysis and electron diffraction confirmed the presence of Mo in the heterostructure untill its total etching. The analogous results took place in the case of Mo (5 nm)/Si (30 nm)-structure. No silicidation were revealed in all the experiments.

The obtained results point to both layer etching and interlayer atom diffusion occuring during ion bombardment of multilayer heterostructures at an energy of 1 keV. It was reconfirmed that lighter atoms diffuse predominantly to the surface and heavier atoms diffuse to the depth of the heterostructure. In the case of a heterostructure with lighter atoms of the lowest layer the structural changes occur at the depth considerably exceeding the mean free ion path.

Design features of the microscope-microanalyser and the affect of ion source magnetic systems do not allow successful observation of the early growth of thin films. Therefore a system for static in situ TEM analysis of film growth initial stages has been created.

SYSTEM FOR STATIC IN SITU TEM ANALYSIS OF ION BEAM SPUTTERED THIN FILMS EARLY GROWTH

A system for static in situ TEM analysis has been implemented on the base of a commercial transmission electron microscope with the resolution of 0.3 nm. In this system an original IBSS was placed inside the arrangement (Fig. 2). The arrangement was incorporated between the shutter mechanism and the specimen movement mechanism

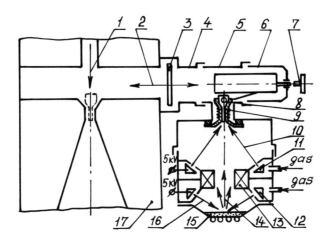

Fig. 2. Schematic of static in situ TEM analysis system: 1 - electron-beam; 2 - specimen movement direction; 3 - shutter; 4 - shutter mechanism; 5 - arrangement for specimen sputtering-heating; 6 - specimen movement mechanism; 7 - stick; 8 - specimen holder; 9 - specimen heating stage; 10 - etching ion-beam; 11 - etching ion source; 12 - sputtering ion source; 13 - magnetic system; 14 - condensed atoms flux; 15 - water-cooled target; 16 - sputtering ion-beam; 17 - microscope column.

without any construction changes of the microscope. The given IBSS also consisted of two axial ion sources of closed-drift type. During all the in situ study the arrangement maintains the common vacuum volume within the microscope. The specimen displacement from the treatment position to the analysis position is carried out by using a stick. To extend the range of the studied processes the arrangement was added by the specimen heating stage up to 700 K. The system operates at the residual gas pressure of $3\mu 10^{-5}$ torr, operational gas pressure of $1\text{-}5\mu 10^{-4}$ torr and provides the generation of inert and reactive gas ion beams with the energy range of 0.5-1.5 keV and current density up to 4 mA/cm^2.

The system of static in situ TEM analysis was applied for investigation of the early growth of In-Sn films ion-beam sputtered on amorphous carbon substrates at temperatures of 300, 400, 573 K and at the deposition rate of 16 nm/min. The target composition was In+5%Sn. The initial growth of quasi-amorphous island films was observed. At the mean thickness of 6 nm the films had a fine-graned polycrystalline structure. At a thickness of 27 nmthe films possessed a continuous grained structure with the incorporation of anomalous large-sized islands in fine-grained matrix.

The relationships of island density and island mean size versus mean film thickness as a function of temperature (Fig. 3) were used for the calculation of surface diffusion energy E_d and surface growth energy E_g according[10,11]. The obtained value of E_d=0.2 eV is in agreement with E_d=0.19±0.03 eV for Au films deposited by ion-beam sputtering onto NaCl[10] but considerably differs from E_d=0.09±0.015 eV for thermaly evaporated Au films[10]. Probably more vigorous surface diffusion for ion-beam sputtered In-Sn films causes the higher adatom kinetic energy than in the case of thermal evaportation. The value of E_g=0.13 eV for ion-beam sputtered In-Sn films is lesser than E_g=0.15 eV for thermaly evaporated Al and Fe films onto NaCl[11]. The island growth for ion-beam sputtered films proceeds slower than in the case of thermal evaportation and it is assumed to be due to the higher island density and smaller diffusion zones. The size of critical

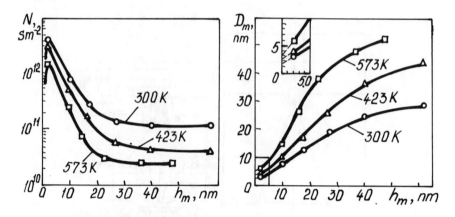

Fig. 3. Island density (a) and mean island size (b) versus mean film thickness for In-Sn - films on carbon as a function of substrate temperature.

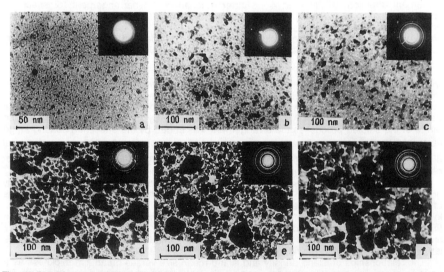

Fig. 4. Growth sequence of In-Sn-O - films on carbon at 573 K: mean film thickness of (a) 2 nm; (b) 4.5 nm; (c) 13 nm; (d) 22 nm; (e) 30 nm; (f) 39 nm.

nucleus of 2-3 nm for ion-beam sputtered In-Sn films was estimated using the extrapolation of the curves of mean island size versus mean film thickness.

The system of static in situ TEM analysis was applied for investigation of early growth of conducting transparent In-Sn-O films . These films were reactively ion-beam sputtered on the amorphous carbon substrates at the temperatures of 300, 423, 573 K and a deposition rate of 7.6 nm/min. The target and operational gas compositions were In+5%Sn and Ar+80%O$_2$ respectively. The quasi-amorphous films growth was observed at 300 K. As shown in Fig. 4 the films had polycrystalline structure of In$_2$O$_3$ with BCC-lattice when deposited on the heated substrates. At the thickness of 10 nm the films became continuous and on their surface was observed the appearance of secondary nucleation centres that developed in the separate large grains with cubic and hexagonal form.

The system of static in situ TEM analysis was applied for the study of early growth of electroluminescent ZnS:Mn films. These films were ion-beam sputtered on the amorphous carbon substrates at temperatures of 300, 423, 573 K and deposition rate of 8.4 nm/min. At the thickness of 7 nm the films had a polycrystalline structure identified as ZnS with FCC-lattice. The films grown at 423 K and below consisted of grains with blurred boundaries. The smeared structure of these films is assumed to be due to the predominance of radiation-stimulated defect generation and atom mixing over the structural ordering and defect annealing[12]. At the thickness of 14 nm the films became continuous and on their surface the appearance of secondary nucleation centres was detected that developed in the separate large-size grains (Fig. 5). The distinction of film growth at 573 K was the formation of pronounced large-grained columnar structure at the thickness more than 33 nm, the appearance of mixed phase with FCC - close-packed lattices and of twin-grains on the plane (111) normal to the electron-beam.

The possibility of specimen heating inside the electron microscope has permitted the study of the recrystallization mechanism of In-Sn-O - films during their vacuum post-annealing. The films were deposited by ion-beam sputtering on the cold carbon substrates. It was established that initial recrystallization processes start at 400 K by the formation of individual grains-spherolytes identified as In_2O_3 with BCC-lattice. The grain concentration gradient directed to the large-size defects of film such as layer breaks and incorporations (Fig. 6) is assumed to be due to the existence of the long-range field and initiating the generation of recrystallization centres in the distant regions. The annealing temperature increase causes the extension of recrystallized area. The extension is due to both the grain-boundary migration and the absorption of the small grains with small-angle boundaries by their larger neighbours. The recrystallization of the film and the formation of axially textured grains-microcrystallines with size of 800-1000 nm identified as In_2O_3 with BCC-lattice takes place at 430 K. The annealing at the higher temperatures results in the slight grain size growth as well as in the appearance of the grain boundariy breaks caused by film intrinsic stresses.

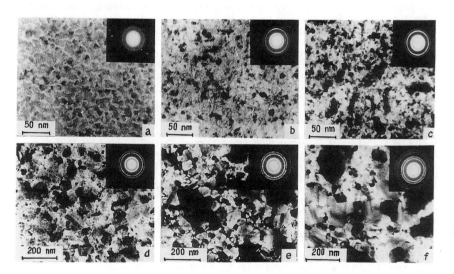

Fig. 5. Growth sequence of ZnS:Mn - films on carbon at 573 K: mean film thickness of (a) 7 nm; (b) 14 nm; (c) 24 nm; (d) 33 nm; (e) 43 nm; (f) 70 nm.

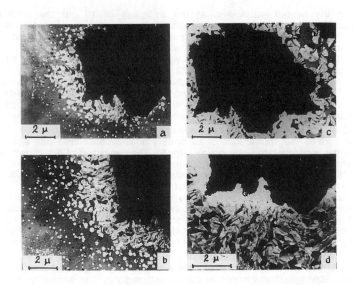

Fig. 6. Recrystallization sequence of In-Sn-O film deposited on carbon at 300 K during vacuum post-annealing. The annealing temperatures are: (a) 400 K; (b) 410 K; (c) 420 K. (d) - electron-beam annealing inside the microscope.

CONCLUSION

The presented results convincingly demonstrate that a system created for dynamic TEM analysis is a powerful analytical tool for the study of radiation-stimulated phase formation and mass transport in multilayered heterostructures. The other developed system of static in situ TEM analysis is indispensable in the investigation of the kinetics of both early growth and recrystallization of vacuum condensates produced by means of ion-beam technique.

Further work will be continued in the direction of the creation of a system for dynamic in situ TEM analysis made on the base of the electron microscope with the specimen side-entry and resolution higher than 0.2 nm.

REFERENCES

1. M.O.Rualt, J.Chaumont, H.Bernas IEEE Trans. Nuclear Sci. NS-30 (2) 1746 (1983).
2. A.Taylor, E.A.Ryan, Nucl. Instr. & Meth. in Phys. Res. B10/11, Pt. 2, 687 (1985).
3. W.A.Isser, Crystal Research and Technology 14 (11) 1393 (1979).
4. H.Poppa, Phil. Mag. 7 (78) 1013 (1962).
5. В.И.Бендиков, О.Л.Тальянская, В.Ф.Рыбалко, Приборы и техника эксперимента 5, 203 (1986).
6. J.-P.Gonchod, P.Ged , D.Bois, France Patent No 2477769, H 01 J,K.
7. K.Heinemann, H.Poppa, J. Vac. Sci. Tehnol. 10 (1) 22 (1973).
8. E.Morita, K.Takayanagi, K.Kobayashi, K.Yagi, G.Honjo, Jap. J. Appl. Phys. 19 (10) 1981 (1980).
9. D.M.Sherman, Th.E.Hutchinson, Rev. Sci. Instrum. 43 (12) 1793 (1972).
10. G.E.Lane, J.C.Anderson, Thin Solid Films 26, 5 (1975).
11. Э.И.Точицкий, Кристаллизация и термообработка тонких пленок (Наука и техника, Минск, 1976).
12. K.-H.Muller, Surf. Sci. Lett. 184, L375 (1987).

SURFACE AND INTERFACE MICROSTRUCTURE OF EPITAXIAL FERROELECTRIC (Ba,Sr)TiO3 THIN FILMS

V. A. ALYOSHIN, E. V. SVIRIDOV, Vl. M. MUKHORTOV, I. N. ZAKHARCHENKO AND
V. P. DUDKEVICH
Institute of Physics, Rostov State University, 194 Stachki Ave.
Rostov-on-Don, 344104 Russia

ABSTRACT

Surface and cross-section relief evolution of ferroelectric
epitaxial (Ba,Sr)TiO3 films rf-sputtered on (001) MgO crystal cleavage surfaces versus the oxygen working gas pressure P and substrate temperature T were studied. Specific features of both three-
dimensional and two-dimensional epitaxy mechanisms corresponding
to   various   deposition   conditions   were   revealed.   Difference
between low and high P-T-value 3D epitaxy was established. The
deposition of films with mirror-smooth surfaces and perfect interfaces is shown to be possible.

INTRODUCTION

Some applications of ferroelectric thin films require the high
quality epitaxial film surface and film-substrate interface. The
dependence of a surface relief of epitaxial $(Ba_{0.85} Sr_{0.15})TiO3$
films rf-sputtered onto (001)MgO cleavage surfaces upon the oxygen working gas pressure P and the substrate temperature T have
been reported in [1]. The P-T diagram corresponded to the variation of the epitaxial film surface relief with P-T values was shown
(Fig. 1). The P values varied in the range from 30 to 150 Pa and

Figure 1.   Microstructure
type diagram illustrating
the influence of sputtering
conditions on relief of epitaxial BST/(001)MgO films:
region 1 - semi-spherical
blocks formation (●),
region 2 - mirror-smooth
epitaxy (o),
region 3 - oriented facet
block growth (●).
The surface relief of thin
films sputtered at conditions corresponded to the boundaries between the regions
was intermediate (φ). [1]

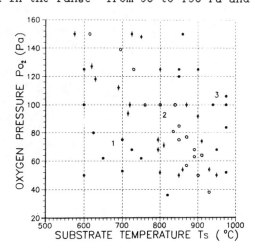

Mat. Res. Soc. Symp. Proc. Vol. 388 © 1995 Materials Research Society

the T values varied from 500 to 1000 °C. Three regions on the microstructure type diagram with the increase of P and/or T were observed: at the lowest P and/or T values the semi-spherical blocks
were formed (region 1 on P-T diagram); the increase of P and/or T
resulted in the mirror-smooth epitaxy (region 2); the further increase of P and/or T led to the oriented facet block growth
(region 3). The diffused boundaries between these regions corresponded to the intermediate surface relief.

The present paper is aimed at understanding the growth mechanisms of epitaxial (Ba, Sr)TiO3/(001)MgO (BST) films.

EXPERIMENTAL PROCEDURE

BST films were rf sputtered onto (001) cleavage faces of MgO
crystals from the stoichiometric targets in an O_2 atmosphere.
The fact of epitaxial growth, block misorientations and orientation relations between the film and the substrate were established by XRD method. The surface and cross-section reliefs (before
and after chemical etching) of thin film - substrate sandwiches
were studied by usage of the electron microscope (Pt/C unfolded
replica technique). To reveal the dislocations both in the MgO
substrate and on the film surface the aqueous HNO3 solution with
the addition of HF was used as an etchant.

RESULTS AND DISCUSSION

Different types of surface reliefs of the epitaxial BST films
deposited in the regions 1, 2, and 3 of the above mentioned P-T
diagram were due to the gradual change in the following mecha-

Figure 2. Surface relief of epitaxial BST/(001)MgO films deposited at the growth conditions corresponding to the regions 1, 2,
and 3 of the P-T diagram, respectively. C/Pt replica. Bar - 1 um.

nisms of nucleation and growth: three-dimensional nucleation, layer-by-layer growth, and three-dimensional nucleation again, respectively (Fig. 2).

At the lowest values of P and T (region 1 on the P-T diagram) the three-dimensional growth of the epitaxial films having semi-spherical grains took place (Fig. 2. 1). The increase in P and/or T (region 2) resulted in the growth of mirror-smooth films (Fig. 2. 2). Spiral steps indicating the presence of screw growth dislocations were observed on the surfaces of these films. Spiral patterns were seen more clear after the chemical etching of the film surface. The thickness of these films at which their continuity may be achieved is not more than the depth resolution of the Auger analysis (3 nm). This evidenced the layer-by-layer growth mechanism (2D epitaxy). The further increase in P and/or T led to the three-dimensional nucleus growth (region 3 of P-T diag-ram). In this case the surfaces of the films were formed with facet blocks (Fig. 2. 3). Thus, the three-dimensional nuclei growth

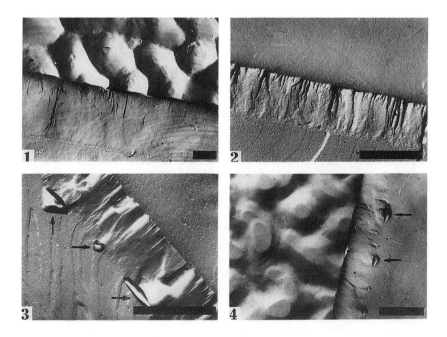

Figure 3. Surface and cross-section relief of epitaxial BST/ (001)MgO films grown at 3D (1, 4) and layer-by-layer (2, 3) epitaxy. The deposition conditions are: 1 - P=60 Pa, T=650 °C (the inner zone of the region 1 of P-T diagram); 2 - 80 Pa, 825 °C (the inner zone of the region 2); 3 - 80 Pa, 875 °C (the high P-T value zone of the region 2); 4 - 100 Pa, 925 °C (the low P-T value zone of the region 3). At the high P-T value deposition conditions the pores in the film-substrate interface are formed (3, 4). Bar - 1 um.

(3D epitaxy) is realized in regions 1 and 3. The difference between the 3D epitaxy mechanisms in these regions is shown below.
    The results of the study of cross-sections of film-substrate
sandwiches formed by different epitaxy mechanisms are illustrated
in Fig. 3. As can be seen, closed pores are formed in the film-
substrate interface during film growth at elevated working gas
pressure and/or substrate temperature (Figs. 3.3 and 3.4; pores
are marked with arrows). Despite the mirror-smooth film surface
relief achieved these macrodefects were observed in the film-
substrate interface region under growth conditions corresponding
to the high P-T value zone of the 2D epitaxy region 2 (Fig. 3.3).
Smooth films fabricated by the conditions of inner zone of the 2D
epitaxy region 2 do not contain pores (Fig. 3.2). The additional
difference between low and high P-T value mechanisms of 3D epitaxy was established: the absence (Fig. 3.1) and formation (Fig. 3.4)
of pores in film-substrate interfaces, respectively.
    To establish the reasons of the pores appearence we examined

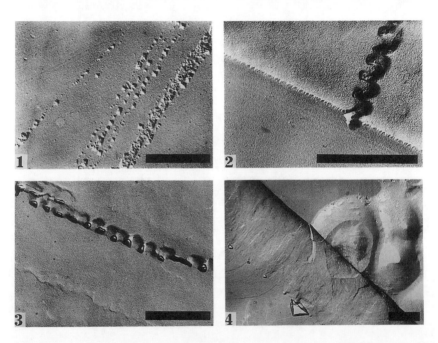

Figure 4.   Dislocation mechanism of the pore formation in the
film-substrate interface.   1 - the chemically etched (001) surface of MgO crystal. 2 - the cross-section of the 35 nm thick BST
film in the vicinity of the microstep on MgO cleavage surface
(inner zone of the region 2). 3 - 35 nm thick BST film surface
relief in the vicinity of the microstep (the high P-T value zone
of region 2). 4 - the cross-section relief of the 2.5 um thick BST
film (under the same growth conditions). C/Pt replica. Bar - 1 um.

first the substrate surface. It was found that the microsteps of cleavage are situated on the surfaces of MgO crystals. After chemical etching the elevated concentration of dislocations formed near and on the microsteps of cleavage surfaces was revealed (Fig. 4.1). In the inner P-T value zone of the smooth epitaxy region 2 the pores were not formed either on the smooth substrate surface or on the microsteps (Fig. 4.2). In the initial step of film deposition in the high P-T value zone of 2D epitaxy region 2 the increased number of pores was observed near microsteps (Fig. 4.3) and with the further film growth the mirror-smooth surface was not formed in this part of film (Fig. 4.4). At some distance from microstep where dislocation concentration was not more than the primary dislocation density in the MgO crystals the smooth film surface was formed over the pores.

Fig. 5 illustrates dynamics of pore formaton in the film-substrate interface by the high P-T value 3D epitaxy. At the stage of nuclei coalescence (average film thickness h = 8 nm) the area of free substrate surface is about 50% of whole surface area (Fig. 5.1). Then, hole formation due to plasma etching the substrate material on the open MgO surface (circles) and the shrink of

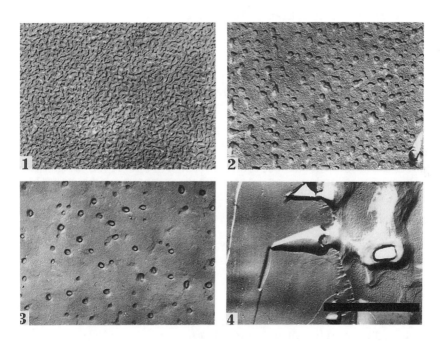

Figure 5.   Dynamics of the pore formation in the film-substrate interface (region 3 of the P-T diagram): 1 - the film surface at the stage of nuclei coalescence. An average film thickness h = 8 nm.   2 - h = 12 nm.   3 - h = 24 nm.   4 - the cross-section of the film of 180 nm in thickness.   C/Pt replica.   Bar - 1 um.

free substrate area take place; hole density is much higher than the initial concentration of dislocations in MgO crystal (Fig. 5.2). After that, the decrease in number of open pores and their further growth at the expense of their widening and deepening into MgO are observed (Fig. 5.3). Cross-section investigation of the film permits to reveal that the growing film did not condenses on the bottoms of the holes but covered them due to the lateral growth and catches micropores in the film-substrate interface (Fig. 5.4). The pores are situated near the boundaries of film blocks.

The increased pore density in the films from region 3 (Fig. 5) allows to assume that an additional dislocations are formed in the substrate on the boundaries with the film nuclei. The absence of pores in the films from region 1 is due to the low P-T values that prevents the substrate etching by plasma. We suppose that the pore influence on growth process is not the main reason of the difference between 3D epitaxy mechanisms in regions 1 and 3. This is confirmed by the possibility of fabricating the pore-free BST block film at growth conditions corresponding to region 3 at the expense of film deposition on the surface of mirror-smooth BST film. Film resputtering seems to play an alternative role in the appearence of P-T high-value region of 3D epitaxy.

CONCLUSION

We suppose that: i) the microholes formation in the substrate at the high P-T values is due to the etching of dislocations by rf-plasma particles, ii) the increased hole concentration in region 3 is due to the resputtering of the substrate material near the nucleus boundaries from the places where the additional dislocations are formed, and iii) the reason of 3D epitaxy in high P-T region 3 is the film resputtering.

Nevertheless, the deposition of transparent films with mirror-smooth surfaces and perfect interfaces is shown to be possible.

Acknowledgements

This work was supported by grant N 95-02-06099a of Russian Basic Research Foundation.

References

1. Z. Surowiak, Y. Nikitin, S. Biryukov, Y. Golovko, V. Mukhortov and V. Dudkevich, Thin Solid Films. 208, 76 (1992).

CONDUCTIVITY OF THE GRANULAR METAL FILMS OBTAINED BY HIGH DOSE ION IMPLANTATION INTO PMMA

V.V.Bazarov, V.Yu.Petukhov, V.A.Zhikharev, I.B.Khaibullin,
Zavoisky Physical Technical Institute of Russian Academy of Science,
Sibirsky Trakt, 10/7, Kazan, 420029,
Russian Federation

ABSTRACT

Thin granular metal films in polymethylmethacrylate(PMMA) have been synthesized by 40 keV Fe^+, Ag^+ or Pb^+ ions implantation with fluencies up to $6*10^{17}$ ion/cm^2. The resistivity of synthesized films was measured in the temperature range from 300K to 5K. The temperature dependence of the resistivity of PMMA implanted with Ag^+, Pb^+ and small fluence Fe^+ obeys the well known law $\ln R \sim (1/T)^{1/2}$. The samples implanted by high fluence Fe+ reveal rather a different behaviour. At low temperature (T<100K) the curves R(T) fit the formulae $\ln R \sim \ln T$. The two mechanisms of conductivity of a granular film are considered: direct tunneling and thermally activated hopping. Combined with the morphology features of films, obtained by high fluence Fe^+ implantation, the above mentioned consideration offers a satisfactory explanation of the observed temperature dependence R(T).

INTRODUCTION

The enhancement of conductivity properties of polymers is an important area of technology since it may give rise to the development of useful and unique devices. An ion implantation is one of the methods of changing a surface layer of an insulating polymer into a conducting material is . The majority of reported experimental results of the investigation of the influence of ion implantation on the transport properties of polymers [1] concerns radiation fluences which are lower than 10^{16}ion/cm^2. In this case an increase of electrical conductivity of polymers has a radiolytic origin. The damages (dangling bonds, additional cross linking, scission of the chains etc.) lead to the formation of diamondlike or graphitic materials in the surface layer of irradiated polymer. The conductivity of such layers is usually described by variable range hopping, temperature dependence of dc resistivity R obeying the relation

$$\ln(R/R_0) = (T_0/T)^n \qquad (1)$$

where n ranges from 1/4 to 1 depending on the structure and dimension of conducting material [2].

At a high dose ion implantation there occurs a creation of new phase (ion synthesis). When metal ions are used as implants a granular metal film emerges just under the polymer surface [3]. The film can give a noticeable contribution to surface conductivity of irradiated samples. Much attention has been given to investigation of the electrical transport in an assembly of metal particles embedded into insulating matrix (see for ex. [4]). The process of transport of a single electron from one grain to another is connected with nonnegligible change in electrostatic

417

energy. This change may be compensated either by the electron while it passes into a lower-energy state in the second grain (direct tunneling) [5] or by phonons (thermally activated hopping) [6]. In many investigations the experimentally observed temperature dependence of dc conductivity of granular metal films corresponds to the formulae (1) with n=1/2. This result is explained if we assume the thermally activatied hopping as the main mechanism of electron transport and take into account the spread of grain sizes. However direct tunneling can contribute noticeably to electron transfer at low temperatures. In this paper the surface conductivity of polymethylmetacrylate (PMMA) after a high dose implantation by metal (Fe^+, Ag^+, Pb^+) ions has been investigated in a wide temperature range.

EXPERIMENT

The PMMA slices (1mm thickness) were implanted by Fe^+ ions at 40keV and by Ag^+ and Pb^+ ions at 30keV with fluences ranging from 10^{15} to $2*10^{17}$ ion/cm^2. To bring out the contribution of metal film to the conductivity an implantation of PMMA by 40keV Ar^+ ions was also carried out. Low ion current densities (<5mkA/cm^2) were used to avoid overheating and destruction of the polymer during the implantation. The surface resistivity was measured by constant voltage method and thermal evaporated Ag electrodes were used. The investigated areas between electrodes had shapes of (5*5)mm^2 squares. Depending on the magnitude of resistance either four-point or two-point methods were applied. The date had accuracies of 3% in surface resistance below 10^9Ohm and 5% in resistance higher than 10^9Ohm. The resistivity was measured at a temperature range from 5K to 300K.

RESULTS AND DISCUSSION

Room temperature resistivity R_0 of PMMA implanted with metal ions doesn't depend on fluence in dose range 10^{16} - $6x10^{17}$ion/cm^2. For different samples and different implants the values of R_0 lie in the interval (0.5-5) MOhm. The temperature dependencies of the resistivity of PMMA implanted by Fe^+, Ag^+, Pb^+ and Ar^+ ions are plotted in Fig.1. One can see that the experimental dependencies R(T) for samples with Ag, Ar, Pb are in good agreement with law(1) with n=1/2. As for the samples implanted by Fe^+ ions, the law (1) was observed only for low dose of implantation. At high fluences (>10^{17}ion/cm^2) a considerable deviation from (1) takes place for temperatures lower than 100K. The curve ln(R(T)) is more gently sloping than ln(R/R$_0$)~(1/T)$^{1/2}$ law. We suppose that such behaviour of resistivity is connected with morphology features of granular Fe-film obtained at a high dose implantation. This supposition is confirmed by TEM study of implanted PMMA samples. Typical electron micrographs for samples implanted by Ag and Fe are shown in Fig.2(a-c). As can be seen in Fig.2c the high dose Fe^+ - implanted layer contains two well distinguished groups of particles (one group - with sizes up to 10nm and the other - with sizes of 60-70nm). It should be marked that large particles were arranged in the form of "lattice" in the layer plane. The space between big particles is sufficiently large and filled with small Fe particles.

It is obvious that in the case of Ar^+ implantation the conductivity is determined only by a carboneouse layer near the polymer surface. Such layers also exist in samples implanted by metal ions. We suppose that for the temperatures higher than 100K the conductivity of implanted

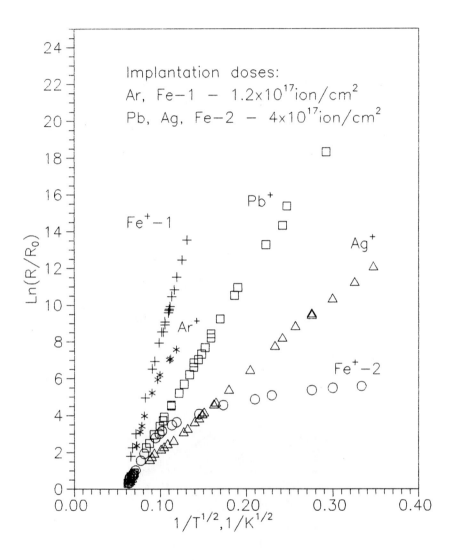

Fig. 1. Temperature dependences of surface resistivities of PMMA implanted by different ions

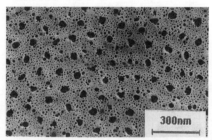

PMMA + Fe$^+$, D = 4x10^{17}ion/cm^2.

Fig.2. TEM micrographs of implanted PMMA.

polymer is a sum of conductivity of the carboneouse layer and that of granular metal film. Because of the high resistivity of the carboneouse layer the resistivity of implanted samples at low (<100K) temperatures is mainly determined by the synthesized metal film. Within this temperature range the direct tunneling of electron from negatively charged particle to a neutral one can contribute to a transition probability comparable with that of thermally activated hopping. At a very low temperature a direct tunneling dominanates. To calculate the conductivity of the film the critical path method [6] was used. At all temperatures the law $\ln(R/R_0) \sim (1/T)^{1/2}$ was obtained for an assembly of particles with a broad distribution of sizes. If the size distribution is narrow (when the difference between charging energies is very small) the thermally activated hopping gives the dependence $\ln(R/R_0) \sim (1/T)^{1/2}$ as in the previous case, while the direct tunneling leads to rather a different law $\ln(R/R_0) \sim \ln T$.

In the case of Ag- (and Pb-) implantation metal films consist of particles with widely spread sizes. Thus the dependence (1) is expected to be observed. Calculations for high dose Fe-implanted samples which take into account the presence of two distinguished groups of particles lead to rather different results. The large difference in charging energies hinders the electron transfer between small and large particles. Combined with the fact that large particles are separated by areas filled with small ones, this leads to an enhancement of contribution of small particles subassembly into resistivity of the film. The size distribution in the small particles subassembly is estimated to be very narrow. Thus at low temperatures where direct tunneling is dominant the resistivity obeys the formulae $\ln(R/R_0) \sim \ln T$. This dependence fits well with the experimental results for high dose Fe- implanted PMMA at T<100K. It is interesting to emphasize that when the temperature decreases the direct tunneling processes effectively rule out the largest particles from electron transport, whereas thermally activated processes effectively "remove" smallest ones.

Thus, the temperature dependence of resistivity PMMA implanted by metal ions has been investigated. And has been demonstrated that an unusual temperature dependence of resistivity $\ln(R/R_0) \sim \ln T$ for high dose Fe- implanted samples is connected with the presence of two groups of particles of different size scales in the synthesized Fe film and with a prevalence of direct tunneling in electron transport at low temperatures.

This work was supported in part by International Science Foundation (Grant NNV000).

REFERENCES

1. M.S.Dresselhaus, B.Wasserman, G.E.Whek, Mat.Res.Soc.Symp.Proc.**27**, 413-422 (1984);
   C.J.Sufield, S.Sugden, C.J.Bedell, P.R.Graves and L.B.Bridwell, Nucl.Instr. and Meth.
   B67, 432 (1992).
2. Yongqiang Wang, L.B.Bridwell, R.E.Giedd, J.Appl.Phys. **73**(1), 474-476 (1993).
3. V.Yu.Petukhov, V.A.Zhikharev, E.P.Zheglov, I.A.Azarov, I.B.Khaibullin,
   Proc. of 27 AMPERE Congress, Kazan, 1994, pp.370-371;
   V.Yu.Petukhov, V.A.Zhikharev, N.R.Khabibullina, I.B.Khaibullin,
   Vysokochistye veschestva, <u>1993,</u> N3, 45-48.
   [High-purity substances, 7, N3, 274-276 (1993)].
4. B.Abeles, Ping Sheng, M.D.Coutts, Y.Arie, Adv.Phys., **24**, 407 (1975).
5. C.J.Adkins, J.Phys.C:Solid State Phys., **15**, 7143-7155 (1982).
6. Ping Sheng, J.Klafter, Phys. Rev., **B27**(4), 2583-2586 (1983).

CONDUCTANCE MEASUREMENTS OF THERMALLY ANNEALED, SI-IMPLANTED QUARTZ

Alex R. Hodges and Gary C. Farlow
Physics Department, Wright State University, Dayton, OH 45435.

ABSTRACT

Preliminary conductance measurements of Si implanted, α-Quartz which had been annealed in Ar to $1000°C$ have been made using a bridge method. The quartz was implanted to a dose expected to yield Si precipitates inside the quartz upon annealing. The measured conductivity, based on a geometry deduced from TRIM calculations and several trans-conductance measurements, is $\sim 2 \times 10^{-4}$ $(\Omega\ m)^{-1}$. This is consistent with large islands of Si in series with an insulating matrix.

BACKGROUND AND SAMPLE PREPARATION

It has been shown that when Si is implanted into quartz and subjected to rapid thermal annealing that crystalline islands or layers of precipitated Si will form at roughly the mean range of the implanted Si.[1] This process is the reverse of the SIMOX process[2] now used in device fabrication. In this paper we describe preliminary measurements of the trans-conductance of such buried 'layers' of Si.

In order to prepare suitable samples it is necessary to implant enough Si to form a layer super-saturated with Si in the vicinity of the mean range. The criterion is that the number density of implanted Si plus the number density of native Si in SiO_2 exceeds the number density of Si atoms in elemental Silicon. The required number densities are listed in Table I.

We chose 120 Kev for implantation since this is high enough to minimize sputtering and small enough to have a rather narrow implanted-ion distribution, thus minimizing the dose required. Other parameters of the implantation are listed in Table II. The Si ion distribution from a TRIM-89[2] simulation of this implantation of Si into quartz gives an effective width of 100 nm. This translates into a minimum required dose (flux) of 2.8×10^{17} Si/cm^2. The actual dose used was $4 \times 10^{17} Si/cm^2$ which corresponds to a

Table I. showing the number
densities of Si in the beginning
and final materials and the net
density needed to obtain the
final result.

	Atomic density of Si (atoms/cc)
Si	5.0×10^{22}
SiO$_2$	2.2×10^{22}
Implanted Si ions	2.8×10^{22}

Table II. showing the implantation parameters for the Si-implanted quartz (* From TRIM SIMULATION)

Energy	120 keV
Flux	$4 \times 10^{17}/cm^2$
Fluence	$2.0 \ \mu A/cm^2$
Temperature of sample	$30 \degree C$
Range	1550 nm*
Straggling	50 nm*

Precipitation occurs during annealing. This can be accomplished by either rapid thermal annealing or ordinary thermal annealing in an oxygen-free ambient. (Annealing in the presence of oxygen will allow the implanted Si to oxidize.) It is important to have a very slow temperature rate of change in the temperature range 550 to $600 \degree C$: Quartz undergoes a structural phase transition at $\sim 575 \degree C$ and the two phases have very nearly the same free energy. If the temperature changes too fast, the quartz will become confused as to which phase it should assume leaving a glassy matrix.

CONDUCTANCE MEASUREMENTS

Contacts to the precipitated Si are made by scribing the implanted substrate with a diamond scribe. The resulting scribed marks are 3 mm long and about 5000 nm deep so that the precipitate layer is exposed. Five hundred (500) nm of Al are evaporated through a mask onto the scribed marks by vapor phase evaporation in a vacuum chamber with pressure not exceeding 1×10^{-6} torr. The contacts were not annealed so some Schottky behavior is expected. A standard probe station was used to make contact to the deposited Al pads.

The resistance between contacts was much higher than expected so that the DC conductance between the contacts was measured with a Wheatstone bridge. The fixed legs of the bridge were constructed of

22 MΩ carbon film resistors. The resistance of each resistor was measured to 3 significant figures before assembly. The variable leg also consisted of carefully measured 22 MΩ resistors and a Leeds and Northrop decade resistance box. The null detector was a Leeds and Northrop optical galvanometer.

RESULTS AND DISCUSSION

The measured resistances and imputed conductivities are listed in Table III. The conductivities are computed from the length of the scribe mark, the average thickness of the Si precipitate layer computed from TRIM simulation data, and the separation of the scribe marks. The conductivities are presented here only for purposes of reference and should not be taken too seriously for reasons discussed below.

That these resistances are representative of a Si precipitate layer is certain because similar Al contacts laid down on regions of the sample having no scribe marks indicate resistances beyond the capacity of the existing bridge circuit to measure. Additionally the RC relaxation time for measurements of the precipitate layer are of the order of seconds and that for the unscribed surface is of the order of many minutes. The conductivity value for SiO_2 is 10^{-21} $\Omega^{-1}m^{-1}$. Conductance through a layer of quartz many square meters in area would be necessary to give the conductance measured on these samples.

Table III. Showing the measurements of resistance and conductivity for various pairs of contacts.

Contact Pair	Separation (mm)	Resistance (10^{-10} Ω)	Conductivity (10^{-4} $\Omega^{-1}m^{-1}$)
1-2	2	11	.59
3-4	4	0.93	1.4
2-3	5	0.90	1.9
1-3	7	0.93	2.5
2-4	9	0.75	1.4
1-4	11	11	3.4

Contact width = 3mm
Conducting layer thickness = 100 nm, computed from TRIM data

It is not clear that the conducting layer is continuous silicon. The conductivity of elemental Si is $2.5 \times 10^3 \ \Omega^{-1} m^{-1}$. The resistances measured would imply a layer of continuous Si of only a few angstroms (ie. a monolayer). That the resistances cluster around either $1 \times 10^{10} \ \Omega$ or $11 \times 10^{10} \ \Omega$ and show no evidence of scaling with the separation is consistent with a large resistance in series with Si precipitates. Islands of Si surrounded by a SiO_2 matrix would account for this effect (assuming the expected precipitate layer thickness). A series circuit model of Si and Quartz indicates that only a 10 to 15 anstrom thickness of Quartz in series with the expected layer of Si is needed to give the measured conductance. Alternatively, it is possible that the contact interfaces themselves are dominating the net resistance between the Al contact pads. At this writing it has not been possible to distinguish between these possibilities.

CONCLUSION

The layer of Si precipitate which forms when SiO_2 is implanted with Si has conductivity at least 8 orders of magnitude larger than SiO_2 itself. The resistance measurements quoted here indicate that, in this case, a precipitate layer formed from a peak concentration of 4×10^{22} Si/cc consists of Si which is electrically in series with a small amount of SiO_2 matrix. In the reported case of visual observation of a continuous layer of Si[1], the peak concentration was nearly 8×10^{22}/cc so that higher doses must be needed to obtain a continuous layer of precipitate Si in Si-implanted SiO_2 than the 1.5 times the theoretical minimumun used in this study..

REFERENCES
1. U. Ramabadran, Howard Jackson, G. C. Farlow, Nucl. Instru. and Methods **B59/60**, (1991) 637.

2. A. M. Ibrahim, A. A. Brezin, Materials Chemistry and Physics **31**, (1992) 285.

PROPERTIES OF LOW-RESISTIVITY UNDOPED INDIUM-OXIDE FILMS GROWN BY REACTIVE ION PLATING AND ELECTROCHROMIC TUNGSTEN-OXIDE FILMS GROWN BY ELECTRON-BEAM EVAPORATION

Y. P. LEE,* J. I. JEONG,** J. H. MOON,**J. H. HONG,** AND J. S. KANG**
*Sunmoon University, Asan, Choongnam, Korea
**Research Institute of Industrial Science and Technology, Pohang, Kyoungbuk, Korea

ABSTRACT

The use of gaseous discharge for ion plating and related techniques have been well known to improve coating properties in several ways. In the arc-induced ion plating (AIIP), the ionization efficiency for the evaporants is so enhanced without any introduction of inert gases that the bias voltage for, and the temperature of the substrate are reduced in the preparation of the coatings. Highly transparent ($\geq$ 90% transmission in the visible range) and highly conductive (resistivity $\cong$ 1.5 x 10^{-4} Ω cm) In-oxide films were deposited at a rate of 500 - 900 Å/min by AIIP of pure In in an O_2 atmosphere of 10^{-4} Torr. Hall-effect measurement revealed that the observed low resistivity is due primarily to the excellent electron mobilty ($\geq$ 70 cm^2/ V sec) with carrier density up to 7 x 10^{20}/cm^3. Electrochromic WO_3 films were also prepared and characterized.

INTRODUCTION

Transparent conducting films of metallic oxides such as In_2O_3, SnO_2, and Sn-doped In_2O_3 (ITO) have been studied for many years because of their many practical applications [1-5]. Among them, ITO films are most popular because of their high conductivity and transparency in the visible region of the spectrum. In the case of undoped In_2O_3 films, limited studies [6-10] have been done because of their low electrical resistivity (about 4 x 10^{-4} Ω cm). Pan and Ma [11] prepared pure In_2O_3 films by evaporating a mixture of In_2O_3-10 wt% In at 320 - 350°C in an O_2 atmosphere, and obtained an average resistivity of 1.8 x 10^{-4} Ω cm. This is the best value obtained to our knowledge.

The electrical and optical properties of In_2O_3 films depend mainly on the preparation method and the oxidation state [12]. We found that the quality of the In_2O_3 films can be improved by using the appropriate deposition method to make the desirable O content in the film. We report remarkable electrical and optical properties of undoped In_2O_3 films prepared by the reactive arc-induced ion

427

plating (AIIP) technique. Films with resistivity of less than $1.5 \times 10^{-4} \, \Omega$ cm and high visible transmission (over 90%) have been readily obtained without any post-deposition annealing. These properties are comparable or superior to the ITO films previously reported, and excellent reproducibility has been achieved.

There are certain materials which are able to change their colors reversibly according to the external stimulus. Electrochromism is one of these chromic phenomena, and is defined to be reversible change in color or transmission according to the applied voltage or current. The electrochromic (EC) films have attracted attention owing to their many practical applications for display devices, building exteriors and interiors, automobile rear-view mirrors, and so on. WO_3 films has been known to have the EC properties [13, 14]. Their color can be changed reversibly between transparency and blue according to the apllied current. This film was prepared by electron-beam evaporation, and the properties such as transmittance, optical density and coloring efficiency were investigated along with characterization by x-ray photoelectron spectroscopy (XPS), x-ray diffraction (XRD) and scanning electron microscopy (SEM).

EXPERIMENTAL

Reactive AIIP was used to prepare In_2O_3 films on microslide glasses (Corning No. 2947). In grains as pure as 99.99% were evaporated using an ordinary thermal source (usually a W boat). The O_2 gas was admitted through a mass flow controller. The details of the experimental system can be found in Ref. 15. The substrate was grounded during the deposition. The distance between the evaporation source and the substrate was 30 cm. The other deposition parameters used in this experiment were as follows ; deposition rate of 300 - 1500 Å/min, substrate temperature of 25 - 300°C, O_2 partial pressure of $(1 - 10) \times 10^{-4}$ Torr, film thickness of 500 - 6000 Å, ionization voltage of 20 - 80 V, ionization current of 1 - 6 A, and filament current of 40 - 60 A. The film thickness was monitored by a quartz crystal sensor calibrated by a stylus profiler.

WO_3 films were deposited by electron-beam evaporation of WO_3 granules as pure as 99.99%, and a self-made EC cell (Fig. 1) was employed to carry out the coloring and bleaching experiment on the prepared EC films. 1-mole solution of H_2SO_4 was used as the electrolyte, and Pt or C as the counterelectrode in the EC cell. The deposition parameters employed in the preparation were as follows ; electron-beam power of 6 kV and 20 mA, deposition rate of 2000 Å/min, substrate temperature of 25 - 300°C, O_2 partial pressure of $(0.05 - 3) \times 10^{-4}$ Torr, film thickness of 4000 - 10000 Å.

The atomic concentrations of In_2O_3 films were measured by AES (Auger electron spectroscopy) and XPS using standard In and In_2O_3 grains, and calculated by the method of Lin et al. [16]. The WO_3 films were measured by XPS using standard WO_3 powder. Resistivity was measured by the four-point probe method. Carrier concentration and Hall mobility were calculated from the

Hall effect measurement by the van der Pauw method. Optical properties of In$_2$O$_3$ and WO$_3$ films were measured by Fourier transform infrared spectroscopy and spectrophotometry, respectively.

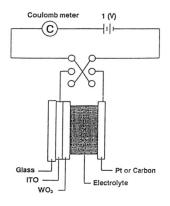

Coulomb meter   1 (V)

Glass — ITO — WO$_3$ — Electrolyte — Pt or Carbon

Fig. 1   Schematic diagram of the EC cell used for the coloring and bleaching experiment.

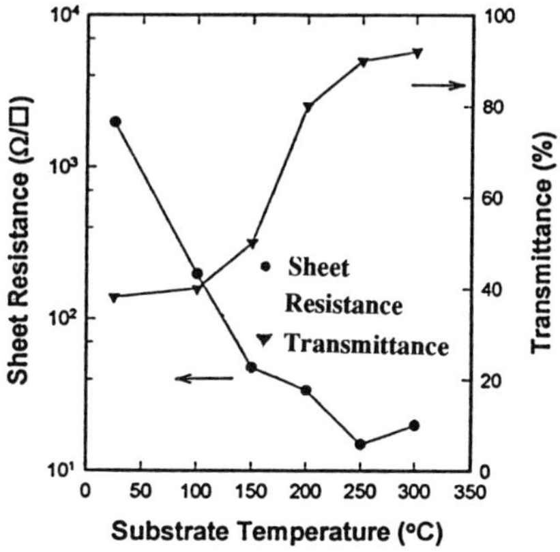

Fig. 2   Change in sheet resistance and visible transmittance of In$_2$O$_3$ films with the substrate temperature.

RESULTS AND DISCUSSION

Figure 2 shows the change in the sheet resistance and visible transmittance of the films with the substrate temperature (T_{sub}). The deposition rate was 500 Å/min, and the film thickness was 1000 Å. The ionization voltage and filament current were fixed at 30 V and 40 A, respectively. The O$_2$ partial pressure was 3 x 10^{-4} Torr. The sheet resistance decreases rapidly with T_{sub}, especially below 150°C, and is saturated at 10 - 20 Ω/□ around 250°C. A transmittance of more than 80% is observed when T_{sub} is above 200°C. We observed very low resistivity at a deposition rate of 500 - 900 Å/min at a T_{sub} of 250°C. This is noticeable compared with films made by the ordinary evaporation method in which a low evaporation rate of a few tens of Å/min is used [10, 17]. The transmittance decreases rapidly above 1000 Å/min, and is 25% at 1200 Å/min. The sheet resistance changes remarkably when the ionization parameters are changed, but the transmittance remains at above 90% (Fig. 3). This fact indicates that the degree of ionization affects the film resistivity greatly. Figure 4 shows the variation of the resistivity, Hall mobility, and carrier concentration with the film thickness. There is a resistivity minimum at a thickness of 200 Å, where the sheet resistance is 6 Ω/□, leading to an average resistivity of 1.2 x 10^{-4}

Ω cm. This result can be ascribed to the high electron mobility together with the high electron concentration.

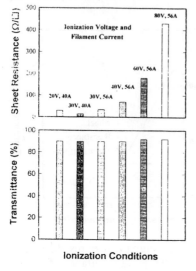

Fig. 3   Change in sheet resistance and
transmittance with
the ionization conditions.

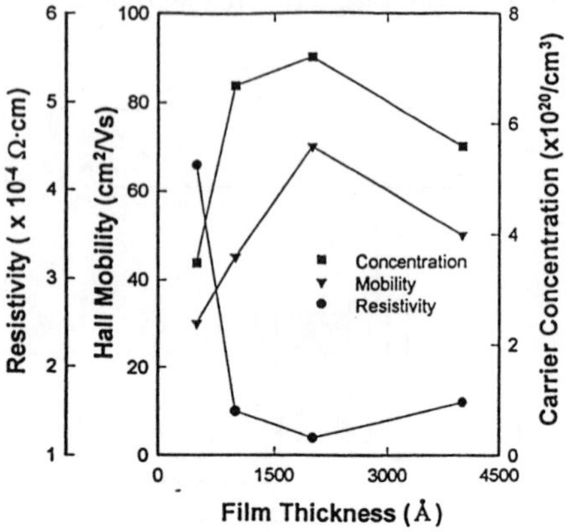

Fig. 4   Variation of resistivity, Hall
mobility, and electron
concentration with the
film thickness.

The results were correlated with the atomic concentration data from AES and XPS.   AES calculations showed that the atomic ratio of O to In (N_O/N_{In}) of the ion-plated films varied from 1.21 to 1.35 according to the deposition conditions, while the rate of the evaporated films varied from 1.23 to 1.29.   These results indicate that, in ion-plated films, more O can be incorporated than in evaporated ones.   There is a resistance minimum when N_O/N_{In} is in the range of 1.29 - 1.31. This result suggests that the proper O content is very important to the film resistivity.

The uncolored EC films of WO_3 showed average visible transmittance of more than 85%.   The colored films with a charge density of 10 mC/cm^2 showed a transmittance of around 55% at a wavelength of 500 nm, and those with a charge density of 50 mC/cm^2 showed a very low transmittance of less than 15%. The color changed from light blue to blue to dark blue with increasing charge density.   The properties turned out to be very stable even after more than tens of coloring/bleaching cycles.   The optical density at 500 nm increased rapidly to be bigger than unity at a charge density of more than 50 mC/cm^2 (Fig. 5).   The transmittance of the self-made EC cell was reduced to less than 50% in 20 s at a current of 3 mA.

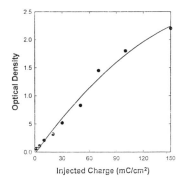

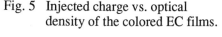

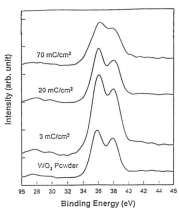

Fig. 5 Injected charge vs. optical
density of the colored EC films.

Fig. 6 XPS spectra of W 4f doublet
for WO_3 powder, and
colored films with
various charge densities.

Both uncolored and colored films showed nearly amorphous structures by XRD. The atomic concentration of the EC films was estimated from the XPS spectra of W $4f_{7/2}$ and O 1s by comparison with standard specimens after removal of the C surface contamination. N_O/N_W was in the range of 2.4 - 2.6 for the uncolored films, while for the colored ones, the ratio was in the range of 2.2 - 2.4. This reduction in the value can be explained by the leakage of oxygen ions from the film during the coloring.   The XPS spectra of W 4f doublets for the colored films showed different lineshapes from that of bulk WO_3 as shown in Fig. 6. This is thought to be due to the additional peaks from W^{+5} and W^{+4} as well as from W^{+6}.

CONCLUSIONS

In oxide films with an electrical resistivity of less than 1.5×10^{-4} Ω cm and of good optical quality were prepared by reactive AIIP of pure In in an O_2 atmosphere of 10^{-4} Torr. The properties are remarkable when we consider that the films were deposited at a high rate of more than 500 Å/min without doping and postannealing.   The film quality was affected greatly by the ionization conditions and the O content in the film.   The observed low resistivity is believed to result from the high electron mobility and high electron concentration, together with the smooth morphology and the appropriate O content.

The EC films of WO_3 which were prepared by electron-beam evaporation showed an average visible transmittance of more than 85% before coloring, and a transmittance of less than 15% and an optical density of around unity at a wavelength of 500 nm after coloring with a charge density of 50 mC/cm^2. A transmittance of less than 50% was achieved in 20 s at a current of 3 mA with the self-made EC cell.

References

[1] G. Cheek, A. Genis, and J. B. Dubow, Appl. Phys. Lett. 35, 495 (1979).
[2] A. Hjortsberg, I. Hamberg, and C. G. Granqvist, Thin Solid Films 90, 323 (1982).
[3] J. C. Manifacier, Thin Solid Films 90, 297 (1982).
[4] K. L. Chopra, S. Major, and D. K. Pandya, Thin Solid Films 102, 1 (1983).
[5] I. Hamberg and C. G. Granqvist, J. Appl. Phys. 60, R123 (1986).
[6] S. Noguchi and H. Sakata, J. Phys. D 13, 1129 (1980).
[7] R. L. Weiher and R. P. Ley, J. Appl. Phys. 37, 299 (1966).
[8] C. E. Wickersham and J. Greene, Phys. Status Solidi A 47, 329 (1978).
[9] J. C. Manifacier, L. Szepessy, J. F. Bresse, M. Perotin, and R. Stuck, Mater. Res. Bull. 14, 163 (1979).
[10] S. Muranaka, Y. Bando, and T. Tanaka, Thin Solid Films 151, 355 (1987).
[11] C. A. Pan and T. P. Ma, Appl. Phys. Lett. 37, 163 (1980).
[12] G. Frank and H. Kostlin, Appl. Phys. A 27, 163 (1982).
[13] S. K. Deb, Philos. Mag. 27, 801 (1969).
[14] G. G. Barna, J. Electron. Mater. 8, 153 (1979).
[15] J. I. Jeong, J. H. Hong, J. S. Kang, H. J. Shin, and Y. P. Lee, J. Vac. Sci. Technol. A 9, 2618 (1991).
[16] A. W. Lin, N. R. Armstrong, and T. Kuwana, Anal. Chem. 49, 1228 (1977).
[17] D. Laser, Thin Solid Films 90, 317 (1982).

OBSERVATION OF THE EARLY STAGES OF
b-AXIS ORIENTED $PrBa_2Cu_3O_{7-x}$ THIN FILM GROWTH BY AFM

GUN YONG SUNG, JEONG DAE SUH, AND SANG-DON JUNG
Electronics and Telecommunications Research Institute,
P.O.Box 106, Yusong, Taejon, 305-600, REP. OF KOREA

ABSTRACT

The initial stages of the growth of b-axis oriented $PrBa_2Cu_3O_{7-x}$ (PBCO) films on $LaSrGaO_4$ (100) substrates were investigated by atomic force microscopy to follow the growth of the thin films. Series of films with thickness ranging between 0.34 nm and 100 nm were prepared under identical pulsed laser deposition conditions. No sprial-topped or flat-topped islands were observed and the scale of the surface roughness was lower than that of the c-axis oriented growth mode. The 300 nm-thick in-plane aligned a-axis oriented YBCO films have the root mean square (RMS) surface roughness of 2 nm. It is considered that the b-axis oriented PBCO films on $LaSrGaO_4$ (100) substrates were nucleated and grown by layer-by-layer like growth mode.

INTRODUCTION

The a-axis oriented $YBa_2Cu_3O_{7-x}$ (YBCO) thin films are important to fabricate the planar Josephson junctions because of the longer coherence length along the a-axis or b-axis. Among the a-axis oriented YBCO films grown by various methods[1~8], in-plane aligned a-axis oriented YBCO thin films on $LaSrGaO_4$ (LSGO) substrate with $PrBa_2Cu_3O_{7-x}$ (PBCO) template have the highest zero resistance temperature($T_{c,zero}$)[6~9]. The a-axis YBCO films with a $T_{c,zero}$ of 83 K have been reported by Inam et al.[5] and these films are formed by inserting an a-axis oriented PBCO thin film as a template layer between the $SrTiO_3$ (100) substrate and the YBCO film. Their low $T_{c,zero}$ was due to the existence of these 90° grain boundaries compared with c-axis YBCO films. Recently, in-plane alignment of the a-axis YBCO films has been realized by using LSGO (100) substrates with PBCO templates, so that the a-axis oriented YBCO films having $T_{c,zero}$ of 87 ~ 90.5 K were obtained.

Since much of a-axis oriented microstructure is determined by the initial formation of the film on the substrate, it is valuable to investigate the early regime of deposition to better understand nucleation and subsequent growth of the film and the creation of microstructural features. Investigation of the surface morphology with scanning tunneling microscopy (STM) and atomic force microscopy (AFM) have revealed a high density of growth spirals on c-axis oriented YBCO thin films[10, 11]. But the STM and AFM studies for a-axis oriented YBCO and PBCO films are very limited. Therefore we have focussed our efforts on the growth mode of PBCO films on LSGO (100) substrate. In this study we report the observation of the growth of b-axis oriented PBCO thin films on LSGO (100) substrates, the in-plane aligned a-axis oriented YBCO thin films on LSGO (100) substrates with b-axis oriented PBCO templates, and the c-axis oriented YBCO thin films on $SrTiO_3$(100) substrates.

EXPERIMENTAL PROCEDURE

PBCO thin films with various thicknesses, ranging from 0.34 nm up to 100 nm, were

Mat. Res. Soc. Symp. Proc. Vol. 388 © 1995 Materials Research Society

prepared by pulsed laser deposition (PLD). Hot-pressed stoichiometric PBCO targets were mounted on a rotating, water-cooled 4-pole target holders. LSGO (100) substrates were bonded onto the heating block using silver paste. The substrates, held at 4 cm from the targets, were pre-annealed at 850 °C under 10^{-6} Torr to remove surface defects including contaminants and scratches created during mechanical polishing. After in-situ pre-annealing, the PBCO films were deposited at 630 °C using a 308 nm XeCl excimer laser operated at repetition rate of 1 Hz and produced 14±4 ns pulses with energy density of 1 J/cm^2. All depositions were carried out in a dynamic oxygen atmosphere of 100 mTorr. The thickness of the films was controlled by scaling the number of laser pulses. Deposition rate of the PLD conditions was 0. 34 nm per single laser pulse.

In-plane aligned a-axis oriented YBCO thin films were grown by two steps PLD process. First, 100 nm-thick PBCO template layers were deposited under above conditions. Second, the substrate temperature was increased at a rate of 4 °C/min to the second step deposition of 750 °C and the 300 nm-thick YBCO films were grown. Immediately following the second step, the chamber was filled with O_2 to a pressure of 500 Torr and the film was cooled down to 500 °C. Following an *in situ* annealing at this temperature for 1 h, the film was cooled to room temperature. In addition, c-axis oriented YBCO thin films on $SrTiO_3(100)$ substrate were prepared to compare their surface morphology and roughness with those of the a-axis oriented YBCO thin films.

After PLD, the films' surfaces were examined using AFM. The observations were made in air at room temperature with Park Scientific Instrument AutoProbe LS equipped with a small field (5 μm x 5 μm) scanner. Images were obtained in the contact mode using Si_3N_4 AFM cantilevers with a large aspect ratio.

RESULTS AND DISCUSSION

Orientation of the 100 nm-thick PBCO films were determined, as the b-axis perpendicular to the substrate surface, by X-ray diffraction (XRD) and transmission electron microscopy[9]. This result reveals the orientation relationship in which the b-axis of the PBCO(a = 0.3870 nm, b = 0.3934 nm, c = 1.1730 nm for $PrBa_2Cu_3O_{6.8}$) films is normal to the substrate surface. The a-axis and c-axis of LSGO are parallel to the surface of LSGO (100) single crystal substrate (tetragonal, a = 0.3843 nm, c = 1.268 nm). The a-axis lattice constant of the substrate is well matched with the a-axis lattice constant rather than the b-axis lattice constant of the PBCO film. So the above orientation relationship can be clearly understood by the lattice matching between the PBCO films and the LSGO (100) substrate. Besides, the b-axis oriented PBCO film is more compatible than the a-axis oriented PBCO film, as a template layer in order to grow a-axis oriented YBCO films. The a-axis and c-axis of PBCO, which are parallel to the surface of b-axis oriented PBCO template layers, are well matched with the b-axis and c-axis of the YBCO (a = 0.3820 nm, b = 0.3880 nm, c = 1.1720 nm) films, respectively. Therefore, it is very important to grow the b-axis oriented PBCO template layer for the subsequent growth of a-axis oriented YBCO films on LSGO (100) substrate.

An AFM image of the surface morphology of LSGO (100) substrate is shown in Fig. 1 (a). Waviness of the substrate surfaces, which has a root mean square (Rms) roughness of 1.3 nm and a mean height of 4 nm, is formed during mechanical polishing and pre-annealing. In the early stages of the PBCO film growth below a thickness of 10 nm (Fig. 1 (b) ~ (f)), the PBCO film initially grew very smoothly, the dominant surface features being substrate waviness, preserved in the film's surface morphology. As shown in Fig. 2, the Rms surface roughness of the film vs. film thickness plots imply quantitatively that initial smoothening of the film surfaces occurred with increasing film thickness. It seems likely that initially deposited species move to the energetically favorable sites, for example, valleys or facets of the wavy surface. Thus, the wavy surface of the substrate would be flatten by the early stages of the PBCO deposition. By a thickness of 9.52 nm the smoothest film surface was obtained, and the observed Rms surface roughness is 0.086 nm, although the surface roughening was followed

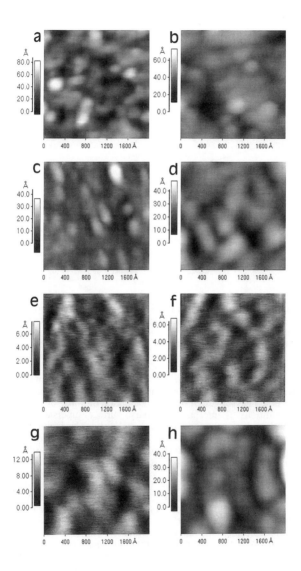

Fig. 1 AFM images of the PBCO films on LSGO (100) substrates with different thicknesses of (a) 0 nm, (b) 0.34 nm, (c) 1.36 nm (d) 4.08 nm (e) 5.44 nm (f) 9.52 nm, (g) 13.6 nm, and (h) 17 nm.

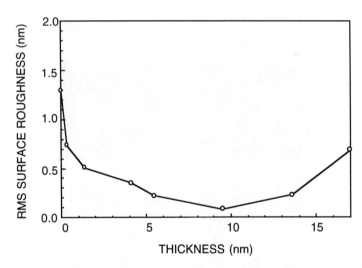

Fig. 2   RMS surface roughness vs. film thickness plots for the surfaces of PBCO
films on LSGO (100) substrates.

with increasing film thickness.

    In contrast to the c-axis oriented YBCO film growth[10,11], no spiral-topped or flat
topped islands were observed at any film thicknesses. This result is not surprising because the
a-axis or b-axis oriented growth mode is quite different from the c-axis oriented growth
mode.[12,13].   Although   c-axis   oriented   two-dimensional   island   growth   is   the
thermodynamically preferred growth mode of YBCO, since it maintains the lowest energy
crystal termination, kinetically it is rather inefficient. A large amount of surface diffusion is
required to grow a layer one unit cell in thickness. However, if an orientation of the nucleus is
determined to be a-axis or b-axis oriented, then practically no surface diffusion would be
necessary. Streiffer et al.[13] suggested that the nucleation of a-axis or b-axis oriented domains
is caused by the conditions of interface chemistry, surface supersaturation, limited surface
diffusion, and Y/Ba disorder or thermodynamic instability. The incoming flux in the plume
during PLD would automatically find itself at the active crystal growth sites. This growth
mode can be continued under the conditions of low temperature deposition, slow deposition
rate, and low oxygen partial pressure. It is considered that the growth conditions of the PBCO
films were sufficiently satisfied with the a-axis or b-axis oriented growth conditions, so that no
islands were observed in AFM images until the films grew to thicknesses of 100 nm.

    Above a thickness of 9.52 nm surface roughening was observed up to a thickness of 100
nm as shown in Fig. 3 and 4. But the scale of the roughness is lower than that of the c-axis
oriented growth, because there is no surface diffusion. Rms surface roughness of 0.6 nm
preserved in the films with thickness ranging between 17 nm to 70 nm as shown in Fig. 1 (h)
and Fig. 3 (b), respectively.   At a thickness of 100 nm the surface roughness increased sharply
to the value of 2 nm. Surprisingly, the surface roughness of the 100 nm-thick PBCO films
were preserved in the 300 nm-thick a-axis oriented YBCO films deposited on the PBCO
films(Fig. 3 (d)). This result means that the growth conditions of the a-axis oriented YBCO or
b-axis oriented PBCO films could be used to reduce the surface roughness to desirable levels.
In the case of 200 nm-thick c-axis oriented YBCO films on $SrTiO_3$(100) substrate the Rms
surface roughness and mean height was 3.5 nm and 18.1 nm, respectively.

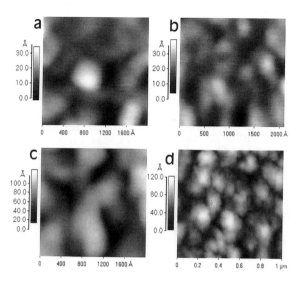

Fig. 3 AFM images of the PBCO films on LSGO (100) substrates with different thicknesses of (a) 34 nm, (b) 68 nm, and (c) 100 nm. (d) AFM image of the a-axis oriented YBCO thin film deposited on the film shown in (c).

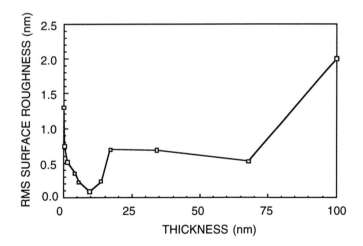

Fig. 4 RMS surface roughness vs. film thickness plots for the surfaces of PBCO films on LSGO (100) substrates.

CONCLUSIONS

We have investigated the initial stages of the growth of b-axis oriented PBCO films on LSGO (100) substrates by AFM to follow the growth of the thin films. In the early stages of the PBCO film growth below a thickness of 10 nm, the PBCO film initially grew very smoothly, the dominant surface features being substrate waviness preserved in the film's surface morphology. By a thickness of 9.52 nm the smoothest film surface was obtained as an observed Rms surface roughness of 0.086 nm, although the surface roughening was followed with increasing film thickness. No spiral-topped or flat topped islands were observed at any film thicknesses and the scale of surface roughness was lower than that of the c-axis oriented growth mode. It is believed that the growth conditions of the PBCO films were sufficiently satisfied with the a-axis or b-axis oriented growth conditions, so that no islands were observed in AFM images until the films grew to a thickness of 100 nm. The 300 nm-thick in-plane aligned a-axis oriented YBCO films having a surface roughness of 2 nm were obtained. We suggest that the growth conditions used in this study are suitable to reduce the surface roughness to desirable levels. It is considered that the b-axis oriented PBCO films on LSGO (100) substrates were nucleated and grown by layer-by-layer-like growth mode.

ACKNOWLEDGEMENTS

We would like to thank Dr. E.-H. Lee for his consistent support and encouragement. This work was supported by Ministry of Information and Communications, Rep. of Korea.

REFERENCES

1. C.B. Eom, A. F. Marshall, J. M. Triscone, B. Wilkens, S. S. Laderman, and T. H. Geballe, Science **251**, 780 (1991).
2. R. Ramesh, C.C. Chang, T.S. Ravi, D.M. Hwang, A. Inam, X.X. Xi, Q. Li, X.D. Wu, and T. Venkatesan, Appl. Phys. Lett. **57**, 1064 (1990).
3. T. Hase, H. Takahashi, H. Izumi, K. Ohata, K. Suzuki, T. Morishita, and S. Tanaka, J. Cry. Growth **115**, 788 (1991).
4. L. Luo, X.D. Wu, R.C. Dye, R.E. Muenchausen, S.R. Foltyn, Y. Coulter, C.J. Maggiore, and T. Inoue, Appl. Phys. Lett. **59**, 2043 (1991).
5. A. Inam, C.T. Rogers, R. Ramesh, K. Remsching, L. Farrow, D. Hart, T. Venkatesan, and B. Wilkens, Appl. Phys. Lett. **57**, 2484 (1990).
6. G.Y. Sung, J.D. Suh, and S. Nahm, in Epitaxial Oxide Thin Films and Heterostructures, edited by D.K. Fork, J.M. Phillips, R. Ramesh, and R.M. Wolf (Mater. Res. Soc. Proc. **341**, Pittsburgh, PA, 1994) pp. 177-182.
7. S. Hontsu, N. Mukai, J. Ishii, T. Kawai, and S. Kawai, Appl. Phys. Lett. **61**, 1134 (1992).
8. Y. Suzuki, D. Lew, A.F. Marshall, M.R. Beasley, and T.H. Geballe, Phy. Rev. B **48**, 10642 (1993).
9. G.Y. Sung and J.D. Suh, Appl. Phys. Lett. (submitted).
10. C. Gerber, D. Anselmetti, J.G. Bednorz, J. Mannhart, and D.G. Schlom, Nature **350**, 279 (1991).
11. M. Hawley, I.D. Raistrick, and J.G. Berry, R.J. Houlton, Science **251**, 1587 (1991).
12. S.J. Pennycook, M.F. Chisholm, D.E. Jesson, R. Feenstra, S. Zhu, X.Y. Zheng, and D.J. Lowndes, Physica C **202**, 1 (1992).
13. S. K. Streiffer, B.M. Lairson, E.M. Zielinski, and J.C. Bravman, Phy. Rev. B **47**, 11431 (1993).

AUTHOR INDEX

440

SUBJECT INDEX

absorption mechanisms (optical), 15, 133
alloys, 97
AlN films, 367, 399
aluminum, 39
  nitride, 367, 399
    BN-AlN, 183
  oxynitride, 311
amorphous
  carbon films (a-C), 281
  diamond-like carbon, 145
  hydrogenated carbon, 305
anion exchange, 259
annealing, 127, 379
anodic vacuum arc, 281
artificially layered thin films, 57
atomic
  beam, 271
  force microscopy (AFM), 79, 103, 115,
    399, 433
  hydrogen, 227
Auger electron spectroscopy, 165, 177, 183,
  293

background gases, 21
BaCuO$_2$, 57
(Ba,Sr)Ti$_3$O, 411
BaTiO$_3$, 45
boron nitride, 165, 183
  BN-AlN, 183
buried liquid layers, 127

C implantation, 195, 241
C$_2$ in plume, 121, 145
cadmium stannate, 51
carbon
  nitride (CN$_x$), 271, 281, 293, 393
  plasmas, 145
cathodic arc, 215, 287
CBr, supersonic jet, 233
chemical
  reactivity, 367
  vapor deposition (CVD), 299, 305
chromium, 305
cluster beams, 207
coating of powders, 215
combined ion beam and molecular beam
  deposition, 241
competitive oxidation, 385
compressive stress, role of, 165
conducting layers, 79, 417, 423
contactless electrochemical etching, 299
copper vapor laser, 121

defects, 15, 195, 349
deposition and growth mechanisms, 139, 165,
  177, 271, 317, 349, 405, 433
  plasma activated CVD, 305

transition metals, 317
diamond(-), 299, 317, 329
  like carbon (DLC), 121, 145, 171
  nucleation, biased enhanced, 177
dielectric
  constant, silicon oxynitride, 115
  susceptibility, 73
di-iodomethane, Cl$_2$H$_2$, 373
dissociative chemisorption, 221, 265
doping, 85, 91, 201, 241
dynamical modeling, 3, 337, 349

ejection of particulates, 127
electrical properties, 51, 57, 73, 79, 91,
  241, 271, 379, 417, 423, 427
electrochromic tungsten-oxide films,
  423
electron(-)
  beam, 323, 427
    evaporation, 423
  cyclotron
    resonance (ECR)(-)
      molecular beam epitaxy
        (ECR-MBE), 259
      plasma, 329, 355
  energy loss spectroscopy (EELS), 287
  impact, ionization and excitation, 133
End of Range (EOR) defect, 195
energetic cluster impact (ECI), 207
epitaxy, 57, 73, 79, 97, 189, 195, 201, 241
  beam-induced crystallization
    (IBIEC), 189
  combined ion beam and molecular
    beam (CIBMBE), 241
  electron cyclotron resonance-molecular
    beam (ECR-MBE), 259
  GaN, 259, 265
  homoepitaxy, 201, 299
  laser-assisted chemical beam
    (CBE), 373
  solid phase growth (SPEG), 189, 195
  TiN, 103
  two vs. three-dimensional, 411
etching
  chemical, 367
  electrochemical, 299
ethyl(trimethylphosphine)gold(I), 323

ferroelectric poling, 79
film roughness/topography, 91, 115, 349, 399,
  411
fluorescence
  planar laser-induced (PLIF), 33
Fourier transform infrared, 393
free-standing diamond plate, 299